Quantum Mechanics

Detailed Historical, Mathematical and Computational Approaches

I0031470

Caio Lima Firme
Organic Chemistry Professor
Researcher in Computational Chemistry
Chemistry Institute/Federal University of Rio Grande do Norte
Brazil

CRC Press
Taylor & Francis Group
Boca Raton London New York

CRC Press is an imprint of the
Taylor & Francis Group, an **informa** business
A SCIENCE PUBLISHERS BOOK

Cover credit: https://unsplash.com/photos/qyhLjwn6Gpc

First edition published 2022
by CRC Press
6000 Broken Sound Parkway NW, Suite 300, Boca Raton, FL 33487-2742

and by CRC Press
4 Park Square, Milton Park, Abingdon, Oxon OX14 4RN

Library of Congress Cataloging-in-Publication Data (applied for)

ISBN: 978-0-367-50633-9 (hbk)
ISBN: 978-0-367-50634-6 (pbk)
ISBN: 978-1-003-05058-2 (ebk)

DOI: 10.1201/9781003050582

Typeset in Times New Roman
by Radiant Productions

To my beloved daughter Ananda França Firme

Preface

In the literature, there are only a couple of excellent books of quantum computational physics and other books with comprehensive mathematical approach for quantum mechanics; and there is only one excellent comprehensive book of the history of the quantum mechanics. Nonetheless, as far as we know, there is no book that encompasses all these three cornerstones of the quantum physics in a detailed manner. When the student learns the quantum mechanics with all these three central components together (historical, mathematical and computational approaches), he/she will have a more proficient learning experience in quantum mechanics. This book intends to present a detailed (but not complete) presentation of historical, mathematical and computational approaches with respect to quantum mechanics. The book is supportive in mathematical and computational topics related to quantum mechanics in order to make the reading straightforward and easy. The historical approach of the book follows the natural path of the chronological sequence of published papers divided in chapters which represent the most important themes of quantum mechanics. By reading the book in its chronological historical sequence, the reader will realize the importance of learning the concepts and facts from the beginning of quantum theory which form important pillars of the theory prior to the advent of Schrödinger's wave mechanics. When encompassing important concepts from Bohr, Sommerfeld, Heisenberg, Pauli, de Broglie and Dirac, it will ensure a better understanding of the whole theory along with those concepts that come with wave mechanics.

The book gives all the mathematical support for learning quantum mechanics: (i) detailed development of the equations for quantum mechanics; (ii) all background information to understand the mathematical approach in quantum mechanics; (iii) chapters of important mathematical issues in quantum mechanics; (iv) and numerical solutions for several mathematical and quantum problems. There are chapters devoted to linear algebra and differential equations along with their numerical solutions in order to provide better support for the mathematical issues in quantum mechanics.

The book is divided into three parts. In the first part, there are chapters introducing Fortran and numerical calculations, series, linear algebra and differential equations and their applications to the quantum mechanics. Chapter one (Fortran) is interrelated with all other chapters of the first part. Chapter two (series) is interrelated with Chapters five to eight of the second part. Chapter three (matrices) is interrelated with Chapters eight, eleven and twelve of the second part. Chapter four (differential equations) is interrelated with Chapter ten of the second part and Chapters fourteen to seventeen of the third part of the book.

Fortran is the most important language for the physical sciences. It is a dinosaur which is permanently modernized and then it is the most important and used language for science and engineering even today. In the first chapter, there are several Fortran source codes to understand the Fortran language with straightforward explanations. In the next three chapters, there is a detailed elucidation of the mathematics for quantum mechanics itself (linear algebra, series and differentiation equations) and for the numerical calculation (integration, finding roots and derivatives) along with several source codes to solve some related problems. There is a detailed support to understand these source codes. All these codes are available in the author's site (https://www.fortran-codes.com/).

In the second part, there are several chapters that follow the timeline of the historical evolution of quantum mechanics from emission spectroscopy towards Heisenberg and Schrödinger's quantum theories. The historical discussion is mainly based on the papers that were published by all important contributors to quantum mechanics accompanied with a mathematical support. The discussion is based on the original scientific papers as basis for the historical approach which gives a better and deeper understanding of the quantum mechanics itself. It also helps understanding the evolution of quantum mechanics as a science and provides several important information to the backbone of quantum mechanics.

In the third part, there are chapters for the Schrödinger's time independent one-particle quantum problems with comprehensive mathematical approach, plus their numerical solutions and Fortran codes. Following the above-mentioned there is a chapter for helium atom using variational method and perturbation theory.

Features: presentation of matrix mechanics and wave mechanics, along with their applications (some of them using both theories altogether), within their historical context which helps to build a more solid understanding of quantum mechanics as a whole—an amalgamation of both theories. Supportive mathematical details for an easy, straightforward reading of the quantum mechanics. Chronological sequence of historical facts based on the original papers which gives a more comprehensive view of quantum mechanics and their most important concepts. Detailed information about the discovery of the spin; wave-matter duality principle, matrix mechanics, important concepts of old quantum theory, wave mechanics and particle statistics. Thorough mathematical and numerical analysis of the most important one-electron and two-electron problems of quantum mechanics. Detailed approach of mathematical and numerical analysis applied to quantum mechanics. Do-it-yourself problems to generate all the illustrations of the book using Fortran or other software. Examples and exercises in most chapters. All code in the author's site: (https://www.fortran-codes.com/).

Contents

Part One
Computational and Mathematical Support

Basics of Fortran

1

1. Notes about Fortran

There are several versions of Fortran (FORmulaTRANslation), from Fortran IV to Fortran 2018 (https://www.fortran.com/). In this book, we have used Fortran-90 to write our source codes (see examples below). Fortran presents a backward compatibility. As a compiled language that results in binary codes after compilation, Fortran has a high performance, fast debugging and testing. It has a huge number of numerical libraries and contains modern features such as OOP (object oriented programming). It also has elegant treatment of arrays and it has its own scalable parallel programming model (Fortran 2008 co-arrays for operation of Cartesian grids). Fortran is a living dinosaur of the scientific programming language and is still evolving. Moreover, the apprentices of scientific programming will see in this chapter that Fortran is very easy to learn. More information can also be found in the link: http://fortranwiki.org/fortran/show/HomePage.

The following paragraphs are related with the use of Fortran in Windows platform. All programs used in this book are free. We have used Force 2.0.9. editor from 'Force Fortran The Force Project' (http://force.lepsch.com/) to write our source codes in .f extension (a type of Fortran file for the source code). In order to transform the source code into executable program we have used MinGW (Minimalist GNU for Windows) compiler (http://www.mingw.org/) along with MSYS (MinimalSystem) program. The MSYS program is used to create a Linux Shell statement line (MinGW Shell) interpreter system (a sort of Linux-based shell within the Windows platform) in order to transform the source code into an executable file and to run the corresponding executable program. See the instructions for installation of the MinGW site.

After installing both programs, one opens the MinGW Shell (located in the main menu of Windows). It opens a window (MINGW32:/) whose path is: C:\MinGW\msys\1.0. Then, one creates the directory \home\user\Fortran where the .f files will be stored. After writing the source code in Force editor, there will three single operations in the (Linux-based) Shell: (i) mv <name-of-source-code>.f <name-of-source-code>.f90; (ii) gfortran<name-of-source-code>.f90 -o <name-of-source-code>; and in case the compiler finds no error in the source code, (iii) <name-of-source-code> type <enter>.

Alternatively, one may install MinGW from the site of GCC (https://gcc.gnu.org/wiki/GFortranBinaries), follow the path "toolchains targeting win64 > personal builds >mingw-builds > 5.3.0 > threads-posix>seh" and unzip the folders in a chosen location. Afterwards, go to the control panel in Windows and follow the path "system and security > system > advanced system settings", in the tab "advanced" click on "environment variables" and modify the PATH variable and add the location of the MinGW bin folder (where there is the file x86_64-w64-mingw32-gfortran.exe). Next, install the software Code::Blocks, the IDE for Fortran from the site of CBFortran (http://cbfortran.sourceforge.net/) and learn how to use it to open existing .f90 files (https://www.youtube.com/watch?v=GFrl8A_JgbE&feature=youtu.be) or to use it for writing new files .f90 (https://www.youtube.com/watch?v=b-9cBnnczNE&feature=youtu.be).

In order to visualize the results from .dat output files, one can use the multiplatform, free GNUPLOT (Willians et al. 2018). The GNUPLOT can be downloaded from http://www.gnuplot.info/

All the source codes of this book can be found in the authors' own website at the link https://www.fortran-codes.com/.

The general-purpose libraries of Fortran can be found in http://fortranwiki.org/fortran/show/Libraries.

The following link (http://fortranwiki.org/fortran/show/Software+repositories) indicates several repositories of Fortran codes, such as : Computer Physics Communications Programs Library (http://cpc.cs.qub.ac.uk/) where you can find several codes for physics problems, e.g., the Hartree-Fock method. Other resources of Fortran libraries are: (http://www.fortranlib.com/freesoft) and (https://people.sc.fsu.edu/~jburkardt/f_src/f_src.html). More specifically, one can find the Fortran codes of Dalton package for quantum chemistry calculation (https://gitlab.com/dalton/dalton/tree/release/2018/DALTON), the Fortran codes of the Valence program (https://github.com/VALENCE-software/VALENCE/tree/master/src) and Fortran codes of the Gamess quantum chemistry package (https://github.com/streaver91/gamess_source_mod).

2. Bits and bytes

Computers use binary (base-2) system. Each bit has 0 or 1 value. From right to left the value increases in the basis 2 and exponent from 0 to n, where n is an integer. For example, the number 21 in decimal system corresponds to 10101 in binary system. In this case, the number 10101 has five bits.

$$21_{(10)} = 2\times10^1 + 1\times10^0$$

$$10101_{(2)} = 1\times2^4 + 0\times2^3 + 1\times2^2 + 0\times2^1 + 1\times2^0$$

$$21_{(10)} = 10101_{(2)}$$

Variables and constants are stored in the hardware as a sequence of 32 or 64 bits called 'word' (natural unit of information). In a 32-bit architecture (e.g., 32-bit processor or 32-software), the bits of a word are grouped in 4 bytes of 8 bits each. This is the integer size or memory address width of four octets in 32-bit computing.

In a 64-bit architecture, the integer size and memory address width have eight octets (8 bytes). Independently of the architecture, one byte is a collection of 8 bits.

As the amount of bits increases the quantity of patterns (or possible combination in that set of bits) increases 'exponentially'. Mathematically, n bits give 2^n patterns. For example, 8 bits correspond to 2^8 or 256 patterns or one byte.

Any number in Fortran can be real (number encompassing all range of numbers except for imaginary numbers, having fractional/decimal part or not), integer (integral number without fractional part) or complex. Then, a rational number (or decimal numbers) has two parts: integral part at the left side of the decimal separator (a dot) and the fractional part at the right side of the decimal separator. In Fortran, a rational number is synonymous with a real number. When a variable is attributed to a real number, it might be represented in Fortran only by the integer part and the decimal separator, e.g., 0. or 10., etc. Fortran stores a complex number a+bi as two real numbers adjacent in memory. For example, the entry: COMPLEX*16 A/(1,2)/, yields 1+2i complex number.

One byte holds 1 typed character. The set of 4 bytes can store integers from –2147483648 to 2147483648, while the set of 8 bytes can store integers from –9223372036854775808 to 9223372036854775808. An integer overflow occurs when a number exceeds this limit.

In ASCII code (American Standard Code for Information Interchange), each typed character is represented by a number stored in 1 byte. The ASCII codifies 128 symbols (alphabet letters, punctuation marks, mathematical symbols and control characters).

The 32-bit computers support a maximum of 4GB (2^{32} bytes) of addressable RAM while 64-bit computers have a theoretical maximum of 18 EB (2^{64} bytes). From now on, let us assume we are working on a 32-software, e.g., a 32-bit Fortran, although modern Fortran is 64-bit (https://www.fortran.com/).

In a 32-bit architecture, an integer or real number has 4 bytes wide, named INTEGER*4 or REAL*4, respectively by Fortran. Then, any number in a 32-bit Fortran is either INTEGER*4 (without dot) or REAL*4.

The most left bit of integer or real number is used for the sign (+ or –) of this number, called s. In a real number, the next 8 bits after the s-bit represent the integer part (named p) and the other 23 bits represent the fractional part, called f.

The default precision of a real number is single precision from REAL*4. In order to increase the precision of a real number it is used REAL*8 or DOUBLE PRECISION along with the suffix d0 or _db after the number. The REAL*16 refers to quadruple precision of real numbers but it works only in 64-bit architecture.

Let us observe the decimal number below in binary and decimal systems.

$$f_{(2)} = .10001100110011001100110$$

$$f_{(2)} = 2^{-1} + 2^{-5} + 2^{-6} + 2^{-9} + 2^{-10} + 2^{-13} + 2^{-14} + 2^{-17} + 2^{-18} + 2^{-21} + 2^{-22}$$

$$f_{(2)} = \tfrac{1}{2} + \tfrac{1}{32} + \tfrac{1}{64} + \tfrac{1}{512} + \tfrac{1}{1024} + \tfrac{1}{8192} + \tfrac{1}{16384} + \tfrac{1}{131072} + \tfrac{1}{262144} + \tfrac{1}{2097152} + \tfrac{1}{4194304}$$

$$f_{(2)} = 0.5 + 0.03125 + 0.015625 + 0.001953 + 0.000976 + \ldots + 0.000000238$$

$$f_{(10)} \approx 0.5498$$

3. Types of scalar data in Fortran

In Fortran, there are two basic types of scalar data: constant data and variable data. A variable might be a letter or a word. In a variable, different values might be assigned, unlike a constant. The constant data may be numerals or alphabetical characters.

There are five types of constant:

(i) integer literal constant (1;0; –99;+100;299456; etc.);

(ii) real literal numbers (0.03; 3.1416; 1792.3; etc);

(iii) complex literal constant (1., 3.2, i.e., 1.0+3.2i; 2.0, 4.78, i.e., 2.0 + 4.78i.), designed by a pair of literal constants separated by a comma;

(iv) String or character literal constant ('Anything is possible') where the apostrophes or quotation marks serve as delimiters;

(v) logical literal constants (.true. and .false.) used to initialize logical variables.

It is also important to note that a number can be written in Fortran in scientific notation by means of the E-notation, where E means 'exponent', i.e., ten raised to the power of number n or 10^n. However, the symbol E can be exchanged into D in Fortran (either E or D can be used). Note that E is used for single precision and D is used for double precision. For example, 6.02 x 10^{23} and 0.55 can be written in Fortran as 6.02D+23 and 55.D-02, respectively.

As to the variables, a value attributed to a variable is stored in the computer memory. If no value is attributed to the variable, the value of the variable will be zero (0).

At the beginning of the source code, one can define whether the scalar variable will be:

- an integer number (e.g., INTEGER :: A,B);
- a real number (e.g., REAL :: C,D);
- a complex number (COMPLEX :: E,F);
- logical (LOGICAL :: matrix);
- or character (CHARACTER :: letter).

```
!name of the program: VARIABLE
integer :: total
real :: average
complex :: cx
logical :: done
character(len=80) :: message
total = 10000
average = 5000
done = .true.
cx = (0.5, -4.0)
message = 'Hello'
write (*,*), total, average, done, cx, message
stop
end
```

There are only two types of logical values: .TRUE. and .FALSE.

It is important to add that the number of decimal places in a real number can be controlled by REAL statement. The REAL*4 or just REAL statement has 7 decimal places while the REAL*8 statement has 16 decimal places. The REAL*8 is a nonstandard notation and it can be replaced by DOUBLE PRECISION statement. When using PARAMETER statement, you can set the value of the variable in the REAL construct. The REAL*8 statement has single precision with 16 decimal places, i.e., it has the precision of REAL*4. The DOUBLE PRECISION statement alone has also single precision. The double precision is only guaranteed when using DOUBLE PRECISION or REAL*8 when there is d0 after the value of the variable. The REAL*16 yields 34 decimal places with single precision or double precision (when there is d0 at the end of the value of the variable) for 32-bit Fortran. Then, d0 has to be used after the value of the variable to guarantee the double precision.

```
!REAL* and double precision program
!Program name: PRECISION
real, parameter :: a=1.12345678901234567890
!real or real*4 has 7 decimal places
double precision :: b,c
real*8 d,e
!real*8 or double precision have 16 decimal places
!real*8 has single precision with 16 decimal places
!double precision has DOUBLE PRECISION when there is d0
real*16 f,g
b=1.12345678901234567890d0
c=1.12345678901234567890
d=1.12345678901234567890
e=1.12345678901234567890d0
f=1.12345678901234567890
g=1.12345678901234567890d0
print *, a !shows 1.1234568 (single precision)
print *, b !shows 1.1234567890123457 (double precision)
print *, c !shows 1.1234568357467651 (single precision)
print *, d !shows 1.1234568357467651 (single precision)
print *, e !shows 1.1234567890123457 (double precision)
print *, f
!shows 1.12345683574676513671875000000000 (single precision)
print *, g
!shows 1.12345678901234569124767403991427 (double precision)
stop
end
```

The intrinsic types (real and integer variable or constant) may be associated with INTEGER-PARAMETER-SELECTED_REAL-KIND construct can also be used to guarantee the double precision. This construct replaces REAL*8 and DOUBLE

PRECISION. The SELECTED_REAL_KIND(p,r) guarantees the decimal precision of at least p digits and exponent range of at least r for the parameter of a real data type. The syntax of this construct is:

INTEGER,PARAMETER :: variable1=SELECTED-REAL_KIND(p,r)

REAL(KIND=variable1) :: variable2

In the next code, one can see that the suffix '_dp' has the same effect as the suffix 'd0'. Both can be used in DOUBLE PRECISION statement or INTEGER-PARAMETER-SELECTED_REAL-KIND construct.

```
!program real_kinds and double precision
!program name:PRECISION2
integer,parameter :: p6 = selected_real_kind(6)
integer,parameter :: p10r100 = selected_real_kind(10,100)
integer,parameter :: p13r200 = selected_real_kind(13,200)
integer,parameter :: r400 = selected_real_kind(r=400)
integer,parameter :: dp = selected_real_kind(15,307)
integer,parameter :: sp = selected_real_kind(6,37)
integer,parameter :: dp2 = selected_real_kind(15,307)
  double precision :: e,f
  real(kind=p6) :: x
  real(kind=p10r100) :: y
  real(kind=r400) :: z
  real(kind=p13r200) :: w
  real(kind=dp) :: a
  real(kind=sp) :: b
  real(kind=dp2) :: c,d
a=1.12345678901234567890
  b=1.12345678901234567890
  c=1.12345678901234567890_dp
  d=1.12345678901234567890d0
  e=1.12345678901234567890d0
  f=1.12345678901234567890_dp
print *, precision(x), range(x) !shows 6 37
  print *, precision(y), range(y) !shows 15 307
  print *, precision(w), range(w) !shows 15 307
  print *, precision(z), range(z) !shows 18 4931
  print *, a !shows 1.1234568357467651 (single precision)
  print *, b !shows 1.1234568
  print *, c !shows 1.1234567890123457 (double precision)
  print *, d !shows 1.1234567890123457 (double precision)
  print *, e !shows 1.1234567890123457 (double precision)
  print *, f !shows 1.1234567890123457 (double precision)
  stop
  end
```

The real value can be rounded to the nearest whole number or real number by using NINT (round to the nearest integer from REAL*4); ANINT (round to the nearest integer but return a real result); IDNINT (round to the nearest integer from REAL*8).

```
!Program name: ROUND
real*4 x
real*8 y
x=1.234e04
y=4.321
print *, x, y, nint(x), idnint(y), anint(x), anint(y)
stop
end
---------------It shows in the screen--------------------------
12340.000 4.3210000991821289 12340
4 12340.000 4.0000000000000000
```

It is a good practice to indicate at the beginning of the source code that all variables have to be declared explicitly, i.e., no variable are implicitly declared. This is done by introducing the statement IMPLICIT NONE at the very beginning of the code.

The maximum value of an integer variable is given by the HUGE function. The KIND statement can be used to determine the number of bytes to be used. The KIND(variable) is used to display the details of the hardware's data during execution of the code.

```
!Name of the program: DATA
!Examples of data
Implicit none
!maximum integer value
Integer :: A
Integer (kind=2) :: B !it uses two bytes
Integer (kind=4) :: C !it uses four bytes
Integer (kind=8) :: D !it uses eight bytes
print *, huge(A), kind(A) !it shows 2147483647
print *, huge(B), kind (B) !it shows 32767 2
Print *, huge(C) !it shows 2147483647 4
Print *, huge(D), kind (D)
!it shows 9223372036854775807 8
stop
end
```

The POINTER statement stores the memory address of an object and contains information such as object, rank and extents. The TARGET statement associates one pointer variable to other target variable by means of the symbol =>. The TARGET statement requires the POINTER statement. The pointer can be re-associated with another variable any time and the pointer can be disassociated by the NULLIFY statement. The function ASSOCIATED can be used to inquire if a pointer is associated with any target.

```
!Name of the program: POINTER
!Example of POINTER and TARGET statements
implicit none
  integer, pointer :: pa
  integer, target :: a
  pa=>a
  pa = 10
  Print *, pa
  Print *, a
  pa = pa + 4
  Print *, pa
  Print *, a
  If (associated(pa, target=a)) then
    Print *, 'pa is associated with a'
  else
    Print *, 'pa is not associated with a'
  End if
  Nullify(pa)
  Print *, pa
  Print *, a
  If (associated(pa, target=a)) then
    Print *, 'pa is associated with a'
  else
    Print *, 'pa is not associated with a'
  End if
  Stop
  End
----------RESULT ON THE SCREEN ----------------
10
10
14
14
pa is associated with a
0
14
pa is not associated with a
```

4. Algorithm structure and Fortran input/output (I/O)

An algorithm encompasses a set of statements to run a determined calculation. It is the initial architecture of a source code. Then, it is an important step prior to its codification into a specific programming language (Fortran, C, Python, etc).

A simple algorithm structure is made up of:: (i) algorithm title; (ii) input parameters (read statement); (iii) arithmetic/logic expression; (iv) output parameters (write statement); (v) end of the algorithm. Let us see the example below:

```
Fahrenheit-Celsius converter
Read Fahrenheit
Celsius ← (Fahrenheit - 32)*5/9
Write Fahrenheit, Celsius
End
```

------FORTRAN Source Code------

```
!Program name: CELSIUS
!Fahrenheit-Celsius converter
  read *, F
  C=(F-32)*5/9
  Print *, 'Fahrenheit= ', F, 'Celsius= ', C
Stop
  End
```

The statement READ * reads the input—a value given by the user—in a free-format input, that can be, for example: 21.0; 21.; +21; 21;2.1E+01. When depicting the value from the PRINT * statement, Fortran decides the precise format of the output. The "*" instructs Fortran to read or to write to the screen.

There are three different ways to determine the size of the array, n, and its elements, a(i), one might give. They are:

```
(i)  READ(*,*) n, (a(i), i=1,n)
(ii) READ(*,*) n
     READ(*,*) (a(i),i=1,n)
(iii) READ(*,*) n
    DO i=1,n
    READ(*,*) a(i)
    END DO
```

The statement WRITE is a general PRINT. Both READ and WRITE can have following forms: READ(unit,format) list; WRITE(unit,format) list. The 'unit' specifies I/O unit to use (5 corresponds to standard-in, 6 is standard-out, and 0 is standard-error). The 'format' is the statement number of the FORMAT statement that specifies the format of the depicted data. The 'list' contains all variables (from READ) or expressions (from WRITE).

The FORMAT statement instructs Fortran how to display the data (integer, real, logical and/or character). It has some field specifications. We use the following convention of symbols:

- w: number of positions to be used
- d: number of digits to the right of the decimal point
- e: number of digits of the exponent part.
- m: minimum of digits to be printed

Important to add that this number of positions is not the precision of the number. The edit descriptors for the variables (which are not case sensitive) are: I,i (for integer), F,f (for real number in decimal form), E,e or D,d (for real number in exponential for), ES, es (for real number in scientific form), L,l (for logical values), A,a (for character). The edit descriptors for the positioning are: X (horizontal positioning), T (tabbing positioning) and /(vertical positioning).

The syntax of the edit descriptors are:

- **I**w or **i**w or **I**w.m or **i**w.m
- **F**w.d or **f**w.d
- **E**w.d or **e**w.d or **D**w.d or **d**w.d
- **ES**w.d or **es**w.d
- **L**w or **l**w
- **A** or **A**w or **a** or **a**w
- n**X** or n**x**
- **T**c or **t**c

Where w, d, e, c and n are numbers
The FORMAT statement can be done in two equivalent forms:
Write (*,'(3i,F8.3)') y, z
Or
Write (*,1) y,z
1 format (3i,f8.3)

See the example below where the variable J is an integer and X and Y variables are real numbers.

```
!Program name: FORMAT
!Use of FORMAT statement
  Real*8 X/1.23/,Y/53./
  Integer a, b, c
  Real*8 d, e, f
  a= -1024
  b= 999
  c= 124438
  d= 3.14159
  e= -96.4
  f= 256.4578E-02
  J=37
Write(6,10) J,X,Y
10 Format(1X,I3/'X=',T7,F5.2/D8.2)
   Print *,
   Write (*,20) 'The area is', X
```

```
20 format (a,f8.5)
   Print *,
   Write (*,30) X,Y,J
30 format (t10,f4.2,/,t10,f6.2,/,t20,i4)
   Print *,
   Write (*,'( I10, I10, I10)') a, b, c
   Write (*,'( 3I10.8)'), a, b, c
   Print *,
   Write (*,'(F7.2, F7.2, F7.2)') d, e ,f
   Write (*,'( 3F7.4)')
    stop
    End
-------------------------It shows on the screen --------------------
37
X = 1.23
0.53D+02
The area is 1.23000
1.23
53.00
 37
–1024 999 124438
–00001024 0000999 00124438
3.14 -96.40 2.56
3.1416 ******** 2.5646
```

The statement CHARACTER is used to associate a ‘character variable’ with a word or sentence as an input (string). Then, one can use this ‘character variable’ to represent the input word/sentence wherever one likes in the source code. For example:

```
!Program name: CHARACTER
!statement CHARACTER to associate a character variable with a sentence
  character*1 LINE(80)
  Print *, 'Write a sentence of 80 characters maximum:'
  Read (5,10) LINE
10 FORMAT (80A1)
  Print *, LINE
  Stop
  End
```

However, there are other two ways to give the size of the variable character:

- CHARACTER, DIMENSION(80) :: line
- CHARACTER(len=80) :: line

There are several functions associated with the CHARACTER statement. They are:

- TRIM(character_variable): returns the value of the character variable (string) without trailing blanks
- INDEX(character_variable,name): it finds the location of the 'name' word in the string of the character variable and it shows the position of the 'name' word or else it gives zero value.

The operator // is the concatenation operator. It concatenates strings in the same line.

```
!Program name:CHARACTER2
Implicit none
character (80) :: text
character (len=6) :: title
character (len=15) :: surname, firstname
character (len=60) :: name, name2
integer :: i
Write (*,*) 'Give your title, first name and surname: '
Read *, title, firstname, surname
name= trim(title)//trim(firstname)//trim(surname)
name2 = title//firstname//surname
Print *, 'Hello,', name
Print *, 'Hello,', name2
Text='Do you believe in the power of love?'
i=index(text,'love')
If (i /= 0) then
print *, 'The word LOVE is at position: ', i
print *, 'in the text: ', text
end if
stop
end
```

5. Conditional structure in Fortran

Conditional structure/expression is a set of performed statements in case certain condition is met. It involves IF-THEN construct, but it can include ELSE in the statements.

In Fortran, in IF-THEN-ELSE construct, the THEN can be omitted after IF (logical expression). The IF statement relates to two expressions by means of the logical expression (or relational operator). There is an equivalent operator that can be used as well.

Relational operator	Meaning	Equivalent operator
.LT. or .lt.	$<$	<
.LE. or .le.	$\leq$	<=
.NE. or .ne.	1	$\neq$
.EQ. or .eq.	$=$	==
.GT. or .gt.	$>$	>
.GE. or .ge.	$\geq$	>=

Some examples:

- IF (i .ne. 0) THEN
- IF (j >= 2) THEN
- IF (answer == 'Y') THEN
- IF (a+b>i*j) THEN

It is possible to use logical expressions (.AND., .OR., .NOT., .EQV., .NEQV.) in IF-THEN-ELSE statement to make a compound expression. For example:

- IF (A .EQ. 7 .OR. A .EQ. -7) THEN
- IF (A(i,j) == MAXVAL(A) .AND. A(i,j) << 10) THEN

First example: The equations below are the quadratic equation and its root(s) that can be two real numbers (if d>0), two imaginary numbers (if d<0) and one real number (if d=0)

$$ax^2 + bx + c = 0$$

$$x = \frac{-b \pm \sqrt{b^2 - 4ac}}{2a} = -\frac{b}{2a} \pm \frac{\sqrt{b^2 - 4ac}}{2a}$$

$$d = b^2 - 4ac, \qquad x - \frac{b}{2a} \pm \frac{\sqrt{d}}{2a}$$

Calculate the roots of aquadratic equation
Read A, B and C
D ← B**2-4*A*C
E ← -B/(2*A)
F ← SQRT(ABS(D))/(2*A)
IF D<0
Write X1←E+Fi and X2←E-Fi
IF D=0
Write X=E
IF D>0
Write X1←E+F and X2←E-F
End

------FORTRAN Source Code------

```
!Program name: QUADRATIC
!Roots of the quadratic equation
  Read *, A, B, C
D = B**2 - 4.*A*C
  E = -B/(2.*A)
  F=sqrt(abs(D))/(2.*A)
if (D .lt. 0.) print *, 'X1=',E, '+i' ,F, &
& 'X2=',E, '-i', F
  if (D .eq. 0) print *, 'X=', E
  if (D .gt. 0) print *, 'X1=', E+F, ' X2=', E-F
  stop
  end
```

Second example: routine to determine whether a number is even or odd. *This routine only works because the variables are integers and the equation n/2 is round to the nearest smaller integer.*

```
!Program name: EVEN-ODD
!Routine to determine whether an input number is even or odd
integer :: n,nh
Print *, "input n: "
Read *, n
nh=n/2
! If you type 1, 3, 5, (odd number), the result is rounded to 0, 1, 2
! If you type 0, 2, 4, (even number), the result is rounded to 0, 1, 2
print *, nh
if (2*nh==n) then
print *, 2*nh, "even"
else
print *, 2*nh, "odd"
! If n=1, nh=0, 2*nh=0, 1≠0, odd
! If n=2, nh=1, 2*nh=2, 2 = 2, even
! If n=3, nh=1, 2*nh=2, 3 ≠ 2, odd
! If n=4, nh=2, 2*nh=4, 4 = 4, even
end if
stop
end
```

6. Repetitive structures (loops) in Fortran

Repetitive structure/expression is a type For-Until loop (or iteration) statement that tells the computer to repeat sections of statements several times. The repetitive structure can have a counter or not. The repetitive structure with a counter is: FOR <control variable>←<initial value> UNTIL <final value>. At each loop, the default increment is a unit (step 1). If the increment is greater than one, then use STEP <value of the increment>

In Fortran, the repetitive structure can be made in three forms: (i) DO loop with a counter; (ii) DO-EXIT construct; (iii) DO-WHILE construct. There can also be nested loops where one loop is inside the other. The nested loops can be named, e.g., outer and inner. In addition, the DO loop can be done in two different ways:
First way:

DO <number>i=<initial_value>,<final_value>,<step>

<number> CONTINUE

Second way:

DO i=<initial_value>,<final_value>,<step>

END DO

First example:

```
! I values and I**2 values - loop
do 1 I=1,20,2
  ISQ=I**2
  print *, 'I= ', I , 'I**2= ', ISQ
1 continue
  stop
 end program
```

Second example:

```
!Program name: SUM
!Summation from 1 to 100
integer::i,sum
sum=0
Do i=1,100
sum=sum+i
End do
Print *,'Sum(1-100)= ',sum
Stop
End
```

Third example:

```
!Program name: EPSILON
!Epsilon - calculation of the precision
  EPS=1.
  do
   EPS=EPS/2
   If (EPS+1 .eq. 1.) exit
  End do
  Print *,'Epsilon= ', EPS
Stop
 End
```

Fourth example:
The exponential (e) of a number (x), e^x, can be given by sum below:

$$e^x \cong sum = 1 + x + \frac{x^2}{2!} + \frac{x^3}{3!} + \frac{x^4}{4!} \cdots \frac{x^n}{n!}$$

Where n is the truncation number. The higher n, the higher the precision of the sum.

```
!Program name: EXPONENTIAL
!Exponential of x
Integer::i,j,n,fact
real::x,sum
Print *, 'Let us calculate Exp(x)'
Print *, 'Input a number for x: '
read *, x
Print *, 'Input a truncation number (n) for the series: '
read *, n
sum=1
outer: do i=1,n
fact=1
   inner: do j=1,i
      fact=fact*j
   End do inner
   sum=sum+(x**i)/REAL(fact)
End do outer
Print *, 'Exp(x)= ', sum
stop
end
```

Fifth example: Calculation of the arithmetic mean (μ) and standard deviation (σ).

$$\mu = \frac{1}{N}\sum_{i=1}^{N} x_i, \qquad \sigma = \sqrt{\frac{1}{N-1}\sum_{i=1}^{N}(x_i - \mu)^2}$$

Where N is the number of elements, x_i, in the sample.

```
!Program name: MEANSIGMA
!Calculation of arithmetic mean and standard deviation
Parameter(IMAX=100)
   REAL*8 X(IMAX),SUM,SSQ,MI,SIGMA
   SUM=0
   N=0
Do 1 I=1,IMAX
   write(0,10)
   10 Format('Input the N values of X (type 000 when finished): ',$)
   Read(5,*) X(I)
     If(X(I) .eq. 000) then
       go to 2
     end if
```

```
  SUM=SUM+X(I)
N=N+1
1 continue
2 MI=SUM/DFLOAT(N)
  SSQ=0
  Do 3 I=1,N
    SSQ=SSQ+(X(I)-MI)**2
3 continue
  SIGMA=DSQRT(SSQ/DFLOAT(N-1))
  Write(6,20) MI,SIGMA
20 format(/'mean= ',F7.3/'standard deviation= ',F5.3)
  stop
  end
```

The EXIT statement is used to leave a loop and CYCLE to skip the rest of the current iteration of a single loop. They are used after IF (logical_expression) in the same line.

The WHILE statement can be used in DO loops meaning that the iteration will continue until one logical expression is met. DO WHILE (logical_expression) is similar to DO (variable=startvalue,stopvalue) IF (.not. logical_expression) *break* or it can be equivalent to DO EXIT statement with opposite logical expression. See the example below where both routines with DO EXIT and DO WHILE (with inverted logical expression) give the same result.

Sixth example: square root

Obs.: See section 10 for ALLOCATABLE, ALLOCATE and DEALLOCATE statements. To sum up DEALLOCATE clears the memory of the final values of variables used in a (previous) DO loop routine so that they can be used in a next DO loop routine with their initial values.

```
!Program name: SQUARE
!Find approximate square root
Implicit none
Real, allocatable :: x, a, b
allocate (x)
allocate (a)
allocate (b)
Write (*,*) 'Enter a number to find its approximate square root: '
Read *, x
!use the first routine with EXIT statement
a=1; b=x/a
do
  a=(a+b)/2.0
  b=x/a
print *, 'New approximation of square root of', x, 'is: ', a
```

```
  if ( abs(a-b) < 1e-10 ) then
     exit
  end if
end do
Print *
Print *, 'Precise square root of', x, 'is: ', sqrt(x)
deallocate (a)
deallocate (b)
!Use the second routine with WHILE STATEMENT
a=1; b=x/a
do while ( abs(a-b) > 1e-10 ) !inverted logical expression
  a=(a+b)/2.0
  b=x/a
  print *, 'New approximation of square root of', x, 'is: ', a
end do
Print *
Print *, 'Precise square root of', x, 'is: ', sqrt(x)
end
```

Seventh example

```
!Program name: CIRCLE
!Area of a circle
Do
  Print *, 'Enter the radius of the circle: '
  Read *, r
  Print *, 'The area is: ', 3.1416*r**2
  Print *, 'Do you want to calculate another area? (Y/N): '
  Read *, resp
  If (resp == N) exit
End do
Stop
end
```

7. Select-case construct in Fortran

The SELECT CASE construct is a type of conditional statement that executes a statement depending on the value of a scalar expression in a CASE statement. It has the general form:

```
[name:] SELECT CASE (variable)
                CASE ('value1', 'value2',...)
                Statement 1
                CASE (low:high)
                Statement 2
                CASE DEFAULT
                Statement 3
END SELECT
```

There can be more than one value in each CASE statement and there can also be minimum to maximum (low:high) range in each CASE statement. The CASE DEFAULT statement means that any value (for the variable) different from the scope of the anterior CASE statements will fall in the statement defined after CASE DEFAULT. Let us see the example below:

```
!Program name: CASE
!Select case construct
character(len=1) :: grade
integer :: mark
Print *, 'Select your grade from A to D: '
Read *, grade
Select case (grade)
  case ('A','a')
     write (*,*) 'Excellent!'
  case ('B','b')
     write (*,*) 'Congratulations'
  case ('C','c')
     write (*,*) 'Well done'
  case ('D','d')
     write (*,*) 'Try again'
  case default
     write (*,*) 'Invalid grade'
End select
Print *, 'Your grade is: ', grade
Print *, 'Select your mark from 0 to 100: '
Read *, mark
Select case (mark)
  case (91:100)
     write (*,*) 'Excellent!'
  case (75:90)
     write (*,*) 'Congratulations'
  case (60:74)
     write (*,*) 'Well done'
  case (:59)
     write (*,*) 'Better try again'
  case default
     write (*,*) 'Invalid mark'
end select
Print *, 'Your mark is: ', mark
stop
end
```

8. Intrinsic and external functions in Fortran

Fortran and other programing languages provide commonly used built-in functions, so-called intrinsic functions. They are:

Intrinsic function	Meaning
ABS(X)	Absolute value of X
SQRT(X)	Square root of X
CONJG (X)	Complex conjugate of X
INT(X)	Rounding off to lower integer
NINT (X)	Rounding off to closer integer
CBRT(X)	Cube root of X
SIGN (X,Y)	Print sign of Y in X
SIN(X)	Sine of X radian
COS(X)	Cosine of X radian
TAN(X)	Tangent of X radian
ASIN(X)	Arc sine of X
ATAN(X)	Arc tangent of X
ACOS(X)	Arc cosine of X
EXP(X)	Exponential of X
MAX(a,b,c,d…)	Maximum value of the series
MIN (a,b,c,d…)	Minimum value of the series
EXPONENT(X)	Exponent part of X
FRACTION (X)	Fraction part of X
LOG(X)	Natural logarithm of X
LOG10(X)	Common logarithm of X
CONJG(X)	Conjugate of complex number X
ERF(X)	Error function: 2/sqrt(pi)* integral from 0 to X of exp(-t*t)dt

Calculate the hypotenuse
Read A and B
HYP ← square root (A**2 + B**2)
Write A, B and HYP
End

------FORTRAN Source Code------

```
!Program name: HYPOTENUSE
!Hypotenuse
read *, A,B
print *, 'A=', A, ' B=', B
hyp = sqrt (A**2 + B**2)
print *, 'hyp=', hyp
stop
end
```

The function SIGN can be used either for scalar variables or array variables (see ahead).

```
!Program name: SIGN
Integer :: A(12)
!ARRAY A in senoidal behavior
A=[1,2,1,-1,-2,-1,1,2,1,-1,-2,-1]
write (*,*) sign(1,A)
! It shows 1 1 1 -1 -1 -1 1 1 1 -1 -1 -1
Write (*,*) sign (1,-2), sign (2,-1)
!It shows -1 -2
Write (*,*) int(3.7), nint(3.7)
!It shows 3 4
write (*,*) int(2.3), nint(2.3)
!It shows 2 2
end
```

There are other important functions that are not in-built. Then, it is needed to write this function. This is the case of the external function. An external function is declared by the FUNCTION statement followed by the name of the variable declaration and an argument between parentheses, FUNCTION <name> (argument). The argument contains the variables of the function. The function subprogram ends with RETURN and END statements. In the example below, X, Y, Z and SUM are dummy arguments used only in the FUNCTION subprogram and they are equivalent to A, B and C, respectively (the variable SUM has no correspondence).

```
!Program name: AVERAGE
!Average of the data A, B and C
Program average
Real A,B,C,AV
Data A,B,C/5.0,2.0,3.0/
AV=AVRAGE(A,B,C)
Print *, 'The numbers ',A,B,C
Print *, 'have average= ',AV
End
REAL function AVRAGE(X,Y,Z)
REAL X,Y,Z,SUM ! dummy arguments
SUM=X+Y+Z
AVRAGE=SUM/3.0
RETURN
end
```

9. Matrices and vectors in Fortran

An array is a data structure to represent matrices and vectors. An array is one list of ordered scalar elements in one dimension (vector) or two dimensions (matrix). The construction of an array always requires a repetitive structure.

First example: Building a vector

```
!Program name: VECTOR
!Building a 5-dimensional vector and sum of its elements
  REAL*8 X(5),sum
  Print *,'Input the five scalar elements of the 5-D vector:'
Read *,X
  sum=0.
  do 1 I=1,5
     sum=sum+X(I)
1 continue
  Print *,'The sum of the five elements of 5-D vector is: ',sum
stop
 end
```

Second example: Building a matrix

```
!Program name: MATRIX
!Building a 2x3 matrix
INTEGER*4 MAT(2,3)
  Do 1 J=1,3
  Do 1 I=1,2
MAT(I,J)=(5*I)+(7*J)
     Print *, MAT(I,J)
1 continue
  Print *, 'Matrix 2x3 (5i+7j): '
  Print *, MAT(1,1),MAT(1,2),MAT(1,3)
  Print *, MAT(2,1),MAT(2,2),MAT(2,3)
  stop
  end
```

Third example: Building a matrix 2

```
!Program name:MATRIX2
  Integer :: n,i,j
Integer, dimension(:,:), allocatable :: A
  Print *, 'Enter the dimension, n, of the square matrix A: '
Read *, n
  Allocate ( A(n,n) )
  Do i=1,n
     Do j=1,n
     Print *, 'A(',i,',',j,')= '
```

```
Read *, A(i,j)
  End do
End do
  Do i=1,n
     Do j=1,n
     Print *, 'A(',i,',',j,')= ', A(i,j)
End do
  End do
Deallocate (A)
stop
end
```

Fourth example:
The summation, S, of product elements, $a_i b_i$, has a general formula below:

$$S = \sum_{i=1}^{n} a_i b_i$$

Where n is the number of elements of both vectors A and B. The Fortran implementation for this sum is:

```
!Program name: SUMMATION
!Summation of product elements, ab, of vectors A and B
  Real*8 S,A(5),B(5)
  Print *, 'Input the five elements of vector A: '
  Read *, A
  Print *, 'Input the five elements of vector B: '
  Read *, B
  S=0
  Do 1 I=1,5
     S=S+A(I)*B(I)
1 continue
  Print *, 'The summation of the product elements, ab, is: ', S
  stop
  end
```

10. Assignment/declaration of an array and dynamic arrays

The dimension(s) of an array can be given by the DIMENSION statement. The number maximum of dimensions is seven. For example:

REAL, DIMENSION(2,3) :: A

INTEGER, DIMENSION(10,20,3) :: B

COMPLEX, DIMENSION(5) :: C

In the example above, the array A has two dimension and 2*3 = 6 elements; the array B has three dimensions and 10*20*3 = 600 elements; the array C has one dimension and 5 elements. The number of elements of an array defines its size.

It is also possible to declare the upper bounds and lower bounds of an array. For example:

REAL, DIMENSION(-10:5) :: D

REAL, DIMENSION(-10:5, -20:-1, 0:1, -1:0, 2, 2, 2) :: E

The size of the array D is 16, from D(-10) to D(5), the upper and lower bounds, respectively. The size of the array E is 16*20*2*2*2*2*2=10240.

The values of the elements of an array may be declared in several ways:

```
A = (/ 1, 2, 3, 4, 5, 6 /)
A=[1,2,4,5,6]
B = A (only when they have the same dimensions)
DATA A /1,2,3,4,5,6/
REAL, DIMENSION(6), PARAMETER :: A = (/ 1, 2, 3, 4, 5, 6 /)
```

It is also possible to declare the values of an array for each row or column in the following way

```
!Program name: ARRAYS
real :: A(4), B(4,4)
integer :: i
A=[1.,-0.5,5.,3.3]
B(1,:)=A ! set the first row be elements of A
B(2,:)=[(B(1,i)**2,i=1,4)] !set the 2nd row be the square of the first row
B(3,:)=0. ! set the third row be zero
B(4,:)=sqrt(abs(A)) ! set the forth row be the square root of A
Print *, A
Print *,
do i=1,4
   print *, B(i,1), B(i,2), B(i,3), B(i,4)
end do
stop
end
```

The result of this code is:

```
1.0000000   -0.50000000   5.0000000   3.3000000
1.0000000   -0.50000000   5.0000000   3.3000000
1.0000000    0.25000000   25.000000   10.889999
0.0000000     0.0000000   0.0000000   0.0000000
1.0000000    0.70710677   2.2360680   1.8165902
```

The PARAMETER attribute permits to include the values of an array in the variable declaration.

```
!Program name: DIMEN
! Dimension and data of arrays A, B, C, D
integer, dimension(6) :: A,B,C,D
A=(/1,2,3,4,5,6/)
B=2*A
Data D/7,8,9,10,11,12/
Print *, A
Print *, B
Print *, D
Do i=1,6
  C=A+i
  Print *, C
end do
stop
end
```

There is another alternative to assign values of an array:

- array ([lowerbound]:[upperbound]) = value.

```
Integer, dimension(10) :: a
a(1:5) = 10 ! a(1) to a(5) assigned 10
a(6:) = 5 ! a(6) to a(10 assigned 5
```

The values of an array (one dimensional array) can be given using implied do syntax. In the example below, the array A will give four values 2*i where i=1 to 4. The array B will give 4 times 6, 24, values where they follow a nested do notation where i belongs to an equivalent inner DO loop and j belongs to an equivalent the outer DO loop.

```
!Program name: DIMEN2
Integer,dimension(4) :: A
Integer,dimension(24) :: B
A=(/ (2*i, i=1,4) /)
B=(/ ((i*j, i=1,4), j=1,6) /)
write (*,*) A
write (*,*) B
stop
end
```

An array can also be assigned by WHERE construct which respect the logical array expression in the WHERE statement. For example:

```
!Program name: DIMEN3
real, dimension(10) :: A=(/ 1,2,3,4,5,6,7,8,9,10 /)
Real, dimension(10) :: B
where (A>5)
  B=1.
Elsewhere
  B=0
end where
write (*,*) A, B
stop
end
```

The UBOUND function gives the value of the upper bound of an array. The LBOUND function gives the value of the lower bound of an array. Remember that arrays of rank 1 such as A(10) or B(0:9) have 10 elements from 1 to 10 or 0 to 9 (if integers), respectively; and that an array of rank 2 such as C(20, 0:19) has 400 elements where first rank is indexed from 1 to 20 and the second rank is indexed form 0 to 19.

```
!Program name: DIMEN4
Integer,dimension (-3:4,2:5) :: i !array of rank 2 from -3 to 4 and from 2 to 5
Integer,dimension (0:6) :: j ! array of rank 1 from 0 to 6
Integer :: k(3,4) ! array of rank 2 from 1 to 3 and from 1 to 4
Write (*,*) ubound (i) ! it shows 4 5
Write (*,*) lbound (i) ! it shows -3 2
Write (*,*) ubound (j) ! it shows 6
write (*,*) lbound (j) ! it shows 0
Write (*,*) ubound (i,1) ! it shows 4
Write (*,*) ubound (i,2) ! it shows 5
Write (*,*) lbound (i,1) ! it shows -3
Write (*,*) lbound (i,2) ! it shows 2
write (*,*) ubound(k) ! it shows 3 4
write (*,*) lbound(k) ! it shows 1 1
Stop
End
```

In a dynamic array the size of the array is unknown during compilation, but it will be given during the execution. They are declared by the statement ALLOCATABLE, followed by the function ALLOCATE and terminated with DEALLOCATE function. In the example below, the order n of the square matrix is defined by the user as well as each of its matrix elements. The symbol: in DIMENSION statement is called pointer.

```
!Program name: DYNAMICARRAY
! Order n of the matrix A and its matrix elements
   Integer :: n,i,j
   Integer, dimension(:,:), allocatable :: A
   Print *, 'Enter the dimension, n, of the square matrix A: '
Read *, n
   Allocate ( A(n,n) )
   Do i=1,n
      Do j=1,n
      Print *, 'A(',i,',',j,')= '
Read *, A(i,j)
      End do
End do
   Do i=1,n
      Do j=1,n
      Print *, 'A(',i,',',j,')= ', A(i,j)
End do
   End do
Deallocate (A)
stop
end
```

The SIZE function gives the amount of the elements of an array.

11. Array operations and intrinsic array functions in Fortran

All mathematical operations for scalar numbers apply to arrays as well, element-by-element.

Likewise scalar numbers, there are also intrinsic functions for arrays. For example:

- MAXVAL(array): returns the maximum value of an array
- MINVAL(array): returns the minimum value of an array
- PRODUCT(array): returns the product of all elements
- SUM(array): returns the sum of all elements of an array
- SIZE(array): returns the number of elements of an array
- TRANSPOSE(A): returns the transpose of the matrix A
- SHAPE(A): returns the shape of the matrix A
- RESHAPE(array,(dimen)): transforms a vector into matrix
- MERGE(A,B,mask): builds a matrix from A and B according to the logical values of the mask (true or false).
- SIGN(X,A): prints sign of each element of array A on X.

- MATMUL(A,B): matrix multiplication for the dimensions (m,k) and (k,n) of the matrices A and B, respectively
- DOT_PRODUCT(A,B): scalar product of vectors A and B of the same dimensions.

```
!Program name: ARRAYS2
real :: A(4), B(4), c(4), D(4,4), E(4,4),F(4,4), G(4,4)
integer :: i
A=[1.,0.5,-3.,5.5]
B=(2*A)/3
c=B**2
D(1,:)=A
D(2,:)=(B*3)/2
D(3,:)=sqrt(C)
D(4,:)=A+B+C
E=D*3
print *, A
Print *,
Print *, B
print *,
Print *, C
print *,
do i=1,4
print *, D(i,1), D(i,2), D(i,3), D(i,4)
end do
print *,
do i=1,4
print *, E(i,1), E(i,2), E(i,3), E(i,4)
end do
print *,
print *, maxval(A), minval(B), product(C), sum(D), size(E)
print *,
F=transpose(E)
do i=1,4
print *, F(i,1), F(i,2), F(i,3), F(i,4)
end do
print *,
print *, dot_product(A,B)
print *,
G=matmul(D,E)
do i=1,4
print *, G(i,1), G(i,2), G(i,3), G(i,4)
end do
stop
end
```

12. Subprograms in Fortran

There are three types of subprograms within a main program: subroutines, external functions and modules. The main program invokes the subroutine/function/module using specific statements for each case.

The statement CALL is a kind of branching transferring the control to the named subroutine. Then, each subroutine has a specific name. The subroutine begins with the statement SUBROUTINE and finishes with the statement RETURN which transfers back the control to the main program.

```
!Program name: SUBROUTINE
!Building a 3x3 matrix and its transpose
INTEGER*4 F(3,3),FT(3,3)
  Do 1 J=1,3
  Do 1 I=1,3
     F(I,J)=(5*I)+(7*J)
     Print *, F(I,J)
1 continue
  Print *, 'Matrix F 3x3 (5i+7j): '
Print *, F(1,1),F(1,2),F(1,3)
Print *, F(2,1),F(2,2),F(2,3)
  Print *, F(3,1),F(3,2),F(3,3)
  Call MATRNS(F,FT)
  Print *, 'Matrix transpose FT 3x3 (5j+7i): '
  Print *, FT(1,1),FT(1,2),FT(1,3)
  Print *, FT(2,1),FT(2,2),FT(2,3)
  Print *, FT(3,1),FT(3,2),FT(3,3)
  stop
  end

  Subroutine MATRNS(F,FT)
Integer*4 F(3,3),FT(3,3)
Do 2 J=1,3
  Do 2 I=1,3
     FT(J,I)= F(I,J)
2 continue
  return
  end
```

In the next subroutine, there are dummy variables (arguments) which are used only in the subprogram. The dummy arguments are internal and replace external variables within a subroutine. In addition, the subprogram can associate each variable of the main program with a dummy variable of the subprogram in order of appearance. One can see in the example below that the variables x, y, z, disc are interchanged to a, b, c and d, respectively. Then, a, b, c and d are called dummy arguments.

```
!Program name: SUBROUTINE2
program subroutine
implicit none
  real :: x, y, z, disc
  x = 1.0
  y = 5.0
  z = 2.0
  call discriminant(x, y, z, disc)
  Print *, "The value of the discriminant is"
  Print *, disc
end program subroutine
subroutine discriminant (a, b, c, d)
implicit none
  ! dummy arguments
  real :: a
  real :: b
  real :: c
  real :: d
  d = b * b - 4.0 * a * c
end subroutine discriminant
```

In the next example, we use the INTERFACE statement to build two matrices (M.N) from two vectors (U,V) and to merge both matrices in another matrix, R.

```
!Program name: SUBROUTINE3
integer :: V(6), U(6), M(2,3), N(2,3), R(2,3)
logical :: T(6), L(2,3)
interface
  subroutine write_array(A)
  integer :: A(:,:)
  end subroutine write_array
  subroutine write_logical_array(A)
  logical :: A(:,:)
  end subroutine write_logical_array
end interface
V=[11,12,13,14,15,16]
U=[1,2,3,4,5,6]
T=[.true., .false., .true., .true., .false., .false.]
M=reshape(V,(/2,3/))
N=reshape(U,(/2,3/))
L=reshape(T,(/2,3/))
R=merge(M,N,L)
```

```
call write_array(M)
call write_array(N)
call write_logical_array(L)
call write_array(R)
end

subroutine write_array(A)
integer :: A(:,:)
do i=lbound(A,1), ubound(A,1)
  write (*,*) (A(i,j), j=lbound(A,2),ubound(A,2))
end do
return
end subroutine write_array

subroutine write_logical_array(A)
logical :: A(:,:)
do i=lbound(A,1), ubound(A,1)
  write (*,*) (A(i,j), j=lbound(A,2),ubound(A,2))
end do
return
end subroutine write_logical_array
```

The INTERFACE statement gives the full knowledge of the types and characteristics of the dummy arguments that are used in the subroutines. If one removes the INTERFACE construct from the program SUBROUTINE3 (above), there will be several errors of the type "procedure <subroutine_name> at (1) with assumed-shape dummy argument 'a' must have an explicit interface" for each CALL statement.

Any external function should be listed in INTERFACE block along with the declaration of its arguments and their types and the type of function value. For each external function used in module or subroutine should have an entry in an interface block. The syntax of the INTERFACE block is shown below. Notice that INTENT(IN) is optional.

```
INTERFACE
  Type FUNCTION name (arg-1, arg-2,...,arg-n)
   Type, INTENT(IN) :: arg-1
   Type, INTENT(IN) :: arg-2
   ................
   Type, INTENT(IN) :: arg-n
   END FUNCTION name
   .............other functions.........
END INTERFACE
```

As stated in a previous section, an external function is a subprogram whose function's variable gives a specific result according to an input. In the example below, the variable 'a' of the main program is associated with the variable of the function 'circle_area' and the input is 'r'.

a = circle_area(r)

The function 'circle_area' has an expression for the area of the circle written in the subprogram 'function circle_area(r)'. It calculates this area according to the input 'r'. The variable 'a' is associated with 'circle_area(r)' in the main program to show the final result.

```
!Name of the program: FUNCTION
  Real :: a, r
  character (len=1) :: answer
  Print *, 'Input the radius of the circle: '
  Read *, r
! Variable function:
  a = circle_area(r)
  Print *, 'The area of a circle with radius ', r, ' is: '
  Print *, a
10 Print *, 'Do you want to calculate another area of a circle (Y/N)?'
  Read *, answer
  If (answer == 'Y') then
     Print *, 'Input the radius of the circle: '
     Read *, r
     a = circle_area(r)
     Print *, 'The area of a circle with radius ', r, ' is: '
     Print *, a
     Go to 10
  else
  print *, 'Ok. Goodbye.'
 end if
Stop
End
! this function computes the area of a circle with radius r
Function circle_area (r)
implicit none
  real :: circle_area
  ! local variables
real :: r
  real :: pi
  pi = 4 * atan (1.0)
area_of_circle = pi * r**2
end function circle_area
```

The explicit function can also be declared from FUNC (variable) argument which is given at the beginning of the code. See the example below:

```
!program name: FUNCTION2
real :: r
func(r) = (4*atan(1.0))*r**2
Print *, 'give the radius of the circle: '
Read *, r
print *, 'the area of the circle is: ', func(r)
stop
end
```

The argument FUNC as well as the subprogram FUNCTION can have more than one variable. See the examples below:

```
!Program name: FUNCTION3
!Euler's method for differential equation dy/dx=-xy
!The exact solution: y=exp(-x^2/2)
!y(n+1)= y(n)+h(-xy)
!Y(n+1) <=> X+h
real :: x, y, h, exact, error
integer :: i
func(x,y)=-x*y
h=0.01
y=1.
do i=0,1
  x=i
  y=y+h*func(x,y)
exact=exp((-(x+h)**2)/2)
error=exact-y
  print *, i, x+h, y, exact, error
end do
stop
end
```

The variables of the explicit function can be declared in the PRINT statement when calling the respective function.

```
!Program name: FUNCTION4
function f(a,b)
integer :: a
real :: b, f
f=b**2-4*a*b
return
end function
program function4
real :: f
print *, 'a=4; b=2.2', ' b^2-4ab= ', f(4, 2.2)
end program
```

The statement INTENT(IN) is used for variables which are not changed in the subprogram/main program while for INTENT(OUT) statement the variables in the subprogram/main program are overwritten.

If an external function or subroutine has to call itself again in the same subprogram, then it is necessary to include the RECURSIVE statement to declare the FUNCTION or SUBROUTINE statements. In addition, in the same line the RESULT statement also has to be used.

- RECURSIVE FUNCTION name (variable_function) RESULT (dummy-argument)

In the example below, the variable 'i' is associated with the dummy argument 'n' and the variable 'f' is associated with the dummy argument 'fac'. The function 'factorial(n)' is called recursively in the line:

- fac = n *factorial (n-1)

But, without the RECURSIVE statement, the code cannot return to the same calculation (within this function) again.

```
!Program name: RECURSIVE
!calculate the factorial of a number n!
program recursive
  integer :: i, f
  character (len=1) :: answer
  Print *, 'Input the value to be factorized: '
  Read *, i
! Variable function:
  f = factorial(i)
  Print *, 'The value of factorial of', i, ' is: '
  Print *, f
10 Print *, 'Do you want to calculate another factorial (Y/N)?'
  Read *, answer
  If (answer == 'Y') then
     Print *, 'Input another value to be factorized: '
     Read *, i
     f = factorial(i)
     Print *, 'The value of the factorial of ', i, ' is: '
     Print *, f
     Go to 10
  else
  print *, 'Ok. Goodbye.'
 end if
end program recursive
! computes the factorial of n (n!)
recursive function factorial (n) result (fac)
! function result
```

```
implicit none
  ! dummy arguments
  integer :: fac
  integer, intent (in) :: n
  select case (n)
     case (0:1)
fac = 1
     case default
fac = n * factorial (n-1) !the function is called recursively here.
  end select
end function factorial
```

The code can be written without dummy arguments and using IF-ELSE construct (see the example below).

```
!Program name: RECURSIVE2
program factorial
integer :: n, f
character (len=1) :: answer
  Print *, 'Input the value to be factorized: '
Read *, n
  f = factorial(n)
Print *, 'The value of factorial of', n, ' is: '
  Print *, f
10 Print *, 'Do you want to calculate another factorial (Y/N)?'
  Read *, answer
  If (answer == 'Y') then
     Print *, 'Input another value to be factorized: '
Read *, n
     f = factorial(n)
Print *, 'The value of the factorial of ', n, ' is: '
     Print *, f
     Go to 10
  else
  print *, 'Ok. Goodbye.'
 end if
end program factorial
! computes the factorial of n (n!)
recursive function factorial (n) result (f)
! function result
```

```
implicit none
  ! dummy arguments
  integer :: f
  integer, intent(in) :: n
  If (n == 0) then
     f = 1
     else
        f = n * factorial (n-1) !the function is called recursively here
  end if
end function factorial
```

The MODULE-USE construct is another type of subprogram, data and interface block and it provides a way for splitting the code in multiple sections. Within the MODULE construct it can contain a subroutine and CONTAINS statement. This statement links the variable values in a subprogram (subroutine within a module construct or not) with the main program and the USE statement makes them available. The MODULE construct can have subroutine and functions subprograms inside. When using FUNCTION statement it has to accompanied by the RESULT statement. See the example below:

```
!Program name: MODULE
!Area of the circle and exponent - module example
module circle_exponent
Implicit none
real, parameter :: pi = 3.1415926536
  real, parameter :: e= 2.7182818284
contains
  subroutine consts()
     print*, "Pi = ", pi
     Print*, 'e= ', e
  end subroutine consts
  function areaCircle(r) result(a)
  implicit none
real::r
     real::a !dummy argument
a = pi * r**2
  end function areaCircle
  function exponent(x) result(epx)
     implicit none
     real::x
     real::epx!dummy argument
epx = e**x
  end function exponent
end module circle_exponent
```

```
!main program
use circle_exponent
Real :: r
Real :: x
   call consts()
   Print *, 'Input the radius of the circle: '
   read *, r
   print*, 'Area of a circle with radius', r, '= ', areaCircle(r)
   Print *, 'Input the exponent of the base e: '
   read *, x
   print*, 'The base e power to', x, '= ', exponent(x)
   Print *, 'Good bye'
stop
end
```

If one removes the MODULE construct from the program above (MODULE), there will be several errors of several distinguished types. Another example of MODULE construct is given in Section 21 for the program BISECTION3. In this program (BISECTION3), one can see that RESULT statement appears without the RECURSIVE statement in the FUNCTION subprogram.

13. Logical variables in Fortran

When it is necessary to store and manipulate logical values explicitly (not using IF conditionals), as in the example below, Fortran provides the logical variables.

In the example below, the code begins declaring two logical variables: *done* and *even*. This code is a nice example of iteration process using logical variables (in that case, variable *done*) and how to use logical variables for a series interchanging the sign of the sum for odd and even values (in that case, using variable *even*). The variable *done* is attributed .FALSE. until it reaches the variable abs (X-XOLD) lower than 0.01 .or. the variable I greater than 100 when the code is stopped. While variable *done* is .FALSE. the conditional of line 1 is not followed and it goes to ELSE statement below.

The logical value in variable *even* plus the 'even= .not. even' argument are used to afford the interchanging type of sum executed at each iteration. There are two types of the sum in the code.

```
X=X-(1./float(I))
X=X+(1./float(I))
```

Notice that X=X-(1./float(I)) is equivalent to X=X-1./float(I) and that X=X+(1./float(I)) is equivalent to X=X+1./float(I). Then, the fraction includes only the number closer to division operator, /, of the numerator when there is a summation in the numerator. The FLOAT statement converts the integer number into a default real value.

For even=.TRUE., the sum X=X+(1./float(I)) is used for the first, third, fifth, and so on, iterations (i.e., I=1, I=3, I=5,...), while the sum X=X-(1./float(I)) is used for the second, fourth, sixth, and so on, iterations (i.e., I=2, I=4, I=6, ...).

X = 0	I = 1	X' = 0 + 1 = 1
X = 1	I = 2	X' = 1–1/2 = 0.5000
X = 0.500	I = 3	X' = 0.500 + 1/3 = 0.8333
X = 0.833	I = 4	X' = 0.833–1/4 = 0.5833

Then, at each first iteration, *even* becomes '.not. even'; in the second iteration, it becomes '.not. .not. even' which yields 'even=.true'.; in the third iteration, *even* becomes '.not. even' again; in the forth iteration, it becomes '.not. .not. even' which yields 'even=.true.' again, and so on.

Just out of curiosity, if even=.FALSE, it would give a reversed operation during the iteration, that is, the sum X=X+(1./float(I)) would be used for the first, third, fifth, and so on, iterations, while the sum X=X-(1./float(I)) would be used for the second, fourth, sixth, and so on, iterations. As a consequence, all the values of X would have negative sign as depicted below:

X = 0	I = 1	X' = 0–1= –1
X = –1	I = 2	X' = –1 + 1/2 = –0.5000
X = –0.500	I = 3	X' = –0.500–1/3 = –0.8333
X = –0.833	I = 4	X' = –0.833+1/4 = –0.5833

The sum in this code has an alternating series with positive sign for odd denominators (I) and negative sign for even denominators. This series is equivalent to the Ln 2. The series ranges from i = 1 to i = infinite. However, in the code below, it is truncated to i = 100. Another truncation is the difference between successive values of X during the iteration. When it is 0.01 or lower, the loop is interrupted.

$$x = 1 - \frac{1}{2} + \frac{1}{3} - \frac{1}{4} + \ldots = \sum_{i=1}^{\infty} \frac{(-1)^{i+1}}{i} = \ln 2$$

```
!Program name: LOGICAL
!Value of Ln 2
  Logical done/.false./, even/.true./ ! first done and subsequent done values are
 false until either condition is met at the end of the code.
  I=0
  X=0
1 If (done) then ! first done and subsequent done values are false and the code
goes to else
      Print *, 'Iteration is over. Ln 2 is: ', X
      stop
```

```
else
     I=I+1
     XOLD=X !first X is zero (even number)
     even= .not. even ! swap even and not even at each iteration
      If (even) X=X-(1./float(I)) ! the second, fourth, sixth,... X values are
    computed here.
      If (.not. even) X=X+(1./float(I)) !the first, third, fifth,... X values are
    computed here.
     Print *, 'XOLD:', XOLD, 'I:', I, 'X:', X
     done=(abs(X-XOLD) .lt. 0.01) .or. (I .ge. 100)
     go to 1
 end if
 end
```

The first lines of the output are:

XOLD: 0.0000000 I: 1 X: 1.0000000

XOLD: 1.0000000 I: 2 X: 0.50000000

XOLD: 0.50000000 I: 3 X: 0.83333331

XOLD: 0.83333331 I: 4 X: 0.58333331

And the last lines of the output are:

XOLD: 0.69827485 I: 98 X: 0.68807077

XOLD: 0.68807077 I: 99 X: 0.69817179

XOLD: 0.69817179 I: 100 X: 0.68817180

Iteration is over. Ln 2 is: 0.68817180

14. Data type construct in Fortran

The TYPE construct is used to represent a library or database consisting of objects/components of different types (called structure). The TYPE statement (*TYPE type_name*) defines the name of the library, i.e., the set of components within the type name (structure). Each member of the library/database is defined by statement TYPE as well whose syntax is: *TYPE(type_name) :: member_number*. The components each member are variables that are declared in the following syntax: *member_number%component="constant-name"*.

First example:

```
!Program name: TYPE
program bookstore
!type declaration
  type Books
     character(len = 50) :: title
     character(len = 50) :: author
     character(len = 150) :: subject
```

```
    integer :: year
  end type Books
!declaring type variables
  type(Books) :: book1
  type(Books) :: book2
!enter the components of the structure
  write(*,*) 'Enter the title of the book 1'
  Read *, book1%title
  Write (*,*) 'Enter the author of the book 1'
  Read *, book1%author
  Write (*,*) 'Enter the subject of the book 1'
  Read *, book1%subject
  Write (*,*) 'Enter the year of the book 1'
  Read *, book1%year
  write(*,*) 'Enter the title of the book 2'
  Read *, book2%title
  Write (*,*) 'Enter the author of the book 2'
  Read *, book2%author
  Write (*,*) 'Enter the subject of the book 2'
  Read *, book2%subject
  Write (*,*) 'Enter the year of the book 2'
  Read *, book2%year
  !display book info
  Print *, book1%title
  Print *, book1%author
  Print *, book1%subject
  Print *, book1%year
  Print *, book2%title
  Print *, book2%author
  Print *, book2%subject
  Print *, book2%year
end program bookstore
```

Second example:

```
!Program name: TYPE2
Program Bookstore
Integer, dimension(:), allocatable :: N
Print *, 'Enter the number of books of the bookstore: '
Read *, n
Allocate ( N(n) )
!type declaration
type Books
  character(len = 50) :: title
```

```
   character(len = 50) :: author
   character(len = 150) :: subject
   integer :: year
end type Books
do i=1,n
!declaring type variables
type(Books) :: booki
end do
do i=1,n
!enter the components of the structure
   write(*,*) 'Enter the title of the book ', i
   Read *, booki%title
   Write (*,*) 'Enter the author of the book ', i
   Read *, booki%author
   Write (*,*) 'Enter the subject of the book ', i
   Read *, booki%subject
   Write (*,*) 'Enter the year of the book ', i
   Read *, booki%year
End do
do i=1,n
   !display book info
   Print *, booki%title
   Print *, booki%author
   Print *, booki%subject
   Print *, booki%year
end do
stop
end program Bookstore
```

15. Data files in Fortran

Fortran allows to create data files (.dat) and to write data files. A data file must have a unit number (from 1–99) and a file name in order to identify this file. The OPEN statement allows to create a new data file and to write data in it using WRITE (number_of_file, *). The statement CLOSE(number_of_file) closes the data file created.

```
!Program name: DATAFILE
Integer, dimension (:,:), allocatable :: a
integer :: i, j, c, l
Print *, 'Give the number of columns of the matrix a: '
Read *, c
Print *, 'Give the number of lines of the matrix a: '
Read *, l
Allocate (a(c,l))
```

```
!Generate the scalar data of the matrix a
do i=1, c
  do j= 1, l
     a(i,j)= i*j
end do
end do
!create data1.dat (replace if it already exists) and write the a(i,j)
open(1,file='data1.dat', status='replace')
do i=1, c
  do j=1, l
  write(1,*) a(i,j)
end do
end do
close(1)
!Display the data recorded in data1.dat
do i=1, c
  do j=1, l
  write(*,*), a(i,j)
end do
end do
Deallocate(a)
stop
end
```

16. Randomization in Fortran

The Fortran subroutine CALL RANDOM_NUMBER(r) generates random numbers in the range (0,1). It gives a pseudorandom sequence (with one or more numbers depending on the loop), i.e., if one starts the sequence at the same point, it will give the same sequence even after recompilation. See the example below. In this case, the use of CALL RANDOM_SEED() at the beginning of the code is optional because it does not change the result.

```
!program name: RANDOM_TEST
implicit none
real :: r, x
integer :: i
call random_seed()
10 call random_number(r)
print *, r
print *,
do i=1,10
  call random_number(x)
  print *, x
end do
stop
end
```

The result is:

```
0.99755955
0.56682467
0.96591532
0.74792767
0.36739087
0.48063689
7.37542510E-02
5.35517931E-03
0.34708124
0.34224379
0.21795171
```

This result is the same every time you run the code. That's why it is a pseudorandom subroutine. The sequence is repeated or is the same once you start the code.

One way to circumvent this limitation by using CALL RANDOM_SEED with its arguments SIZE and SEED and the subroutine CALL DATE_AND_TIME.

The subroutine CALL DATE_AND_TIME generates a one-dimensional array with eight elements. The first three elements or values are related to the date in the form yyyy mm dd, i.e., the year, month and day. The fourth value is the time difference with respect to Coordinated Universal Time (UTC) in minutes. The next values correspond to the CPU clock time: hour from 0 to 23 (value 5); minutes from 0 to 59 (value 6); seconds from 0 to 59 (value 7); and milliseconds from 0 to 999 (value 8).

In the code below, every time you run the code, it gives a different number. The array dzt has 8 elements from the eight values of the subroutine CALL DATE_AND_TIME. The dynamic vector 'array' has 15 elements because the argument PUT in the subroutine CALL RANDOM_SEED imposes a minimum size for the array (12 at minimum). The line 'array(8)=dzt(1)' picks the first value of the subroutine CALL DATE_AND_TIME, i.e., the year for the seed of the CALL RANDOM_SEED. Then, every time you run the code, it depicts the same value. On the other hand, the line 'array(8)=dzt(8)' picks the eighth value of the subroutine CALL DATE_AND_TIME, i.e., the milliseconds of CPU clock for the seed of the CALL RANDOM_SEED. Then, every time you run the code, it is depicted a different value.

```
!program name: RANDOM
integer :: dzt(8)
integer, dimension(:), allocatable :: array
real :: r

allocate(array(15))
call date_and_time(values=dzt)
!print *, dzt !OPTIONAL

!program name: RANDOM
integer :: dzt(8)
integer, dimension(:), allocatable :: array
real :: r
```

```
allocate(array(15))
call date_and_time(values=dzt)
!print *, dzt !OPTIONAL
array(8) = dzt(1) !It picks the first value (year) of CALL DATE_AND_TIME
!print *, array !OPTIONAL
call random_seed(put=array)
call random_number(r)
print *,r
array(8) = dzt(8) !It picks the eighth value (milliseconds) of CALL DATE_
AND_TIME
!print *, array !OPTIONAL
call random_seed(put=array)
call random_number(r)
print *,r
stop
end
```

Exercise

By knowing that the wave function of a particle in ring can be described by the function cosine of mθ where m is an integer and θ is the angle in radian unit (Chapter sixteen), write a Fortran program that plots this function from $\theta = 0$ to $\theta = 2\pi$ with n (grid) points. Use the argument grid (equivalent to the number of points of a function). This code must contain:

1. User´s entry for m where m=0, ±1, ±2, ...
2. User´s number of points, grid. The grid should be 100 ≤ grid ≤ 300.
3. The interval, h, between the points of the function is given by: 2π/grid.
4. Output file (.dat) in the format to be read in GNUPLOT having the data of the wave function, that is, a column with n points (grid) of x=theta (0 ≤ theta ≤ 2π) and n points (grid) of y=cos(theta*m). This block should be at the end of the program. See the example below.

Note: The free program GNUPLOT can be downloaded from http://www.gnuplot.info/

Tips:

(1) Example of DO loop block containing the x and y variables:

```
do i = 0, grid
  x(i) = i * h
  y(i) = cos (m*i)
  write (*,*) i, 'X(i)=', x(i),'Y(i)=', y(i)
end do
```

(2) Example of Fortran block to export data to generate a plot in GNUPLOT:

```
open(7,file=fileout,status='replace')
do i=1,grid
  write (7,'(f7.3,3e16.8,f12.6)') x(i), y(i)
  End do
  close(7)
```

References cited

Williams, T., Kelley, C., Merritt, E.A. et al. 2018. Gnuplot 5.2—An interactive plotting program.

Basics of Numerical Calculation and Series

2

1. Introduction

There are three basic numerical operations in computational physics: quadrature (or numerical integration), numerical differentiation and finding the roots. In addition, the Taylor and MacLaurin series and the differentiation processes due to their use in differential equations and perturbation theory (next sections) are very important for several numerical calculations. At the end of this chapter we see some basics of derivatives. We also present some information about Fourier series and Fourier transform, which are not relevant to our numerical calculations, but they are important to quantum mechanics and are related to series method.

2. Power series method

A power series is an expansion of a function into an infinite sum of terms in which the variable x has the form of an infinitesimal polynomial. It is used to represent a determined function. For example, the power series of the function e^x is:

$$e^x = 1 + x + \frac{x^2}{2!} + \frac{x^3}{3!} + \frac{x^4}{4!} + \frac{x^5}{5!} + \ldots$$

$$Example: e^2 = 7.389056$$

$$e^2 = 1 + 2 + \frac{2^2}{2!} \qquad value = 5$$

$$e^2 = 1 + 2 + \frac{2^2}{2!} + \frac{2^3}{3!} \qquad value = 6.333$$

$$e^2 = 1 + 2 + \frac{2^2}{2!} + \frac{2^3}{3!} + \frac{2^4}{4!} + \frac{2^5}{5!} \qquad value = 7.266$$

$$e^2 = 1 + 2 + \frac{2^2}{2!} + \frac{2^3}{3!} + \frac{2^4}{4!} + \frac{2^5}{5!} + \frac{2^6}{6!} + \frac{2^7}{7!} \qquad value = 7.3809$$

Many functions can be represented as a power series of the general formula:

$$f(x) = \sum_{n=0}^{\infty} a_n x^n, \qquad n = 0, 1, 2, 3, \ldots$$

Where n is an integer number and a_n is a constant. Then, we have:

$$f(x) = a_0 + a_1x + a_2x^2 + a_3x^3 + a_4x^4 + a_5x^5 + ...$$

Let us assume a partial sum, $S_n(x)$, below:

$$S_k(x) = a_0 + a_1x + a_2x^2 + a_3x^3 + a_4x^4 + ... + a_kx^k$$

If, for some x, the limit given below exists, then the power series terminates or converges.

$$\lim_{n\to\infty} S_k(x) = \lim_{n\to\infty} \sum_{n=0}^{k} a_kx^k$$

When the power series does not correspond to the limit above for any value of x, then the power series diverges or does not terminate. For example, the series below converges to exp(x).

$$\sum_{n=0}^{\infty} \frac{1}{k!} = 1 + x + \frac{x^2}{2} + \frac{x^3}{6} + ... \Rightarrow e^x$$

Then, it is important to know about convergent and divergent series. If a series has a finite sum, it is called convergent. Otherwise, it is called divergent. This is a very important classification since divergent series cannot be used in numerical calculation. Let us define S_n as the partial sum of the n terms of the series. The letter n can be any integer.

$$S_1 = a_1$$

$$S_2 = a_1 + a_2$$

$$S_3 = a_1 + a_2 + a_3$$

...

$$S_n = a_1 + a_2 + a_3 + \cdots + a_n$$

As n increases, the partial sum may increase without any limit where the series diverges, or it may oscillate as in the series (1-2+3-4+5-...) which can converge as in the telescoping series or it may have a more complicated behavior. When the partial sum approaches a limiting value, the series converges. Then, when the series converges, we have the following condition:

$$\lim_{n\to\infty} S_n = S \therefore converges!$$

Where S is the sum of the series. The difference between S and S_n is called remainder, R_n.

$$R_n = S - S_n$$

The power series method is powerful tool for solving differential equations with non-constant coefficients from Hermite, Legendre, Laguerre differential equations (Chapter four) and the energy solution for one-particle quantum harmonic oscillator (Chapter fifteen). In order to solve these differential equations, we need to find the power series of the first two derivatives of the function represented as a power series as well.

$$f(x) = \sum_{n=0}^{\infty} a_n x^n, \qquad n = 0,1,2,3,...$$

Its first derivative is given below:

$$f'(x) = \sum_{n=1}^{\infty} n a_n x^{n-1}$$

Let us do the power equalization for the expressions above, i.e., we have to change the previous summation in order to begin with n=0. In fact, for the summation of the first derivative, f'(x), if n starts at 0 or at 1 does not change the result since for n=0 the second the corresponding term is zero. Then:

$$\sum_{n=1}^{\infty} n a_n x^n = \sum_{n=0}^{\infty} n a_n x^n$$

Its second derivative is:

$$f''(x) = \sum_{n=2}^{\infty} n(n-1) a_n x^{n-2}$$

Let us do the power equalization so that n starts at zero.

$$\begin{aligned}
&for: n(n-1)\\
&n = 2 \rightarrow 2(2-1) = 2\\
&n = 3 \rightarrow 3(3-1) = 6\\
&for: (n+2)(n+1)\\
&n = 0 \rightarrow (0+2)(0+1) = 2\\
&n = 1 \rightarrow (1+2)(1+1) = 6\\
&then: n(n-1)\big|_{n=2} = (n+2)(n+1)\big|_{n=0}
\end{aligned}$$

$$finally: \sum_{n=2}^{\infty} n(n-1) = \sum_{n=0}^{\infty} (n+2)(n+1)$$

Hence:

$$f''(x) = \sum_{n=0}^{\infty} (n+2)(n+1) a_{n+2} x^n$$

The procedure of the power equalization above will be used to solve any differential equation with power series method. Please, be aware that the procedure of power equalization is of the power series method and is different from that of the Frobenius method (see Chapter four).

The function *f*(x) which can be described by a series is called generating function, $G(a_n, x)$ given by:

$$G(a_n; x) = \sum_{n=0}^{\infty} a_n x^n$$

"A generating function is a device somewhat similar to a bag. Instead of carrying many little objects detachedly, which could be embarrassing, we put them all in a bag, and then we have only one object to carry, the bag" George Pólya once said.

3. Fourier series and Euler's formula

Fourier proved that any continuous function could be represented by an infinite sum of sine and cosine waves. In the Fourier series, the sines and cosines are used to represent periodic functions. On the other hand, the non-periodic functions are represented by Fourier integral or transform (next section).

Consider the particle P moving in a ring (see Chapter fifteen) as an example of periodic function. The motion of the particle P in the ring is given by the independent coordinates x and y:

$$x = A\cos\theta = A\cos\omega \cdot t$$

$$y = A\sin\theta = A\sin\omega \cdot t$$

Where ω is the angular velocity, A is the amplitude and t is time.

We can use a single equation of the motion of P encompassing x and y coordinates shown below which can be rewritten as a Euler's formula (see its derivation ahead).

$$z = x + iy = A\cos\omega \cdot t + iA\sin\omega \cdot t$$

$$Euler: z = Ae^{i\omega t}$$

We want to expand a given periodic function in a series of sines and cosines. Firstly, we start with functions of period 2π and, then, the functions of sin x and cos x or sin nx and cos nx (n being an integer) have period 2π. Let us suppose a given function of period 2π, *f*(x), represented by the following equation:

$$f(x) = \tfrac{1}{2}a_0 + a_1\cos x + a_2\cos 2x + a_3\cos 3x + \dots$$
$$+ b_1\sin x + b_2\sin 2x + b_3\sin 3x + \dots$$

Then, we have to find the formulas of the coefficients a_n and b_n for the above expansion. The derivation of the following equations can be found elsewhere (Boas, 2006). These coefficients are used to represent square waves, for example.

$$a_n = \frac{1}{\pi}\int_{-\pi}^{\pi} f(x)\cos(nx)\,dx$$

$$b_n = \frac{1}{\pi}\int_{-\pi}^{\pi} f(x)\sin(nx)\,dx$$

For even or odd functions, the coefficients formulas for a_n and b_n are different from those used for square waves.

$$f(x): odd \begin{cases} b_n = \dfrac{2}{P}\int_0^P f(x)\sin\dfrac{n\pi x}{P}dx \\ a_n = 0 \end{cases}$$

$$f(x): even \begin{cases} a_n = \dfrac{2}{P}\int_0^P f(x)\cos\dfrac{n\pi x}{P}dx \\ b_n = 0 \end{cases}$$

Where the interval of length P is the period of the Fourier series.

An even function presents symmetry along the vertical axis. An odd function has another type of symmetry, i.e., an odd function has symmetry with respect to the origin.

$$f(x) \Rightarrow odd : f(x) = -f(-x)$$

$$Examples : f(x) = \frac{1}{x}, \qquad f(x) = x^3, \qquad f(x) = x^5$$

$$f(x) \Rightarrow even : f(x) = f(-x)$$

$$Examples : f(x) = x^2, \qquad f(x) = |x|, \qquad f(x) = \sqrt{|x|}$$

The general Fourier series for even-odd functions is:

$$f(x) = \frac{a_0}{2} + \sum_{n=1}^{N} \left[a_n \cos\left(\frac{2\pi n x}{P}\right) + b_n \sin\left(\frac{2\pi n x}{P}\right)\right]$$

Let us now apply the Fourier series to a square wave. A square wave has a non-sinusoidal periodic waveform in which the amplitude alternates between fixed minimum and maximum values with the same interval between minimum and maximum. Let us imagine the minimum in *f*(x) as 0 and the maximum as 1 and the interval between minimum and maximum is π.

$$f(x) = \begin{cases} 0, & -\pi < x < 0 \\ 1, & 0 < x < \pi \end{cases}$$

This square wave can be represented by the following Fourier series (Boas, 2006):

$$f(x) = \frac{1}{2} + \frac{2}{\pi}\left(\frac{\sin x}{1} + \frac{\sin 3x}{3} + \frac{\sin 5x}{5} + \ldots\right)$$

The Fourier series can also be represented in the complex form. Since the sines and cosines can be represented in terms of the complex exponential according to Euler's formula, then we have:

$$\sin nx = \frac{e^{inx} - e^{-inx}}{2i}$$

$$\cos nx = \frac{e^{inx} + e^{-inx}}{2}$$

The Euler's formula is

$$e^{ix} = \cos x + i \sin x$$

The proof for Euler's equation is given here. Let us firstly prove the Euler´s formula from the summation below:

$$e^x = 1 + x + \frac{x^2}{2!} + \frac{x^3}{3!} + \frac{x^4}{4!} \cdots \frac{x^n}{n!}$$

Then, by replacing x into ix, we have:

$$e^{ix} = 1 + \frac{ix}{1!} + \frac{(ix)^2}{2!} + \frac{(ix)^3}{3!} + \frac{(ix)^4}{4!} + \frac{(ix)^5}{5!} + \frac{(ix)^6}{6!} + \ldots$$

$$e^{ix} = 1 + \frac{ix}{1!} - \frac{x^2}{2!} - \frac{ix^3}{3!} + \frac{x^4}{4!} + \frac{ix^5}{5!} - \frac{x^6}{6!} + \ldots$$

$$e^{ix} = \left(1 - \frac{x^2}{2!} + \frac{x^4}{4!} - \frac{x^6}{6!} + \ldots\right) + i\left(\frac{x}{1!} - \frac{x^3}{3!} + \frac{x^5}{5!} + \ldots\right)$$

$$e^{ix} = \cos x + i \sin x$$

As a consequence, one continuous function can be represented as:

$$f(x) = c_0 + c_1 e^{ix} + c_{-1} e^{-ix} + c_2 e^{2ix} + c_{-2} e^{-2ix} + \ldots$$

The Fourier series in the complex form becomes

$$c_n = \frac{1}{2P} \int_{-P}^{P} f(x) e^{-in\pi x/P} dx$$

$$f(x) = \sum_{n=-\infty}^{n=+\infty} c_n e^{inx}$$

To sum up, the Fourier series changes the time domain (represented by the period T of a wave) of a continuous function into its frequency domain of sine-cosine form. Remember that the period, T, is the time taken to complete a cycle of a wave vibration and that the period is the inverse of the frequency of the vibration. See the Table 2.1.

Table 2.1: Time domain (original *f*(x)) and frequency domain (Fourier series) of some continuous functions.

Time domain (original *f*(x))	Frequency domain (Fourier series)
$f(x) = A\left\|\sin\left(\frac{2\pi}{T}x\right)\right\|$ $0 \le x < T, \quad T = \frac{1}{\nu}$	$a_0 = \frac{4A}{\pi}, \quad b_n = 0$ $a_n = \begin{cases} \frac{-4A}{\pi}\frac{1}{n^2-1}, & n = even \\ 0, & n = odd \end{cases}$
$f(x) = \begin{cases} A\sin\left(\frac{2\pi}{T}x\right), & 0 \le x < T \\ 0, & T/2 \le x < T \end{cases}$	$a_0 = \frac{2A}{\pi} \therefore b_n = \begin{cases} \frac{A}{2}, & n = 1 \\ 0, & n > 1 \end{cases}$ $a_n = \begin{cases} \frac{-2A}{\pi}\frac{1}{1-n^2}, & n = even \\ 0, & n = odd \end{cases}$
$f(x) = \frac{Ax}{T}, \quad 0 \le x < T$	$a_0 = A$ $a_n = 0$ $b_n = -\frac{A}{n\pi}$

4. Fourier transform

Whereas the Fourier series is used to represent periodic functions, the Fourier transform is used to represent non-periodic functions and a continuous spectrum of waves of different frequencies.

As an example of a set of waves of different frequencies, imagine a classic music concert in which the sounds consist of a fundamental and a complement of overtones or harmonics of frequencies 2, 3, 4, ... times the frequency of the fundamental, i.e., a superposition of sine waves with multiple integer frequencies of the fundamental wave. Another example of spectrum of waves of different frequencies is the sunlight having a spectrum from infrared to ultraviolet wavelengths.

The Fourier integrals can be represented as:

$$g(\alpha)=\frac{1}{2\pi}\int_{-\infty}^{\infty} f(x)e^{-i\alpha x}dx$$

$$f(x)=\int_{-\infty}^{\infty} g(\alpha)e^{i\alpha x}d\alpha$$

Let us represent a nonperiodic function below

$$f(x)=\begin{cases}1 \therefore -1<x<1\\ 0 \therefore |x|>1\end{cases}$$

with Fourier transform (Boas 2006).

$$g(\alpha)=\frac{1}{2\pi}\int_{-\infty}^{\infty} f(x)e^{-i\alpha x}dx=\frac{1}{2\pi}\int_{-1}^{1} e^{-i\alpha x}dx$$

$$g(\alpha)=\frac{1}{2\pi}\frac{e^{-i\alpha x}}{-i\alpha}\Bigg|_{-1}^{1}=\frac{1}{\pi\alpha}\frac{e^{-i\alpha}-e^{i\alpha}}{-2i}=\frac{\sin\alpha}{\pi\alpha}$$

$$f(x)=\frac{2}{\pi}\int_{0}^{\infty}\frac{\sin\alpha\cos\alpha x}{\alpha}d\alpha$$

5. Taylor series and MacLaurin series

The Taylor series is a power series to represent a function as an infinite sum of the terms calculated from the function's derivative at a single point. The Taylor series expansion of a function about the point x=a is given by:

$$f(x+a)=f(a)+\frac{x}{1!}f'(a)+\frac{x^2}{2!}f''(a)+\frac{x^3}{3!}f'''(a)+...+\frac{x^n}{n!}f^{(n)}(a)+R_n(x)$$

$$f(x+a)=\sum_{n=0}^{\infty}\frac{x^n}{n!}f^{(n)}(a)$$

or:

$$f(x-a)=f(a)-\frac{x}{1!}f'(a)+\frac{x^2}{2!}f''(a)-\frac{x^3}{3!}f'''(a)+\ldots+R_n(x)$$

$$f(x-a)=\sum_{n=0}^{\infty}\frac{(-1)^n x^n}{n!}f^{(n)}(a)$$

The term $R_n(x)$ is the error involved in the approximation, called remainder term. As the number of terms in the series increase, the error of remainder term decreases. In the infinite of the sum, the $R_n(x)=0$.
An alternative formula for the Taylor series is:

$$f(x)=f(a)+\frac{(x-a)}{1!}f'(a)+\frac{(x-a)^2}{2!}f''(a)+\frac{(x-a)^3}{3!}f'''(a)+\ldots$$

$$\ldots+\frac{(x-a)^n}{n!}f^{(n)}(a)+R_n(x)$$

$$f(x)=\sum_{n=0}^{\infty}\frac{(x-a)^n}{n!}f^{(n)}(a)$$

Then, any function can be represented as:

$$f(x)=T_n(x)+R_n(x)$$

$$T_n(x)=\sum_{i=0}^{n}\frac{(x-a)^i}{i!}f^{(i)}(a)$$

Where $T_n(x)$ is the nth degree Taylor polynomial of the function $f(x)$.

The special case where the expansion of a function occurs about the point x=0 is called MacLaurin series obeying the condition that all derivatives at x=0 exists. The function $(1+x)^a$ can be represented in MacLaurin series, where the power rule of the derivative of x^n, i.e., $d(x^n)/dy= n\,x^{n-1}$, was used.

$$f(x)=(1+x)^a$$

$$f'(x)=a(1+x)^{a-1}$$

$$f''(x)=a(a-1)(1+x)^{a-2}$$

$$f'''(x)=a(a-1)(a-2)(1+x)^{a-3}$$

etc.

When x=0 (condition for MacLaurin series), we have:

$$f(0)=1, f'(0)=a, f''(0)=a(a-1), f'''(0)=a(a-1)(a-2)$$

Then, the MacLaurin series for $(1+x)^a$ is:

$$(1+x)^a=1+ax+\frac{a(a-1)}{2!}x^2+\frac{a(a-1)(a-2)}{3!}x^3+\ldots$$

Let us find the Taylor series for $y = 1/x^2$ at $x = -1$.

$$f^{(0)}(x)=\frac{1}{x^2}, f^{(1)}(x)=-\frac{2}{x^3}, f^{(2)}(x)=\frac{6}{x^4}, f^{(3)}(x)=-\frac{24}{x^5}$$

$$f^{(n)}(x)=-\frac{(-1)^n(n+1)!}{x^{n+2}}$$

$$x=-1, f^{(0)}(-1)=\frac{1}{(-1)^2}=1, f^{(1)}(-2)=-\frac{2}{(-1)^3}=2, f^{(2)}(-1)=\frac{6}{(-1)^4}=6$$

$$f^{(n)}(-1)=-\frac{(-1)^n(n+1)!}{(-1)^{n+2}}=(n+1)!$$

$$\frac{1}{x^2}=\sum_{n=0}^{\infty}\frac{f^{(n)}(-1)}{n!}(x+1)^n=\sum_{n=0}^{\infty}\frac{(n+1)!}{n!}(x+1)^n=\sum_{n=0}^{\infty}(n+1)(x+1)^n$$

By using the same rational we get the Taylor series for $y = 1/x$ at $x = 1$.

$$f^{(0)}(x)=\frac{1}{x^1}, f^{(1)}(x)=-\frac{1}{x^2}, f^{(2)}(x)=\frac{2}{x^3}, f^{(3)}(x)=-\frac{6}{x^4}$$

$$f^{(n)}(x)=-\frac{(-1)^n n!}{x^{n+1}}$$

$$x=1, f^{(0)}(1)=\frac{1}{(1)^2}=1, f^{(1)}(1)=-\frac{1}{(1)^3}=-1, f^{(2)}(1)=2, f^{(3)}(1)=-6$$

$$f^{(n)}(1)=-\frac{(-1)^n n!}{(1)^{n+1}}=(-1)^n n!$$

$$\frac{1}{x}=\sum_{n=0}^{\infty}\frac{f^{(n)}(1)}{n!}(x-1)^n=\sum_{n=0}^{\infty}\frac{(-1)^n n!}{n!}(x-1)^n=\sum_{n=0}^{\infty}(-1)^n(x-1)^n$$

6. The differentiation process

Let us assume the gravity displacement for falling objects. The gravity acceleration is g (9.8 ms^{-2}) due to the gravity force on the falling object which falls at velocity v which increases with time t along the displacement y. The equations for the velocity and displacement of the falling object are given below along with the hypothesis of zero initial velocity (as it was dropped from the sky to the ground).

$$v = gt + v_i, \quad if: v_i = 0 \rightarrow v = gt$$

$$y = \frac{gt^2}{2} + v_i t, \quad if: v_i = 0 \rightarrow y = \frac{gt^2}{2}$$

The average velocity and the instantaneous velocity are given below:

$$\bar{v} = \frac{\Delta y}{\Delta t}$$

$$v_{inst} = \lim_{\Delta t \to 0}\frac{\Delta y}{\Delta t} = \lim_{\Delta t \to 0}\frac{f(t+\Delta t)-f(t)}{\Delta t}$$

The process of taking the limit of the above equation is called differentiation.

In a more general perspective, let us assume the function where y varies according to x variable respecting the function $f(x)$ and we want to obtain the differentiation of this function.

$$y = f(x)$$

$$\frac{dy}{dx} = \lim_{\Delta x \to 0}\left(\frac{\Delta y}{\Delta x}\right) = \lim_{\Delta x \to 0}\left(\frac{f(x+\Delta x) - f(x)}{\Delta x}\right)$$

Let us now assume the function $y = 1/x$ and we want to obtain the differentiation of this function.

$$y = f(x) = \frac{1}{x}, \qquad \lim_{\Delta x \to 0}\left(\frac{\Delta y}{\Delta x}\right)$$

$$\Delta y = f(x+\Delta x) - f(x) = \frac{1}{x+\Delta x} - \frac{1}{x} = \frac{x-(x+\Delta x)}{x(x+\Delta x)} = \frac{-\Delta x}{x(x+\Delta x)}$$

$$\frac{\Delta y}{\Delta x} = \frac{\frac{-\Delta x}{x(x+\Delta x)}}{\Delta x} = \frac{-1}{x(x+\Delta x)}$$

$$\lim_{\Delta x \to 0}\left(\frac{\Delta y}{\Delta x}\right) = \lim_{\Delta x \to 0}\left(\frac{-1}{x(x+\Delta x)}\right) = -\frac{1}{x^2}$$

The differentiation process gives the derivative of a determined function, dy/dx. Actually, there are three different ways to obtain the derivative of a given function: the forward finite-difference, the backward finite difference and central finite difference.

$$forward: \frac{dy}{dx} = \lim_{\Delta x \to 0}\left(\frac{f(x+\Delta x) - f(x)}{\Delta x}\right)$$

$$backward: \frac{dy}{dx} = \lim_{\Delta x \to 0}\left(\frac{f(x) - f(x-\Delta x)}{\Delta x}\right)$$

$$central: \frac{dy}{dx} = \lim_{\Delta x \to 0}\left(\frac{f(x+\Delta x) - f(x-\Delta x)}{2\Delta x}\right)$$

7. Finite difference approximation

By knowing that dx is an infinitesimal variation of the x variable, we have the following approximation:

$$\lim_{\Delta x \to 0} \Delta x \approx dx$$

Then, the derivative of a given function can be obtained from this approximation in three different ways:

$$\frac{dy}{dx} \approx \left(\frac{f(x+dx) - f(x)}{dx} \right)$$

$$\frac{dy}{dx} \approx \left(\frac{f(x) - f(x-dx)}{dx} \right)$$

$$\frac{dy}{dx} \approx \left(\frac{f(x+dx) - f(x-dx)}{2dx} \right)$$

Let us take the expression of the Taylor series below:

$$f(x+a) = f(a) + \frac{x}{1!} f'(a) + \frac{x^2}{2!} f''(a) + \frac{x^3}{3!} f'''(a) + \ldots + R_n(x)$$

$$f(x+dx) = f(x) + \frac{dx}{1!} f'(x) + \frac{dx^2}{2!} f''(x) + \frac{dx^3}{3!} f'''(x) + \ldots + R_n(x)$$

Let us pass $f(x)$ to the left side of the equation and divide by dx:

$$\frac{f(x+dx) - f(x)}{dx} = f'(x) + \frac{dx}{2!} f''(x) + \frac{dx^2}{3!} f'''(x) + \ldots + R_n(x)$$

Then, a numerical expression for the

$$\frac{dy}{dx} \approx \left(\frac{f(x+dx) - f(x)}{dx} \right)$$

8. Numerical quadrature or integration: Simpson's rule

The numerical integration or quadrature is used to calculate the approximate solution for the definite integral of the function $f(x)$ between the intervals a and b (b > a). This can be done by separating the definite integral into quadratures of the integral of $f(x)$ between –h and +h interval in order to obtain a precise numerical integration.

$$\int_a^b f(x)dx = \int_a^{a+2h} f(x)dx + \int_{a+2h}^{a+4h} f(x)dx + \ldots + \int_{b-2h}^{b} f(x)dx$$

The Taylor series can be used to provide a nearly exact interpolation of the function $f(x)$.

$$f(x) = f(x_1) + f'(x_1)(x - x_1) + \frac{f''(x_1)}{2}(x - x_1)^2 + \\ + \frac{f'''(x_1)}{6}(x - x_1)^3 + \frac{f^{iv}(x_1)}{24}(x - x_1)^4 \therefore x_1 = \frac{a+b}{2}$$

The integration of the Taylor series above gives:

$$\int_a^b f(x)dx = \int_a^b \begin{bmatrix} f(x_1) + f'(x_1)(x - x_1) + \frac{f''(x_1)}{2}(x - x_1)^2 + \\ + \frac{f'''(x_1)}{6}(x - x_1)^3 + \frac{f^{iv}(x_1)}{24}(x - x_1)^4 \end{bmatrix} dx$$

By knowing that $f(x_1)$, $f'(x_1)$, $f''(x_1)$, $f'''(x_1)$ and $f^{iv}(x_1)$ are constants, then we have:

$$\int_a^b f(x)dx = \left[\begin{array}{l} f(x_1)x + \frac{f'(x_1)}{2}(x-x_1)^2 + \frac{f''(x_1)}{6}(x-x_1)^3 + \\ + \frac{f'''(x_1)}{24}(x-x_1)^4 + \frac{f^{iv}(x_1)}{120}(x-x_1)^5 \end{array} \right]_a^b$$

Where

$$\int_a^b x^n dx = \left[\frac{x^{n+1}}{n+1} \right]_a^b$$

Then:

$$\int_a^b f(x)dx = f(x_1)[b-a] + \frac{f'(x_1)}{2}\left[(b-x_1)^2 - (a-x_1)^2\right] +$$

$$+\frac{f''(x_1)}{6}\left[(b-x_1)^3 - (a-x_1)^3\right] + \frac{f'''(x_1)}{24}\left[(b-x_1)^4 - (a-x_1)^4\right] +$$

$$+\frac{f^{iv}(x_1)}{120}\left[(b-x_1)^5 - (a-x_1)^5\right]$$

Let:

$$x_1 = \frac{a+b}{2}, \qquad h = \frac{b-a}{2}$$

Then, the value of x_1 is (a + h) or (b – h):

$$x_1 = a + \frac{b-a}{2} = a+h$$

$$x_1 = b + \frac{a-b}{2} = b-h$$

Hence, we can replace the values of x_1 in the equations below:

$$x_1 - a = (a+h) - a = h$$

$$Hence: (a - x_1) = -h$$

$$and: (b - x_1) = b - (b-h) = h$$

Let us substitute, $(b-a) = 2h$, $(a-x_1) = -h$ and $(b-x_1) = h$ into the last definite integral above to obtain:

$$\int_a^b f(x)dx = 2hf(x_1) + \frac{f'(x_1)}{2}\left[h^2 - (-h)^2\right] +$$

$$+\frac{f''(x_1)}{6}\left[(h)^3 - (-h)^3\right] + \frac{f'''(x_1)}{24}\left[(h)^4 - (-h)^4\right] +$$

$$+\frac{f^{iv}(x_1)}{120}\left[(h)^5 - (-h)^5\right]$$

Since $(-h)^2 = h^2$ and $(-h)^4 = h^4$, then:

$$\int_a^b f(x)dx = 2hf(x_1) + \frac{f''(x_1)}{6}\left[h^3 + h^3\right] + \frac{f^{iv}(x_1)}{120}\left[h^5 + h^5\right]$$

$$\int_a^b f(x)dx = 2hf(x_1) + h^3\frac{f''(x_1)}{3} + h^5\frac{f^{iv}(x_1)}{60}$$

By knowing that:

$$f'(x) \approx \frac{f(x+\delta_x) - f(x)}{\delta_x}$$

Then:

$$f''(x) \approx \frac{f'(x+\delta_x) - f'(x)}{\delta_x} =$$

$$f''(x) \approx \frac{1}{\delta_x}\left(\frac{\left[f(x+2\delta_x) - f(x+\delta_x)\right]}{\delta_x} - \frac{f(x+\delta_x) - f(x)}{\delta_x}\right)$$

Let

$x_1 = x + \delta_x$, $\quad x = a$, $\quad x + 2\delta_x = b$, $\quad h = \delta_x$

Then:

$$f''(x) \approx \frac{1}{h}\left(\frac{\left[f(b) - f(x_1)\right]}{h} - \frac{f(x_1) - f(a)}{h}\right)$$

$$f''(x) \approx \frac{f(b) - 2f(x_1) - f(a)}{h^2}$$

Now, replace the $f''(x)$ into the last definite integral above:

$$\int_a^b f(x)dx = 2hf(x_1) + \frac{h^3}{3}\left[\frac{f(b) - 2f(x_1) + f(a)}{h^2}\right] + h^5\frac{f^{iv}(x_1)}{60}$$

$$\int_a^b f(x)dx = 2hf(x_1) + \frac{h}{3}\left[f(b) - 2f(x_1) + f(a)\right] + h^5\frac{f^{iv}(x_1)}{60}$$

$$\int_a^b f(x)dx \approx \frac{h}{3}\left[f(b) + 4f(x_1) + f(a)\right]$$

$$\int_a^b f(x)dx \approx \frac{b-a}{6}\left[f(b) + 4f\left(\frac{a+b}{2}\right) + f(a)\right]$$

The numerical integration above is known as Simpson's rule. In more general terms, the Simpson´s rule is written as:

$$\int_{x_0}^{x_2} f(x)dx \approx \frac{h}{3}\left(f_2 + 4f_0 + f_1\right)$$

Higher-order quadrature formulas can be derived by retaining more terms of the Taylor series used to interpolate the function *f*(x) between a and b (mesh points). The numerical integration from quadratic polynomial used to interpolate yields:

$$\int_{x_0}^{x_4} f(x)dx \approx \frac{2h}{45}(7f_0 + 32f_1 + 12f_2 + 32f_3 + 7f_4)$$

Which is known as Bode's rule.

We depict below the code for the Simpson´s rule for the integration of the function *f*(x) = exp(x) in the interval from 0 to 1.

```
!Program name: SIMPSON
!Integration of a function from Simpson's rule
!Integral of the function exp(x) from 0 to 1 interval
!the exact integration is exp(1)-1=1.1718282
!h=(b-a)/2
!x1=a+h
! S=h/3[f(b)+4f(x1)+f(a)}
! n: number of interval (even number)
!S=integral
Real :: fx, exact, integral, a, b
Integer :: i, n
exact=exp(1.)-1
write (*,*) "Exact integration of exp(x) from 0 to 1= ", exact
a=0
b=1
n=2
call simpson(a,b,h,fx1,n,integral)
print *, n, h, fx1, integral
stop
end
subroutine simpson(a,b,h,fx1,n,integral)
real :: fa, fb, fx1, integral, h, a, b
integer :: n, i
fa=exp(a)
fb=exp(b)
h=(b-a)/dfloat(n)
fx1=exp(a+h)
integral = (fa+fb+4.0*fx1)*h/3.0
return
end subroutine simpson
```

9. Mean of a function and Monte-Carlo integration

The mean of a function, $<f(x)>$, is the average value of the function *f*(x) over its interval (a,b).

$$\langle f(x)\rangle = \frac{1}{b-a}\int_a^b f(x)dx$$

The mean of a function can also be written as:

$$\langle f(x)\rangle = \lim_{N\to\infty}\frac{1}{N}\sum_{i=1}^{N} f(x_i)$$

We can use both equations to obtain the integrand I

$$I = \int_a^b f(x)dx$$

See below:

$$I = \int_a^b f(x)dx = b-a\langle f(x)\rangle$$

$$I = \int_a^b f(x)dx \approx \frac{b-a}{N}\sum_{i=1}^{N} f(x_i)$$

$$\lim_{N\to\infty}\frac{b-a}{N}\sum_{i=1}^{N} f(x_i) = \int_a^b f(x)dx$$

The Monte Carlo method uses suitable random numbers to solve a problem. The real power of Monte Carlo is for the evaluation of multi-dimensional integrals, but we will consider the simplest case of one-dimensional integral as shown above.

The Fortran subroutine CALL RANDOM_NUMBER(r) generates random numbers in the range (0,1). Then, we have to multiply this random number to scale it over the base width of the integral of interest.

In order to estimate the error of the Monte Carlo integration, we use the formula below:

$$\sigma_I = (b-a)\cdot\sqrt{\frac{1}{N}\left[\frac{1}{N}\sum_{i=1}^{N} f_i^2 - \left(\frac{1}{N}\sum_{i=1}^{N} f_i\right)^2\right]}$$

Let us take the example of the following integral:

$$I = \int_0^1 \frac{1}{1+x^2}dx = 0.78540$$

```
! Program name: MC-INTEGRATION
! INTEGRATION BY MONTE CARLO METHOD
! INTEGRAL BY MEAN FUNCTION
! EXAMPLE FUNCTION f= 1/(1+x**2)
! In the interval (0,1) the integral is 0.78540
Program monte_carlo
real :: f, f2, integrand, integral, sigma, a, b, x
integer :: i, n
data exact/.78540/
Print *, "Enter the grid of the integration, n: "
Read *, n
If (n .eq. 0) stop
Print *, "Enter the lower limit and upper limit of integration, a, b:"
Read *, a, b
! For the integral in the example, a=0 and b=1
f= 0.0
f2=0.0
call random_seed()
do i=1,n
   call random_number(x)
   x=x*b
   f= f+integrand(x)
   f2=f2+(integrand(x)**2)
! print *, x, f, f2 !OPTIONAL VISUALIZATION
End do
f=f/n
f2=f2/n
integral=(b-a)*f
sigma=(b-a)*sqrt((f2-f**2)/n)
Print *,
Print *, integral, sigma, "error= ", exact-integral
End program monte_carlo
function integrand(x) result(func)
   Implicit none
Real :: func, x
func= 1./(1.+x**2)
End function integrand
```

The results of Monte Carlo integration (I_{MC}), sigma and error for each value of N is given below. Notice that, except for N = 10, when increasing the value of N, the error decreases.

$for : N = 10$

$I_{MC} = 0.78922778 \quad \sigma = 0.05420838 \quad error = -0.0038278102$

$for : N = 50$

$I_{MC} = 0.75583577 \quad \sigma = 0.0226997659 \quad error = -0.0295642018$

$for : N = 100$

$I_{MC} = 0.76436228 \quad \sigma = 0.0162020326 \quad error = -0.0210376978$

$for : N = 500$

$I_{MC} = 0.77315885 \quad \sigma = 0.00713236118 \quad error = -0.0122411251$

$for : N = 1000$

$I_{MC} = 0.77808177 \quad \sigma = 0.00512474775 \quad error = -0.0073181986$

$for : N = 10000$

$I_{MC} = 0.78349620 \quad \sigma = 0.00161808426 \quad error = -0.0019037723$

Let us now use Monte-Carlo integration algorithm for the double integral. The double integral gives us the area of a determined tridimensional surface. For the general double integral below, it can be written as the average function multiplied by the two limit differences.

$$\int_a^b \left(\int_c^d f(x,y)dy \right) dx = (b-a)(d-c)\langle f(x,y) \rangle$$

$$\int_a^b \left(\int_c^d f(x,y)dy \right) dx \approx (b-a)(d-c)\frac{1}{N}\sum_{i=1}^{N} f_i\left(x_i, y_i\right)$$

The formula for sigma is:

$$\sigma_I = (d-c)(b-a) \cdot \sqrt{\frac{1}{N}\left[\frac{1}{N}\sum_{i=1}^{N} f_i^2 - \left(\frac{1}{N}\sum_{i=1}^{N} f_i \right)^2 \right]}$$

Let us consider the following double integral;

$$\int_0^{5/4} \left[\int_0^{5/4} \left(4 - x^2 - y^2\right) dy \right] dx = 4.62239$$

The code is very similar to the previous one:

```
! Program name: MC-INTEGRATION2
! INTEGRATION BY MONTE CARLO METHOD
! INTEGRAL BY MEAN FUNCTION
! EXAMPLE FUNCTION f(x,y)=4-x^2-y^2
! In the interval (0,5/4) and (0,5/4) the integral is 4.62239
Program monte_carlo
real :: f, f2, integrand, integral, sigma, a, b, x, c, d, y
integer :: i, n
data exact/4.62239/
Print *, "Enter the grid of the integration, n: "
Read *, n
If (n .eq. 0) stop
Print *, "Enter the lower limit and upper limit of first integration, a, b:"
Read *, a, b ! In the suggested problem a=2 and b=5/4
Print *, "Enter the lower limit and upper limit of second integration, c, d:"
Read *, c, d ! In the suggested problem a=0 and d=5/4
f= 0.0
f2=0.0
call random_seed()
do i=1,n
  call random_number(x)
  x=x*b
  call random_number(y)
  y=y*d
  f= f+integrand(x,y)
  f2=f2+(integrand(x,y)**2)
 ! print *, x, f, f2
End do
f=f/n
f2=f2/n
integral=(b-a)*(d-c)*f
sigma=(b-a)*(d-c)*sqrt((f2-f**2)/n)
Print *,
Print *, integral, sigma, "error= ", exact-integral
End program monte_carlo
function integrand(x,y) result(func)
  Implicit none
Real :: func, x, y
func= 4-x**2-y**2
End function integrand
```

The results are:

$$for : N = 100$$
$$I_{MC} = 4.4867902 \quad \sigma = 0.11153496 \quad error = -0.13559961$$
$$for : N = 1000$$
$$I_{MC} = 4.5845666 \quad \sigma = 0.033094299 \quad error = -0.0378232002$$
$$for : N = 10000$$
$$I_{MC} = 4.6197872 \quad \sigma = 0.0010365607 \quad error = -0.0026025772$$

10. Integration by midpoint rule

Let us assume a function $f(x)$ over the interval (x_0, x_1) where $x_1 = x_0 + h$. The midpoint rule is the area of the length h times the height $f(x_0 + h/2)$:

$$MP = h \times f\left(x_0 + \tfrac{h}{2}\right)$$

$$\int_{x_0}^{x_1} f(x)dx \approx h \times f\left(x_0 + \tfrac{h}{2}\right)$$

Now consider that the interval (a,b) which is divided into m equal m intervals of the same width h (b – a/m) by using equally spaced node n_k, where k = 1,2,3,..., m. The composite midpoint rule is:

$$h = \frac{b-a}{m}$$

$$MP = h\sum_{k=1}^{m} f\left(a + \left(k - \tfrac{1}{2}\right)h\right) \therefore k = 1,2,3...,m$$

$$\int_{a}^{b} f(x)dx \approx h\sum_{k=1}^{m} f\left(a + \left(k - \tfrac{1}{2}\right)h\right)$$

Let us find the following definite integral with the midpoint rule:

$$\int_{0}^{10} \left(2 + \cos\left(2\sqrt{x}\right)\right)dx$$

```
!Program name: MIDPOINT
!Integration of function: 2+cos(2*sqrt(x))
!Interval: 0 to 10
! I=20.13035
integer :: k, a, m
real :: h, i, x, n, b
a=0
b=10.
m=20
i=0.
```

```
h=(b-a)/m
print *, h
do k=1,m
  n=a+(k-0.5)*h
  x=n
  i=i+h*func(x)
  print *, k, n, func(x), i
end do
print *, i
stop
end
function func(x) result(f)
real :: f
f=2+cos(2*sqrt(x))
end function func
```

See the stability of the midpoint rule for the Runge function below. We get the exact value of the integrand of this function.

```
!Program name: MIDPOINT_STABILITY
!Integration of function: 1/1+25x^2
!Interval: -3 to 3
! I=0.60169208
integer :: k, a, m
real :: h, i, x, n, b
a=-3
b=3.
m=20
i=0.
h=(b-a)/m
print *, h
do k=1,m
  n=a+(k-0.5)*h
  x=n
  i=i+h*func(x)
  print *, k, n, func(x), i
end do
print *, i
stop
end
function func(x) result(f)
real :: f
f=1/(1+25*x**2)
end function func
```

11. Numerical differentiation

The numerical differentiation is the mathematical procedure to estimate the derivative of a determined function.

$$f'(x) = \lim_{h\to 0} \frac{f(x+h)-f(x)}{h}$$

Which can be numerically approximated to

$$f'(x) \approx \frac{f(x+h)-f(x)}{h} + O(h^2)$$

$$f'(x) \approx \frac{f(x+h)-f(x-h)}{2h} + O(h^2)$$

Where h is a small positive or negative change. As h grows smaller, the numerical integration becomes more precise (see more details in Chapter One).

Next, it is depicted the code for the derivative of the function $f(x) = x^n$. Note that the value of h is given by formula: h = sqrt(eps)*x, where eps is the precision of the calculation (5.421E-20).

$$h = x\sqrt{eps}$$

```
!Program name: SLOPE
!derivative (slope) of a function f(x) = x**n
real *16 xph, x, slope, dx, h
integer :: n
Print *, 'Derivative (slope) of the function x^n'
Print *, 'Give the value of x: '
Read *, x
Print *, 'Give the value of n: '
Read *, n
call epsilon(eps)
print *, 'eps: ', eps
Print *, 'sqrt(eps): ', sqrt(eps)
h=sqrt(eps)*x
Print *, 'h: ', h
xph=x+h
dx=xph-x
Print *, 'xph: ', xph
slope=(xph**n-x**n)/dx
Print *, 'slope=(xph**n-x**n)/dx'
write (*,*) 'slope: ', slope
stop
end
subroutine epsilon(eps)
real :: eps
```

```
!Epsilon - calculation of the precision
  eps=1.
  do
     eps=eps/2
     If (eps+1 .eq. 1.) exit
  End do
  return
  End subroutine epsilon
```

12. Finding roots of a function: bisection method

There are three basic methods to find the roots of a function: bisection, secant and Newton-Raphson. The process of finding the roots of a given function *f*(x) means to find the values of x where *f*(x) = 0.

In the bisection method, the function is bisected (or divided into small sections) and one uses the recursive process to find the root(s) by increasing step-by-step the value of x, according to a given interval dx, from an initial trial value of x. The recursive process ends when the tolerance (i.e., a given value approaching zero) is reached for a given x in the function.

Let us use the function $f(x) = x^3 - 9$, where the root is 2.080084 with single precision. In the code below the trial value for x is 1.0, the tolerance is 10^{-6} and the initial step size is 0.5. In this first code, the number of bisections is indefinite.

```
!Program name: BISECTION1
!Function: x^3-9
!Initial x value is 1.0
!Tolerance 1E-06
!initial Interval dx=0.5
!The number of bisections is indefinite
!Root is 2.08008382
Real :: x, dx, tolx
integer :: iter
y(x)=x**3-9
fold=y(x) ! fold=-9.
tolx=1.E-06
x=1.
dx=0.5
iter=0
write (*,5) "iter", "x", "y(x)", "fold", "y(x)*fold", "dx"
5 format (a5X,a10X,a20X,a20X,a15X,a15X)
10 continue
iter=iter+1
  x=x+dx
  print *,iter, x, y(x), fold, y(x)*fold, dx
```

```
        if ((fold*y(x)) .lt. 0.) then
           x=x-dx
           dx=dx/2
        end if
   if (abs(dx) .gt.tolx) goto 10
stop
   end
```

The result of the code BISECTION1 is shown Table 2.1. One can see that when fold*y(x) < 0 the value of dx decreases by half. The recursive procedure stops when dx = tolx.

Table 2.2: Iteration process of the BISECTION1 code.

iter	x	**y(x)**	**fold**	**fold*y(x)**	**dx**
0	1.50000	-5.6250000	-9.0	50.625000	0.50000000
1	2.00000	-1.0000000	-9.0	9.0000000	0.50000000
2	2.50000	6.6250000	-9.0	-59.625000	0.50000000
3	2.25000	2.3906250	-9. 0	-21.515625	0.25000000
4	2.12500	0.59570313	-9.0	-5.3613281	0.12500000
5	2.06250	-0.22631836	-9.0	2.0368652	6.25000000E-02
6	2.12500	0.59570313	-9.0	-5.3613281	6.25000000E-02
7	2.09375	0.17855835	-9.0	-1.6070251	3.12500000E-02
8	2.07812	-2.54020E-02	-9.0	0.22861862	1.56250000E-02
9	2.09375	0.17855835	-9.0	-1.6070251	1.56250000E-02
10	2.085937	7.6196193E-02	-9.0	-0.68576574	7.81250000E-03
11	2.0820313	2.5301754E-02	-9.0	-0.22771579	3.90625000E-03
12	2.0800781	-7.396191E-05	-9.0	6.65657E-04	1.95312500E-03
13	2.0820313	2.5301754E-02	-9.0	-0.22771579	1.95312500E-03
14	2.0810547	1.260794E-02	-9.0	-0.11347148	9.76562500E-04
15	2.0805664	6.2655019E-03	-9.0	-5.638951E-02	4.88281250E-04
16	2.0803223	3.0953981E-03	-9.0	-2.785858E-02	2.44140625E-04
17	2.0802002	1.5106251E-03	-9.0	-1.359562E-02	1.22070313E-04
18	2.0801392	7.1830832E-04	-9.0	-6.464774E-03	6.10351563E-05
19	2.0801086	3.2216741E-04	-9.0	-2.899506E-03	3.05175781E-05
20	2.0800934	1.2410129E-04	-9.0	-1.116911E-03	1.52587891E-05
21	2.0800858	2.5069326E-05	-9.0	-2.256239E-04	7.62939453E-06
22	2.0800819	-2.444638E-05	-9.0	2.2001746E-04	3.81469727E-06
23	2.0800858	2.5069326E-05	-9.0	-2.256239E-04	3.81469727E-06
24	2.0800838	3.1144796E-07	-9.0	-2.803031E-06	1.90734863E-06

Next, we present a different code for the same calculation where the number of bisections is initially established (iter=100). In this case, the last iteration is 24.

```
!Program name: BISECTION2
!Function: x^3-9
!Initial x value is 1.0
!Tolerance 1E-06
!initial Interval dx=0.5
!The number of bisections is 100
!Root is 2.08008382
Real :: x, dx, y, fold
Real :: tolx
integer :: iter, i
func(x)=x**3-9
fold=func(x)
tolx=1.E-06
x=1.
dx=0.5
iter=100
write (*,5) "iter", "x", "y(x)", "fold", "fold*y(x)", "dx"
5 format (a5X,a10X,a20X,a20X,a15X, a15X)
do i=0,iter
  x=x+dx
y=func(x)
  print *,i, x, y, fold, fold*y, dx
     if ((fold*y) .lt. 0.) then
     x=x-dx
  dx=dx/2
  end if
  if (abs(dx) .gt.tolx) cycle
  if (abs(dx) .lt. tolx) exit
end do
  stop
  end
```

The next bisection algorithm took lesser steps (19 steps) to converge to the tolerance value. In this code, there is no dx interval.

```
!Program name: BISECTION3
!Function: x^3-9
!Initial x value is 1.0
!Tolerance 1E-06
!initial Interval dx=0.5
!The number of bisections is 100
!Root is 2.08008382
module finding_roots
implicit none
Real :: xl, xr
contains
  subroutine init()
!Root between x=0 and x=3
  xl=0.
xr=3.
  write (*,5) "iter", "x", "y(x)", "fl", "fl*y(x)"
5 format (a5X,a10X,a20X,a20X,a15X)
end subroutine init
  function func(x) result(y)
  implicit none
Real :: x
Real :: y !dummy argument
  y=x**3-9
  end function func
  subroutine bisect(xl,xr,x,func,f)
     implicit none
integer :: i
real :: xl, xr, x, func, toly
real :: fl, fr, f !dummy arguments
  call init()
fl=func(xl)
fr=func(xr)
  toly=1.E-06
  do i=1,100
     x=0.5*(xl+xr)
     f=func(x)
if (f*fl .lt. 0) then
xr=x
fr=f
     else
     xl=x
fl=f
```

```
        end if
        Write (*,*) i, x, f, fl, f*fl
        if (abs(f) .lt. toly) exit
    end do
    end subroutine bisect
end module finding_roots
use finding_roots
call bisect(xl,xr,x,func,f)
print *, x, func(x)
stop
end
```

The result of this last code is shown in Table 2.3.

Table 2.3: Iteration process of the BISECTION3 code.

ite	x	y(x)	fl	fl*y(x)
1	1.5000000	-5.6250000	-5.6250000	31.640625
2	2.2500000	2.3906250	-5.6250000	-13.447266
3	1.8750000	-2.4082031	-2.4082031	5.7994423
4	2.0625000	-0.22631836	-0.22631836	5.12199998E-02
5	2.1562500	1.0252991	-0.22631836	-0.23204401
6	2.1093750	0.38558578	-0.22631836	-8.72651413E-02
7	2.0859375	7.6196193E-02	-0.22631836	-1.72445979E-02
8	2.0742188	-7.5915634E-02	-7.5915634E-02	5.76318381E-03
9	2.0800781	-7.3961913E-05	-7.3961913E-05	5.47036461E-09
10	2.0830078	3.8007479E-02	-7.3961913E-05	-2.81110579E-06
11	2.0815430	1.8953358E-02	-7.3961913E-05	-1.40182669E-06
12	2.0808105	9.4363503E-03	-7.3961913E-05	-6.97930545E-07
13	2.0804443	4.6803569E-03	-7.3961913E-05	-3.46168150E-07
14	2.0802612	2.3029884E-03	-7.3961913E-05	-1.70333436E-07
15	2.0801697	1.1144608E-03	-7.3961913E-05	-8.24276611E-08
16	2.0801239	5.2023644E-04	-7.3961913E-05	-3.84776833E-08
17	2.0801010	2.2313398E-04	-7.3961913E-05	-1.65034173E-08
18	2.0800896	7.4585215E-05	-7.3961913E-05	-5.51646506E-09
19	2.0800838	3.1144796E-07	-7.3961913E-05	-2.30352872E-11
2.0800838		3.11447963E-07		

13. Finding roots of a function: Newton-Raphson method

The Newton-Raphson method is based on the tangent line equation.

Suppose a function $f(x)$ in which we intend to find the tangent line at the point p $(x_n, f(x_n))$. Let us consider a nearby point q $(x_n + h, f(x_n + h))$ on that curve. The segment uniting these points is the secant line. The slope of this secant line (the ratio y/x) is:

$$\frac{f(x_n + h) - f(x_n)}{h}$$

As the value of h becomes smaller and smaller, the point q approaches p and the secant line tends to become the tangent line when h is infinitesimally small, the slope assumes a certain value k. The coordinates of the tangent line (x_t,y_t) can be found as:

$$\frac{y_t - f(x_n)}{(x_t - x_n)} = k = f'(x_n)$$

$$then:$$

$$y_t = f(x_n) + f'(x_n)(x_t - x_n)$$

Let us now choose the coordinates of the tangent line (x_t,y_t) as the root of the function *f*(x) , that is, at x_{n+1} and $y_t = 0$.

$$for: y_t = 0 \therefore x_t = x_{n+1}$$

$$0 = f(x_n) + f'(x_n)(x_{n+1} - x_n)$$

The value of x_{n+1} is the guess for the root of the function:

$$x_{n+1} = x_n - \frac{f(x_n)}{f'(x_n)}$$

The Newton-Raphson method has a faster convergence than bisection method, but not always converge if the initial point x_0 is not within the neighborhood of the root or if the in any iteration point it finds a stationary point or it finds a k-cycle. The Newton-Raphson method is only possible when the derivative of the function is known.

14. Finding roots of a function: secant method

The secant method derives from Newton-Raphson method. It has an intermediate efficiency between bisection and Newton-Raphson method. On the other hand, there is no need to find the derivative of function in the secant method.

In the secant method the derivative of the function is substituted by the backward finite difference equation:

$$f'(x_n) = \frac{f(x_n) - f(x_{n-1})}{x_n - x_{n-1}}$$

Then, the iterative equation becomes:

$$x_{n+1} = x_n - \frac{f(x_n)}{f'(x_n)}$$

$$x_{n+1} = x_n - f(x_n)\frac{x_n - x_{n-1}}{f(x_n) - f(x_{n-1})}$$

Let:

$$n+1=n, \quad n=n-1, \quad n-1=n-2$$

$$x_n = x_{n-1} - f(x_{n-1})\frac{x_{n-1}-x_{n-2}}{f(x_{n-1})-f(x_{n-2})}$$

And then:

$$x_n = \frac{x_{n-1}\left(f(x_{n-1})-f(x_{n-2})\right)-f(x_{n-1})x_{n-1}+f(x_{n-1})x_{n-2}}{f(x_{n-1})-f(x_{n-2})}$$

$$x_n = \frac{f(x_{n-1})x_{n-1}-f(x_{n-2})x_{n-1}-f(x_{n-1})x_{n-1}+f(x_{n-1})x_{n-2}}{f(x_{n-1})-f(x_{n-2})}$$

$$x_n = \frac{f(x_{n-1})x_{n-2}-f(x_{n-2})x_{n-1}}{f(x_{n-1})-f(x_{n-2})}$$

The last equation is the secant equation. Unlike Newton-Raphson, two initial guesses (instead of just one) are needed in the secant method. They are x_{n-1} and x_{n+1} or x_{n-2} and x_{n-1}. Provided they are close to the root of the equation (upper and lower value than x_n), the convergence is as fast as the Newton-Raphson method.

During the iteration process of the secant method, we have to obtain the intermediate value between x_{n-1} and x_{n+1}. It is:

$$\frac{f(x_{n-1})x_{n-2}-f(x_{n-2})x_{n-1}}{f(x_{n-1})-f(x_{n-2})} \equiv \frac{f(x_{n+1})x_{n-1}-f(x_{n-1})x_{n+1}}{f(x_{n+1})-f(x_{n-1})}$$

$$x_n = \frac{x_{n-1}f(x_{n+1})-x_{n+1}f(x_{n-1})}{f(x_{n+1})-f(x_{n-1})}$$

For the equation $x^3 + x - 3$, the secant method finds the root after eight iterations. For the equation $x^3 - 9$, the secant method finds the root in six iterations (for $x_1 = 0.5$ and $x_2 = 3.0$), that is three times more efficient than the bisection method.

```
!Program name: SECANT
!Function: x^3+x-3
!Initial x values are 0 and 3.0
!Tolerance 1E-06
!Root=1.2134118
Real :: x0, x1, x2, xn, c
Real :: tolx
integer :: i
f(x)=x**3+x-3
tolx=1.E-06
x1=0.
x2=3.
```

```
write (*,5) "iter", "x1", "x2","y1","y2","c","xn-x0"
5 format (a5X,a10X,a20X,a20X,a15X, a15X,a15X,a15x)
do i=0,100
x0=(x1*f(x2)-x2*f(x1))/(f(x2)-f(x1))
c=f(x1)*f(x0)
      x1=x2
      x2=x0
xn=(x1*f(x2)-x2*f(x1))/(f(x2)-f(x1))
  print *,i, x1, x2, f(x1), f(x2), c, xn-x0
      if (c == 0.) exit
      if (abs(xn-x0) .gt.tolx) cycle
      if (abs(xn-x0) .lt. tolx) exit
end do
  Print *, "The root of the equation is: ", x0
stop
end
```

15. Numerov method

The Numerov method is used to calculate numerically the ordinary second order differential equation such as the one-particle quantum harmonic oscillator (Chapter 14) and radial wave function of the hydrogen atom (Chapter 17):

$$-\frac{\hbar^2}{2m}\frac{d^2\psi}{dx^2}+V(x)\psi(x)=E\psi(x)$$

$$\frac{d^2\psi}{dx^2}=\frac{2m}{\hbar^2}(V(x)-E)\psi(x)$$

$$K^2(x)=\frac{2m}{\hbar^2}(E-V(x))$$

$$\frac{d^2\psi}{dx^2}+K^2(x)\psi(x)=0$$

Any smooth function can be rewritten as a sum of infinite polynomial terms. Then, let us use the general Taylor series (see previous section):

$$f(x+a)=f(a)+\frac{x}{1!}f'(a)+\frac{x^2}{2!}f''(a)+\frac{x^3}{3!}f'''(a)+...+R_n(x)$$

to represent the wave function $\psi(x)$:

$$\psi(x+h)=\psi(x)+h\psi'(x)+\frac{h^2}{2!}\psi''(x)+\frac{h^3}{3!}\psi'''(x)+\frac{h^4}{4!}\psi^{iv}(x)+...$$

Where h is a small increment. Observe that h and x were interchanged in their positions in the equation.

The Taylor series can also be written in the following way (see previous section):

$$\psi(x-h)=\psi(x)-h\psi'(x)+\frac{h^2}{2!}\psi''(x)-\frac{h^3}{3!}\psi'''(x)+\frac{h^4}{4!}\psi^{iv}(x)+...$$

By summing both Taylor series equations for the wave function, we have:

$$\psi(x+h)=\psi(x)+h\psi'(x)+\frac{h^2}{2!}\psi''(x)+\frac{h^3}{3!}\psi'''(x)+\frac{h^4}{4!}\psi^{iv}(x)+$$
$$+\frac{h^5}{5!}\psi^{v}(x)+O(h^6)$$
$$+\left(\begin{array}{l}\psi(x-h)=\psi(x)-h\psi'(x)+\frac{h^2}{2!}\psi''(x)-\frac{h^3}{3!}\psi'''(x)+\frac{h^4}{4!}\psi^{iv}(x)\\ -\frac{h^5}{5!}\psi^{v}(x)+O(h^6)\end{array}\right)$$

In both series above, they were truncated to the fifth term, then $O(h^6)$ represents the error of the series approximation, or the remainder term (see previous section). The result of this sum yields only even terms:

$$\psi(x+h)+\psi(x-h)=2\psi(x)+h^2\psi''(x)+\frac{h^4}{12}\psi^{iv}(x)+O(h^6)$$

Let us make the term ψ''(x) explicit:

$$\psi(x+h)+\psi(x-h)-2\psi(x)-\frac{h^4}{12}\psi^{iv}(x)-O(h^6)=h^2\psi''(x)$$

$$\psi''(x)=\frac{\psi(x+h)+\psi(x-h)-2\psi(x)-\frac{h^4}{12}\psi^{iv}(x)-O(h^6)}{h^2}$$

$$\psi''(x)=\frac{\psi(x+h)+\psi(x-h)-2\psi(x)}{h^2}-\frac{h^2}{12}\psi^{iv}(x)-O(h^4)$$

Then, we found the expression from Taylor series for the second derivative of the wave function.

Let us rearrange the above equation and omit the error of the series $O(h^4)$:

$$\frac{\psi(x+h)+\psi(x-h)-2\psi(x)}{h^2}=\psi''(x)+\frac{h^2}{12}\psi^{iv}(x)$$

$$\frac{\psi(x+h)+\psi(x-h)-2\psi(x)}{h^2}=\left(1+\frac{h^2}{12}\frac{d^2}{dx^2}\right)\psi''(x)$$

Then, we use the following operator:

$$1+\frac{h^2}{12}\frac{d^2}{dx^2}$$

On the equation:

$$\psi''(x)+K^2(x)\psi(x)=0$$

And we have the following result:

$$\left(1+\frac{h^2}{12}\frac{d^2}{dx^2}\right)\psi''(x)=-K^2(x)\psi(x)-\frac{h^2}{12}\frac{d^2}{dx^2}\left[K^2(x)\psi(x)\right]$$

$$\psi''(x)+\frac{h^2}{12}\psi^{iv}(x)=-\frac{h^2}{12}\frac{d^2}{dx^2}\left[K^2(x)\psi(x)\right]-K^2(x)\psi(x)$$

Let us replace the above equation on the previous expression for the second derivative of the wave function:

$$\psi''(x)=\frac{\psi(x+h)+\psi(x-h)-2\psi(x)}{h^2}-\frac{h^2}{12}\psi^{iv}(x)-O(h^4)$$

$$\psi''(x)+\frac{h^2}{12}\psi^{iv}(x)=\frac{\psi(x+h)+\psi(x-h)-2\psi(x)}{h^2}-O(h^4)$$

Then, we have:

$$-\frac{h^2}{12}\frac{d^2}{dx^2}\left[K^2(x)\psi(x)\right]-K^2(x)\psi(x)=$$
$$=\frac{\psi(x+h)+\psi(x-h)-2\psi(x)}{h^2}-O(h^4)$$

Multiply the above equation by h^2:

$$-\frac{h^4}{12}\frac{d^2}{dx^2}\left[K^2(x)\psi(x)\right]-h^2K^2(x)\psi(x)=$$
$$=\psi(x+h)+\psi(x-h)-2\psi(x)-O(h^6)$$

$$\psi(x+h)+\psi(x-h)-2\psi(x)+\frac{h^4}{12}\frac{d^2}{dx^2}\left[K^2(x)\psi(x)\right]$$
$$+h^2K^2(x)\psi(x)=0$$

Let us now find the expression for the first term of the left side of the above equation by using the relation we have found for the second derivative of the wave function, but we omit the last two terms

$$\psi''(x)=\frac{\psi(x+h)+\psi(x-h)-2\psi(x)}{h^2}-\frac{h^2}{12}\psi^{iv}(x)-O(h^4)$$

$$\psi''(x)\approx\frac{\psi(x+h)+\psi(x-h)-2\psi(x)}{h^2}$$

Then, we have:

$$\frac{d^2}{dx^2}\left[K^2(x)\psi(x)\right]\approx$$
$$\approx\frac{K^2(x+h)\psi(x+h)+K^2(x-h)\psi(x-h)-2K^2(x)\psi(x)}{h^2}$$

By replacing the above equation in

$$\psi(x+h)+\psi(x-h)-2\psi(x)+\frac{h^4}{12}\frac{d^2}{dx^2}\left[K^2(x)\psi(x)\right]$$
$$+h^2K^2(x)\psi(x)=0$$

We have:

$$\frac{h^4}{12}\left(\frac{K^2(x+h)\psi(x+h)+K^2(x-h)\psi(x-h)-2K^2(x)\psi(x)}{h^2}\right)+$$
$$+\psi(x+h)+\psi(x-h)-2\psi(x)+h^2K^2(x)\psi(x)=0$$

By rearranging the last equation, we obtain the Numerov equation for the wave function:

$$\psi(x+h)=\frac{2\left[1-\tfrac{5}{12}h^2K^2(x)\right]\psi(x)-\left[1+\tfrac{1}{12}h^2K^2(x-h)\right]\psi(x-h)}{1+\tfrac{1}{12}h^2K^2(x+h)}$$

For convenience, let us introduce the array f_n defined as:

$$f_n=1+\frac{h^2}{12}K^2(x)$$
$$f_{n+1}=1+\frac{h^2}{12}K^2(x+h)$$
$$f_{n-1}=1+\frac{h^2}{12}K^2(x-h)$$
$$\psi(x+h)=\psi_{n+1},\qquad \psi(x-h)=\psi_{n-1}$$

Then, the Numerov formula is written as:

$$\psi_{n+1}=\frac{(12-10f_n)\psi_n-f_{n-1}\psi_{n-1}}{f_{n+1}}$$

The above equation can be applied to any ordinary second order differential. As we omitted the O(h^4), the numerical error is O(h^4).

References cited

Koonin, S.E. and Meredith, D.C. 1990. Computational Physics. Fortran version. CRC Press. Boca Raton.
Boas, M.L. 2006. Mathematical methods in the physical sciences. John Wiley & Sons, Inc. Third edition, Hoboken.

Linear Algebra for Quantum Mechanics

3

1. Matrix and matrix multiplication

A matrix is used to arrange real or complex scalars (or functions) in a rectangular array of m,n dimensions where, m is the dimension of the row (m-rows) and n is the dimension of the column (n-column). The matrix is designated as m x n matrix or matrix of orders m,n. The order is always: number of rows x number of columns. The elements of the matrix are designated by i and j indexes.

$$\mathbf{A} = \begin{bmatrix} a_{11} & a_{12} & \cdots & a_{1n} \\ a_{21} & a_{22} & \cdots & a_{2n} \\ \vdots & \vdots & \cdots & \vdots \\ a_{n1} & a_{2n} & \cdots & a_{nn} \end{bmatrix}$$

A very important property to the quantum matrix mechanics is that the commutative property is not applied to matrices (see section 39 – commutators). In chapter nine we see that the linear momentum and position matrices do not commute and this brings important consequences to the quantum angular momentum, uncertainty principle among other consequences.

When both matrices are not square matrices (whose rows and columns have the same number), probably the inverted order of the product provides no matrix at all. Even when both matrices are square matrices, the matrix **AB** is mostly different from matrix **BA**.

The matrix product **AB** (in this order) is defined only if: the number of columns of **A** is equal to the number of rows of B. The dimensions of this matrix product is the number of rows of A and the number of columns of **B**. Likewise, the product **BA** is defined only if: the number of columns of **B** is equal to the number of rows of **A**.

$$\mathbf{A}_{n\times m} \cdot \mathbf{B}_{m\times p} = \mathbf{C}_{n\times p}$$

$$However: \mathbf{B}_{m\times p} \cdot \mathbf{A}_{n\times m} = impossible$$

$$\mathbf{A}_{n\times n} \cdot \mathbf{B}_{n\times n} \neq \mathbf{B}_{n\times n} \cdot \mathbf{A}_{n\times n}$$

$$\mathbf{C} = \mathbf{A} \cdot \mathbf{B} \therefore c_{ij} = \sum_{k=1}^{n} a_{ik} b_{kj}$$

Let us see the example below:

$$\mathbf{F} = \begin{bmatrix} 1 & 4 & 7 \\ 2 & 5 & 8 \\ 3 & 6 & 9 \end{bmatrix}, \quad \mathbf{G} = \begin{bmatrix} 7 & 4 & 1 \\ 8 & 5 & 2 \\ 9 & 6 & 3 \end{bmatrix}, \quad \mathbf{FG} = \begin{bmatrix} 102 & 66 & 30 \\ 126 & 81 & 36 \\ 150 & 96 & 42 \end{bmatrix}$$

In Section 19, it is depicted the Fortran implementation for the matrix multiplication. Try out the matrices F and G above.

2. Trace of a matrix

The trace of a square matrix (matrix of order n), A_n, is the sum of its matrix elements on the main diagonal of A, $a_{k,k}$.

$$\mathbf{Tr\ A} = a_{11} + a_{22} + \ldots + a_{nn} = \sum_{i=1}^{n} a_{ii}$$

3. Transpose of a matrix

The transpose of the matrix $\mathbf{A}_{nxm}$, $\mathbf{A}^T$, is a new matrix where its rows are the columns of the original matrix and vice versa. In the case where the transpose of the matrix is similar to the original matrix, the former is called symmetric matrix. The diagonal elements, $a_{k,k}$, of both matrices are the same

$$\mathbf{A} = \begin{bmatrix} a_{11} & a_{12} & \cdots & a_{1n} \\ a_{21} & a_{22} & \cdots & a_{2n} \\ \vdots & \vdots & \vdots & \vdots \\ a_{n1} & a_{n2} & \cdots & a_{nn} \end{bmatrix}, \mathbf{A}^T = \begin{bmatrix} a_{11} & a_{21} & \cdots & a_{n1} \\ a_{12} & a_{22} & \cdots & a_{n2} \\ \vdots & \vdots & \vdots & \vdots \\ a_{1n} & a_{2n} & \cdots & a_{nn} \end{bmatrix} (a_{ij}) = (a_{ji})^T$$

4. Symmetric matrix and orthogonal matrix

A symmetric matrix, **A**, is a square matrix which respects the relation below:

$$\mathbf{A} = \mathbf{A}^T$$

$$\mathbf{A}^T\mathbf{A} = \mathbf{A}^2$$

A square matrix A is said to be orthogonal when:

$$\mathbf{A}\mathbf{A}^T = \mathbf{A}^T\mathbf{A} = \mathbf{I}$$

Where **I** is the unit (or identity) matrix of the same order whose diagonal elements are unitary values and non-diagonal elements are zero.

5. The determinant of a matrix

The concept of the determinant came from the resolution of the system of linear equations. The determinant of order n derives from a square array of n^2 elements and results in a scalar number. The determinant is named as D_A or det(**A**), where **A** is related to the square array of n^2 elements.

$$D_A = \begin{vmatrix} a_{11} & a_{12} & \cdots & a_{1n} \\ a_{21} & a_{22} & \cdots & a_{2n} \\ \vdots & \vdots & \vdots & \vdots \\ a_{n1} & a_{n2} & \cdots & a_{nn} \end{vmatrix}$$

There are two methods to obtain the scalar value of a determinant: the one given in the next Section 6 (the diagonal product of an upper triangular matrix) and the Laplace expansion (Section 7). The triangular matrix has 0 elements below the main diagonal of the matrix. It is possible to transform any matrix into an upper triangular matrix by the Gaussian elimination method.

6. Gaussian elimination method and the determinant of a matrix

The Gaussian elimination method is used to transform any matrix into upper triangular matrix (row echelon form). This method is based on three row operations:

(i) swapping two rows: $(E_i)\leftrightarrow(E_j)$;

(ii) multiplying a row by a nonzero number: $(\lambda E_i)\rightarrow(E_i)$;

(iii) adding a multiple of one row to another row: $(E_i+\lambda E_j)\rightarrow(E_i)$.

These operations do not change the determinant of the matrix (and also the system of linear equations). The diagonal product of an upper triangular matrix gives the determinant of this matrix, where a zero element is not permitted.
In the example below:

(1)→(2): –3E1+E4

(2)→(3): –2E2+E3 // 6E2+E4

(3)→(4): –1/3E3+E4

$$\det(\mathbf{A}) = \underset{(1)}{\begin{vmatrix} 1 & 2 & 0 & 0 \\ 0 & 1 & 0 & 2 \\ 0 & 2 & 3 & 0 \\ 3 & 0 & 1 & 0 \end{vmatrix}} = \underset{(2)}{\begin{vmatrix} 1 & 2 & 0 & 0 \\ 0 & 1 & 0 & 2 \\ 0 & 2 & 3 & 0 \\ 0 & -6 & 1 & 0 \end{vmatrix}} = 40$$

$$\det(\mathbf{A}) = \underset{(3)}{\begin{vmatrix} 1 & 2 & 0 & 0 \\ 0 & 1 & 0 & 2 \\ 0 & 0 & 3 & -4 \\ 0 & 0 & 1 & 12 \end{vmatrix}} = \underset{(4)}{\begin{vmatrix} 1 & 2 & 0 & 0 \\ 0 & 1 & 0 & 2 \\ 0 & 0 & 3 & -4 \\ 0 & 0 & 0 & 40/3 \end{vmatrix}} = 40$$

7. Laplace expansion and determinant of a matrix

The Laplace expansion is a method to obtain the determinant of a matrix based on cofactors minor matrix. The cofactors, $\mathbf{C}_{ij}$, of the square matrix **A** are $(-1)^{i+j}$ times the determinant of the submatrix $\mathbf{A}_{ij}$, $D(\mathbf{A}_{ij})$, obtained from **A** by deleting ith rows and jth columns of A. The $D(\mathbf{A}_{ij})$ is called minor M_{ij} of element a_{ij} of a determinant D obtained by deleting row i and column j. The cofactors of **A** form a new matrix called cofactor matrix, C, whose elements are:

$$C_{ij} = (-1)^{i+j} \cdot D(\mathbf{A}_{ij})$$

$$C_{ij} = (-1)^{i+j} \cdot M_{ij}$$

The determinant, D, of a square matrix of order n can be obtained by the expansion along any row i or by the expansion of any column j according to Laplace expansion.

$$D = a_{i1}C_{i1} + a_{i2}C_{i2} + \ldots + a_{in}C_{in} = \sum_{j=1}^{n} a_{ij}C_{ij}$$

$$D = a_{1j}C_{1j} + a_{2j}C_{2j} + \ldots + a_{nj}C_{nj} = \sum_{i=1}^{n} a_{ij}C_{ij}$$

Let us find the determinant of the matrix **A** below:

$$\mathbf{A} = \begin{bmatrix} 1 & 2 & 3 \\ -2 & 1 & 2 \\ 3 & -1 & -1 \end{bmatrix}$$

$$\det(\mathbf{A}) = 1 \times \begin{vmatrix} 1 & 2 \\ -1 & -1 \end{vmatrix} - 2 \times \begin{vmatrix} -2 & 2 \\ 3 & -1 \end{vmatrix} + 3 \times \begin{vmatrix} -2 & 1 \\ 3 & -1 \end{vmatrix}$$

$$\det(\mathbf{A}) = 1 \times (-1 + 2) - 2 \times (2 - 6) + 3 \times (2 - 3) = 6$$

8. Adjugate/adjunct/adjoint matrix

The adjugate or adjunct matrix is the transpose of the cofactor matrix, **C**, of the square matrix **A**. Sometimes, one can find also the term adjoint of **A**, Â, to refer to the adjunct matrix

$$adj(\mathbf{A}) = \mathbf{C}^T$$

9. Inverse matrix

If **A** and **B** are square matrices of the same order, **B** is the inverse matrix of **A**, $\mathbf{A}^{-1}$, if:

$$\mathbf{A} \cdot \mathbf{B} = \mathbf{B} \cdot \mathbf{A} = \mathbf{I}$$

Where **I** is the unit (or identity) matrix of the same order whose diagonal elements are unitary values and non-diagonal elements are zero. The inverse matrix can be obtained by:

$$\mathbf{A}^{-1} = \frac{adj(\mathbf{A})}{\det \mathbf{A}}$$

Where det A is the determinant of matrix **A**.

10. Gauss-Jordan elimination and inverse matrix

The Gauss-Jordan method (or elimination) is a method to solve a system of linear equations based on the property of the inverse matrix of order n:

$$\mathbf{A}\cdot\mathbf{A}^{-1} = \mathbf{I}$$

$$\mathbf{A}^{-1} = \begin{pmatrix}\mathbf{x}_1 & \mathbf{x}_2 & \cdots & \mathbf{x}_k & \cdots & \mathbf{x}_n\end{pmatrix} \therefore \mathbf{x} = \textbf{column vector}$$

$$\mathbf{I} = \begin{pmatrix}\mathbf{e}_1 & \mathbf{e}_2 & \cdots & \mathbf{e}_k & \cdots & \mathbf{e}_n\end{pmatrix} \therefore \mathbf{e} = \textbf{column vector}$$

$$\textbf{Then: } \mathbf{A}\mathbf{x}_k = \mathbf{e}_k \therefore (k = 1, 2, \ldots, n)$$

Where the set of all equations $\mathbf{Ax}_k = \mathbf{e}_k$ represents a system of linear equations (see Section 22) which can be solved by the Gauss-Jordan elimination method which uses the same arithmetic rules of the Gaussian elimination method in order to transform the augmented matrix $(\mathbf{A}|\mathbf{I})$ into $(\mathbf{I}|\mathbf{A}^{-1})$.

$$\mathbf{A} = \begin{bmatrix} 1 & 2 & 3 \\ -2 & 1 & 2 \\ 3 & -1 & -1 \end{bmatrix}$$

$$(\mathbf{A}|\mathbf{I}) = \left[\begin{array}{ccc|ccc} 1 & 2 & 3 & 1 & 0 & 0 \\ -2 & 1 & 2 & 0 & 1 & 0 \\ 3 & -1 & -1 & 0 & 0 & 1 \end{array}\right] \therefore$$

$$(\mathbf{I}|\mathbf{A}^{-1}) = \left[\begin{array}{ccc|ccc} 1 & 0 & 0 & \frac{1}{6} & -\frac{1}{6} & \frac{1}{6} \\ 0 & 1 & 0 & \frac{2}{3} & -\frac{5}{3} & -\frac{4}{3} \\ 0 & 0 & 1 & -\frac{1}{6} & \frac{7}{6} & \frac{5}{6} \end{array}\right]$$

11. Properties of orthogonal matrix

An orthogonal matrix is a type of square matrix where its inverse is equal to its transpose. If **A** is an orthogonal matrix, then:

$$\mathbf{A}^T = \mathbf{A}^{-1}$$

$$\mathbf{A}\cdot\mathbf{A}^T = \mathbf{I}$$

$$\det \mathbf{A} = \pm 1$$

$$\det(\mathbf{A}^T\cdot\mathbf{A}) = \det \mathbf{A}^T \times \det \mathbf{A} = (\det \mathbf{A})^2 = 1$$

If a 3 x 3 matrix, A, is orthogonal, then it represents three orthogonal vectors, **a**, **b**, and **c** whose coefficients in x, y and z directions (a_i, b_i, c_i, where i = 1,2,3) are the matrix elements in each row. As a consequence, the scalar products of the vectors **a**, **b**, and **c** are:

$$\mathbf{A} = \begin{bmatrix} a_1 & a_2 & a_3 \\ b_1 & b_2 & b_3 \\ c_1 & c_2 & c_3 \end{bmatrix}$$

$$\mathbf{a} = a_1 x + a_2 y + a_3 z, \qquad \mathbf{a} \cdot \mathbf{a} = \sqrt{a_1^2 + a_2^2 + a_3^2} = 1$$

$$\mathbf{b} = b_1 x + b_2 y + b_3 z, \qquad \mathbf{b} \cdot \mathbf{b} = \sqrt{b_1^2 + b_2^2 + b_3^2} = 1$$

$$\mathbf{c} = c_1 x + c_2 y + c_3 z, \qquad \mathbf{c} \cdot \mathbf{c} = \sqrt{c_1^2 + c_2^2 + c_3^2} = 1$$

$$\mathbf{a} \cdot \mathbf{b} = 0 \therefore \mathbf{b} \cdot \mathbf{c} = 0, \qquad \mathbf{c} \cdot \mathbf{a} = 0$$

The inverse of an orthogonal matrix is equal to its transpose. Let **O** be an orthogonal matrix:

$$\mathbf{O}^{-1} = \mathbf{O}^T$$

12. Antisymmetry in matrices and permutation matrix

A permutation is an operation that changes two rows or two columns of a matrix which gives the minus determinant of the former matrix. Supposing the 2 × 2 matrix **A** has determinant D, if two rows or two columns of a matrix **A** are interchanged, then the determinant of the second matrix is –D. This is the antisymmetry property of the determinants. The transformation of matrix **A** into matrix **B** occurs by the permutation matrix, **P**.

$$\mathbf{A} = \begin{bmatrix} a_{11} & a_{12} \\ a_{21} & a_{22} \end{bmatrix}, \qquad \det \mathbf{A} = D$$

$$If \quad \mathbf{B} = \begin{bmatrix} a_{21} & a_{22} \\ a_{11} & a_{12} \end{bmatrix}, then \quad \det \mathbf{B} = -D$$

$$\det \mathbf{A} = a_{11}a_{22} - a_{21}a_{12}$$

$$\det \mathbf{B} = a_{21}a_{12} - a_{11}a_{22}$$

The permutation matrix is derived from an identity matrix where the unit diagonal elements of the former are reordered in the latter. The multiplication of the permutation matrix over matrix **A** gives the matrix **B** (where the rows are interchanged).

$$\mathbf{I}_2 = \begin{bmatrix} 1 & 0 \\ 0 & 1 \end{bmatrix}, \qquad \mathbf{P} = \begin{bmatrix} 0 & 1 \\ 1 & 0 \end{bmatrix}$$

$$\begin{bmatrix} 0 & 1 \\ 1 & 0 \end{bmatrix} \cdot \begin{bmatrix} a_{11} & a_{12} \\ a_{21} & a_{22} \end{bmatrix} = \begin{bmatrix} a_{21} & a_{22} \\ a_{11} & a_{12} \end{bmatrix}$$

13. Antisymmetric wave function and the Slater determinant

An alternating function has an antisymmetric property that by interchanging two variables it gives the value of the function multiplied by –1 (see Chapter thirteen).

$$f(x_1, x_2, x_3, \ldots, x_n) = -f(x_2, x_1, x_3, \ldots, x_n)$$

When the matrix elements are functions, $f_i(x_i)$, the second (or fourth or sixth and so on) term of the determinant of this matrix, D, is an alternating function of its previous term of the determinant. The determinant of order n has n! permutations, i.e., operations of changing two rows or two columns in the matrix that give the minus determinant of the orginal matrix.

$$D_{2x2} = \begin{vmatrix} f_1(x_1) & f_1(x_2) \\ f_2(x_1) & f_2(x_2) \end{vmatrix} = f_1(x_1)f_2(x_2) - f_1(x_2)f_2(x_1)$$

$$D_{3x3} = \begin{vmatrix} f_1(x_1) & f_1(x_2) & f_1(x_3) \\ f_2(x_1) & f_2(x_2) & f_2(x_3) \\ f_3(x_1) & f_3(x_2) & f_3(x_3) \end{vmatrix} = \begin{cases} f_1(x_1)f_2(x_2)f_3(x_3) - f_1(x_1)f_2(x_3)f_3(x_2) \\ f_1(x_2)f_2(x_3)f_3(x_1) - f_1(x_2)f_2(x_1)f_3(x_3) \\ f_1(x_3)f_2(x_1)f_3(x_2) - f_1(x_3)f_2(x_2)f_3(x_1) \end{cases}$$

If the matrix element, $f_i(x_i)$, is a spin-orbital, then the determinant is called Slater determinant, although it was firstly used by Heisenberg and Dirac (see Chapter thirteen). Alternating functions are important to construct the antisymmetric wave function.

14. Properties of determinants and the Slater determinant

If two rows or two columns of a determinant, D, are equal, then D = 0. Pauli exclusion principle states where two electrons are in the same orbital, then they cannot have the same spin-orbital. Then, in the Slater determinant, two rows or two spin-orbitals cannot be the same, i.e., $f_1 \neq f_2$, otherwise, the Slater determinant (which represents the set of molecular orbital, MO, wave functions) is zero.

To sum up, the properties of the determinants for square matrices (including comments on the Slater determinants, i.e., the MO wave functions), we have:

(i) If any row of column of the matrix has only zero elements, its determinant is zero (there cannot exist one orbital with zero elements, even LUMO orbital has to have non zero elements, because otherwise the Slater determinant is zero).

(ii) The determinant of a matrix, det(**A**), gives the same determinant with minus sign, –det(Â), after the operation $(E_i) \leftrightarrow (E_k)$, with $i \neq k$. This corresponds to the antisymmetric property of a fermion (previous section). That is: *det(**A**)* = *–det(Â)*, where Â: $(E_i) \leftrightarrow (E_k)$

(iii) If a matrix has two rows or two columns with the same elements, then its determinant is zero.

(iv) If a matrix is obtained after the operation $(\lambda E_i) \rightarrow (E_i)$, then: *det(Â)* = *λdet(**A**)*, where Â: $(\lambda E_i) \rightarrow (E_i)$.

(v) If a matrix is obtained from the operation $(E_i+\lambda E_k) \rightarrow (E_i)$, with $i \neq k$, then: *det(Â)* = *det(**A**)*, where Â: $(E_i+\lambda E_k) \rightarrow (E_i)$. The Slater determinant is the same after changing two orbitals by $(E_i+\lambda E_k) \rightarrow (E_i)$.

(vi) If **A** and **B** are square matrices of the same order, then *det(**AB**)* = *det(**A**)det(**B**)*

(vii) The determinant of the transpose of a matrix is the same as the determinant of the original matrix, *det (**A**ᵗ)* = *det(**A**)*. Then, the MO orbitals can be in rows or columns of the Slater determinant.

(viii) If the inverse of a matrix exists, then its determinant is: $det(A^{-1}) = 1/det(A)$.

(ix) If **A** is an upper triangular or lower triangular or diagonal triangular, then: $\det(\mathbf{A}) = a_{11}a_{22}a_{33}...a_{nn}$

15. Eigenvector/eigenvalues and matrix diagonalization

Let the matrix **A** have the following properties below.

$$\mathbf{A} = \begin{bmatrix} 10 & -4 \\ 12 & -4 \end{bmatrix}$$

$$\begin{bmatrix} 10 & -4 \\ 12 & -4 \end{bmatrix}\begin{bmatrix} 1 \\ 2 \end{bmatrix} = 2\begin{bmatrix} 1 \\ 2 \end{bmatrix}$$

$$\begin{bmatrix} 10 & -4 \\ 12 & -4 \end{bmatrix}\begin{bmatrix} 2 \\ 3 \end{bmatrix} = 4\begin{bmatrix} 2 \\ 3 \end{bmatrix}$$

The column matrices $[1\ 2]^T$ and $[2\ 3]^T$ are known as eigenvectors, υ, and the scalars 2 and 4 are known as eigenvalues, λ. By considering υ as the matrix of the eigenvectors and λ as the matrix of the eigenvalues, we have the eigenvalue equation below:

$$\mathbf{A}\upsilon = \lambda\upsilon$$

Indeed, **A** is the linear operator matrix in the equation above but it can also be represented by **O**, from operator. One can also represent υ using other symbols as long as the relation above exists.

All set of eigenvalue equations below

$$\mathbf{OA}_k = \lambda_k \mathbf{A}_k, \quad k = 1,2,\ldots n$$

has one equation that encompasses all these equations. It is:

$$\mathbf{OA} = \mathbf{AD}$$

$$\mathbf{O} = \begin{bmatrix} o_{1,1} & o_{1,2} & \cdots & o_{1,n} \\ o_{2,1} & o_{2,2} & \cdots & o_{2,n} \\ \vdots & \vdots & \vdots & \vdots \\ o_{n,1} & o_{n,2} & \cdots & o_{n,n} \end{bmatrix} \therefore$$

$$\mathbf{A} = (A_1, A_2, \ldots A_n) = \begin{bmatrix} a_{1,1} & a_{1,2} & \cdots & a_{1,n} \\ a_{2,1} & a_{2,2} & \cdots & a_{2,n} \\ \vdots & \vdots & \vdots & \vdots \\ a_{n,1} & a_{n,2} & \cdots & a_{n,n} \end{bmatrix}$$

$$\mathbf{D} = \begin{bmatrix} \lambda_1 & 0 & \cdots & 0 \\ 0 & \lambda_2 & \cdots & 0 \\ \vdots & \vdots & \vdots & \vdots \\ 0 & 0 & \cdots & \lambda_n \end{bmatrix}$$

Where **O** is the linear operator matrix, **A** is the matrix representing the set of column eigenvectors of **O**, and **D** is the diagonal matrix whose diagonal elements are the set of eigenvalues of **O**.

The process to find the diagonal eigenvalue matrix from the matrix A and linear operator **O** is called matrix diagonalization. It is only possible by knowing the matrix **A**, the set of eigenvectors of the linear operator, and the matrix operator **O**.

$$\mathbf{OA} = \mathbf{AD}$$

$$\mathbf{D} = \mathbf{A}^{-1}\,\mathbf{OA}$$

Let us use the following example:

$$\mathbf{O} = \begin{bmatrix} 1 & 3 \\ 2 & 2 \end{bmatrix}; \mathbf{A} = \begin{bmatrix} 1 & 3 \\ 1 & -2 \end{bmatrix}$$

$$\mathbf{OA} = \mathbf{AD}, \qquad \mathbf{D} = \mathbf{A}^{-1}\mathbf{OA}$$

$$\begin{bmatrix} 2/5 & 3/5 \\ 1/5 & -1/5 \end{bmatrix}\begin{bmatrix} 1 & 3 \\ 2 & 2 \end{bmatrix}\begin{bmatrix} 1 & 3 \\ 1 & -2 \end{bmatrix} = \begin{bmatrix} 4 & 0 \\ 0 & -1 \end{bmatrix}$$

One method to obtain the matrix diagonalization from a symmmetric matrix is called Jacobi method (see Section 38).

16. Similarity of the matrices

Two square matrices, **A** and **B**, of order n, are similar, **A** ~ **B**, when the following rules are satisfied:

a. If **A** ~ **B**, then det(**A**) = det(**B**); tr(**A**) = tr(**B**); and
b. **AP** = **PB**, then $\mathbf{B} = \mathbf{P}^{-1}\mathbf{AP}$, where **P** is a nonsingular matrix.

17. Decomposition LU and spectral decomposition of a matrix.

The square matrix **A** of order n can be decomposed in the product of two matrices **L** and **U**. The matrix **L** is a triangular inferior matrix and the matrix **U** is a triangular superior matrix.

$$\mathbf{A} = \begin{bmatrix} a_{11} & a_{12} & \cdots & a_{1n} \\ a_{21} & a_{22} & \cdots & a_{2n} \\ \vdots & \vdots & \vdots & \vdots \\ a_{n1} & a_{n2} & \cdots & a_{nn} \end{bmatrix}$$

$$\mathbf{LU} = \begin{bmatrix} 1 & 0 & \cdots & 0 \\ l_{21} & 1 & \cdots & 0 \\ \vdots & \vdots & \vdots & \vdots \\ l_{n1} & l_{n2} & \cdots & 1 \end{bmatrix}\begin{bmatrix} u_{11} & u_{12} & \cdots & u_{1n} \\ 0 & u_{22} & \cdots & u_{2n} \\ \vdots & \vdots & \vdots & \vdots \\ 0 & 0 & \cdots & u_{nn} \end{bmatrix}$$

$$\mathbf{A} = \mathbf{LU}$$

See the example below:

$$\begin{bmatrix} 1 & -3 & 2 \\ -2 & 8 & -1 \\ 4 & -6 & 5 \end{bmatrix} = \begin{bmatrix} 1 & 0 & 0 \\ -2 & 1 & 0 \\ 4 & 3 & 1 \end{bmatrix} \begin{bmatrix} 1 & -3 & 2 \\ 0 & 2 & 3 \\ 0 & 0 & -12 \end{bmatrix}$$

If the square matrix **A** of order n has a set of eigenvalues, λ_i (i = 1,2,...,n), represented by the diagonal matrix $\mathbf{\Lambda}$, and a set of eigenvectors, υ_i (i = 1,2,...,n), represented by the matrix **V** (where columns represent the eigenvectors υ_i), then the matrix **A** can be decomposed as

$$\mathbf{A} = \mathbf{V}\mathbf{\Lambda}\mathbf{V}^{-1}$$

Which is known as the spectral decomposition of **A**.

18. Transpose matrix in Fortran

In the chart below, we depict one example of transforming the matrix $\mathbf{F}_{3x3}$ into its transpose **FT**. The basic procedure to install the Fortran editor, Fortran compiler and how to understand the fundamentals of algorithm and Fortran language is given in Chapter One.

```
!name of the program: TRANSPOSE
!Building a 3x3 matrix and its transpose
  INTEGER*4 F(3,3),FT(3,3)
  Print *, 'Enter the matrix elements column by column: '
  Read *, F
  Print *, 'Matrix F 3x3: '
  Print *, F(1,1),F(1,2),F(1,3)
  Print *, F(2,1),F(2,2),F(2,3)
  Print *, F(3,1),F(3,2),F(3,3)
  Call MATRNS(F,FT)
  Print *, 'Matrix transpose FT 3x3: '
  Print *, FT(1,1),FT(1,2),FT(1,3)
  Print *, FT(2,1),FT(2,2),FT(2,3)
  Print *, FT(3,1),FT(3,2),FT(3,3)
  stop
  end

  Subroutine MATRNS(F,FT)
Integer*4 F(3,3),FT(3,3)
Do 2 J=1,3
  Do 2 I=1,3
     FT(J,I)= F(I,J)
2 continue
return
end
```

19. Matrix multiplication in Fortran

In the chart below, it is depicted the Fortran implementation for matrix product between two square matrices of order 3. It can be easily modified for any order. Try out the matrices given in the first section, **F** and **G**.

```
!name of the program:MULT
!Building two 3x3 matrices F, G and their multiplication FG
  INTEGER*4 F(3,3),G(3,3), FG(3,3)
 Print *, 'Enter the matrix elements column by column for F: '
  Read *, F
  Print *, 'Matrix F 3x3: '
  Print *, F(1,1),F(1,2),F(1,3)
  Print *, F(2,1),F(2,2),F(2,3)
  Print *, F(3,1),F(3,2),F(3,3)
 Print *, 'Enter the matrix elements column by column for G: '
  Read *, G
  Print *, 'Matrix G 3x3: '
  Print *, G(1,1),G(1,2),G(1,3)
  Print *, G(2,1),G(2,2),G(2,3)
  Print *, G(3,1),G(3,2),G(3,3)
  Call MATMPY(F,G,FG)
  Print *, 'Matrix multiplication FG 3x3: '
  Print *, FG(1,1),FG(1,2),FG(1,3)
  Print *, FG(2,1),FG(2,2),FG(2,3)
  Print *, FG(3,1),FG(3,2),FG(3,3)
  stop
  end

  Subroutine MATMPY(F,G,FG)
 Integer*4 F(3,3),G(3,3),FG(3,3)
  Do 2 J=1,3
  Do 2 I=1,3
     FG(I,J)=0
     Do 3 K=1,3
 FG(I,J)= FG(I,J)+F(I,K)*G(K,J)
 3 continue
 2 continue
  return
 end
```

Another Fortran code is shown below. In this case, it is possible to choose any types of matrices and it is used the intrinsic function MATMUL.

```
!name of the program:MULT2
! matrix multiplication AxB=C
integer , allocatable :: A(:,:), B(:,:), C(:,:)
integer :: m, n, o, i, j
Print *, "Enter the dimensions of matrix A(m,n)"
Read *, m, n
allocate(A(n,m))
Print *, "Enter the dimension o of matrix B(n,o)"
Read *, o
allocate(B(n,o))
allocate(C(m,o))
Do i=1,m
  Do j=1,n
     Print *, 'A(',i,',',j,')= '
Read *, A(i,j)
  End do
end do
do i=1,m
write (*,*) (A(i,j),j=1,n)
end do
print *,
Do i=1,n
  Do j=1,o
Print *, 'B(',i,',',j,')= '
Read *, B(i,j)
  End do
end do
do i=1,n
write (*,*) (B(i,j),j=1,o)
end do
print *,
C=matmul(A,B)
do i=1,m
write (*,*) (C(i,j),j=1,o)
end do
print *,
  stop
  end
```

20. Gaussian operation in Fortran

The source code presents one operation of the Gaussian elimination method: swapping the first two rows.

```
!program name: GAUSS
!Gauss elimination a matrix of order n
!Input dimension of the matrix and its elements
!This program does one Gauss operation: swapping two rows
Integer :: n,i,j
  Integer, dimension(:,:), allocatable :: A,elem
  Print *, 'Enter the dimension, n, of the square matrix A: '
Read *, n
Allocate( A(n,n) )
Allocate( elem(n,n) )
  Do i=1,n
      Do j=1,n
      Print *, 'A(',i,',',j,')= '
Read *, A(i,j)
      End do
End do
  Do i=1,n
      Do j=1,n
      Print *, 'A(',i,',',j,')= ', A(i,j)
End do
  End do
  ! sawpping first and second rows
Do j=1,n
elem(2,j)=A(1,j)
elem(1,j)=A(2,j)
  end do
do i=3,n
            do j=1,n
elem(i,j)=A(i,j)
end do
  end do
      Print *, 'The matrix, B, after one Gauss operation is: '
Do i=1,n
            Do j=1,n
Print *, 'B(',i,',',j,')= ', elem(i,j)
End do
  end do
  Deallocate (A)
Deallocate (elem)
stop
end
```

21. Determinant of a square matrix of order 2 or 3 in Fortran

```
!programname:DETERMINANT
! Program for square matrices of order 2 or 3
!---------------------------------
Integer :: n,i,j
real,dimension(:,:),allocatable :: A
integer::k,l,s
integer::nh,m
real,dimension(:,:,:),allocatable::minor
real,dimension(:),allocatable::det
real :: lap_det
  Print *, 'Enter the dimension, n, of the square matrix A: '
Read *, n
Allocate( A(n,n) )
  Do i=1,n
     Do j=1,n
     Print *, 'A(',i,',',j,')= '
Read *, A(i,j)
     End do
End do
  Do i=1,n
write (*,*) (A(i,j),j=1,n)
End do
!--------------------------------------
  print *,
allocate(minor(n,n-1,n-1))
  do k=1,n-1
     i=1+k
     do l=1,n-1
j=1+l
minor(1,k,l)=A(i,j)
end do
end do
  do k=1,n-1
     i=1+k
     j=1
     do l=1,n-1
minor(2,k,l)=A(i,j)
j=l+2
     end do
```

```
end do
do k=1,n-1
      i=1+k
      do l=1,n-1
j=l
minor(3,k,l)=A(i,j)
end do
end do
  do s=1,n
  Do k=1,n-1
write (*,*) (minor(s,k,l),l=1,n-1)
End do
  print *,
end do
!-----------------------------------
allocate(det(n))
If(n==3) then
do i=1,n
det(i)= (minor(i,1,1)*minor(i,2,2))-(minor(i,1,2)*minor(i,2,1))
print *, det(i)
  end do
else
do i=1,n
det(i)=minor(i,1,1)
end do
end if
!--------------------------
lap_det=0.
  do i=1,n
      m=i
nh=m/2
      if(2*nh==m) then
      sign=-1
      else
      sign=1
      end if
lap_det=lap_det+(sign*A(1,i))*det(i)
      print *, lap_det, 2*nh, sign
  end do
print *,
print *, lap_det
!-------------------------------------------------
stop
end
```

22. System of non-homogeneous linear equations

A linear equation has two or more variables ($x_1, x_2, \ldots, x_n$) along with their respective non-zero coefficients ($a_1, a_2, \ldots, a_n$), as parameters of the equation, plus a zero or non-zero element b.

$$a_1x_1 + a_2x_2 + \ldots + a_nx_n + b = 0$$

A system of non-homogeneous linear equations is a collection with two or more linear equations whose variables are the same in all the linear equations of the system.

$$a_{11}x_1 + a_{12}x_2 + \ldots + a_{1n}x_n + b_1 = 0$$
$$a_{21}x_1 + a_{22}x_2 + \ldots + a_{2n}x_n + b_2 = 0$$
$$a_{31}x_1 + a_{32}x_2 + \ldots + a_{3n}x_n + b_3 = 0$$
$$\vdots$$
$$a_{n1}x_1 + a_{n2}x_2 + \ldots + a_{nn}x_n + b_n = 0$$

The matrix form of the system of the non-homogeneous linear equations is:

$$\mathbf{A} = \begin{bmatrix} a_{11} & a_{12} & \cdots & a_{1n} \\ a_{21} & a_{22} & \cdots & a_{2n} \\ \vdots & \vdots & \vdots & \vdots \\ a_{n1} & a_{n2} & \cdots & a_{nn} \end{bmatrix}, \quad \mathbf{X} = \begin{bmatrix} x_1 \\ x_2 \\ \vdots \\ x_n \end{bmatrix}, \quad \mathbf{B} = \begin{bmatrix} b_1 \\ b_2 \\ \vdots \\ b_n \end{bmatrix}$$

$$\mathbf{AX} + \mathbf{B} = 0$$

Where X is the solution column matrix of the linear equations.

23. Solutions of system of non-homogeneous linear equations

One solution for the system of linear equations is:

$$\mathbf{X} = \mathbf{A}^{-1}\mathbf{B}$$

This is a consequence of some properties of the identity matrix:

$$\mathbf{I} = \mathbf{A}^{-1}\mathbf{A}$$

$$\mathbf{IX} = \mathbf{X}$$

Then, the multiplication of both sides of previous equation by A^{-1} gives:

$$\mathbf{A}^{-1}(\mathbf{AX}) = \mathbf{A}^{-1}\mathbf{B}$$

$$(\mathbf{A}^{-1}\mathbf{A})\mathbf{X} = \mathbf{A}^{-1}\mathbf{B}$$

$$\mathbf{IX} = \mathbf{A}^{-1}\mathbf{B}$$

$$\mathbf{X} = \mathbf{A}^{-1}\mathbf{B}$$

Another method to solve the system of linear equations is the Cramer's rule:

$$D = \begin{vmatrix} a_{11} & a_{12} & \cdots & a_{1n} \\ a_{21} & a_{22} & \cdots & a_{2n} \\ \vdots & \vdots & \vdots & \vdots \\ a_{n1} & a_{n2} & \cdots & a_{nn} \end{vmatrix}$$

$$x_1 = \frac{D_1}{D}, \qquad x_2 = \frac{D_2}{D}, \qquad x_n = \frac{D_n}{D}$$

$$where \qquad D_n = \begin{vmatrix} a_{11} & a_{12} & \cdots & b_1 \\ a_{21} & a_{22} & \cdots & b_2 \\ \vdots & \vdots & \vdots & \vdots \\ a_{n1} & a_{n2} & \cdots & b_n \end{vmatrix}$$

D_1 is obtained from the replacement of the first column by the column matrix **B**. D_2 is the obtained from the replacement of the second column by the column matrix **B**, and so on.

A third method to solve the system of linear equations is to apply backward substitution after Gaussian elimination (Section 6). The augmented matrix $\hat{\mathbf{A}}$ is:

$$\hat{A} = [A, b] = \begin{bmatrix} a_{11} & a_{12} & \cdots & a_{1n} & b_1 \\ a_{21} & a_{22} & \cdots & a_{2n} & b_2 \\ \vdots & \vdots & \vdots & \vdots & \vdots \\ a_{n1} & a_{n2} & \cdots & a_{nn} & b_n \end{bmatrix} = \begin{bmatrix} a_{11} & a_{12} & \cdots & a_{1n} & a_{1,n+1} \\ a_{21} & a_{22} & \cdots & a_{2n} & a_{2,n+1} \\ \vdots & \vdots & \vdots & \vdots & \vdots \\ a_{n1} & a_{n2} & \cdots & a_{nn} & a_{n,n+1} \end{bmatrix}$$

After Gaussian elimination procedures, it becomes:

$$\hat{A} = \begin{bmatrix} a_{11} & a_{12} & \cdots & a_{1n} & a_{1,n+1} \\ 0 & a_{22} & \cdots & a_{2n} & a_{2,n+1} \\ \vdots & \vdots & \vdots & \vdots & \vdots \\ 0 & 0 & \cdots & a_{nn} & a_{n,n+1} \end{bmatrix}$$

Which yields a triangular linear system:

$$\begin{array}{ccccccccc} a_{11}x_1 & + & a_{12}x_2 & + & \cdots & + & a_{1n}x_n & = & a_{1,n+1} \\ & & a_{22}x_2 & + & \cdots & + & a_{2n}x_n & = & a_{2,n+1} \\ & & & & & & \vdots & & \vdots \\ & & & & & & a_{2n}x_n & = & a_{n,n+1} \end{array}$$

Where:

$$x_n = \frac{a_{n,n+1}}{a_{nn}}$$

$$x_{n-1} = \frac{a_{n-1,n+1} - a_{n-1,n}x_n}{a_{n-1,n-1}}$$

And so on. The general solution is:

$$x_i = \frac{a_{i,n+1} - \sum_{j=i+1}^{n} a_{ij} x_j}{a_{ii}} \quad \therefore i = n-1, n-2, \ldots 2, 1$$

The procedure fails if at the ith step the pivot element a_{ii} is zero. This is known as Jacobi method. See Section 37.

24. System of homogeneous linear equations

When all b_n elements are zero, the system of linear equations is called homogeneous.

$$a_{11}x_1 + a_{12}x_2 + \ldots + a_{1n}x_n = 0$$
$$a_{21}x_1 + a_{22}x_2 + \ldots + a_{2n}x_n = 0$$
$$a_{31}x_1 + a_{32}x_2 + \ldots + a_{3n}x_n = 0$$
$$\vdots$$
$$a_{nn}x_1 + a_{n2}x_2 + \ldots + a_{nn}x_n = 0$$

The matrix equation of the homogeneous system of linear equations is:

$$\mathbf{AX = 0, \quad B = 0}$$

The above equation is called homogeneous matrix equation and it has at least the zero solution, i.e., X = 0 or a trivial solution. Some homogeneous system of linear equations have only trivial solution and other have trivial and non-trivial (non-zero) solutions.

25. Secular homogeneous linear equations

The secular homogeneous equations have a different type of element b in each equation of the system: a common parameter in all equations of the system, λ, multiplied by the x^{ith} variable of the row i, i.e., λx_i.

$$a_{11}x_1 + a_{12}x_2 + \ldots + a_{1n}x_n = \lambda x_1$$
$$a_{21}x_1 + a_{22}x_2 + \ldots + a_{2n}x_n = \lambda x_2$$
$$a_{31}x_1 + a_{32}x_2 + \ldots + a_{3n}x_n = \lambda x_3$$
$$\vdots$$
$$a_{nn}x_1 + a_{n2}x_2 + \ldots + a_{nn}x_n = \lambda x_n$$

By passing the right side of each equation above to the left side, then we have:

$$(a_{11} - \lambda)x_1 + a_{12}x_2 + \ldots + a_{1n}x_n = 0$$
$$a_{21}x_1 + (a_{22} - \lambda)x_2 + \ldots + a_{2n}x_n = 0$$
$$a_{31}x_1 + a_{32}x_2 + (a_{33} - \lambda)x_3 + \ldots + a_{3n}x_n = 0$$
$$\vdots$$
$$a_{nn}x_1 + a_{n2}x_2 + \ldots + (a_{nn} - \lambda)x_n = 0$$

The secular homogeneous equations have non-trivial solution only if the secular determinant below (the determinant of the coefficients) is zero.

$$D = \begin{vmatrix} (a_{11}-\lambda) & a_{12} & a_{13} & \cdots & a_{1n} \\ a_{21} & (a_{22}-\lambda) & a_{23} & \cdots & a_{2n} \\ a_{31} & a_{32} & (a_{33}-\lambda) & \cdots & a_{3n} \\ \vdots & \vdots & \vdots & \vdots & \vdots \\ a_{n1} & a_{n2} & a_{n3} & \cdots & (a_{nn}-\lambda) \end{vmatrix} = 0$$

Then, one has to find the n roots of the secular equations—the n solutions for the λ parameter—in order to find the non-trivial solutions for the secular equations. See Section 34.

26. Complex numbers and properties of the complex unit

The complex numbers are an extension of the real numbers where there is an element which represents the square root of –1, called the imaginary unit, i. A complex number is represented by Z composed of a real part, a, the imaginary part, bi.

$$Z = a + bi, \quad i = \sqrt{-1}$$

Any power of the complex unit belongs to the set {i, –i, 1, –1}.

$$i = \sqrt{-1}$$

$$i^2 = -1$$

$$i^3 = -i$$

$$i^4 = 1$$

$$i^{-1} = \frac{1}{i} = \frac{i}{i^2} = -i$$

27. Complex conjugate and Euler's formula

The complex conjugate of Z is the one where the sign of the imaginary part is switched. The complex conjugate is represented by Z^*.

$$z^* = a - bi$$

The product of the complex number and its conjugate, ZZ*, gives a real number, $|Z|^2$

$$zz^* = |z|^2 = (a+bi)(a-bi) = a^2 + b^2$$

Any complex number can be represented in the Euler's formula (see Chapter two).

$$Z = a + bi = re^{i\theta}$$

$$Z^* = a - bi = re^{-i\theta}$$

$$e^{i\theta} = \cos\theta + i\sin\theta$$

Where r and θ are the radial and angular terms of the polar coordinate.

28. Hermitian matrix and unitary matrix

Hermitian operator or self-adjoint operator in matrix algebra is called Hermitian matrix, A, when it is equal to its conjugate transpose, $\mathbf{A}^{H}$ or $\mathbf{A}^{\dagger}$ (A dagger).

$\mathbf{A} = \mathbf{A}^{H} = \mathbf{A}^{\dagger}$, **A is Hermetian**

$$\mathbf{A}^{H} = \mathbf{A}^{\dagger} = \left(\mathbf{A}^{*}\right)^{T} = \left(\mathbf{A}^{T}\right)^{*}$$

Let us suppose A as a complex matrix, then its conjugate transpose is:

$$\mathbf{A} = \begin{bmatrix} a_{11} & a_{12} & a_{13} \\ a_{21} & a_{22} & a_{23} \\ a_{31} & a_{32} & a_{33} \end{bmatrix}, \quad \mathbf{A}^{H} = \begin{bmatrix} a_{11}^{*} & a_{21}^{*} & a_{31}^{*} \\ a_{12}^{*} & a_{22}^{*} & a_{32}^{*} \\ a_{13}^{*} & a_{23}^{*} & a_{33}^{*} \end{bmatrix}$$

A Hermitian matrix is a square complex matrix whose main diagonal elements are real and its off-diagonal elements are complex where the diagonally-opposite entries are complex conjugates. For example:

$$\begin{bmatrix} m & a-ib & c-id \\ a+ib & n & e-if \\ c+id & e+if & o \end{bmatrix} \therefore m,n,o \in \mathbb{R}$$

Let us find the conjugate transpose for the matrix below from one example the example below:

$$\mathbf{A} = \begin{bmatrix} 3+i & 2+3i & -1 \\ 1-2i & 2 & 2i \end{bmatrix}$$

$$\mathbf{A}^{*} = \begin{bmatrix} 3-i & 2-3i & -1 \\ 1+2i & 2 & -2i \end{bmatrix}$$

$$\left(\mathbf{A}^{*}\right)^{T} = \begin{bmatrix} 3-i & 1+2i \\ 2-3i & 2 \\ -1 & -2i \end{bmatrix}$$

$$\mathbf{A} \neq \left(\mathbf{A}^{*}\right)^{T}$$

In this case, the conjugate transpose, $(\mathbf{A}^{*})^{T}$ or $(\mathbf{A}\dagger)$ is different from the matrix **A**. Let us take now another example:

$$\mathbf{A} = \begin{bmatrix} 3 & 2+3i \\ 2-3i & 1 \end{bmatrix}$$

$$\mathbf{A}^{*} = \begin{bmatrix} 3 & 2-3i \\ 2+3i & 1 \end{bmatrix}$$

$$\left(\mathbf{A}^{*}\right)^{T} = \begin{bmatrix} 3 & 2+3i \\ 2-3i & 1 \end{bmatrix}$$

$$\mathbf{A} = \left(\mathbf{A}^{*}\right)^{T} = \mathbf{A}^{H}$$

In this last case, the conjugate transpose, $(\mathbf{A}^*)^T$, is equivalent to the matrix **A** and this matrix is Hermitian.

A Hermitian matrix **A** of NxN order has N linearly independent and orthogonal eigenvectors and all Hermitian matrices are diagonalizable. For a Hermitian matrix, all eigenvalues are positive.

The Hermitian matrix **A** can be written as:

$$\mathbf{A} = \mathbf{P\Lambda P}^{-1}$$

Where **P** is a nonsingular column matrix and **Λ** is the diagonal matrix. A non-singular matrix has a non-zero determinant.

As a consequence of the properties of the Hermitian matrix, the diagonal matrix, **Λ**, can be given from the similarity transformation of **A**:

$$\mathbf{\Lambda} = \mathbf{P}^{-1}\,\mathbf{AP}$$

A square matrix **A** is diagonalizable whenever **A** is similar to its diagonal matrix and it has a complete set of eigenvectors.

In addition, the Hermitian matrix, **A**, is used in the decomposition of the unitary matrix **U** to obtain the triangular matrix **T** (Schur's theorem).

$$\mathbf{T} = \mathbf{U}^{-1}\mathbf{AU}$$

The unitary matrix is a complex square matrix whose its conjugate transpose is also its inverse matrix and it has determinant as a unit. The properties of a unitary matrix are:

$$\mathbf{U}^{-1} = \mathbf{U}^{\dagger}$$

$$\mathbf{U}^{\dagger}\mathbf{U} = \mathbf{I}$$

$$|\det(\mathbf{U})| = 1$$

Where **I** is the identity matrix. The identity matrix is a diagonal matrix where all diagonal elements are 1.

The general expressions of the 2 x 2 unitary matrix is:

$$\mathbf{U} = \begin{bmatrix} a & b \\ -e^{i\varphi}b^* & e^{i\varphi}a^* \end{bmatrix}, \quad |a|^2 + |b|^2 = 1$$

If **v** is an arbitrary eigenvector of the Hermitian matrix **A**, then:

$$\mathbf{v}^{\dagger}(\mathbf{Av}) = \lambda\mathbf{v}^{\dagger}\mathbf{v}$$

The Jacobi method is produced by splitting the Hermitian matrix, **A**, into:

$$Jacobi: \mathbf{A} = \mathbf{D} + \mathbf{N}$$

Where **D** is the diagonal part of **A** and **N** is the matrix containing the off-diagonal elements of the matrix **A**.

In the Gauss-Seidel method, the splitting of the Hermitian matrix is:

$$Gauss-Seidel: \mathbf{A} = (\mathbf{D} - \mathbf{L}) - \mathbf{U}$$

Where **D** is the diagonal part of **A** and –**L** and –**U** contain the elements below and above the diagonal of **A**, respectively.

29. Vector space and basis set

A vector space is a set of vectors. In a simplified definition, a vector is an entity represented by an arrow having a specific magnitude and direction (with angle ϕ) while a scalar has no direction. Examples of vectors are velocity and force. However, functions can also be vectors.

A vector might be 2-dimensional, 3-dimensional (in real space) or n-dimensional (n > 3). Since vector might have n-dimensions (n > 3), a vector might be considered as an abstract mathematical object.

A vector in a real space, **A**, has three elements, **A**(a_1,a_2,a_3), that is, a 3-dimensional vector with ϕ direction, and it can be represented by row matrix or column matrix. In addition, any n-dimensional vector can be represented as row or column vector.

$$\varphi = \frac{1}{\sqrt{\pi}} e^{-\sqrt{a_1^2 + a_2^2 + a_3^2}}$$

$$\mathbf{A} = (a_1, a_2, a_3)$$

$$\mathbf{A} = \begin{bmatrix} a_1 & a_2 & a_3 \end{bmatrix}$$

A column vector can be represented alternatively by:

$$\mathbf{A} = \begin{bmatrix} a_1 \\ a_2 \\ a_3 \end{bmatrix} = \begin{bmatrix} a_1 & a_2 & a_3 \end{bmatrix}^T$$

A basis set is a set of linearly independent vectors (basis vectors) which spans a vector space and it is used to construct a vector upon a linear combination of the basis of the vector space. Any other vector in this vector space is linearly dependent on the basis set and it may be expressed as a linear combination of the basis vectors.

For example, a unit vector, **u**, has a unit value of magnitude in one direction exclusively (x, y or z) and zero value in another two directions. Then, there are three unit vectors, $\mathbf{u_x}$, $\mathbf{u_y}$ and $\mathbf{u_z}$. This set of three orthogonal unit vectors is a type of basis set. Below, it is depicted the vector **A** as a linear combination of the three unit vectors, $\mathbf{u_x}$, $\mathbf{u_y}$ and $\mathbf{u_z}$, as a basis vectors.

$$\mathbf{u_x} = \begin{bmatrix} 1 & 0 & 0 \end{bmatrix}$$

$$\mathbf{u_y} = \begin{bmatrix} 0 & 1 & 0 \end{bmatrix}$$

$$\mathbf{u_z} = \begin{bmatrix} 0 & 0 & 1 \end{bmatrix}$$

$$\mathbf{A} = (a_1, a_2, a_3) = \sum_{i=1}^{3} a_i u_i$$

The basis vector can be abbreviated as $<\mathbf{n}|$ as a row vector or $|\mathbf{n}>$ as a column vector. For a n-dimensional basis vector, the vector ϕ can be represented as:

$$\phi = c_1 n_1 + c_2 n_2 + \ldots c_n n_n = \begin{bmatrix} c_1 & c_2 & \cdots & c_n \end{bmatrix} \begin{bmatrix} n_1 \\ n_2 \\ \vdots \\ n_n \end{bmatrix}$$

This is the method used to obtain molecular orbitals form atomic orbitals as basis set {n}, where ϕ represents a determined molecular orbital, MO, and $c_1, c_2, \ldots, c_n$ are the coefficients of this MO. This method is also used to obtain the wave function by means of the superposition principle (see Chapter ten) where the basis set is a set of eigenvectors. It is also used to obtain the guess function (from its expansion) in the variational method (see Chapter eighteen).

30. Matrix multiplications of vectors

There are two types of multiplication of vectors: the dot product (or scalar product) and the cross product (or vector product).
The dot product is an operation between two vectors that yields a scalar amount.

$$\mathbf{A}\cdot\mathbf{B} = \|A\|\|B\|\cos\theta$$

Where $\|A\|$ is the module (or the norm) of vector **A** and θ is the angle between vectors **A** and **B**.
Then, the angle between two vectors is:

$$\cos\theta = \frac{A\cdot B}{\|A\|\|B\|}$$

When the dot product is zero, the vectors are orthogonal ($\theta = 90°$).

The dot product can be represented in terms of matrix algebra as the multiplication of the row vector **A** and the column vector **B**, $\mathbf{B}^T$, which is the transpose of vector **B**.

$$\mathbf{A}\cdot\mathbf{B} = \mathbf{A}\mathbf{B}^T = \begin{bmatrix} a_1 & a_2 & a_3 \end{bmatrix}\begin{bmatrix} b_1 \\ b_2 \\ b_3 \end{bmatrix} = a_1b_1 + a_2b_2 + a_3b_3$$

The inner (scalar) product of two complex vectors is:

$$\mathbf{A}\cdot\mathbf{B} = \mathbf{A}^\dagger\mathbf{B} = \begin{bmatrix} a_1^* & a_2^* & a_3^* \end{bmatrix}\begin{bmatrix} b_1 \\ b_2 \\ b_3 \end{bmatrix}$$

The cross product is an operation between two vectors which yields another vector in the direction of **n**, the unit vector perpendicular to the plane determined by vectors **A** and **B**.

$$\mathbf{A}\times\mathbf{B} = \left(\|A\|\|B\|\sin\theta\right)\mathbf{n}$$

The cross product can be represented in terms of matrix algebra, where **i**, **j** and **k** are unit vectors for the vector basis **n**.

$$\mathbf{A}\times\mathbf{B} = \begin{bmatrix} \mathrm{i} & \mathrm{j} & \mathrm{k} \\ a_1 & a_2 & a_3 \\ b_1 & b_2 & b_3 \end{bmatrix}$$

$$\mathbf{A}\times\mathbf{B} = (a_2b_3\mathbf{i} + a_3b_1\mathbf{j} + a_1b_2\mathbf{k}) - (a_3b_2\mathbf{i} + a_1b3\mathbf{j} + a_2b_1\mathbf{k})$$
$$\mathbf{A}\times\mathbf{B} = (a_2b_3 - a_3b_2)\mathbf{i} + (a_3b_1 - a_1b_3)\mathbf{j} + (a_1b_2 - a_2b_1)\mathbf{k}$$
$$\mathbf{A}\times\mathbf{B} = \begin{bmatrix} a_2 & a_3 \\ b_2 & b_3 \end{bmatrix}\mathbf{i} - \begin{bmatrix} a_1 & a_3 \\ b_1 & b_3 \end{bmatrix}\mathbf{j} + \begin{bmatrix} a_1 & a_2 \\ b_1 & b_2 \end{bmatrix}\mathbf{k}$$

Vector (or cross) product is important for representing the components of the angular momentum (see Chapter sixteen).

31. Normalization of the vector and orthonormality

A normalized vector has a unit length. Let us suppose the normalized vector **A** $(a_1,a_2,...,a_n)$. Then, we have:

$$\mathbf{A}^T\mathbf{A} = 1$$

$$\begin{bmatrix} a_1 & a_2 & \cdots & a_n \end{bmatrix} \begin{bmatrix} a_1 \\ a_2 \\ \vdots \\ a_n \end{bmatrix} = 1$$

$$\mathbf{A}^T\mathbf{A} = (a_1^2 + a_2^2 + ... + a_n^2) = 1$$

For a not normalized vector **B** $(b_1,b_2,...,b_n)$, we have:

$$\mathbf{B}^T\mathbf{B} \neq 1$$

In order to normalize this vector, we have to assume $\mathbf{B}^2$ has to be a unit. Then, we find the normalization constant, N:

$$\mathbf{B}^T\mathbf{B} = \mathbf{B}^2 = b_1^2 + b_2^2 + ... + b_n^2 = 1$$

$$N = \frac{1}{norm}$$

$$norm = \|B\| = \sqrt{b_1^2 + b_2^2 + ... + b_n^2}$$

Where $||\mathbf{B}||$ is the norm of the vector **B**. Then, the normalized vector, $\mathbf{B}_N$, is

$$\mathbf{B}_N = \frac{1}{\sqrt{b_1^2 + b_2^2 + ... + b_n^2}}(b_1, b_2, ..., b_n)$$

If a pair of vectors is orthonormal, they are orthogonal and they are unit vectors. Let us suppose a pair of 2-dimensional orthonormal vectors $\mathbf{U}(u_1, u_2)$ and $\mathbf{V}(v_1,v_2)$. Then, we have:

$$u_1v_1 + u_2v_2 = 0$$

$$\sqrt{u_1^2 + u_2^2} = 1$$

$$\sqrt{v_1^2 + v_2^2} = 1$$

The orthogonalization of basis vectors in quantum chemistry, i.e., the atomic orbitals, facilitates the integral calculations. When the basis set is not orthogonal, one can use Lowdin or Schmidt methods of orthogonalization. Usually, more than two atomic basis vectors are used and one vector, e.g., orbital 1s, is chosen to be fixed while all other vectors are orthogonalized with respect to it. This is the Schmidt orthogonalization.

The dot product, or overlap, between members of an orthonormal set of vector is represented by the Kronecker delta.

$$\phi_j^T \cdot \phi_k = \langle \phi_j | \phi_k \rangle = \delta_{jk} = 0$$

$$\phi_j^T \cdot \phi_j = \langle \phi_j | \phi_j \rangle = \delta_{jj} = 1$$

Where $< \phi|$ is the row vector and $|\phi >$ is the column vector.

32. Eigenvector in vector space

An eigenvector is a vector that changes only a scalar factor after a linear transformation of an operator acting on it. An operator can be represented by a matrix, O, acting on a vector space V in the vector **A** to yield another vector **B** which is a scalar, λ, of the vector **A**. Below, the linear transformation in a generic vector space with n elements.

$$\mathbf{OA} = \mathbf{B} = \lambda \mathbf{A}$$

$$\begin{bmatrix} o_{1,1} & o_{1,2} & \cdots & o_{1,n} \\ o_{2,1} & o_{2,2} & \cdots & o_{2,n} \\ \vdots & \vdots & \vdots & \vdots \\ o_{n,1} & o_{n,2} & \cdots & o_{n,n} \end{bmatrix} \begin{bmatrix} a_1 \\ a_2 \\ \vdots \\ a_n \end{bmatrix} = \begin{bmatrix} b_1 \\ b_2 \\ \vdots \\ b_n \end{bmatrix} = \lambda \begin{bmatrix} a_1 \\ a_2 \\ \vdots \\ a_n \end{bmatrix}$$

Next, it is shown the linear transformation in a real space where n = 3.

$$\mathbf{OA} = \mathbf{B} = \lambda \mathbf{A}$$

$$\begin{bmatrix} o_{1,1} & o_{1,2} & o_{1,3} \\ o_{2,1} & o_{2,2} & o_{2,3} \\ o_{3,1} & o_{3,2} & o_{3,3} \end{bmatrix} \begin{bmatrix} a_1 \\ a_2 \\ a_3 \end{bmatrix} = \begin{bmatrix} b_1 \\ b_2 \\ b_3 \end{bmatrix} = \lambda \begin{bmatrix} a_1 \\ a_2 \\ a_3 \end{bmatrix}$$

The equation above is called an eigenvalue equation, the operator **O** is a linear operator, the scalar λ is called eigenvalue and the vector **A** is called eigenvector. Important to add that not all operators are linear, i.e., they act on a vector under a linear transformation (an eigenvalue equation).

The linear operator of order n yields a set of n eigenvalues and n eigenvectors, one eigenvalue for each eigenvector.

Two important properties of eigenvectors are:

(1) If the linear operator **O** is a real, symmetric matrix, the eigenvectors of different eigenvalues ($\lambda j \neq \lambda_k$) are orthogonal, that is:

$$\mathbf{A}_j^T \mathbf{A}_k = 0$$

$$\begin{bmatrix} a_1 & a_2 & \cdots & a_n \end{bmatrix}_j \begin{bmatrix} a_1 \\ a_2 \\ \vdots \\ a_n \end{bmatrix}_k = 0$$

Then, eigenvectors from different eigenvalues, are orthogonal, that is, their dot product is zero (see previous section).

$$\mathbf{A}_j^T \cdot \mathbf{A}_k = \|\mathbf{A}_j\| \|\mathbf{A}_k\| \cos\theta = 0$$

(2) As for the eigenvectors of the same eigenvalues (degenerate eigenvalues) they are orthogonal or can be made orthogonal.

Two important vector spaces are: the Euclidean vector space (whose basis set is real) and the Hermitian vector space (whose basis set is complex). One important difference between these vector spaces lies in the inner (dot) product:

$$Eucledian: \langle \mathbf{A} | \mathbf{B} \rangle = a_1 b_1 + a_2 b_2 + \ldots + a_n b_n$$
$$Hermitian: \langle \mathbf{A} | \mathbf{B} \rangle = a_1^* b_1 + a_2^* b_2 + \ldots + a_n^* b_n$$

33. Unitary transformation and linear transformations

A transformation in linear algebra converts a n-dimensional vector space into m-dimensional vector space, where n ≥ m. A matrix multiplication is a type of transformation. For example, one n x n matrix multiplying a n-dimensional vector to give a n-dimensional row vector.

$$\begin{bmatrix} n_1 & n_2 & \cdots & n_n \end{bmatrix} \begin{bmatrix} c_{1,1} & c_{1,2} & \cdots & c_{1,n} \\ c_{2,1} & c_{2,2} & \cdots & c_{2,n} \\ \vdots & \vdots & \vdots & \vdots \\ c_{n,1} & c_{n,2} & \cdots & c_{n,n} \end{bmatrix} = \begin{bmatrix} \phi_1 & \phi_2 & \cdots & \phi_n \end{bmatrix}$$

A unitary transformation occurs by means of a unitary matrix, **U**, in a Hermitian vector space. An orthogonal transformation occurs by means of an orthogonal matrix, **O**, in a Eucledian vector space. Both of them preserve the normalization and orthogonality of vectors. A rotation of a vector in a 2-dimensional real (Eucledian) vector space is a type of orthogonal transformation.

For example, let us rotate the 2-dimensional vector **V**(x,y) from α to β angle in the (x,y) vector space to form the vector V'(x',y'). The modulus of **V** ||**V**||, v, is preserved during the rotation.

$$\|\mathbf{V}\| = v$$

$$\mathbf{V}(x, y) \rightarrow \mathbf{V}'(x', y')$$

$$\alpha \rightarrow \beta \therefore \beta > \alpha$$

$$x' = v \cdot \cos(\alpha + \beta) = v \cdot \cos\alpha \cos\beta - v \sin\alpha \sin\beta$$

$$y' = v \cdot \sin(\alpha + \beta) = v \cdot \sin\alpha \cos\beta + v \cos\alpha \sin\beta$$

$$x = v \cdot \cos\alpha$$

$$y = v \sin\alpha$$

$$\begin{bmatrix} x' \\ y' \end{bmatrix} = \begin{bmatrix} \cos\alpha & -\sin\alpha \\ \sin\alpha & \cos\alpha \end{bmatrix} \begin{bmatrix} x \\ y \end{bmatrix}$$

Another important transformation is called linear transformation. In the pair of linear equations below, the coordinate of the 2-dimensional vector changes from (x,y) to (x',y') by the action of four coefficients a_1, b_1, a_2 and b_2.

$$x' = a_1 x + b_1 y$$

$$y' = a_2 x + b_2 y$$

Which can be described in the matrix form:

$$\begin{pmatrix} x' \\ y' \end{pmatrix} = \begin{pmatrix} a_1 & b_1 \\ a_2 & b_2 \end{pmatrix} \begin{pmatrix} x \\ y \end{pmatrix}$$

$$\mathbf{r}' = \mathbf{Ar}$$

The equations above are matrix and vector forms of the linear transformation where **r** and **r'** are column vectors/matrices representing the coordinates before and after the transformation of the coordinates and **A** is the operator (a linear operator) or the transformation matrix which transforms **r** into **r'**.

Should an operator **O** be linear, then it must satisfy the eigenvalue equation which is a linear transformation. Besides eigenvectors, one can use eigenfunctions *f*(x). Then:

$$Of(x) = \lambda f(x)$$

As a consequence, all quantum operators must perform linear transformations.

A linear transformation occurs by means of a linear operator **A** (that can be a square matrix) acting on an eigenvector **X** to yield a new eigenvector **Y** that is λ**X**, where λ is a scalar.

$$\mathbf{AX} = \mathbf{Y} = \lambda\mathbf{X}$$

$$\begin{bmatrix} a_{11} & a_{12} & \cdots & a_{1n} \\ a_{21} & a_{22} & \cdots & a_{2n} \\ \vdots & \vdots & \ddots & \vdots \\ a_{n1} & a_{n2} & \cdots & a_{nn} \end{bmatrix} \begin{bmatrix} x_1 \\ x_2 \\ \vdots \\ x_n \end{bmatrix} = \begin{bmatrix} y_1 \\ y_2 \\ \vdots \\ y_n \end{bmatrix} = \begin{bmatrix} \lambda x_1 \\ \lambda x_2 \\ \vdots \\ \lambda x_n \end{bmatrix}$$

For a linear operator 3x3 matrix A acting on a tridimensional vector, the linear transformation is:

$$\mathbf{A}_{3,3}\mathbf{X}_3 = \mathbf{Y}_3$$

$$\begin{pmatrix} a_{11} & a_{12} & a_{13} \\ a_{21} & a_{22} & a_{23} \\ a_{31} & a_{32} & a_{33} \end{pmatrix} \times \begin{pmatrix} x_1 \\ x_2 \\ x_3 \end{pmatrix} = \begin{pmatrix} x_1a_{11} + x_2a_{12} + x_3a_{13} \\ x_1a_{21} + x_2a_{22} + x_3a_{23} \\ x_1a_{31} + x_2a_{32} + x_3a_{33} \end{pmatrix} = \begin{pmatrix} \lambda x_1 \\ \lambda x_2 \\ \lambda x_3 \end{pmatrix} = \begin{pmatrix} y_1 \\ y_2 \\ y_3 \end{pmatrix}$$

The square 2 x 2 matrix **A** (depicted below) and the vector **X** (depicted below) is an eigenvector with eigenvalue 1.

$$\mathbf{A} = \begin{bmatrix} 2 & 1 \\ 1 & 2 \end{bmatrix}, \mathbf{X} = \begin{bmatrix} 3 \\ -3 \end{bmatrix}$$

$$\mathbf{AX} = \begin{bmatrix} 2 & 1 \\ 1 & 2 \end{bmatrix}\begin{bmatrix} 3 \\ -3 \end{bmatrix} = \begin{bmatrix} 2\cdot 3 + 1\cdot(-3) \\ 1\cdot 3 + 2\cdot(-3) \end{bmatrix} = \begin{bmatrix} 3 \\ -3 \end{bmatrix} = 1\cdot\begin{bmatrix} 3 \\ -3 \end{bmatrix}$$

On the other hand, the vector **Y** below is not an eigenvector for the linear operator **A**. There is no eigenvalue associated with **Y** from the linear operator **A**.

$$\mathbf{A} = \begin{bmatrix} 2 & 1 \\ 1 & 2 \end{bmatrix}, \mathbf{Y} = \begin{bmatrix} 0 \\ 1 \end{bmatrix}$$

$$\mathbf{AY} = \begin{bmatrix} 2 & 1 \\ 1 & 2 \end{bmatrix}\begin{bmatrix} 0 \\ 1 \end{bmatrix} = \begin{bmatrix} 2\cdot 0 + 1\cdot 1 \\ 1\cdot 0 + 2\cdot 1 \end{bmatrix} = \begin{bmatrix} 1 \\ 2 \end{bmatrix}$$

Let us see whether the differential operator, D_x or d/d_x, is linear or not.

$$\frac{d}{dx}x^n = nx^{n-1}, \qquad x^{n-1} = x^n \cdot x^{-1}$$

$$\frac{d}{dx}x^n = \frac{n}{x}x^n, \qquad \lambda = n/x$$

Then, the differential operator is a linear operator. Other simple cases of linear operators are the identity operator, I, and the multiplication operator, **X**.

$\mathbf{I}f(x) = f(x)$

$\mathbf{X}f(x) = xf(x)$

34. Eigenvalues-eigenvectors in system of secular homogeneous linear equations

In the eigenvalue equation from matrix algebra below, it can be rewritten in the following way:

$$\mathbf{OA} = \lambda\mathbf{A}$$

$$(\mathbf{O} - \lambda\mathbf{I})\mathbf{A} = \mathbf{0}$$

Where **O** is the linear operator in a square matrix, **A** is a column vector, λ is a number and **I** is the unit matrix.

From a generic square matrix of order n for the linear operator **O** and a eigenvector **A** represented by column matrix of order n, the above equation yields a system of and they (or characteristic polynomial):

$$(o_{11}-\lambda)a_1+o_{12}a_2+o_{13}a_3\ldots+a_{1n}a_n=0$$
$$o_{21}a_1+(o_{22}-\lambda)a_2+o_{23}a_3\ldots+o_{2n}a_n=0$$
$$o_{31}a_1+o_{32}a_2+(o_{33}-\lambda)a_3+\ldots+o_{3n}a_n=0$$
$$\vdots$$
$$o_{nn}a_1+o_{n2}a_2+o_{n3}a_3\ldots+(o_{nn}-\lambda)a_n=0$$

The secular equations have trivial solution (**A** = 0) and they can have a non-trivial solution for

$$\det(\mathbf{O}-\lambda\mathbf{I})=0$$

Giving n eigenvalues (λ_1, λ_2, ...λ_n) as the non-zero solution set or eigenvalue spectrum of the eigenvector **A**.
The **O**-λ**I** is the singular matrix and the linear system is defined as:

$$(\mathrm{O}-\lambda\mathrm{I})\mathrm{A}=0$$
$$\mathrm{A}\neq 0$$

Where **A** is the eigenvector (a column matrix).

When the coefficients of the eigenvector are unknown, then, for each λ_k as solution, there is a corresponding A_k eigenvector, i.e:, there are n eigenvectors and n eigenvalues as solution set.

$$\mathbf{OA}_k=\lambda_k\mathbf{A}_k,\quad k=1,2,\ldots n$$

See the example below:

$$\mathbf{O}=\begin{bmatrix}-2 & 1 & 1\\ -11 & 4 & 5\\ -1 & 1 & 0\end{bmatrix},\quad \mathbf{A}=\begin{bmatrix}\mathbf{x}\\ \mathbf{y}\\ \mathbf{z}\end{bmatrix}$$

$$\begin{bmatrix}-2 & 1 & 1\\ -11 & 4 & 5\\ -1 & 1 & 0\end{bmatrix}\cdot\begin{bmatrix}\mathbf{x}\\ \mathbf{y}\\ \mathbf{z}\end{bmatrix}=\lambda\begin{bmatrix}\mathbf{x}\\ \mathbf{y}\\ \mathbf{z}\end{bmatrix}$$

The elements **x**, **y** and **z** are column eigenvectors.

It yields the secular equation below:

$$-2x + y + z = \lambda x$$
$$-11x + 4y + 5z = \lambda y$$
$$-x + y + 0z = \lambda z$$

$$\begin{bmatrix} (-2-\lambda) & y & z \\ -11 & (4-\lambda) & 5 \\ -x & y & (-\lambda)z \end{bmatrix} \begin{matrix} 0 \\ =0 \\ 0 \end{matrix}$$

$$\det(\mathbf{O}-\lambda\mathbf{I}) = 0, \qquad \mathbf{D} = \begin{vmatrix} -2-\lambda & 1 & 1 \\ -11 & 4-\lambda & 5 \\ -1 & 1 & -\lambda \end{vmatrix} = 0$$

Whose solution set of the eigenvalues is {–1, 1, 2}. Each eigenvalue yields a corresponding eigenvector as solution. By using the expression:

$$(\mathrm{O}-\lambda\mathrm{I})\mathrm{A} = 0$$

Let us begin with λ_1 and eigenvector **x**:

$$\lambda_1 = -1,$$

$$\begin{bmatrix} -1 & 1 & 1 \\ -11 & 5 & 5 \\ -1 & 1 & 1 \end{bmatrix} \cdot \begin{bmatrix} x_1 \\ x_2 \\ x_3 \end{bmatrix} = \begin{bmatrix} 0 \\ 0 \\ 0 \end{bmatrix}, \qquad \mathrm{x} = \begin{bmatrix} x_1 \\ x_2 \\ x_3 \end{bmatrix}$$

Then, we have:

$$x_1 = x_2 + x_3$$
$$-11(x_2 + x_3) + 5x_2 + 5x_3 = 0$$
$$-6x_2 - 6x_3 = 0$$
$$x_2 = -x_3$$

The trivial solution is **x** = [0,0,0]. If $x_3 = 1$, $x_2 = -1$ and $x_1 = 0$.
For λ_2 and eigenvector **y:**

$$\lambda_2 = 1,$$

$$\begin{bmatrix} -3 & 1 & 1 \\ -11 & 3 & 5 \\ -1 & 1 & -1 \end{bmatrix} \cdot \begin{bmatrix} y_1 \\ y_2 \\ y_3 \end{bmatrix} = \begin{bmatrix} 0 \\ 0 \\ 0 \end{bmatrix}, \qquad \mathrm{y} = \begin{bmatrix} y_1 \\ y_2 \\ y_3 \end{bmatrix}$$

Then, we have:

$$-3y_1 + y_2 + y_3 = 0$$
$$-11y_1 + 3y_2 + 5y_3 = 0$$
$$-y_1 + y_2 - y_3 = 0$$
$$y_1 = y_2 - y_3$$
$$-3(y_2 - y_3) + y_2 + y_3 = 0$$
$$-2y_2 + 4y_3 = 0, \quad y_2 = 2y_3$$
$$y_1 = y_3$$
$$-11y_3 + 3(2y_3) + 5y_3 = 0, \quad y_3 = 0$$

The trivial solution is **y** = [0,0,0]. If $y_3 = 1$, $y_2 = 2$ and $y_1 = 1$.
For λ_3 and eigenvector **z:**

$$\lambda_3 = 2$$
$$\begin{bmatrix} -4 & 1 & 1 \\ -11 & 2 & 5 \\ -1 & 1 & -2 \end{bmatrix} \cdot \begin{bmatrix} z_1 \\ z_2 \\ z_3 \end{bmatrix} = \begin{bmatrix} 0 \\ 0 \\ 0 \end{bmatrix}, \qquad z = \begin{bmatrix} z_1 \\ z_2 \\ z_3 \end{bmatrix}$$

Then, we have:

$$-4z_1 + z_2 + z_3 = 0, \quad 4z_1 = z_2 + z_3$$
$$-z_1 + z_2 - 2z_3 = 0, \quad z_1 = z_2 - 2z_3$$
$$4(z_2 - 2z_3) = z_2 + z_3, \quad 4z_2 - 8z_3 = z_2 + z_3$$
$$3z_2 - 9z_3 = 0, \quad z_2 = 3z_3$$
$$z_1 = z_2 - 2z_3, \quad z_1 = 3z_3 - 2z_3 = z_3$$
$$-11z_3 + 6z_3 + 5z_3 = 0, \quad z_3 = 0$$

The trivial solution is **z** = [0,0,0]. If $z_3 = 1$, $z_2 = 3$ and $z_1 = 1$.
The matrix **A** is made up with the columns from the eigenvectors **x**, **y** and **z**.

$$\mathbf{x} = \begin{bmatrix} 0 \\ -1 \\ 1 \end{bmatrix}, \mathbf{y} = \begin{bmatrix} 1 \\ 2 \\ 1 \end{bmatrix}, \quad \mathbf{z} = \begin{bmatrix} 1 \\ 3 \\ 1 \end{bmatrix}, \quad \mathbf{A} = [\mathbf{x} \quad \mathbf{y} \quad \mathbf{z}]$$

$$\mathbf{O} = \begin{bmatrix} -2 & 1 & 1 \\ -11 & 4 & 5 \\ -1 & 1 & 0 \end{bmatrix}, \mathbf{A} = \begin{bmatrix} 0 & 1 & 1 \\ -1 & 2 & 3 \\ 1 & 1 & 1 \end{bmatrix}, \mathbf{D} = \begin{bmatrix} -1 & 0 & 0 \\ 0 & 1 & 0 \\ 0 & 0 & 2 \end{bmatrix}$$

Where **D** is the diagonal matrix whose diagonal elements are the set of eigenvalues of **O**. As we will see in Section 8:

OA = AD

$$\begin{bmatrix} -2 & 1 & 1 \\ -11 & 4 & 5 \\ -1 & 1 & 0 \end{bmatrix}\begin{bmatrix} 0 & 1 & 1 \\ -1 & 2 & 3 \\ 1 & 1 & 1 \end{bmatrix} = \begin{bmatrix} 0 & 1 & 2 \\ 1 & 2 & 6 \\ -1 & 1 & 2 \end{bmatrix}$$

$$\begin{bmatrix} 0 & 1 & 1 \\ -1 & 2 & 3 \\ 1 & 1 & 1 \end{bmatrix}\begin{bmatrix} -1 & 0 & 0 \\ 0 & 1 & 0 \\ 0 & 0 & 2 \end{bmatrix} = \begin{bmatrix} 0 & 1 & 2 \\ 1 & 2 & 6 \\ -1 & 1 & 2 \end{bmatrix}$$

From the matrix **O** below

$$\mathbf{O} = \begin{bmatrix} 10 & -4 \\ 12 & -4 \end{bmatrix} \therefore \det(\mathbf{O} - \lambda\mathbf{I}) = \begin{bmatrix} 10-\lambda & -4 \\ 12 & -4-\lambda \end{bmatrix}$$

$$\det(\mathbf{O} - \lambda\mathbf{I}) = (10-\lambda)(-4-\lambda) - 12(-4) = 0$$

$$\lambda^2 - 6\lambda + 8 = 0, \quad \lambda = \{2, 4\}$$

Two important features of the linear matrix operator, **O**, are:

$$Tr(\mathbf{O}) = \sum_{i=1}^{n} \lambda_i$$

$$\det(\mathbf{O}) = \prod_{i=1}^{n} \lambda_i$$

Then, from the last example above, trace of **O** is 2 + 4 = 6 and determinant of **O** is 2 x 4 = 8.

Since the eigenvalues of a Hermitian operator **O***nxn* are real, they can be ordered as:

$$\lambda_1 \geq \lambda_2 \geq \cdots \geq \lambda_n$$

35. Characteristic polynomial and the characteristic equation

The characteristic polynomial, p(λ), is:

$$p(\lambda) = \det(\mathrm{O} - \lambda\mathrm{I})$$

which can be written as:

$$p(\lambda) = (-1)^n \lambda^n + o_1\lambda^{n-1} + o_2\lambda^{n-2} + \ldots + o_{n-1}\lambda + o_n$$

And the characteristic equation is:

$$\lambda^n + c_1\lambda^{n-1} + c_2\lambda^{n-2} + \ldots + c_{n-1}\lambda + c_n = 0 \therefore c_i = (-1)^n o_i$$

Where:

$$trace(\mathrm{O}) = \lambda_1 + \lambda_2 + \ldots\lambda_n = -c_1$$

$$\det(\mathrm{O}) = \lambda_1\lambda_2\lambda_n = (-1)^n c_n$$

The characteristic equation can also be written as:

$$|O - \lambda I| = (\lambda_1 - \lambda)(\lambda_2 - \lambda)\cdots(\lambda_N - \lambda) = 0$$
$$|O - \lambda I| = \det(O - \lambda I)$$

For linear operator matrices, O, which are diagonal or triangular, the characteristic equation is straightforward because their diagonal elements are the eigenvalues of a given eigenequation

$$O = \begin{bmatrix} 1 & 0 & 0 \\ 0 & 2 & 0 \\ 0 & 0 & 3 \end{bmatrix}$$
$$|O - \lambda I| = (1 - \lambda)(2 - \lambda)(3 - \lambda)$$

$$O = \begin{bmatrix} 1 & 0 & 0 \\ 3 & 2 & 0 \\ 1 & 2 & 3 \end{bmatrix}$$
$$|O - \lambda I| = (1 - \lambda)(2 - \lambda)(3 - \lambda)$$

For the example below, let us consider the first form of the characteristic equation.

$$O = \begin{bmatrix} 2 & 0 & 0 \\ 0 & 3 & 4 \\ 0 & 4 & 9 \end{bmatrix}$$

$$|O - \lambda I| = \left\| \begin{bmatrix} 2 & 0 & 0 \\ 0 & 3 & 4 \\ 0 & 4 & 9 \end{bmatrix} - \lambda \begin{bmatrix} 1 & 0 & 0 \\ 0 & 1 & 0 \\ 0 & 0 & 1 \end{bmatrix} \right\| = \begin{vmatrix} 2-\lambda & 0 & 0 \\ 0 & 3-\lambda & 4 \\ 0 & 4 & 9-\lambda \end{vmatrix}$$

$$|O - \lambda I| = (2-\lambda)\cdot\left[(3-\lambda)(9-\lambda) - 16\right] = -\lambda^3 + 14\lambda^2 - 35\lambda + 22$$

36. Jordan matrix form and convergence

The square matrix is diagonalizable only if it has linearly independent eigenvectors. Not all matrices are diagonalizable and they are called defective. However, non-diagonalizable matrix can have a near diagonalizable matrix called Jordan normal form of the matrix.

A square complex matrix **A** is similar to a block diagonal matrix **J**:

$$J = P^{-1}AP$$

$$A = J = \begin{bmatrix} J_1 & & \\ & \ddots & \\ & & J_p \end{bmatrix}, \quad J_i = \begin{bmatrix} \lambda_i & 1 & & \\ & \lambda_i & \ddots & \\ & & \ddots & 1 \\ & & & \lambda_i \end{bmatrix}$$

For the solutions of a system of linear equations:

$$\mathrm{x}(k+1) = \mathrm{Ax}(k)$$

$$\mathrm{x}(k) = \mathrm{A}^k \mathrm{x}(0), \quad k = 1,2,3,...$$

Where the initial vector x(0) is known. Instead of finding x(k) for any finite k, the real problem is to understand the limiting solution:

$$\lim_{k\to\infty} x(k) \equiv \lim_{k\to\infty} A^k = 0, \quad A^k \to 0$$

The above condition is only met when the spectral radius of A, $\rho(A)$, is less than one. The spectral radius of a square matrix is the largest value of its eigenvalues.

$$\rho(A) = \max\{|\lambda_1|,...,|\lambda_n|\}$$

The spectral radius is closely related to the behavior of the convergence of the power sequence of a matrix.
Let (v,λ) be an eigenvector-eigenvalue pair of matrix **A**. Then:

$$\mathbf{A}^k v = \lambda^k v$$

$$0 = \left(\lim_{k\to\infty} \mathbf{A}^k\right) v$$

$$0 = \left(\lim_{k\to\infty} \mathbf{A}^k v\right)$$

$$0 = \lim_{k\to\infty} \lambda^k v$$

$$0 = v \lim_{k\to\infty} \lambda^k$$

$$for: v \neq 0$$

$$\lim_{k\to\infty} \lambda^k = 0$$

It is easy to see that:

$$\mathbf{A}^k = \mathbf{V}\mathbf{J}^k\mathbf{V}^{-1}$$

$$\mathbf{J}^k = \begin{bmatrix} \lambda_1^k & 0 & 0 & \cdots & 0 \\ 0 & \lambda_2^k & 0 & \cdots & 0 \\ \vdots & \cdots & \ddots & \cdots & \vdots \\ 0 & \cdots & 0 & \lambda_{s-1}^k & \\ 0 & \cdots & \cdots & 0 & \lambda_s^k \end{bmatrix}$$

$$Then: \lim_{k\to\infty} \mathbf{A}^k = \lim_{k\to\infty} \mathbf{J}^k = 0$$

37. Jacobi method for non-homogeneous linear equations

The Jacobi method is used to solve the system of non-homogeneous linear equations (see Section 23). It is an iterative method to obtain the eigenvectors (x_1, x_2, ..., x_n) of a square linear system which is strictly diagonally dominant. In a strictly diagonally dominant matrix:

$$|a_{ii}| \geq \sum_{j\neq i} |a_{ij}|, \quad \text{for all i}$$

For example, the system of non-homogeneous linear equations below cannot be solved by Jacobi method.

$$\mathbf{A} = \begin{bmatrix} 1 & -2 & 3 \\ -3 & 9 & 1 \\ 2 & -1 & -7 \end{bmatrix}, \quad \mathbf{B} = \begin{bmatrix} -1 \\ 2 \\ 3 \end{bmatrix}, \quad \mathbf{X}_0 = \begin{bmatrix} 0 \\ 0 \\ 0 \end{bmatrix}$$

As shown in the previous section, the square system of linear equations:

$$a_{11}x_1 + a_{12}x_2 + \ldots + a_{1n}x_n + b_1 = 0$$
$$a_{21}x_1 + a_{22}x_2 + \ldots + a_{2n}x_n + b_2 = 0$$
$$a_{31}x_1 + a_{32}x_2 + \ldots + a_{3n}x_n + b_3 = 0$$
$$\vdots$$
$$a_{nn}x_1 + a_{n2}x_2 + \ldots + a_{nn}x_n + b_n = 0$$

Can be described in the matricial form:

$$\mathbf{Ax} = \mathbf{b}$$

$$\mathbf{A} = \begin{bmatrix} a_{11} & a_{12} & \cdots & a_{1n} \\ a_{21} & a_{22} & \cdots & a_{2n} \\ \vdots & \vdots & \vdots & \vdots \\ a_{n1} & a_{n2} & \cdots & a_{nn} \end{bmatrix}, \quad \mathbf{x} = \begin{bmatrix} x_1 \\ x_2 \\ \vdots \\ x_n \end{bmatrix}, \quad \mathbf{b} = \begin{bmatrix} b_1 \\ b_2 \\ \vdots \\ b_n \end{bmatrix}$$

In the Jacobi method, the Hermetian matrix, **A**, is decomposed into:

$$\mathbf{A} = \mathbf{D} + \mathbf{N}$$

$$\mathbf{D} = \begin{bmatrix} a_{11} & 0 & \cdots & 0 \\ 0 & a_{22} & \cdots & 0 \\ \vdots & \vdots & \vdots & \vdots \\ 0 & 0 & \cdots & a_{nn} \end{bmatrix}, \quad \mathbf{N} = \begin{bmatrix} 0 & a_{12} & \cdots & a_{1n} \\ a_{21} & 0 & \cdots & a_{2n} \\ \vdots & \vdots & \vdots & \vdots \\ a_{n1} & a_{n2} & \cdots & 0 \end{bmatrix}$$

Where **D** is the diagonal part of **A** and **N** is the matrix containing the off-diagonal elements of the matrix **A**.
The iteration matrix, **H**, is:

$$\mathrm{H} = \mathrm{D}^{-1}\mathrm{N}$$

$$\mathrm{d} = \mathrm{D}^{-1}\mathrm{b}$$

For an initial vector, x(0), a linear stationary iteration is:

$$\mathrm{x}(k) = \mathrm{Hx}(k-1) + \mathrm{d} \therefore k = 1, 2, 3, \ldots$$

The solution set, the column matrix x, is obtained iteratively from the matricial equation below:

$$\mathrm{x}^{(k+1)} = \mathrm{D}^{-1}\left(\mathrm{b} - \mathrm{Nx}^{(k)}\right)$$

Where $x^{(k)}$ is the kth iteration of x and $x^{(k+1)}$ is the k+1 iteration of x.

The formula for the solution of each element is (see Section 23):

$$x_i^{(k+1)} = \frac{1}{a_{ii}}\left(b_i - \sum_{j\neq i} a_{ij}x_j^{(k)}\right), i = 1,2,3,\ldots,n$$

That is:

$$x_1^{(k+1)} = \frac{1}{a_{11}}\left(b_1 - a_{12}x_2^{(k)} - a_{13}x_3^{(k)} - \ldots - a_{1n}x_n^{(k)}\right)$$

$$x_2^{(k+1)} = \frac{1}{a_{22}}\left(b_2 - a_{21}x_1^{(k)} - a_{23}x_3^{(k)} - \ldots - a_{2n}x_n^{(k)}\right)$$

$$\vdots$$

$$x_n^{(k+1)} = \frac{1}{a_{nn}}\left(b_n - a_{n1}x_1^{(k)} - a_{n2}x_2^{(k)} - \ldots - a_{n,n-1}x_{n-1}^{(k)}\right)$$

The convergence condition is when the spectral radius of the iteration matrix is less than one (previous section)

$$\rho\left(\mathbf{D}^{-1}\mathbf{N}\right) < 1$$

However, for practical reasons, the iterative process is interrupted when the difference between **X**(i) and **X** is small.

$$\lim_{k\to\infty}\left\|\mathbf{x}^{(k)} - \mathbf{x}\right\| = 0$$

For the linear system:

$$2x_1 + x_2 = 11$$

$$5x_1 + 7x_2 = 13$$

$$\mathbf{A} = \begin{bmatrix} 2 & 1 \\ 5 & 7 \end{bmatrix}, \quad \mathbf{b} = \begin{bmatrix} 11 \\ 13 \end{bmatrix}$$

An dx(0) = (1,1), we have:

$$\mathbf{x}^{(k+1)} = -\mathbf{D}^{-1}\mathbf{N}x^k + \mathbf{D}^{-1}\mathbf{b}$$

$$\mathbf{D}^{-1} = \begin{bmatrix} 1/2 & 0 \\ 0 & 1/7 \end{bmatrix}, \quad \mathbf{N} = \begin{bmatrix} 0 & 1 \\ 5 & 0 \end{bmatrix}$$

$$-\mathbf{D}^{-1}\mathbf{N} = \begin{bmatrix} 1/2 & 0 \\ 0 & 1/7 \end{bmatrix} \cdot \begin{bmatrix} 0 & -1 \\ -5 & 0 \end{bmatrix} = \begin{bmatrix} 0 & -1/2 \\ -5/7 & 0 \end{bmatrix}$$

$$\mathbf{D}^{-1}b = \begin{bmatrix} 1/2 & 0 \\ 0 & 1/7 \end{bmatrix} \cdot \begin{bmatrix} 11 \\ 13 \end{bmatrix} = \begin{bmatrix} 11/2 \\ 13/7 \end{bmatrix}$$

$$\mathbf{x}^{(1)} = -\mathbf{D}^{-1}\mathbf{N}\mathbf{x}^{(0)} + \mathbf{D}^{-1}\mathbf{b}$$

$$x^{(0)} = [1 \quad 1]$$

$$x_1^{(1)} = \frac{1}{a_{11}}\left(b_1 - a_{12}x_2^{(0)}\right) \qquad x_2^{(1)} = \frac{1}{a_{22}}\left(b_2 - a_{21}x_1^{(0)}\right)$$

$$x_1^{(1)} = \frac{1}{2}\left(11 - 1(1)\right) = 5.0 \qquad x_2^{(1)} = \frac{1}{7}\left(13 - 5(1)\right) = 1.14$$

$$x_1^{(2)} = \frac{1}{2}\left(11 - 1(1.14)\right) = 4.93 \qquad x_2^{(2)} = \frac{1}{7}\left(13 - 5(5.0)\right) = -1.71$$

$$x_1^{(3)} = \frac{1}{2}\left(11 - 1(-1.71)\right) = 6.35 \qquad x_2^{(3)} = \frac{1}{7}\left(13 - 5(4.93)\right) = -1.66$$

$$x_1^{(32)} = 7.1111107 \qquad x_2^{(32)} = -3.2222219$$

The final result after 32 iterations is:

$$x = \begin{bmatrix} 7.1111107 \\ -3.2222219 \end{bmatrix}$$

```
!Programname: JACOBI
Integer :: i, j, k, n
Real :: A(2,2), B(2)
Real :: X0(2), X(2), tol, norm, sigma
A(1,:)=[2,1]
A(2,:)=[5,7]
B=(/11,13/)
X0=(/1,1/)
k=0
tol=1.0E-06
10 do i=1,2
  sigma=0
  do j=1,2
If (j .ne. i) sigma=sigma+A(i,j)*X0(j)
  end do
X(i)=(B(i)-sigma)/A(i,i)
end do
k=k+1
print *, k, (X(i), i=1,2)
norm=abs(X(1)-X0(1))
If (abs(X(2)-X0(2)) .gt. norm) then
norm=abs(X(2)-X0(2))
if (norm .lt. tol) go to 20
end if
do i=1,2
X0(i)=X(i)
end do
go to 10
20 stop
end
!THE FINAL RESULT IS: k = 32 7.1111107 –3.2222219
```

Let us solve the following non-homogeneous linear equations:

$$A = \begin{bmatrix} 5 & -2 & 3 \\ -3 & 9 & 1 \\ 2 & -1 & -7 \end{bmatrix}, \quad B = \begin{bmatrix} -1 \\ 2 \\ 3 \end{bmatrix}, \quad X_0 = \begin{bmatrix} 0 \\ 0 \\ 0 \end{bmatrix}$$

After 13 iterations (k = 13), the solution is:
k = 13 0.186119888 0.33123031 –0.422271289

```
!Programname: JACOBI2
Integer :: i, j, k, n
Real, allocatable :: A(:,:)
Real, allocatable :: X0(:), X(:), B(:)
Real :: tol, norm, sigma
Print *, "Enter the dimension of the system of non-homogeneous equations"
Read *, n
allocate(A(n,n))
Allocate (X0(n))
Allocate (X(n))
Allocate (B(n))
Do i=1,n
  Do j=1,n
Print *, 'A(',i,',',j,')= '
Read *, A(i,j)
End do
end do
do i=1,n
write (*,*) (A(i,j),j=1,n)
end do
print *,
do i=1,n
  Print *, 'B(',i,')='
Read *, B(i)
end do
write (*,*) (B(j),j=1,n)
Print *,
do i=1,n
 Print *, 'X0(',i,')='
Read *, X0(i)
end do
write (*,*) (X0(j),j=1,n)
Print *,
k=0
```

```
tol=1.0E-06
10 do i=1,n
  sigma=0
  do j=1,n
If (j .ne. i) sigma=sigma+A(i,j)*X0(j)
end do
  X(i)=(B(i)-sigma)/A(i,i)
end do
k=k+1
print *, k, (X(i), i=1,n)
norm=abs(X(1)-X0(1))
do i=1,n
If (abs(X(i)-X0(i)) .gt. norm) then
norm=abs(X(i)-X0(i))
if (norm .lt. tol) go to 20
end if
end do
do i=1,n
X0(i)=X(i)
end do
go to 10
20 stop
end
!THE FINAL RESULT IS: k = 13 0.186119888 0.33123031 –0.422271289
```

38. Jacobi method of diagonalization

The diagonalization procedure gives the eigenvalues of a eigenvector matrix. The Jacobi method is one of the most used to obtain the eigenvalues. This method is based on the successive plane rotations of the real, symmetric matrix. This successive plane rotations eliminate the off-diagonal elements. In the kth step, the following relation exists:

$$\mathbf{A}^{(k+1)} = \mathbf{O}^{T(k)}\mathbf{A}^{(k)}\mathbf{O}^{(k)}$$

Where $\mathbf{A}^{(k)}$ is the kth plane rotation of the matrix **A** and $\mathbf{O}^{(k)}$ is the orthogonal matrix in the kth step.

Remember from Section 33, the rotation transformation of vector V:

$$V(x,y) \to V'(x',y')$$

$$\alpha \to \beta, \quad \beta > \alpha$$

$$\begin{bmatrix} x' \\ y' \end{bmatrix} = \begin{bmatrix} \cos\alpha & -\sin\alpha \\ \sin\alpha & \cos\alpha \end{bmatrix} \begin{bmatrix} x \\ y \end{bmatrix}$$

In the step k + 1, there is one rotation (with an angle α) in the plane r,s. Then, the $\mathbf{O}^{(k+1)}$ has the following matrix:

$$\mathbf{O}^{(k+1)} = \begin{bmatrix} 1 & 0 & \cdots & 0 & \cdots & 0 & \cdots & 0 & 0 \\ 0 & 1 & \cdots & 0 & \cdots & 0 & \cdots & 0 & 0 \\ \vdots & \vdots & \vdots & \vdots & \vdots & \vdots & \vdots & \vdots & \vdots \\ 0 & 0 & \cdots & \cos\alpha & \cdots & -\sin\alpha & \cdots & 0 & 0 \\ \vdots & \vdots & \vdots & \vdots & \vdots & \vdots & \vdots & \vdots & \vdots \\ 0 & 0 & \cdots & \sin\alpha & \cdots & \cos\alpha & \cdots & 0 & 0 \\ \vdots & \vdots & \vdots & \vdots & \vdots & \vdots & \vdots & \vdots & \vdots \\ 0 & 0 & \cdots & 0 & \cdots & 0 & \cdots & 0 & 1 \end{bmatrix} \begin{matrix} \\ \\ \\ \cdots r \\ \\ \cdots s \\ \\ \\ \end{matrix}$$

$$\begin{matrix} \vdots & \qquad & \vdots \\ r & & s \end{matrix}$$

The elements of this orthogonal matrix are:

$\mathbf{O}_{rr} = \mathbf{O}_{ss} = \cos\alpha$

$\mathbf{O}_{rs} = \mathbf{O}_{sr} = -\sin\alpha$

$\mathbf{O}_{ii} = \mathbf{O}_{jj} = 1, \quad i = j \neq r \neq s$

$\mathbf{O}_{ij} = \mathbf{O}_{ji} = 1, \quad \mathbf{i\ or\ j \neq r\ or\ s}$

In order to eliminate the off-diagonal elements of the eigenvector matrix A at each step k+1, the angle of rotation (α) must be chosen in such a way that the r,s element of matrix **A** in the k step must equal to zero in the step k+1. Then, we apply the following equation:

$$\mathbf{A}^{(k+1)} = \mathbf{O}^{T(k)}\mathbf{A}^{(k)}\mathbf{O}^{(k)}$$

Due to the zero values of the orthonormal matrix, the above multiplication can be reduced to 2 x 2 matrices, **O'** and **A'**.

$$A^{(k)} = \begin{bmatrix} a_{11} & a_{12} & a_{13} & \cdots & a_{1n} \\ a_{21} & a_{rr} & a_{rs} & \cdots & a_{2n} \\ a_{31} & a_{sr} & a_{ss} & \cdots & a_{3n} \\ \vdots & \vdots & \vdots & \cdots & \vdots \\ a_{n1} & a_{n2} & a_{n3} & \cdots & a_{nn} \end{bmatrix}$$

$$O' = \begin{bmatrix} \cos\alpha & -\sin\alpha \\ \sin\alpha & \cos\alpha \end{bmatrix}$$

$$O'^{T} = \begin{bmatrix} \cos\alpha & \sin\alpha \\ -\sin\alpha & \cos\alpha \end{bmatrix}$$

$$A'^{(k)} = \begin{bmatrix} a_{rr} & a_{rs} \\ a_{sr} & a_{ss} \end{bmatrix} \therefore a_{rs} = a_{sr}$$

$$A'^{(k+1)} = \begin{bmatrix} \cos\alpha & \sin\alpha \\ -\sin\alpha & \cos\alpha \end{bmatrix} \cdot \left(\begin{bmatrix} a_{rr} & a_{rs} \\ a_{sr} & a_{ss} \end{bmatrix} \cdot \begin{bmatrix} \cos\alpha & -\sin\alpha \\ \sin\alpha & \cos\alpha \end{bmatrix} \right)$$

step – by – step :

$$\begin{bmatrix} a_{rr} & a_{rs} \\ a_{sr} & a_{ss} \end{bmatrix} \cdot \begin{bmatrix} \cos\alpha & -\sin\alpha \\ \sin\alpha & \cos\alpha \end{bmatrix} = \begin{bmatrix} a_{rr}\cos\alpha + a_{rs}\sin\alpha & -a_{rr}\sin\alpha + a_{rs}\cos\alpha \\ a_{sr}\cos\alpha + a_{ss}\sin\alpha & -a_{sr}\sin\alpha + a_{ss}\cos\alpha \end{bmatrix}$$

$$\begin{bmatrix} \cos\alpha & \sin\alpha \\ -\sin\alpha & \cos\alpha \end{bmatrix} \cdot \begin{bmatrix} a_{rr}\cos\alpha + a_{rs}\sin\alpha & -a_{rr}\sin\alpha + a_{rs}\cos\alpha \\ a_{sr}\cos\alpha + a_{ss}\sin\alpha & -a_{sr}\sin\alpha + a_{ss}\cos\alpha \end{bmatrix} =$$

$$A'^{(k+1)} = \begin{bmatrix} a_{rr}^{(k+1)} & a_{rs}^{(k+1)} \\ a_{sr}^{(k+1)} & a_{ss}^{(k+1)} \end{bmatrix}$$

$$a_{rr}^{(k+1)} = a_{rr}\cos^2\alpha + a_{rs}\sin\alpha\cos\alpha + a_{ss}\cos\alpha\sin\alpha + a_{sr}\cos\alpha\sin\alpha$$

$$a_{sr}^{(k+1)} = -a_{rr}\cos\alpha\sin\alpha - a_{rs}\sin^2\alpha + a_{sr}\cos^2\alpha + a_{ss}\cos\alpha\sin\alpha$$

$$a_{rs}^{(k+1)} = -a_{rr}\cos\alpha\sin\alpha + a_{rs}\cos^2\alpha - a_{sr}\sin^2\alpha + a_{ss}\cos\alpha\sin\alpha$$

$$a_{ss}^{(k+1)} = a_{rr}\sin^2\alpha - a_{rs}\cos\alpha\sin\alpha - a_{sr}\cos\alpha\sin\alpha + a_{ss}\cos^2\alpha$$

Since the matrix A is symmetric, $a_{sr} = a_{rs}$ everywhere. Then:

$$a_{sr} = a_{rs}$$

$$a_{rr}^{(k+1)} = a_{rr}\cos^2\alpha + 2a_{rs}\sin\alpha\cos\alpha + a_{ss}\cos\alpha\sin\alpha$$

$$a_{sr}^{(k+1)} = -a_{rr}\cos\alpha\sin\alpha + a_{rs}\left(\cos^2\alpha - \sin^2\alpha\right) + a_{ss}\cos\alpha\sin\alpha$$

$$a_{ss}^{(k+1)} = a_{rr}\sin^2\alpha - 2a_{rs}\cos\alpha\sin\alpha + a_{ss}\cos^2\alpha$$

For the diagonalization process, the $a_{sr}^{(k+1)} = a_{rs}^{(k+1)}$ must be zero. Then, we have:

$$-a_{rr}\cos\alpha\sin\alpha + a_{rs}\left(\cos^2\alpha - \sin^2\alpha\right) + a_{ss}\cos\alpha\sin\alpha = 0$$

This equation yields to:

$$\tan 2\alpha = \frac{2a_{rs}}{a_{rr} - a_{ss}}, \qquad \alpha = \tfrac{1}{2}\tan^{-1}\left[\frac{2a_{rs}}{a_{rr} - a_{ss}}\right]$$

The calculation begins with a guess angle. It is customary to choose α in the range ±45. There are two algorithms to determine the element a_{rs} in the previous equation (to find α). Nonetheless, the main part of the code is the same for both cases:

```
!n=dimension of the square matrices
! Enter the matrix elements (A(i,j)) of the symmetric matrix A
Do i=1,n
  Do j=1,n
If (i .eq. j) O(i,j)=1
if (i .ne. j) O(i,j)=0
  end do
end do
OT=transpose(O)
  alfa=0.5*atan(2*A(i,j)/(A(i,i)-A(j,j)))
```

```
    O(i,i)=cos(alfa)
    O(j,j)=cos(alfa)
    O(i,j)=-sin(alfa)
    O(j,i)=sin(alfa)
    OT(i,i)=cos(alfa)
    OT(j,j)=cos(alfa)
  OT(i,j)=sin(alfa)
    OT(j,i)=-sin(alfa)
  M=matmul(A,O)
    A=matmul(OT,M)
    Do k=1,n
       Do l=1,n
  If (k .eq. l) O(k,l)=1
  if (k .ne. l) O(k,l)=0
       end do
    end do
    OT=transpose(O)
  !Start a new iteration until the final step
```

The following algorithm is limited for some types of symmetrical matrices where the first off-diagonal element, A(1,2), has the greatest value.

The plane passing through the element A(1,2) is chosen to rotate and then the elements of the matrix $A^{(0)}$ that corresponds to the chosen plane are inserted in the tangent equation above which yields the angle. Afterwards, the first orthogonal matrix $\mathbf{O}^{(1)}$ is calculated. Then, the matrix $\mathbf{A}^{(1)}$ is obtained.

$$\mathbf{A}^{(1)} = \mathbf{O}^{T(1)}\mathbf{A}^{(0)}\mathbf{O}^{(1)}$$

A second rotation is done at k = 2 by choosing another plane to rotate, and the second orthogonal matrix is calculated in order to obtain $\mathbf{A}^{(1)}$. Then, the matrix $\mathbf{A}^{(2)}$ is obtained.

$$\mathbf{A}^{(2)} = \mathbf{O}^{T(2)}\mathbf{A}^{(1)}\mathbf{O}^{(2)}$$

The iteration process stops when all off-diagonal elements of the matrix A are zero. As the end, for example, at the step n, the eigenvector matrix is the product of all orthogonal matrices.

$$\mathbf{O} = \mathbf{O}^{(1)}\mathbf{O}^{(2)}\cdots\mathbf{O}^{(k-1)}\mathbf{O}^{(k)}\cdots\mathbf{O}^{(n-1)}\mathbf{O}^{(n)}$$

```
! Program name:JACOBI-EIGEN
real, allocatable :: A(:,:), O(:,:),OT(:,:), M(:,:)
real :: alfa
integer :: i, j, k, l
Print *, "Enter the dimension of the symmetric matrix A"
Read *, n
Allocate(A(n,n))
```

```
Allocate(O(n,n))
Allocate(OT(n,n))
Allocate(M(n,n))
Do i=1,n
  Do j=1,n
     Print *, 'A(',i,',',j,')= '
Read *, A(i,j)
  End do
end do
do i=1,n
write (*,*) (A(i,j),j=1,n)
end do
print *,
Do i=1,n
  Do j=1,n
If (i .eq. j) O(i,j)=1
if (i .ne. j) O(i,j)=0
end do
end do
OT=transpose(O)
do i=1,n-1
    do j=i+1,n
       alfa=0.5*atan(2*A(i,j)/(A(i,i)-A(j,j)))
       O(i,i)=cos(alfa)
       O(j,j)=cos(alfa)
       O(i,j)=-sin(alfa)
       O(j,i)=sin(alfa)
       OT(i,i)=cos(alfa)
       OT(j,j)=cos(alfa)
OT(i,j)=sin(alfa)
       OT(j,i)=-sin(alfa)
M=matmul(A,O)
     A=matmul(OT,M)
     Do k=1,n
        Do l=1,n
If (k .eq. l) O(k,l)=1
if (k .ne. l) O(k,l)=0
        end do
     end do
     OT=transpose(O)
  end do
```

```
end do
do i=1,n
write (*,*) (A(i,j),j=1,n)
end do
print *
stop
end
```

For the matrix **A**:

$$\mathbf{A} = \begin{bmatrix} 0.6532 & 0.2165 & 0.0031 \\ 0.2165 & 0.4105 & 0.0052 \\ 0.2165 & 0.031 & 0.2132 \end{bmatrix}$$

The eigenvalue matrixis:

$$A^{(k=3)} = \begin{bmatrix} 0.78008908 & 2.72800826E-05 & -1.12739292E-06 \\ 2.72952857E-05 & 0.28378111 & -7.65449426E-10 \\ -1.12760961E-06 & 2.44833265E-10 & 0.21302973 \end{bmatrix}$$

If the first off-diagonal element, A(1,2), does not have the greatest value, then one has to pick the off-diagonal element with the greatest value to begin the calculation. All the rest of the procedure is similar to that previously described.

39. Commutators

Matrices and operators can be elements of commutators. A commutator is defined as:

$$[\mathbf{A},\mathbf{B}] = \mathbf{AB} - \mathbf{BA}$$

As already mentioned in Section 1 of this chapter, matrices do not commute, except for the cases discussed in the chapters involving matrix mechanics.

When **AB** = **BA**, the matrices or operators **A** and **B** commutate. The commutator has the following properties or commutator's identities:

$$[\mathbf{A},\mathbf{A}] = 0$$

$$[\mathbf{A},\mathbf{B}] = -[\mathbf{B},\mathbf{A}]$$

$$[\mathbf{A}+\mathbf{B},\mathbf{C}] = [\mathbf{A},\mathbf{C}] + [\mathbf{B},\mathbf{C}]$$

$$[\mathbf{A},\mathbf{B}+\mathbf{C}] = [\mathbf{A},\mathbf{B}] + [\mathbf{A},\mathbf{C}]$$

$$[\mathbf{A},\mathbf{BC}] = [\mathbf{A},\mathbf{B}]\mathbf{C} + \mathbf{B}[\mathbf{A},\mathbf{C}]$$

$$[\mathbf{AB},\mathbf{C}] = \mathbf{A}[\mathbf{B},\mathbf{C}] + [\mathbf{A},\mathbf{C}]\mathbf{B}$$

$$[\mathbf{AB},\mathbf{CD}] = \mathbf{A}[\mathbf{B},\mathbf{C}]\mathbf{D} + [\mathbf{A},\mathbf{C}]\mathbf{BD} + \mathbf{CA}[\mathbf{B},\mathbf{D}] + \mathbf{C}[\mathbf{A},\mathbf{D}]\mathbf{B}$$

When the operators or matrices are of the same type they commutate even for different components. For example, the linear momentum and position operators, p and r, respectively, can have two or three components for two-dimensional and three-dimensional spaces. Then, we have the following commutator's properties:

$$[p_x, p_y] = 0, [p_x, p_z] = 0, [p_y, p_z] = 0$$
$$[x, y] = 0, [x, z] = 0, [y, z] = 0$$

40. Examples

1. Give a literal example of a linear transformation.

Solution: A linear transformation transforms a vector into another one which is obtained from an operation matrix operator, **A**, acting on a basis function (a vector **X**). The matrix **A** changes the modulus of **X**, but it does not change its direction. Then, **X** is an eigenvector of **A**. The resulting vector, **Y**, is the outcome of the linear operation of **A** on **X**, giving the equation: **AX** = λ**X**, where λ**X** = **Y**. The objective of the linear transformation is to obtain the eigenvalues λ_n of the operation of **A** on **X**.

$$\mathbf{A}_{nn} \cdot \mathbf{X}_n = \mathbf{Y}_n = \lambda_n \cdot \mathbf{X}_n$$

$$\begin{vmatrix} a_{11} & a_{12} & a_{13} \\ a_{21} & a_{22} & a_{23} \\ a_{31} & a_{32} & a_{33} \end{vmatrix} \cdot \begin{vmatrix} x_1 \\ x_2 \\ x_3 \end{vmatrix} = \begin{vmatrix} a_{11}x_1 + a_{12}x_2 + a_{13}x_3 \\ a_{21}x_1 + a_{22}x_2 + a_{23}x_3 \\ a_{31}x_1 + a_{32}x_2 + a_{33}x_3 \end{vmatrix} = \begin{vmatrix} \lambda_1 x_1 \\ \lambda_2 x_2 \\ \lambda_3 x_3 \end{vmatrix}$$

2. By knowing that the spin projection operators S_z, S_y and S_x are 2 x 2 square matrices operators that affects a measurement of the spin in the z, y, and x directions, respectively, yielding the two states of the spin in each of these directions. Find their commutative properties.

Data:

$$\mathbf{S}_x = \frac{\hbar}{2}\sigma_x = \frac{\hbar}{2}\begin{bmatrix} 0 & 1 \\ 1 & 0 \end{bmatrix}$$

$$\mathbf{S}_y = \frac{\hbar}{2}\sigma_y = \frac{\hbar}{2}\begin{bmatrix} 0 & -i \\ i & 0 \end{bmatrix}$$

$$\mathbf{S}_z = \frac{\hbar}{2}\sigma_z = \frac{\hbar}{2}\begin{bmatrix} 1 & 0 \\ 0 & -1 \end{bmatrix}$$

Solution: Let us find the commutative property between S_x and S_y

$$[S_x, S_y] = \mathbf{S}_x\mathbf{S}_y - \mathbf{S}_y\mathbf{S}_x = \frac{1}{4}\hbar^2 \begin{bmatrix} 0 & 1 \\ 1 & 0 \end{bmatrix}\begin{bmatrix} 0 & -i \\ i & 0 \end{bmatrix} - \frac{1}{4}\hbar^2 \begin{bmatrix} 0 & -i \\ i & 0 \end{bmatrix}\begin{bmatrix} 0 & 1 \\ 1 & 0 \end{bmatrix}$$

$$[S_x, S_y] = \frac{1}{4}\hbar^2 \begin{bmatrix} i & 0 \\ 0 & -i \end{bmatrix} - \frac{1}{4}\hbar^2 \begin{bmatrix} -i & 0 \\ 0 & i \end{bmatrix} = \frac{1}{4}\hbar^2 \left\{ \begin{bmatrix} i & 0 \\ 0 & -i \end{bmatrix} - \begin{bmatrix} -i & 0 \\ 0 & i \end{bmatrix} \right\}$$

$$[S_x, S_y] = \frac{1}{4}\hbar^2 \begin{bmatrix} i-(-i) & 0 \\ 0 & -i-i \end{bmatrix} = \frac{1}{4}\hbar^2 \begin{bmatrix} 2i & 0 \\ 0 & -2i \end{bmatrix} = \frac{1}{2}\hbar^2 \begin{bmatrix} i & 0 \\ 0 & -i \end{bmatrix}$$

$$[S_x, S_y] = \frac{i}{2}\hbar^2 \begin{bmatrix} 1 & 0 \\ 0 & -1 \end{bmatrix} = i\hbar\mathbf{S}_z$$

3. Demonstrate that **AP** = **PD** to **AP** = **DP** is an eigenvalue equation being $\mathbf{P}_{2x2}$ an inversible matrix and $\mathbf{D}_{2x2}$ a diagonal matrix.

Solution:

Let us designate the matrices **D** and **P** as:

$$\mathbf{D} = \begin{bmatrix} \lambda_1 & 0 \\ 0 & \lambda_2 \end{bmatrix}, \quad \mathbf{P} = \begin{bmatrix} v_1 & w_1 \\ v_2 & w_2 \end{bmatrix} = [\mathbf{V} \quad \mathbf{W}]$$

Then, we have:

$$\mathbf{AP} = \mathbf{A}[\mathbf{V} \quad \mathbf{W}] = [\mathbf{AV} \quad \mathbf{AW}]$$

$$\mathbf{PD} = \begin{bmatrix} v_1 & w_1 \\ v_2 & w_2 \end{bmatrix} \begin{bmatrix} \lambda_1 & 0 \\ 0 & \lambda_2 \end{bmatrix} = \begin{bmatrix} v_1\lambda_1 & w_1\lambda_2 \\ v_2\lambda_1 & w_2\lambda_2 \end{bmatrix} = [\lambda_1\mathbf{V} \quad \lambda_2\mathbf{W}]$$

$$[\mathbf{AV} \quad \mathbf{AW}] = [\lambda_1\mathbf{V} \quad \lambda_2\mathbf{W}]$$

$$\mathbf{A}[\mathbf{V} \quad \mathbf{W}] = \mathbf{D}[\mathbf{V} \quad \mathbf{W}]$$

4. Let T be a linear transformation matrix given below. Find the three associated eigenvalues λ_n.

$$\mathbf{T} = \begin{pmatrix} 1 & 0 & 0 \\ 0 & 1 & 2 \\ 0 & 2 & 1 \end{pmatrix}$$

Solution: The linear transformation matrix operates on the vector **X** to yield the λ_n **X**.

$$\textit{Let}: \mathbf{TX} = \lambda_n\mathbf{X}$$

$$(\mathbf{T} - \lambda_n\mathbf{I}_n)\mathbf{X} = 0$$

$$\textit{Then}: \det(\mathbf{T} - \lambda_n\mathbf{I}_n) = 0$$

$$\textit{Hence}:$$

$$\begin{bmatrix} 1 & 0 & 0 \\ 0 & 1 & 2 \\ 0 & 2 & 1 \end{bmatrix} - \lambda_n \begin{bmatrix} 1 & 0 & 0 \\ 0 & 1 & 0 \\ 0 & 0 & 1 \end{bmatrix} = \begin{bmatrix} 1-\lambda_n & 0 & 0 \\ 0 & 1-\lambda_n & 2 \\ 0 & 2 & 1-\lambda_n \end{bmatrix}$$

$$\det \begin{bmatrix} 1-\lambda_n & 0 & 0 \\ 0 & 1-\lambda_n & 2 \\ 0 & 2 & 1-\lambda_n \end{bmatrix} = 0$$

$$(1-\lambda_n)\left[(1-\lambda_n)^2 - 4\right] = 0$$

$$S = \{1, -1, 3\}$$

$$\lambda_n = \begin{bmatrix} 1 \\ -1 \\ 3 \end{bmatrix}$$

5. Given the matrix **A** below, find its conjugate transpose $(\mathbf{A}^*)^T$ or $(\mathbf{A}^\dagger)$.

$$\mathbf{A} = \begin{bmatrix} 3+i & 5 & -2i \\ 2-2i & i & -7-13i \end{bmatrix}$$

Solution:

$$\mathbf{A}^* = \begin{bmatrix} 3-i & 5 & 2i \\ 2+2i & -i & -7+13i \end{bmatrix}$$

$$\left(\mathbf{A}^*\right)^T = \begin{bmatrix} 3-i & 2+2i \\ 5 & -i \\ 2i & -7+13i \end{bmatrix}$$

$$\mathbf{A} \neq \mathbf{A}^\dagger$$

Differential Equations for Quantum Mechanics

4

1. Introduction

The differential equations are used to formulate many laws of physics and engineering. The most general form of an ordinary equation is a set of coupled first order equations. It is possible to transform a higher order differential equation into a set of coupled first order differential equations.

Let us suppose the second Newton's law of motion (***F*** = *m**a***), where **F** and **a** are vectors, of an object of mass m moving in x direction and the linear momentum, ***p*** = *m**v***, where **p** and **v** are also vectors.

$$m\frac{d^2x}{dt^2} = F(x) \qquad m\frac{dx}{dt} = p(t)$$

The second order differential equation of the second Newton's law of motion can be converted into two first order differential equations.

$$F(x) = \frac{d\left[p(t)\right]}{dt}, \qquad p(t) = m\frac{dx}{dt}$$

The terms first order or second order are the degree of the differential equation which is given by the highest power of the highest-order derivative in the equation.

Nonetheless, many equations in Physics are in the form of linear, inhomogeneous/homogeneous second-order equations. **When S(x) is zero, the equation is called homogeneous and when the S(x) is nonzero, the equation is called inhomogeneous.**

$$\frac{d^2y}{dx^2} + k(x)y = S(x)$$

Where k is a real function with positive sign (oscillatory wave number) or negative sign (exponential growth or decay).

The differential equations can be ordinary (ODE: ordinary differential equation) or partial (PDE: partial differential equation). *The ordinary differential equations have only one independent variable and partial differential equations have two or more independent variables.*

One differential equation is an eigenvalue equation when the solutions satisfying the boundary conditions exist only for particular values of a constant wavenumber, k, as a set of $\{k_n$ where n is an integer$\}$.

2. First-order differential equation

A first-order differential equation is linear when it has the form:

$$\frac{dy}{dx} + k(x)y = S(x)$$

When S(x) = 0, it is called homogeneous and the equation is separable. The term homogenous means that every term in the equation has the variable y(x). The solution of the linear homogeneous first-order differential equation is given below:

$$\frac{dy}{dx} + k(x)y = 0$$

$$\frac{dy}{dx} = -k(x)y$$

$$\frac{dy}{y} = -k(x)dx$$

$$\ln y = -\int k(x)dx + c$$

$$y = ae^{-\int k(x)dx}$$

Where a is an arbitrary constant. When a = 0, y = 0 (trivial solution).

When S(x) ≠ 0, it is called inhomogeneous linear equation. If we take the function F(x):

$$F(x) = e^{\int k(x)dx}$$

Which has the following characteristic:

$$\frac{dF(x)}{dx} = F(x)k(x)$$

Since:

$$\frac{dF(x)}{dx} = \frac{d\left(e^{\int k(x)dx}\right)}{dx}, \qquad u = \int k(x)dx$$

$$\frac{d(e^u)}{du}\frac{du}{dx}$$

$$\frac{d(e^u)}{du} = e^u = F(x), \qquad \frac{d(\int k(x)dx)}{dx} = k(x)$$

$$\frac{dF(x)}{dx} = F(x)k(x)$$

Then, multiplying F(x) by each term of the inhomogeneous equation gives:

$$F(x)\frac{dy}{dx} + F(x)k(x)y = F(x)S(x)$$

The second term of the left side of the above equation is:

$$F(x)k(x) = \frac{dF(x)}{dx}$$

$$F(x)k(x)y = \frac{dF(x)}{dx}y$$

Then, the left side of the equation becomes:

$$F(x)\frac{dy}{dx} + \frac{dF(x)}{dx}y = \frac{d}{dx}\left[F(x)y\right]$$

Hence, we have:

$$\frac{d}{dx}\left[F(x)y\right] = F(x)S(x)$$

And the general solution becomes:

$$F(x)y = \int F(x)S(x)dx + c$$

Let us take the linear inhomogeneous equation below as example:

$$\frac{dy}{dx} - 2y = 5e^{3x} \therefore k(x) = -2$$

$$\int k(x)dx = -2x$$

$$F(x) = e^{\int k(x)dx} = e^{-2x}$$

The solution from the general solution is:

$$e^{-2x}y = \int e^{-2x}(5e^{3x})dx + c = 5\int e^{x}dx + c = 5e^{x} + c$$

$$y = e^{2x}\left(5e^{x} + c\right) = 5e^{3x} + ce^{2x}$$

In the example above, we performed indefinite integration and we obtained a general solution having an unknown constant c. In order to obtain a particular solution, we have to know the boundary conditions (or initial conditions) of the system, where the initial values of x and y are known. Let us use the boundary condition y(0) = 7 in the example above to find its particular solution.

$$y = 5e^{3x} + ce^{2x}, \quad x = 0, \quad y = 7$$

$$y = 5e^{3(0)} + ce^{2(0)} = 5 + c = 7, \quad c = 2$$

$$y = 5e^{3x} + 2e^{2x}$$

There is another method to solve the first-order differential equation, that is, to use separable variables. The first-order differential equation with separable variables has the form:

$$A(x)dx + B(y)dy = 0$$

Let us see the example below:

$$\frac{dy}{dx}\ln x - \frac{y}{x} = 0$$

After some algebraic procedures, we have:

$$\frac{dy}{dx}\ln x = \frac{y}{x}, \qquad dy\ln x = dx\frac{y}{x}$$

$$\frac{1}{y}dy\ln x = dx\frac{1}{x}, \qquad \frac{1}{y}dy = dx\frac{1}{x\ln x}$$

$$\frac{1}{y}dy - \frac{1}{x\ln x}dx = 0$$

$$A(x) = -\frac{1}{x\ln x}, \qquad B(y) = \frac{1}{y}$$

Then, by integrating the above equation, we have the general solution:

$$\int\frac{1}{y}dy - \int\frac{1}{x\ln x}dx = 0$$

$$u = \ln x, \qquad \frac{du}{dx} = \frac{1}{x}, \qquad du = \frac{dx}{x}$$

$$\int\frac{1}{y}dy - \int\frac{1}{u}\frac{1}{x}dx = 0$$

$$\int\frac{1}{y}dy - \int\frac{1}{u}du = 0$$

$$(\ln y + c') - (\ln u + c") = 0$$

$$(\ln y + c') - (\ln(\ln x) + c") = 0$$

$$e^{(\ln y + c')} - e^{(\ln(\ln x) + c")} = 0$$

$$k'y - k"\ln x = 0$$

$$y = c\ln x \therefore c = k"/k'$$

3. Second-order differential equation: classification

General linear second-order differential equations have the form given below. The first is called inhomogeneous and the second is called homogeneous. *The terms p(x), k(x) and S(x) are the coefficients of the equation. These terms might be constant or variables.* When they are constants, the solutions can be expressed in terms of elementary functions.

$$\frac{d^2y}{dx^2} + p(x)\frac{dy}{dx} + k(x)y = S(x)$$

and

$$\frac{d^2y}{dx^2} + p(x)\frac{dy}{dx} + k(x)y = 0$$

Again, the word homogenous means that every term in the equation has the variable y(x).

The second-order differential equations with constant coefficients have three classes: elliptical, hyperbolic and parabolic. Let us take another general second-order differential equation:

$$A\frac{d^2\phi}{dx^2}+B\frac{d^2\phi}{dxdy}+C\frac{d^2\phi}{dy}=F\left(x,y,\phi,\frac{d\phi}{dx},\frac{d\phi}{dy},\ldots\right)$$

or

$$Au_{xx}+Bu_{xy}+Cu_{yy}+Du_x+Eu_y+Hu=0$$

$$u_{xx}=\frac{d^2\phi}{dx^2}\therefore u_x=\frac{d\phi}{dx}\therefore u=\phi\therefore F=Du_x+Eu_y+Hu$$

Where A, B, C, D, E, H are the coefficients of the second-order differential equation. The coefficients can be constants or functions.

If $B^2–4AC > 0$	Hyperbolic equation	2 real characteristics
If $B^2–4AC = 0$	Parabolic equation	1 real characteristic
If $B^2–4AC < 0$	elliptic equation	no real characteristic

For example, the Laplace's equation is a case of elliptic second-order differential equation.

$$\nabla^2\varphi(x,y)=0$$

$$\frac{\partial^2\varphi(x,y)}{\partial x^2}+\frac{\partial^2\varphi(x,y)}{\partial y^2}=0$$

$$A=1,\quad B=0,\quad C=1$$

$$0^2-4(1)(1)=-4$$

$$x^2+y^2=1\Rightarrow ellipse$$

Let us take another example:

$$\frac{d\varphi^2}{dx^2}-\frac{d\varphi^2}{dy^2}=1$$

$$A=1,\quad B=0,\quad C=1$$

$$0^2-4(1)(-1)=4$$

$$x^2-y^2=1\Rightarrow hyperbola$$

4. Second-order differential equation with constant coefficients: general and particular solutions

For the general second-order differential equation given below:

$$\frac{d^2y}{dx^2}+a\frac{dy}{dx}+by=0$$

The general solution is:

$$y = e^{\lambda x}$$

Then, by substituting y in the former equation, we have:

$$\frac{d}{dx}e^{\lambda x} = \lambda e^{\lambda x}, \quad \frac{d^2}{dx^2}e^{\lambda x} = \lambda^2 e^{\lambda x}$$

$$\lambda^2 e^{\lambda x} + a\lambda e^{\lambda x} + be^{\lambda x} = 0$$

$$e^{\lambda x}\left(\lambda^2 + a\lambda + b\right) = 0$$

$$\lambda_1 = \tfrac{1}{2}\left(-a + \sqrt{a^2 - 4b}\right), \lambda_2 = \tfrac{1}{2}\left(-a - \sqrt{a^2 - 4b}\right)$$

The equation:

$$\left(\lambda^2 + a\lambda + b\right) = 0$$

is called characteristic equation and λ_1 and λ_2 are the roots of the characteristic equation. The possible solutions:

$$y_1 = e^{\lambda_1 x}, \quad y_2 = e^{\lambda_2 x}$$

occur when the discriminant $a^2 - 4b > 0$.
And the general solution is:

$$y = c_1 y_1 + c_2 y_2$$

$$y = c_1 e^{\lambda_1 x} + c_2 e^{\lambda_2 x}$$

Where c_1 and c_2 are the arbitrary constants.

When the discriminant $a^2 - 4b = 0$, there is only one real root for the characteristic equation.

$$\lambda = \tfrac{1}{2}\left(-a - \sqrt{0}\right) = -a/2$$

Although there is only one root, it is possible to have two solutions:

$$y_1 = e^{\lambda x} = e^{-ax/2}$$

The second solution is linearly independent of the first:

$$y_2 = xy_1 = xe^{-ax/2}$$

The general solution is:

$$y = \left(c_1 + xc_2\right)e^{-ax/2}$$

When the discriminant $a^2 - 4b < 0$, there are two imaginary roots for the characteristic equation. Let us take the example:

$$\frac{d^2 y}{dx^2} - 2\frac{dy}{dx} + 2y = 0$$

The characteristic equation and corresponding solutions are:

$$\lambda = \frac{1}{2}\left(2 \pm \sqrt{4-8}\right)$$

$$\lambda_1 = 1+i, \qquad \lambda_2 = 1-i = \lambda_1^*$$

$$y_1 = e^{(1+i)x}, \qquad y_2 = e^{(1-i)x}$$

The general solution is:

$$y = e^x\left(c_1 e^{ix} + c_2 e^{-ix}\right)$$

As stated in chapter three, any complex number, Z, can be represented in Euler's formula.

$$Z = a + bi = re^{i\theta}$$

$$Z^* = a - bi = re^{-i\theta}$$

$$e^{i\theta} = \cos\theta + i\sin\theta$$

Then, the general solution can be written as:

$$y = e^x\left[c_1\left(\cos x + i\sin x\right) + c_2\left(\cos x - i\sin x\right)\right]$$

$$y = e^x\left[\left(c_1 + c_2\right)\cos x + i\left(c_1 - c_2\right)\sin x\right]$$

$$y = e^x\left[d_1 \cos x + d_2 \sin x\right]$$

Let us see the case where the second-order differential equation has a real parameter, ω^2, in the equation below. Note that a = 0 since y' = 0.

$$y'' + \omega^2 y = 0$$

In that case, the general solution is:

$$a = 0, \qquad b = \omega^2$$

$$\lambda = \frac{1}{2}\left(0 \pm \sqrt{0 - 4\omega^2}\right) = \pm i\omega$$

$$\lambda_1 = i\omega, \qquad \lambda_2 = -i\omega$$

$$y_1 = e^{i\omega x}, \qquad y_2 = e^{-i\omega x}$$

$$y = c_1 e^{i\omega x} + c_2 e^{-i\omega x}$$

$$y = d_1 \cos\omega x + d_2 \sin\omega x$$

If ω is a constant, we have the classical harmonic oscillator as the corresponding differential equation. If ω is a parameter to be determined, an additional condition has to be established in order to find the general solution.

If we have a periodic boundary condition, e.g., at each λ interval, the function y equals the previous value without λ, in a cyclic condition.

$$y(x + n\lambda) = y(x), \qquad n = 0, \pm1, \pm2, \pm3...$$

Then, we have:

$$y(x+\lambda) = c_1 e^{i\omega(x+\lambda)} + c_2 e^{-i\omega(x+\lambda)}$$
$$y(x+\lambda) = c_1 e^{i\omega x} e^{i\omega\lambda} + c_2 e^{-i\omega x} e^{-i\omega\lambda}$$
$$y(x) = c_1 e^{i\omega x} + c_2 e^{-i\omega x}$$
$$If : e^{i\omega\lambda} = 1, \qquad e^{-i\omega\lambda} = 1$$
$$then : y(x+\lambda) = y(x)$$

This is the case when $\omega\lambda = 2\pi n$. Then:

$$\omega_n = \frac{2\pi n}{\lambda}$$
$$y_n = d_1 \cos\left(\frac{2\pi n x}{\lambda}\right) + d_2 \sin\left(\frac{2\pi n x}{\lambda}\right)$$

To sum up, for the general second-order differential equation given below

$$\frac{d^2 y}{dx^2} + a\frac{dy}{dx} + by = 0$$

Whose discriminant is:

$$a^2 - 4b$$

We have the following chart:

Discriminant	General solution
If $a^2 - 4b > 0$	$y = c_1 e^{\lambda_1 x} + c_2 e^{\lambda_2 x}$
If $a^2 - 4b = 0$	$y = e^{-ax/2}(c_1 + c_2 x)$
If $a^2 - 4b < 0$	$y = e^{x}\left(c_1 e^{ix} + c_2 e^{-ix}\right)$ $y = e^{x}\left[d_1 \cos x + d_2 \sin x\right]$ Or (for $b = \omega^2$) $y = c_1 e^{i\omega x} + c_2 e^{-i\omega x}$ $y = d_1 \cos \omega x + d_2 \sin \omega x$

Let us take the example of a body of mass, m, attached at one extreme of a massless spring with constant k while the other extreme of the spring is fixed in the wall. This is the classical harmonic oscillator.

$$F = -kx$$
$$F = m\frac{d^2 x}{dt^2} = -kx$$
$$\frac{d^2 x}{dt^2} + \frac{k}{m} x = 0$$

Let the initial velocity at the limit distance, l, be 0, as expected. The discriminant is $a^2 - 4b < 0$, then, we have

$$x(t) = l \sin \omega t + l \cos \omega t$$

In this case, we see that the parameter ω in the differential equation represents the angular frequency. The vibrational frequency, ν, is also given below.

$$\omega^2 = \frac{k}{m}$$

$$\omega = \sqrt{\frac{k}{m}}$$

$$\nu = \frac{\omega}{2\pi}, \qquad \nu = \frac{1}{2\pi}\sqrt{\frac{k}{m}}$$

The potential energy for the classical harmonic oscillator is given according to the following relation:

$$F = -\frac{dV}{dx}$$

$$F = -kx$$

$$-\frac{dV}{dx} = -kx, \qquad dV = kxdx$$

$$\int dV = \int kxdx, \qquad V = \frac{1}{2}kx^2 + c$$

At the equilibrium position (x = 0), where the force of the spring acting on the body is zero, we have V = 0. Then, we find:

$$F = -\frac{dV}{dx}$$

$$For: x = 0, \qquad V = 0, \qquad c = 0$$

$$V = \frac{1}{2}kx^2$$

If we take the case of the stretching vibration of a diatomic molecule. The Morse potential for this system is:

$$V = BDE \cdot \left[1 - e^{-a(R-R_e)}\right]^2$$

$$V \approx BDE \cdot \left[a^2(R-R_e)^2 - a^3(R-R_e)^3 + \ldots\right]$$

Where R is the distance between the nuclei, R_e is the equilibrium distance (Where V = 0), a is a constant, and BDE is the bond dissociation energy. For small displacements, as in the ground state, the potential energy can be approximated to:

$$V = a^2 BDE \cdot (R-R_e)^2$$

Then, the force acting on the nuclei is:

$$F = -\frac{dV}{dR} = \frac{d\left[a^2 BDE\left(R - R_e\right)^2\right]}{R} = -2a^2 BDE\left(R - R_e\right)$$

$$k = 2a^2 BDE, \quad x = \left(R - R_e\right), \quad F = -kx$$

Then, the stretching vibration of a molecule is type of approximate simple harmonic.

5. Power-series method for second-order homogenous equation

The power-series method can be used for first-order, second-order, or higher-order homogeneous differential equations. The power series represent, at least, one particular solution for these cases and it is represented as:

$$y(x) = a_0 + a_1 x + a_2 x^2 + a_3 x^3 + \ldots = \sum_{n=0}^{\infty} a_n x^n$$

Let us see the power-series method for the first-order differential equation below:

$$\frac{dy}{dx} + y = 0$$

By knowing that above general formula one possible solution, then we have the result below. In the fifth line below we perform the power equalization (see Chapter two).

$$y(x) = \sum_{n=0}^{\infty} a_n x^n$$

$$y'(x) = \sum_{n=1}^{\infty} n a_n x^{n-1}$$

$$y'(x) + y(x) = 0$$

$$\sum_{n=1}^{\infty} n a_n x^{n-1} + \sum_{n=0}^{\infty} a_n x^n = 0$$

$$\sum_{n=0}^{\infty} (n+1) a_{n+1} x^n + \sum_{n=0}^{\infty} a_n x^n = 0$$

$$\sum_{n=0}^{\infty} \left[(n+1) a_{n+1} + a_n\right] x^n = 0$$

$$then: (n+1) a_{n+1} + a_n = 0$$

As a consequence, the recursion formula (or recurrence relation) is:

$$\frac{a_{n+1}}{a_n} = -\frac{1}{n+1}$$

$$let: n = 0$$

$$\frac{a_1}{a_0} = -\frac{1}{0+1} = -1$$

$$let: n = 1$$

$$\frac{a_2}{a_1} = -\frac{1}{1+1} = -\frac{1}{2}$$

$$let: n = 2$$

$$\frac{a_3}{a_2} = -\frac{1}{2+1} = -\frac{1}{3}$$

And so on. Let us now express all the coefficients in terms of a_0.

$$\frac{a_1}{a_0}\frac{a_2}{a_1}\frac{a_3}{a_2}\cdots\frac{a_n}{a_{n-1}} = \frac{a_n}{a_0}$$

$$\frac{a_n}{a_0} = -1\times\left(-\frac{1}{2}\right)\times\left(-\frac{1}{3}\right)\times\left(-\frac{1}{4}\right)\ldots\times\left(-\frac{1}{n}\right) = \frac{(-1)^n}{n!}$$

$$a_n = a_0\frac{(-1)^n}{n!}$$

$$y(x) = \sum_{n=0}^{\infty} a_0\frac{(-1)^n}{n!}x^n$$

Another solution is:

$$y(x) = ce^{-x}$$

We know that:

$$e^{-x} = \sum_{n=0}^{\infty}\frac{(-1)^n}{n!}x_n$$

The last equation is the MacLaurin series (see chapter two). The constant a_0 is arbitrary and can be obtained by knowing the boundary conditions.

Let us now solve the following second-order differential equation from the power-series method:

$$\frac{d^2y}{dx^2} + y = 0$$

Then, we see the result below where we also perform the power equalization in the fifth line.

$$y = \sum_{n=0}^{\infty} a_n x^n$$

$$y'' = \sum_{n=2}^{\infty} n(n-1)a_n x^{n-2}$$

$$y'' + y = 0$$

$$\sum_{n=2}^{\infty} n(n-1)a_n x^{n-2} + \sum_{n=0}^{\infty} a_n x^n = 0$$

$$\sum_{n=0}^{\infty}(n+2)(n+1)a_{n+2}x^n + \sum_{n=0}^{\infty}a_n x^n = 0$$

$$\sum_{n=0}^{\infty}\left[(n+2)(n+1)a_{n+2} + a_n\right]x^n = 0$$

$$then: (n+2)(n+1)a_{n+2} + a_n = 0$$

As a consequence, the recursion formula is:

$$a_{n+2} = -\frac{a_n}{(n+2)(n+1)}$$

Then, for even values of n:

$$a_2 = -\frac{a_0}{2} = -\frac{a_0}{2\times 1}$$

$$a_4 = -\frac{a_2}{12} = -\frac{1}{4\times 3}\times\left(-\frac{a_0}{2\times 1}\right) = +\frac{a_0}{4!}$$

$$a_6 = -\frac{a_4}{30} = -\frac{1}{6\times 5}\times\frac{a_0}{4!} = -\frac{a_0}{6!}$$

For the odd values of n:

$$a_3 = -\frac{a_1}{6} = -\frac{a_1}{2\times 3}$$

$$a_5 = -\frac{a_3}{20} = -\frac{1}{5\times 4}\times\left(-\frac{a_1}{2\times 3}\right) = +\frac{a_1}{5!}$$

$$a_7 = -\frac{a_5}{42} = -\frac{1}{7\times 6}\times\frac{a_1}{5!} = -\frac{a_1}{7!}$$

Then, we have:

$$y = \left(a_0 + a_2x^2 + a_4x^4 + \ldots\right) + \left(a_1x + a_3x^3 + a_5x^5 + \ldots\right)$$

$$y = \left(a_0 - \frac{a_0}{2!}x^2 + \frac{a_0}{4!}x^4 - \frac{a_0}{6!}x^6 + \ldots\right) + \left(a_1x - \frac{a_1}{3!}x^3 + \frac{a_1}{5!}x^5 - \frac{a_1}{7!}x^7 + \ldots\right)$$

$$y = a_0\left(1 - \frac{x^2}{2!} + \frac{x^4}{4!} - \frac{x^6}{6!} + \ldots\right) + a_1\left(x - \frac{x^3}{3!} + \frac{x^5}{5!} - \frac{x^7}{7!} + \ldots\right)$$

The power series in the brackets are expansions of cos x and sin x, respectively. Then:

$$y = a_0\cos x + a_1\sin x$$

The Fuchs' theorem establishes that for the second-order differential equation with non-constant coefficients of the form:

$$y'' + P(x)y' + Q(x)y = G(x)$$

can have a series solution about $x = a$ if the $x = a$ is a regular singular point or ordinary point of the differential equation. An ordinary point is one where P(x), Q(x) and

G(x) are analytic at x = a. **A regular singular point** of the second-order differential equation is one at which the second-order differential equation can be written as

$$y = \sum_{n=0}^{\infty} a_n (x-a)^{n+s}, \qquad a_0 \neq 0$$

or

$$y = y_0 \ln(x-a) + \sum_{n=0}^{\infty} b_n (x-a)^{n+r}$$

For some real s or r, where y_0 is the solution of the first kind.

An ordinary point at $x = x_0$ occurs when P(x) and Q(x) are finite there, i.e., when P(x) and Q(x) are analytic there. A regular singular point occurs when P(x) or Q(x) at $x = x_0$ is not analytic there, but $(x - x_0)P(x)$ and $(x - x_0)^2Q(x)$ are analytic at that point. An irregular singular point occurs when at least one of $(x - x_0)P(x)$ and $(x - x_0)$ Q(x) is not analytic at $x = x_0$.

6. Frobenius method

The Frobenius method is used to second-order differential equations with non-constant coefficients of the form:

$$x^2 y'' + P(x)xy' + Q(x)y = 0$$

Where x^2, p(x)x, and q(x) are non-constant coefficients and x = 0 is the regular singular point.

Let us define: (i) singularity occurs at a point at which it fails to be well-behaved, for example, not having differentiability at this point; (ii) singular point is a point at which some coefficient has a singularity; (iii) regular singular point is where the growth of solutions is bounded by an algebraic function.

Let us divide the above equation by x^2. Then:

$$y'' + \frac{P(x)}{x} y' + \frac{Q(x)}{x^2} y = 0$$

We see that this equation is undefined and will not yield solutions at x = 0 or near 0. But, the Frobenius method enables to create a power series as a solution provided that xP(x) and $x^2Q(x)$ are well-behaved or analytic at x = 0 (i.e., they have power series solutions that converge near x = 0). If these conditions are met, then one can find a power series as a solution in the form:

$$y = x^r \sum_{n=0}^{\infty} a_n x^n = \sum_{n=0}^{\infty} a_n x^{n+r}$$

Where $a_0 \neq 0$ and r is a real or integer constant to be determined and it is called the indicial parameter. In this case, the parameter r belongs to the recursion formula. Then, besides finding the recursion formula or recurrence (as it was done in the previous section), we have to determine the values of the exponent r in order to obtain the exact values of a_n.

The first step is to substitute the above power series in the differential equation to be solved and we get the recursion formula. After, we have to derive the indicial equation (that is, the equation to obtain the parameter r) by setting n = 0. Then, we find the value(s) of r that enable to solve the differential equation for the condition that a_0 is non zero. Afterwards, we substitute the value(s) of r into the recursion formula to obtain the coefficients (a_1, a_2, a_3,...) for each solution and we obtain equations of y_1 and y_2. Finally, we substitute y_1 and y_2 to obtain the general solution. Let us find the solution for the equation below:

$$2x^2y''+7x(x+1)y'-3y=0$$

$$y''+\frac{7x(x+1)}{2x^2}y'-\frac{3}{2x^2}y=0$$

$$P(x)=\frac{7x(x+1)}{2x^2}, \quad Q(x)=-\frac{3}{2x^2}$$

Both xP(x) and $x^2Q(x)$ are well behaved at $x_0 = 0$ and there will be power series near 0. Then, we have the following values of y, y' and y":

$$y=\sum_{n=0}^{\infty}a_nx^{n+r}$$

$$y'=\sum_{n=0}^{\infty}(n+r)a_nx^{n+r-1}$$

$$y''=\sum_{n=0}^{\infty}(n+r)(n+r-1)a_nx^{n+r-2}$$

Notice that in the equations above we do not change the lower limit of the summation as we did it in the power series method (last section), where n = 0, n = 1 and n = 2 for y, y' and y", respectively. In the case above, n = 0 for y, y' and y" because of the presence of the factor x^r. The value of the indicial parameter, r, is chosen so that the first non zero term in the expansion is a_0.

Let us substitute these equations into the differential equation.

$$2x^2y''+7x^2y'+7xy'-3y=0$$

$$2x^2\sum_{n=0}^{\infty}(n+r)(n+r-1)a_nx^{n+r-2}+7x^2\sum_{n=0}^{\infty}(n+r)a_nx^{n+r-1}+$$

$$+7x\sum_{n=0}^{\infty}(n+r)a_nx^{n+r-1}-3\sum_{n=0}^{\infty}a_nx^{n+r}=0$$

$$x:inside$$

$$2\sum_{n=0}^{\infty}(n+r)(n+r-1)a_nx^{n+r}+7\sum_{n=0}^{\infty}(n+r)a_nx^{n+r+1}+$$

$$+7\sum_{n=0}^{\infty}(n+r)a_nx^{n+r}-3\sum_{n=0}^{\infty}a_nx^{n+r}=0$$

$$power: equalization$$

$$2\sum_{n=0}^{\infty}(n+r)(n+r-1)a_n x^{n+r} + 7\sum_{n=1}^{\infty}(n+r-1)a_{n-1}x^{n+r} +$$

$$+7\sum_{n=0}^{\infty}(n+r)a_n x^{n+r} - 3\sum_{n=0}^{\infty}a_n x^{n+r} = 0$$

Now, we have all the summation terms at the same power (n + r) and then let us find the recursion formula of a_n.

$$\left[2(n+r)(n+r-1)+7(n+r)-3\right]a_n + 7(n+r-1)a_{n-1} = 0$$

$$a_n = \frac{-7(n+r-1)a_{n-1}}{\left[2(n+r)-1\right]\left[(n+r)+3\right]}$$

Let us set n = 0 in order to find the value(s) of r so that a_0 is non zero. Note that the second term with n = 0 does not appear in the equation below:

$$n = 0$$

$$2\sum_{n=0}^{0}(0+r)(0+r-1)a_0 x^{0+r} + 7\sum_{n=1}^{\infty}(n+r-1)a_{n-1}x^{n+r} +$$

$$+7\sum_{n=0}^{0}(0+r)a_0 x^{0+r} - 3\sum_{n=0}^{0}a_0 x^{0+r} = 0$$

$$\left[2r(r-1)+7r-3\right]a_0 = 0 \therefore \left(2r^2+5r-3\right)a_0 = 0$$

$$see: a_0 \neq 0$$

$$then: 2r^2+5r-3 = 0$$

The above equation is the indicial equation. The values of r are $r = 1/2$ and $r = -3$. Let us now find the values of a_n according to the values of r in the recursion formula. Note that the values of n below are n ≥ 1 because of the lower limit of the second term of the summation (see above).

$$r = \frac{1}{2}$$

$$a_n = \frac{-7\left(n+\frac{1}{2}-1\right)a_{n-1}}{\left[2\left(n+\frac{1}{2}\right)-1\right]\left[\left(n+\frac{1}{2}\right)+3\right]} = \frac{-7\left(n-\frac{1}{2}\right)a_{n-1}}{\left[2\left(n+\frac{1}{2}\right)-1\right]\left[\left(n+\frac{7}{2}\right)\right]}$$

$$a_n = \frac{-7(2n-1)a_{n-1}}{2n(2n+7)}, \qquad n \geq 1$$

Let us find a_1 and a_2.

$$a_1 = -\frac{7}{18}a_0, \qquad a_2 = -\frac{21}{44}a_1 = \frac{147}{792}a_0$$

And get the first solution:

$$y_1 = a_0 x^{1/2}\left(1 - \frac{7}{18}x + \frac{147}{792}x^2 + \ldots\right)$$

Now, let us find a_1 and a_2 for r = –3.

$$r = \tfrac{1}{2}$$

$$a_n = \frac{-7(n-3-1)a_{n-1}}{[2(n-3)-1][(n-3)+3]} = \frac{-7(n-4)a_{n-1}}{n(2n-7)} \therefore n \geq 1$$

$$a_1 = -\frac{21}{5}a_0 \therefore a_2 = -\frac{7}{3}a_1 = \frac{49}{5}a_0$$

And the second solution is:

$$y_2 = a_0 x^{-3}\left(1 - \frac{21}{5}x + \frac{49}{5}x^2 + \ldots\right)$$

The general solution is:

$$y = c_1 y_1 + c_2 y_2$$

7. Second-order differential equations with non-constant coefficients

The second-order differential equations in previous sections have constant coefficients. However, many important differential equations have non-constant coefficients. Some of these equations are:

Hermite equation: $y'' - 2xy' + 2my = 0$

Laguerre equation: $xy'' + (1-x)y' + \lambda y = 0$

Legendre equation: $(1-x^2)y'' - 2xy' + l(l+1)y = 0$

The solution of these expressions can be given as Maclaurin series:

$$y(x) = \sum_{n=0}^{\infty} a_n x^n$$

8. The Hermite equation

The Hermite equation appears in the one-particle quantum harmonic oscillator problem (See chapter fourteen). The Hermite equation is:

$$y'' - 2xy' + 2my = 0$$

Where n is a non-negative integer. We solve this equation using power series method.

$$y(x) = \sum_{n=0}^{\infty} a_n x^n$$

$$y'(x) = \sum_{n=1}^{\infty} n a_n x^{n-1}$$

$$y''(x) = \sum_{n=2}^{\infty} n(n-1) a_n x^{n-2}$$

We see that in the second summation, if n starts at 0 or at 1 does not change the result since for n = 0 the second corresponding term is zero.

$$2xy' = 2x\sum_{n=1}^{\infty} a_n n x^{n-1} = 2\sum_{n=1}^{\infty} a_n n x^n$$

$$as: \sum_{n=1}^{\infty} a_n n x^n = \sum_{n=0}^{\infty} a_n n x^n$$

$$then: 2\sum_{n=1}^{\infty} a_n n x^n = \sum_{n=0}^{\infty} 2a_n n x^n$$

In the expansion of the y", let us change n to start at zero.

$$for: n(n-1)$$
$$n = 2 \rightarrow 2(2-1) = 2$$
$$n = 3 \rightarrow 3(3-1) = 6$$
$$for: (n+2)(n+1)$$
$$n = 0 \rightarrow (0+2)(0+1) = 2$$
$$n = 1 \rightarrow (1+2)(1+1) = 6$$
$$then: n(n-1)\big|_{n=2} = (n+2)(n+1)\big|_{n=0}$$

As a consequence:

$$y'' = \sum_{n=2}^{\infty} n(n-1)a_n x^{n-2} = \sum_{n=0}^{\infty} (n+2)(n+1)a_{n+2} x^n$$

The expansion for the last term of the Hermite equation is:

$$2my = \sum_{n=0}^{\infty} 2m a_n x^n$$

Let us substitute these expansions in the Hermite equation:

$$\sum_{n=0}^{\infty} (n+2)(n+1)a_{n+2} x^n - \sum_{n=0}^{\infty} 2n a_n x^n + \sum_{n=0}^{\infty} 2m a_n x^n = 0$$

$$\sum_{n=0}^{\infty} \left[(n+2)(n+1)a_{n+2} - 2n a_n + 2m a_n\right] x^n = 0$$

The recurrence relation is:

$$(n+2)(n+1)a_{n+2} - 2n a_n + 2m a_n = 0$$

$$a_{n+2} = \frac{2(n-m)}{(n+2)(n+1)} a_n, \qquad n = 0,1,2,3,4...$$

If m is odd, the initial value conditions are: $a_0 = 0$ and $a_1 = 1$. If m is even, the initial value conditions are: $a_0 = 1$ and $a_1 = 0$.

$$m \rightarrow odd, \quad a_0 = 0, \quad a_1 = 1$$
$$m \rightarrow even, \quad a_0 = 1, \quad a_1 = 0$$

In order to obtain the Hermite polynomials, we have to set the boundary conditions and the value of m. Let us begin with m = 3

$$initial:$$
$$m = 3$$
$$y(0) = a_0 = 0$$
$$y'(0) = a_1 = 1$$

In this case, all even coefficients will be zero due to the relation a_{n+2} and a_n in the recursion equation, that is, a_2 is related to a_0, a_4 is related to a_2 and so on.

$$for: a_0 = 0$$
$$a_0 = a_2 = a_4 = a_6 = \ldots = 0$$

For the odd coefficients, we have:

$$a_3 = a_{1+2}\Big|_{n=1} = \frac{2(1-3)}{(2+1)(1+1)} = -\frac{2}{3}$$

$$a_5 = a_{3+2}\Big|_{n=3} = \frac{2(3-3)}{(2+3)(1+3)}a_3 = 0$$

$$then: H_3(n) = x - \frac{2}{3}x^3$$

Let us now set m = 4. In this case, the initial conditions are:

$$initial:$$
$$m = 4$$
$$y(0) = a_0 = 1$$
$$y'(0) = a_1 = 0$$

In this case, all odd coefficients will be zero due to the relation a_{n+2} and a_n in the recursion equation, that is, a_3 is related to a_1, a_5 is related to a_3 and so on.

$$for: a_1 = 0$$
$$a_1 = a_3 = a_5 = a_7 = \ldots = 0$$

For the even coefficients, we have:

$$a_2 = a_{0+2}\Big|_{n=0} = \frac{2(0-4)}{(2+0)(1+0)} = -\frac{8}{2} = -4$$

$$a_4 = a_{2+2}\Big|_{n=2} = \frac{2(2-4)}{(2+2)(1+2)}a_2 = \left(-\frac{4}{12}\right)(-4) = \frac{4}{3}$$

$$a_6 = a_{4+2}\Big|_{n=4} = \frac{2(4-4)}{(2+4)(1+4)}a_4 = 0$$

$$then: H_4(n) = 1 - 4x^2 + \frac{4}{3}x^4$$

9. The Laguerre equation

The Laguerre equation is:

$$xy''+(1-x)y'+\lambda y=0$$

Where λ is a positive integer. Some authors use another term for y:

$$y=L_n(x)$$

Let us use the Frobenius method to solve this equation.

$$\lambda y=\lambda\sum_{n=0}^{\infty}a_n x^{r+n}$$

$$(1-x)y'=(1-x)\sum_{n=0}^{\infty}(r+n)a_n x^{r+n-1}$$

$$xy''=x\sum_{n=0}^{\infty}(r+n)(r+n-1)a_n x^{r+n-2}$$

Remember that in Frobenius method, n = 0 for y, y' and y'' because of the presence of the factor x^r. The value of the indicial parameter, r, is chosen so that the first non zero term in the expansion is a_0.

Let us insert the coefficients in the summations:

$$\lambda y=\sum_{n=0}^{\infty}\lambda a_n x^{r+n}$$

$$(1-x)y'=\sum_{n=0}^{\infty}(r+n)a_n x^{r+n-1}-\sum_{n=0}^{\infty}(r+n)a_n x^{r+n}$$

$$xy''=\sum_{n=0}^{\infty}(r+n)(r+n-1)a_n x^{r+n-1}$$

And substitute in the Laguerre equation:

$$\sum_{n=0}^{\infty}(r+n)(r+n-1)a_n x^{r+n-1}+\sum_{n=0}^{\infty}(r+n)a_n x^{r+n-1}$$

$$-\sum_{n=0}^{\infty}(r+n)a_n x^{r+n}+\sum_{n=0}^{\infty}\lambda a_n x^{r+n}$$

Let us rearrange the equation

$$\sum_{n=0}^{\infty}\left[(r+n)(r+n-1)+(r+n)\right]a_n x^{r+n-1}-\sum_{n=0}^{\infty}\left[\lambda-(r+n)\right]a_n x^{r+n}=0$$

$$\sum_{n=0}^{\infty}\left[\left(r^2+2rn+n^2-(r+n)\right)+(r+n)\right]a_n x^{r+n-1}+$$

$$+\sum_{n=0}^{\infty}\left[\lambda-(r+n)\right]a_n x^{r+n}=0$$

$$\sum_{n=0}^{\infty}(r+n)^2 a_n x^{r+n-1}+\sum_{n=0}^{\infty}\left[\lambda-r-n\right]a_n x^{r+n}=0$$

In order to obtain the recursion equation, we have to change one of the a_n terms into a_{n-1} and equalize the powers.

$$\left[\lambda - r - n\right] a_{n-1} x^{r+n}\Big|_{n=0} = \left[\lambda - r - n + 1\right] a_{n-1} x^{r+n-1}\Big|_{n=1}$$

then:

$$\sum_{n=0}^{\infty} (r+n)^2 a_n x^{r+n-1} + \sum_{n=1}^{\infty} \left[\lambda - r - n + 1\right] a_{n-1} x^{r+n-1} = 0$$

$$\sum_{n=0}^{\infty} (r+n)^2 a_n x^{r+n-1} = r^2 a_0 x^{r-1} + \sum_{n=1}^{\infty} (r+n)^2 a_n x^{r+n-1}$$

then:

$$r^2 a_0 x^{r-1} + \sum_{n=1}^{\infty} (r+n)^2 a_n x^{r+n-1} + \sum_{n=1}^{\infty} \left[\lambda - r - n + 1\right] a_{n-1} x^{r+n-1} = 0$$

$$r^2 a_0 x^{r-1} + \sum_{n=1}^{\infty} \left[(r+n)^2 a_n + (\lambda - r - n + 1) a_{n-1}\right] x^{r+n-1} = 0$$

Thus, the indicial equation is:

$$r^2 = 0, \quad r_1 = 0, \quad r_2 = 0$$

The recurrence relation is:

$$r^2 = 0$$

$$0^2 a_0 x^{r-1} + (r+n)^2 a_n + (\lambda - r - n + 1) a_{n-1} = 0$$

$$r = 0$$

$$n^2 a_n + (\lambda - n + 1) a_{n-1} = 0$$

$$a_n = \frac{(n - \lambda - 1) a_{n-1}}{n^2}, \qquad n \geq 1$$

Let us find the values of a_n:

$$a_1 = -\frac{\lambda}{1^2} a_0$$

$$a_2 = \frac{1-\lambda}{2^2} a_1 = \left(\frac{1-\lambda}{2^2}\right)\left(-\frac{\lambda}{1^2} a_0\right) = -\frac{\lambda(1-\lambda)}{(2!)^2} a_0$$

$$a_3 = \frac{2-\lambda}{3^2} a_2 = \left(\frac{2-\lambda}{3^2}\right)\left(-\frac{\lambda(1-\lambda)}{(2!)^2} a_0\right) = -\frac{\lambda(1-\lambda)(2-\lambda)}{(3!)^2} a_0$$

$$a_n = \frac{\prod_{k=0}^{n-1} (k - \lambda)}{(n!)^2}$$

Thus, the solution to Laguerre equation is:

$$y = a_0 \left(1 + \sum_{n=1}^{\infty} \frac{\prod_{k=0}^{n-1}(k-\lambda)}{(n!)^2} x^n \right)$$

The first few Laguerre polynomials are:

$$L_0(x) = 1$$
$$L_1(x) = 1 - x$$
$$L_2(x) = 1 - 2x + x^2/2$$
$$L_3(x) = 1 - 3x + 3x^2/2 - x^3/6$$
$$L_4(x) = 1 - 4x + 3x^2 - 2x^3/3 + x^4/24$$

10. The associated Laguerre equation

The associated Laguerre equation is:

$$xy'' + (\alpha + 1 - x)y' + \lambda y = 0$$

Some authors use k instead of α for the second index and j instead of λ for the first index. When $\alpha = 0$, the above equation reduces to the Laguerre equation. Some authors use a different notation for y:

$$y = L_\lambda^\alpha(x) \quad or \quad L_j^k(x)$$

The point x = 0 is a regular singular point. Then, we can use the Frobenius method to find the solution.

$$\lambda y = \lambda \sum_{n=0}^{\infty} a_n x^{r+n}$$

$$(\alpha + 1 - x)y' = (\alpha + 1 - x)\sum_{n=0}^{\infty}(r+n)a_n x^{r+n-1}$$

$$xy'' = x\sum_{n=0}^{\infty}(r+n)(r+n-1)a_n x^{r+n-2}$$

By replacing these terms in the associated Laguerre equation, we have (where the variable x of the first term was incorporated in the summation):

$$\sum_{n=0}^{\infty}(r+n)(r+n-1)a_n x^{r+n-1} + (\alpha + 1 - x)\sum_{n=0}^{\infty}(r+n)a_n x^{r+n-1} +$$
$$+\lambda \sum_{n=0}^{\infty} a_n x^{r+n} = 0$$

The recurrence relation for r = 0 (see previous section for similar derivation) is:

$$\frac{a_{n+1}}{a_n} = \frac{n-\lambda}{(n+1)(n+\alpha+1)}$$

The first associated Laguerre polynomials are:

$$L_0^0(x) = L_0(x) = 1$$
$$L_1^0(x) = L_1(x) = 1 - x$$
$$L_1^1(x) = -2x + 4$$
$$L_0^1(x) = 1$$
$$L_2^0(x) = L_2(x)$$
$$L_2^1(x) = 3x^2 - 18x + 18$$
$$L_2^2(x) = 12x^2 - 96x + 144$$
$$L_1^2(x) = -6x + 18$$
$$L_0^2(x) = 2$$

The associated Laguerre polynomials have Rodrigues' formula:

$$L_n^k(x) = \frac{e^x x^{-k}}{n!}\frac{d^n}{dx^n}\left(e^{-x}x^{n+k}\right)$$

Whereas the Rodrigues' formula for the Laguerre polynomials is:

$$L_n(x) = \frac{e^x}{n!}\frac{d^n}{dx^n}\left(x^n e^{-x}\right)$$

The associated Laguerre formula can be related to Laguerre polynomials via:

$$L_n^k(x) = (-1)^k \frac{d^k}{dx^k} L_{n+k}(x)$$

The general solution for the associated Laguerre equation is:

$$L_n^k(x) = \sum_{m=0}^{n} (-1)^0 \frac{(n+k)!}{(n-m)(k+m)!m!} x^m$$

The associated Laguerre polynomial is not self-adjoint, but it can be made self-adjoint by multiplying by a factor [a weighting function, w(x)]: $e^{-x}x^k$.

$$x\frac{d^2 L_j^k(x)}{dx^2} + (k+1-x)\frac{dL_j^k(x)}{dx} + jL_j^k(x) = 0$$

$\times e^{-x}x^k$:

$$e^{-x}x^{k+1}\frac{d^2 L_j^k(x)}{dx^2} + (k+1-x)e^{-x}x^k\frac{dL_j^k(x)}{dx} + je^{-x}x^k L_j^k(x) = 0$$

The **self-adjoint condition of a differential equation** is that in which the derivative of the function multiplying the first derivative equals the function multiplying the second derivative. In the case above, we see that:

$$\frac{d}{dx}\left(e^{-x}x^{k+1}\right) = -e^{-x}x^{k+1} + (k+1)e^{-x}x^k$$
$$\frac{d}{dx}\left(e^{-x}x^{k+1}\right) = (k+1-x)e^{-x}x^k$$

We can see that this modified associated Laguerre equation:

$$e^{-x}x^{k+1}\frac{d^2L_j^k(x)}{dx^2}+(k+1-x)e^{-x}x^k\frac{dL_j^k(x)}{dx}+je^{-x}x^kL_j^k(x)=0$$

is self-adjoint.

In addition, for distinct values of j, the corresponding associated Laguerre polynomials are orthogonal within the limits of integration (between zero and infinite) for j ≠ l. The orthogonality of the polynomial is an important condition for obtaining the normalization constant N of the wave function.

$$\int_0^\infty e^{-x}x^kL_j^k(x)L_l^k(x)dx=0$$

$$L_j^k(x)p_0L_l^k(x)'-L_l^k(x)p_0L_n^k(x)'\Big|_0^\infty=0$$

$$p_0(x)=e^{-x}x^{k+1}$$

Here, the function $p_0(x)$ is the coefficient of the second derivative term which is zero at both limits.

The generating function of the self-adjoint associated Laguerre equation is:

$$G(x,z)=\sum_{j=0}^{\infty}z^jL_j^k(x)=\frac{e^{xz/(z-1)}}{(1-z)^{k+1}}$$

Let us evaluate the product G(x,t)G(x,s) multiplied by the weight function x^ke^{-x} and expand the functions G(x,t) and G(x,s) in powers of t and s.

$$\sum_{n,m=0}^{\infty}t^ns^m\int_0^\infty e^{-x}x^kL_n^k(x)L_m^k(x)dx$$

$$\int_0^\infty e^{-x}x^kL_n^k(x)L_m^k(x)dx=\int_0^\infty e^{-x}x^k\frac{e^{xt/(t-1)}}{(1-t)^{k+1}}\frac{e^{xs/(s-1)}}{(1-s)^{k+1}}dx$$

$$\Rightarrow\int_0^\infty x^k\frac{1}{(1-t)^{k+1}}\frac{1}{(1-s)^{k+1}}\exp\left[\frac{-x(1-st)}{(1-t)(1-s)}\right]dx$$

$$\Rightarrow\frac{1}{(1-t)^{k+1}}\frac{1}{(1-s)^{k+1}}\int_0^\infty x^k\exp\left[\frac{-x(1-st)}{(1-t)(1-s)}\right]dx$$

$$v=\frac{x(1-st)}{(1-t)(1-s)}$$

$$x=\frac{(1-t)(1-s)}{(1-st)}v,\qquad dx=\frac{(1-t)(1-s)}{(1-st)}dv$$

$$\Rightarrow\frac{(1-t)^k(1-s)^k(1-t)(1-s)}{(1-t)^{k+1}(1-s)^{k+1}(1-st)^k(1-st)}\int_0^\infty v^ke^{-v}dv$$

$$see:\int_0^\infty x^n e^{-ax}dx = \frac{n!}{a^{n+1}}$$

$$\Rightarrow \frac{1}{(1-st)^{k+1}}\int_0^\infty v^k e^{-v}dv = \frac{k!}{(1-st)^{k+1}}$$

$$\sum_{n,m=0}^{\infty} t^n s^m \left(\frac{k!}{(1-st)^{k+1}}\right) = \frac{(k+n)!}{n!}, \qquad for: n = m$$

In the Appendix it is depicted the derivation of a different type of the associated Laguerre equation in order to become an associated Laguerre equation. The former is the equation obtained from the radial wave function of the hydrogen atom (see chapter seventeen).

11. The Legendre equation

The Legendre equation occurs in problems with axial symmetry involving the Laplace operator, ∇^2, expressed in terms of spherical polar coordinates where the variable x is replaced by cos θ, and $-1 \le x \le 1$. The constant j represents a real number in the Legendre equation. The parameter j is a real number and any solution to this equation is called Legendre function. The Legendre equation can be rearranged into the Legendre differential where L is the Legendre operator.

Note: Most books use parameter l instead of j (or other letter such as n). In this section, we opted for using j instead of l in order to avoid typographical confusion with number 1 (one). In the next section (associated Legendre equation), we changed j into l because the number of equations is much smaller.

$$(1-x^2)y''-2xy'+j(j+1)y=0$$

$$\frac{d}{dx}\left[(1-x^2)\frac{dy}{dx}\right]+j(j+1)y=0$$

$$L=\frac{d}{dx}\left[(1-x^2)\frac{d}{dx}\right]+j(j+1)$$

$$Ly=0$$

Let us check whether the Fuchs' theorem is satisfied or not for the Legendre equation. Firstly, we have to change the form of the Legendre equation to that used in Fuchs' theorem:

$$y''-\frac{2x}{(1+x)(1-x)}y'+\frac{j(j+1)}{(1+x)(1-x)}y=0$$

We see that x = 0 is a regular point of the second-order differential equation and that x = −1 and x = 1 are regular singular points. Then, the Legendre differential equation has regular singular points at −1, +1 and ∞. The solutions of interest correspond to the interval $-1 \le x \le +1$, i.e., $0 \le \theta \le \pi$. The Legendre equation can be solved using the power-series method.

$$y(x)=\sum_{n=0}^{\infty}a_nx^n$$

$$y'(x)=\sum_{n=1}^{\infty}na_nx^{n-1}$$

$$y''(x)=\sum_{n=2}^{\infty}n(n-1)a_nx^{n-2}$$

Let us substitute these expansions in the respective terms of the Legendre equation:

$$2xy'=2x\sum_{n=1}^{\infty}na_nx^{n-1}=\sum_{n=0}^{\infty}2na_nx^n$$

$$\left(1-x^2\right)y''=\left(1-x^2\right)\sum_{n=2}^{\infty}n(n-1)a_nx^{n-2}$$

$$\left(1-x^2\right)y''=\sum_{n=2}^{\infty}n(n-1)a_nx^{n-2}-x^2\sum_{n=2}^{\infty}n(n-1)a_nx^{n-2}$$

$$\left(1-x^2\right)y''=\sum_{n=2}^{\infty}n(n-1)a_nx^{n-2}-\sum_{n=0}^{\infty}n(n-1)a_nx^n$$

$$power: equalize$$

$$n=2\rightarrow 2(2-1)=2,\qquad n=0\rightarrow(0+2)(0+1)=2$$

$$n=3\rightarrow 3(3-1)=6,\qquad n=1\rightarrow(1+2)(1+1)=6$$

$$then:\sum_{n=2}^{\infty}n(n-1)=\sum_{n=0}^{\infty}(n+2)(n+1)$$

$$so:\left(1-x^2\right)y''=\sum_{n=0}^{\infty}(n+2)(n+1)a_{n+2}x^n-\sum_{n=0}^{\infty}n(n-1)a_nx^n$$

$$\left(1-x^2\right)y''=\sum_{n=0}^{\infty}\left[(n+2)(n+1)a_{n+2}-n(n-1)a_n\right]x^n$$

Let us substitute the expansions in the Legendre equation.

$$\left(1-x^2\right)y''-2xy'+j(j+1)y=0$$

$$y=\sum_{n=0}^{\infty}a_nx^n,\qquad 2xy'=\sum_{n=0}^{\infty}2na_nx^n$$

$$\left(1-x^2\right)y''=\sum_{n=0}^{\infty}\left[(n+2)(n+1)a_{n+2}-n(n-1)a_n\right]x^n$$

$$\sum_{n=0}^{\infty}\left[(n+2)(n+1)a_{n+2}-n(n-1)a_n\right]x^n-$$

$$-\sum_{n=0}^{\infty}2na_nx^n+j(j+1)\left[\sum_{n=0}^{\infty}a_nx^n\right]=0$$

For every $n \geq 0$, we have:

$$(n+2)(n+1)a_{n+2} - n(n-1)a_n - 2na_n + j(j+1)a_n = 0$$

We obtain the recurrence relation:

$$a_{n+2} = -\frac{(j-n)(j+n+1)}{(n+1)(n+2)}a_n$$

This recurrence relation can also be written as:

$$a_{n+2} = \frac{[n(n+1) - j(j+1)]}{(n+1)(n+2)}a_n$$

According to recursion equation, bear in mind that:

$$a_2 = a_{0+2}, \qquad n = 0$$
$$a_4 = a_{2+2}, \qquad n = 2$$
$$a_6 = a_{4+2}, \qquad n = 4$$

The even values of a_n are given below.

$$a_2 = -\frac{j(j+1)}{1\cdot 2}a_0$$
$$a_4 = -\frac{(j-2)(j+3)}{3\cdot 4}a_2 = (-1)^2\frac{j(j-2)(j+1)(j+3)}{4!}a_0$$
$$a_{2n} = (-1)^n\frac{j(j-2)\cdots(j-2n+2)(j+1)(j+3)\cdots(j+2n-1)}{(2n)!}a_0$$

The odd values of a_n are:

$$a_3 = -\frac{(j-1)(j+2)}{2\cdot 3}a_1$$
$$a_5 = -\frac{(j-3)(j+4)}{4\cdot 5}a_3 = (-1)^2\frac{(j-1)(j-3)(j+2)(j+4)}{5!}a_1$$
$$a_{2n+1} = (-1)^n\frac{(j-1)(j-3)\cdots(j-2n+1)(j+2)(j+4)\cdots(j+2n)}{(2n+1)!}a_1$$

The general solution of the Legendre equation in the interval $(-1,1)$ is:

$$y = c_1y_1 + c_2y_2$$

where:

$$y_1 = 1 - \frac{j(j+1)}{2!}x^2 + \frac{(j-2)(j+1)(j+3)}{4!}x^4 - \cdots$$

$$y_1 = 1 - \sum_{n=1}^{\infty}\frac{j(j-2)\cdots(j-2n+2)(j+1)(j+3)\cdots(j+2n-1)}{(2n)!}x^{2n}$$

and

$$y_2 = x - \frac{(j-1)(j+2)}{3!}x^3 + \frac{(j-3)(j-1)(j+2)(j+4)}{5!}x^5$$

$$y_2 = x + \sum_{n=1}^{\infty}(-1)^n \frac{(j-1)(j-3)\cdots(j-2n+1)(j+2)(j+4)\cdots(j+2n)}{(2n+1)!}x^{2n+1}$$

From the d'Alembert's ratio test both y_1 and y_2 series converge if $|x| < 1$ and diverge if $|x| > 1$, unless j is an integer. For the constant j as an even integer, y_1 reduces to polynomial with only even powers of x and y_2 diverges. For the constant j as an odd integer, y_2 reduces to a polynomial with only odd powers of x and y_1 diverges. The general solution is then rewritten as Legendre Polynomials, Pj(x).

$$P_j(x) = c_j \begin{cases} y_1(x), & j \to even \\ y_2(x), & j \to odd \end{cases}$$

The value of cj is chosen for $P_j(1) = 1$. As a consequence, $P_j(-1) = (-1)^j$.
Let us find the expressions of y and $P_j(x)$ for first four values of j:

$$j = 0, \quad y = 1, \quad P_0(x) = 1$$

$$j = 1, \quad y = x, \quad P_1(x) = x$$

$$j = 2, \quad y = 1 - 3x^2, \quad P_2(x) = \tfrac{1}{2}(3x^2 - 1)$$

$$j = 3, \quad y = \frac{2}{6}(3x - 5x^3), P_3(x) = \tfrac{1}{2}(5x^3 - 3x)$$

These equations are the $P_j(x)$ Legendre polynomials. Since the Legendre differential equation is second-order, it will have a second solution (for each n value) that is not polynomial designated as $Q_j(x)$. The $Q_j(x)$ are known as the Legendre functions of the second kind and they are not appropriate to replace $P_j(x)$ polynomials.

The Legendre Polynomials can be defined as the coefficients in a formal expansion in powers of t of the generating function (see Chapter two) below:

$$G(x;t) = \frac{1}{\sqrt{(1 - 2xt + t^2)}} = \sum_{j=0}^{\infty} P_j(x)t^j$$

When we set x = 1, we have the following generating function:

$$G(1;t) = \frac{1}{\sqrt{(1-t)^2}} = \frac{1}{1-t} = 1 + t + t^2 + t^3 + \ldots$$

where :

$$G(1;t) = P_0(1) + P_1(1)t + P_2(1)t^2 + \ldots$$

We see that $P_j(1) = 1$ for all values of j.

Let us check that the Legendre polynomial is the coefficient of the series expansion of the generating function above. Firstly, note that:

$$t\frac{\partial^2}{\partial t^2}t^{j+1} = j(j+1)t^j$$

$$then:$$

$$t\frac{\partial^2}{\partial t^2}\left[tG(x;t)\right] = \sum_{j=0}^{\infty} j(j+1)P_j(x)t^j$$

If the polynomials $P_j(x)$ are the solutions of the Legendre equation, then we have:

$$\sum_{j=0}^{\infty}\left[\left(1-x^2\right)\frac{\partial^2}{\partial x^2} - 2x\frac{\partial}{\partial x} + j(j+1)\right]P_j(x)t^j = 0$$

By rewriting the above equation by using the previous equation for the term for j(j + 1), we have:

$$\left(1-x^2\right)\frac{\partial^2 G(x;t)}{\partial x^2} - 2x\frac{\partial G(x;t)}{\partial x} + t\frac{\partial^2\left[tG(x;t)\right]}{\partial x^2} = 0$$

One can easily note that the above equation is an identity.

The generating function can be used to obtain the recurrence formula of the Legendre polynomials. Let us differentiate G(x;t) with respect to t.

$$G(x,t) = \left(1-2xt+t^2\right)^{-1/2}$$

$$u = 1-2xt+t^2$$

$$\frac{\partial G(x,t)}{\partial t} = \frac{\partial G(x,t)}{\partial u}\frac{\partial u}{\partial t} = -\frac{1}{2}\frac{(2t-2x)}{\left(1-2xt+t^2\right)^{3/2}}$$

$$\frac{\partial G(x,t)}{\partial t} = \frac{x-t}{\left(1-2xt+t^2\right)^{3/2}}$$

$$\frac{\partial G(x;t)}{\partial t} = \sum_{j=0}^{\infty} jP_j(x)t^{j-1}$$

$$\frac{x-t}{\left(1-2xt+t^2\right)^{3/2}} = \sum_{j=0}^{\infty} jP_j(x)t^{j-1}$$

Let us rearrange the above equation:

$$G(x;t) = \left(1-2xt+t^2\right)^{-1/2}$$

$$\frac{x-t}{\left(1-2xt+t^2\right)^{3/2}} = \sum_{j=0}^{\infty} jP_j(x)t^{j-1}$$

$$(x-t)\left(1-2xt+t^2\right)^{-3/2} = \sum_{j=0}^{\infty} jP_j(x)t^{j-1}$$

$$\left(x-t\right)G(x,t)\left(1-2xt+t^2\right)^{-1}=\sum_{j=0}^{\infty} jP_j(x)t^{j-1}$$

$$\left(x-t\right)G(x,t)=\left(1-2xt+t^2\right)\sum_{j=0}^{\infty} jP_j(x)t^{j-1}$$

$$\left(x-t\right)G(x,t)-\left(1-2xt+t^2\right)\sum_{j=0}^{\infty} jP_j(x)t^{j-1}=0$$

$$G(x,t)=\sum_{j=0}^{\infty} P_j(x)t^j$$

$$\left(x-t\right)\sum_{j=0}^{\infty} P_j(x)t^j+\left(-1+2xt-t^2\right)\sum_{j=0}^{\infty} jP_j(x)t^{j-1}=0$$

$$x\sum_{j=0}^{\infty} P_j(x)t^j-\sum_{j=0}^{\infty} P_j(x)t^{j+1}-\sum_{j=0}^{\infty} jP_j(x)t^{j-1}$$
$$+2x\sum_{j=0}^{\infty} jP_j(x)t^j-\sum_{j=0}^{\infty} jP_j(x)t^{j+1}=0$$

$$x\sum_{j=0}^{\infty} P_j(x)t^j-\sum_{j=0}^{\infty} P_j(x)t^{j+1}-\sum_{j=0}^{\infty} jP_j(x)t^{j-1}$$
$$+2xj\sum_{j=0}^{\infty} P_j(x)t^j-j\sum_{j=0}^{\infty} P_j(x)t^{j+1}=0$$

$$\left(2j+1\right)x\sum_{j=0}^{\infty} P_j(x)t^j-\left(j+1\right)\sum_{j=0}^{\infty} P_j(x)t^{j+1}-\sum_{j=0}^{\infty} jP_j(x)t^{j-1}=0$$

$$\sum_{j=0}^{\infty}\left[\left(2j+1\right)xP_j(x)t^j-\left(j+1\right)P_j(x)t^{j+1}-jP_j(x)t^{j-1}\right]=0$$

By combining the coefficients of t^j from the individual terms of the equation above, we obtain for each n the recurrence formula:

$$\left[\left(2j+1\right)xP_j(x)-jP_{j-1}(x)-\left(j+1\right)P_{j+1}(x)\right]t^j=0$$
$$\left(2j+1\right)xP_j(x)-jP_{j-1}(x)-\left(j+1\right)P_{j+1}(x)=0$$
$$\left(2j+1\right)xP_j(x)=jP_{j-1}(x)+\left(j+1\right)P_{j+1}(x)$$

The equation above gives an efficient way to obtain $P_j(x)$ from the initial values $P_0(x) = 1$ and $P_1(x) = x$. Let us rearrange the equation in terms of the Pj_{-1} and Pj_{-2}

$$\left(2j-1\right)xP_{j-1}(x)=\left(j-1\right)P_{j-2}(x)+jP_j(x)$$
$$P_j(x)=\frac{\left(2j-1\right)xP_{j-1}(x)-\left(j-1\right)P_{j-2}(x)}{j}$$

Another way to express the Legendre polynomials is the form of the Rodrigues' formula:

$$P_j(x)=\frac{1}{2^j j!}\frac{d^j}{dx^j}\left(x^2-1\right)^j$$

The Legendre equation is self-adjoint. The Legendre polynomials are eigenfunctions of a particular self-adjoint operator. The coefficient of $P_j(x)$, $(1-x^2)$, vanishes at the interval $[-1,1]$. Then, its solutions of different j will be orthogonal in this interval (see Chapter three). That is:

$$\int_{-1}^{1} P_j(x)P_k(x)dx = 0, \qquad j \neq k$$

$$x = \cos\theta$$

$$\int_{0}^{\pi} P_j(\cos\theta)P_k(\cos\theta)\sin\theta d\theta = 0$$

Nonetheless, the definition of $P_j(x)$ does not guarantee they are normalized. One way to establish the normalization is by squaring the generating function of the $P_j(x)$.

$$G(x;t) = \frac{1}{\sqrt{\left(1-2xt+t^2\right)}} = \sum_{j=0}^{\infty} P_j(x)t^j$$

$$square:$$

$$\left(1-2xt+t^2\right)^{-1} = \left[\sum_{j=0}^{\infty} P_j(x)t^j\right]^2$$

Let us now integrate from x = –1 to x = 1 and remember that the cross terms vanish due to the orthogonality. Then:

$$\int_{-1}^{1} \frac{dx}{\left(1-2xt+t^2\right)} = \sum_{j=0}^{\infty} t^{2j} \int_{-1}^{1} \left[P_j(x)\right]^2 dx$$

Let us now integrate the left term of the above equation:

$$\int_{-1}^{1} \frac{dx}{\left(1-2xt+t^2\right)} = ?$$

$$y = 1-2tx+t^2, \qquad dy = -2tdx$$

$$if: x=-1 \Rightarrow y = 1+2t+t^2 = \left(1+t\right)^2$$

$$if: x=1 \Rightarrow y = 1-2t+t^2 = \left(1-t\right)^2$$

$$Then:$$

$$\int_{-1}^{1} \frac{dx}{\left(1-2xt+t^2\right)} = \frac{1}{2t} \int_{(1-t)^2}^{(1+t)^2} \frac{dy}{y} = \frac{1}{t}\ln\left(\frac{1+t}{1-t}\right)$$

$$power-series:$$

$$\frac{1}{t}\ln\left(\frac{1+t}{1-t}\right) = 2\sum_{j=0}^{\infty} \frac{t^{2j}}{2j+1}$$

Let us now equalize with the second term (or right term) of the previous integral:

$$2\sum_{j=0}^{\infty}\frac{t^{2j}}{2j+1}=\sum_{j=0}^{\infty}t^{2j}\int_{-1}^{1}\left[P_j(x)\right]^2 dx$$

$$then:$$

$$\int_{-1}^{1}\left[P_j(x)\right]^2 dx=\frac{2}{2j+1}$$

Then, the orthonormality condition for the Legendre polynomials are:

$$\int_{-1}^{1}P_j(x)P_k(x)dx=\frac{2}{2j+1}\delta_{jk}$$

The normalization constant (see chapter ten), N_j, is:

$$N_j=\sqrt{\frac{2j+1}{2}}$$

Any normalized wave function (an eigenfunction), ψ, has the properties:

$$\int_{-\infty}^{\infty}\psi_n^*\psi_m d\tau=\delta_{nm}$$

$$\delta_{nm}=\begin{cases}1 \therefore n=m\Rightarrow normalization\\ 0\therefore n\neq m\Rightarrow orthogonalization\end{cases}$$

12. The associated Legendre equation

The Legendre equation is a special type of the Legendre associated equation where m = 0. The associated Legendre equation is then a generalized version of the Legendre equation.

$$\left(1-x^2\right)y''-2xy'+\left[l(l+1)-\frac{m^2}{1-x^2}\right]y=0$$

The associated Legendre equation has three regular singular points at x = –1, 1 and ∞. This equation appears in the applications of the Laplace operator, ∇^2, or the Helmholtz equation expressed in terms of spherical coordinates (not exhibiting symmetry about the polar axis). Likewise the Legendre equation, the variable x is the cosine of the polar angle in spherical coordinates and then $-1\leq x\leq 1$. The point x = 0 is an ordinary point and it can be obtained the solution for the associated Legendre equation from power series method in a similar way used to the Legendre equation. The associated Legendre equation can be written in the form:

$$\frac{d}{dx}\left[\left(1-x^2\right)\frac{d}{dx}P_l^m(x)\right]+\left[l(l+1)-\frac{m^2}{1-x^2}\right]P_l^m(x)=0$$

$$where: y=P_l^m(x)$$

or:

$$\left(1-x^2\right)P''(x)-2xP'(x)+\left[l(l+1)-\frac{m^2}{1-x^2}\right]P(x)=0$$

This equation has nonzero nonsingular solutions in the interval [–1,1] only if l and m are integers where $0 \le m \le l$. When m is even, the solution is a polynomial and when m = 0 and l is an integer, the solution is identical to the Legendre polynomials.

Let us eliminate the troublesome factor $1 - x^2$ from the denominator by making the following substitution:

$$P=\left(1-x^2\right)^{m/2}p$$

Hence, we differentiate P with respect to x.

$$y=\left(1-x^2\right)^{m/2}, \qquad u=1-x^2$$

$$\frac{dy}{du}=\frac{m}{2}\left(1-x^2\right)^{m/2-1}$$

$$\frac{du}{dx}=-2x$$

$$\frac{dy}{du}\frac{du}{dx}=-mx\left(1-x^2\right)^{m/2-1}$$

$$P'=\left(1-x^2\right)^{m/2}p'-mx\left(1-x^2\right)^{m/2-1}p$$

Let us differentiate the first part of the above equation:

$$\frac{d}{dx}yp'=?$$

$$y=\left(1-x^2\right)^{m/2}$$

$$\frac{dy}{du}\frac{du}{dx}=-mx\left(1-x^2\right)^{m/2-1}$$

$$\frac{d}{dx}yp'=\left(1-x^2\right)^{m/2}p''-mx\left(1-x^2\right)^{m/2-1}p'$$

And let us differentiate the second part of the previous equation (P'):

$$\frac{d}{dx}wzp=?$$

$$w=-mx$$

$$\frac{dw}{dx}=-m$$

$$z=\left(1-x^2\right)^{m/2-1}$$

$$\frac{dz}{du}=\frac{m}{2}-1\left(1-x^2\right)^{m/2-2}$$
$$\frac{du}{dx}=-2x$$
$$\frac{dz}{du}\frac{du}{dx}=\left(-mx+2x\right)\left(1-x^2\right)^{m/2-2}$$
$$\frac{d}{dx}wzp=-mx\left(1-x^2\right)^{m/2-1}p'-m\left(1-x^2\right)^{m/2-1}p+$$
$$+mx\left(-mx+2x\right)\left(1-x^2\right)^{m/2-2}p$$
$$\frac{d}{dx}wzp=-mx\left(1-x^2\right)^{m/2-1}p'-m\left(1-x^2\right)^{m/2-1}p+$$
$$+\left(2m-m^2\right)x^2\left(1-x^2\right)^{m/2-2}p$$

Now, let us add both terms:

$$\frac{d}{dx}yp'+\frac{d}{dx}wzp=\left(1-x^2\right)^{m/2}p''-mx\left(1-x^2\right)^{m/2-1}p'$$
$$-mx\left(1-x^2\right)^{m/2-1}p'-m\left(1-x^2\right)^{m/2-1}p+$$
$$+\left(2m-m^2\right)x^2\left(1-x^2\right)^{m/2-2}p$$
$$P''=\left(1-x^2\right)^{m/2}p''-2mx\left(1-x^2\right)^{m/2-1}p'+$$
$$+\left[-m\left(1-x^2\right)^{m/2-1}+\left(2m-m^2\right)x^2\left(1-x^2\right)^{m/2-2}\right]p$$

By replacing P, P' and P'' in the associated Legendre equation, we have:

$$\left(1-x^2\right)p''-2x\left(m+1\right)p'+\left[l(l+1)-m(m+1)\right]p=0$$

By differentiating the above equation, we obtain a *similar differential equation*

$$-2x(p')'+(1-x^2)(p')''-2(m+1)p'-2(m+1)x(p')'+$$
$$+\left[l(l+1)-m(m+1)\right]p'=0$$
$$\textit{simplification}:$$
$$(1-x^2)(p')''-2x(m+2)(p')'+\left[-2(m+1)+l(l+1)-m(m+1)\right]p'=0$$
$$\textit{Then}:$$
$$(1-x^2)(p')''-2x(m+2)(p')'+\left[l(l+1)-(m+2)(m+1)\right]p'=0$$

This means that if P(x) is a solution for m = 0 (the Legendre polynomial), then the differentiation of P(x) for a positive integer m combining with the equation $y = (1 - x^2)^{m/2}$ is the solution of the associated Legendre equation. Therefore, the general solution of the associated Legendre equation:

$$P_l^m(x)=(-1)^m\left(1-x^2\right)^{m/2}\frac{d^m}{dx^m}P_l(x)$$

When m = 0, $P_l^0 = P_l$. The factor $(-1)^m$ is sometimes omitted. This factor is called Condon-Shortley phase and it is a sign convention that enables to deal with the momentum ladder operators.

By using Rodrigues' formula for P_n^m, we find the solution of the associated Legendre equation:

$$p_l^m(x) = \frac{(-1)^m}{2^l l!}(1-x^2)^{m/2}\frac{d^{l+m}}{dx^{l+m}}(x^2-1)^l$$

$$p_l^{-m}(x) = (-1)^m \frac{(l-m)!}{(l+m)!} p_l^m(x)$$

Hence, the associated Legendre functions are:

$$l=0, m=0 \therefore\ p_0^0(x) = \frac{(-1)^0}{2^0 0!}(1-x^2)^{0/2}\frac{d^0}{dx^0}(x^2-1)^0 = 1$$

$$l=1, m=1, \quad p_1^1(x) = \frac{(-1)^1}{2^1 1!}(1-x^2)^{1/2}\frac{d^2}{dx^2}(x^2-1)^1$$

$$l=1, m=1, \quad p_1^1(x) = -(1-x^2)^{1/2}$$

$$l=1, m=0, \quad p_1^0(x) = \frac{(-1)^0}{2^1 1!}(1-x^2)^{0/2}\frac{d^1}{dx^1}(x^2-1)^1 = x$$

$$l=1, m=-1, \quad p_1^{-1}(x) = (-1)^1 \frac{(1-1)!}{(1+1)!}\left[-(1-x^2)^{1/2}\right]$$

$$l=1, m=-1, \quad p_1^{-1}(x) = \frac{1}{2}(1-x^2)^{1/2}$$

$$l=2, m=1, \quad p_2^1(x) = \frac{(-1)^1}{2^2 2!}(1-x^2)^{1/2}\frac{d^3}{dx^3}(x^2-1)^2$$

$$l=2, m=1, \quad p_2^1(x) = -3x(1-x^2)^{1/2}$$

$$l=2, m=2, \quad p_2^2(x) = 3(1-x^2)$$

$$l=2, m=0, \quad p_2^0(x) = \tfrac{1}{2}(3x^2-1)$$

The plots of the above polynomials are shown in Fig. 4.1. The first few associated Legendre functions in radian unit are:

$$P_1^1(x) = -\sin\theta$$

$$P_2^1(x) = -3\cos\theta\sin\theta$$

$$P_2^2(x) = 3\sin^2\theta$$

By using the Frobenius method, the recurrence formula for r = 0 is (see exercise section):

$$a_{n+2} = a_n\left[\frac{n^2+(2m+1)n-l(l+1)+m(m+1)}{(n+1)(n+2)}\right]$$

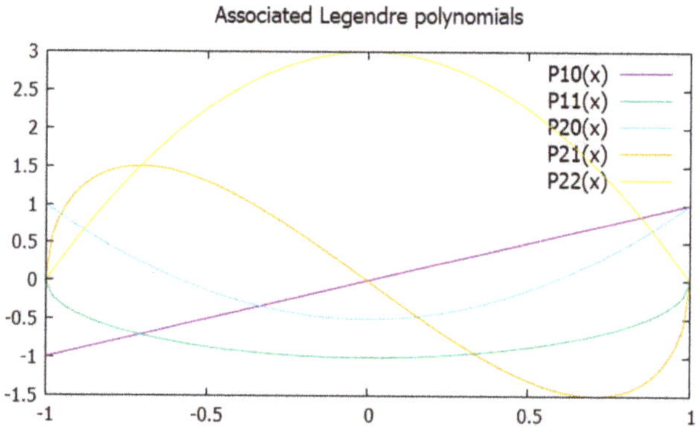

Fig. 4.1: Plots of some associated Legendre polynomials.

The recurrence formula leads to a power series that is divergent at –1 and +1. To avoid the divergence, the numerator in the recurrence formula must become zero for some non-negative integer n. We can verify that a zero numerator is obtained for $n = l - m$, a condition that is only met if l is an integer at least as large as m. Then, the regular solutions to the associated Legendre equation depend on the integer indices l and m.

The orthogonality relation for the associated Legendre functions is:

$$\int_{-1}^{1} P_l^m(x) P_k^m(x) dx = \frac{2}{2l+1} \frac{(l+m)!}{(l-m)!} \delta_{lk}, \delta_{lk} \begin{cases} = 0 \therefore l = k \\ = 1 \therefore l \neq k \end{cases}$$

Where δ_{lk} is the Kronecker delta. The proof of the orthogonality relation for the associated Legendre functions is given below in seven steps.

(1) Let k ≥ l and let us use the Rodrigues' formula to represent the associated Legendre functions in the previous equation:

$$K_{kl}^m = \int_{-1}^{1} \left\{ \begin{array}{l} \left[\frac{(-1)^m}{2^k k!} (1-x^2)^{m/2} \frac{d^{k+m}}{dx^{k+m}} (x^2-1)^k \right] \\ \left[\frac{(-1)^m}{2^l l!} (1-x^2)^{m/2} \frac{d^{l+m}}{dx^{l+m}} (x^2-1)^l \right] \end{array} \right\} dx$$

$$K_{kl}^m = \frac{1}{2^{k+l} k! l!} \int_{-1}^{1} \left\{ \left[(1-x^2)^m \frac{d^{k+m}}{dx^{k+m}} (x^2-1)^k \right] \left[\frac{d^{l+m}}{dx^{l+m}} (x^2-1)^l \right] \right\} dx$$

(2) Let us use the integration by parts l+m times:

$$\int_{-1}^{1} uv' dx = uv\Big|_{-1}^{1} - \int_{-1}^{1} vu' dx$$

$$u = (1-x^2)^m \frac{d^{k+m}}{dx^{k+m}} (x^2-1)^k$$

$$v' = \frac{d^{l+m}}{dx^{l+m}} (x^2-1)^l$$

At each integration by parts, the product uv vanishes because $(1 - x^2)$ and $(x^2 - 1)$ in the interval $[-1,1]$ will be zero.

$$\int_{-1}^{1} uv'dx = uv\Big|_{-1}^{1} - \int_{-1}^{1} vu'dx$$

$$\left[\left(1-x^2\right)^m \frac{d^{k+m}}{dx^{k+m}}\left(x^2-1\right)^k\right]\left[\frac{d^{l+m}}{dx^{l+m}}\left(x^2-1\right)^l\right]\Bigg|_{-1}^{1} = 0$$

$$\left(1-x^2\right)^m\Big|_{-1}^{1} = 0 \therefore \left(x^2-1\right)^l\Big|_{-1}^{1} = 0$$

Then, the first term of the integration by parts is zero. As a consequence, the integration becomes:

$$K_{kl}^{m} = \frac{(-1)^{l+m}}{2^{k+l}\,k!\,l!}\int_{-1}^{1}\left(x^2-1\right)^l \frac{d^{l+m}}{dx^{l+m}}\left[\left(1-x^2\right)^m \frac{d^{k+m}}{dx^{k+m}}\left(x^2-1\right)^k\right]dx$$

(3) By knowing that general Leibniz rule states that:

$$(uv)^{(n)} = \sum_{j=0}^{n}\binom{n}{j}u^{(n-j)}v^{(j)}$$

$$\binom{n}{j} = \frac{n!}{j!(n-j)!}$$

We use this rule in the second part of the previous integration:

$$\frac{d^{l+m}}{dx^{l+m}}\left[\left(1-x^2\right)^m \frac{d^{k+m}}{dx^{k+m}}\left(x^2-1\right)^k\right] =$$

$$= \sum_{j=0}^{l+m}\binom{l+m}{j}\frac{d^j}{dx^j}\left(\left(1-x^2\right)^m\right)\frac{d^{l+k+2m-j}}{dx^{l+k+2m-j}}\left(\left(x^2-1\right)^k\right)$$

(4) The leftmost derivative in the sum above is non-zero only when $j \le 2m$ (recall that $m \le l$) and the rightmost derivative is non-zero only when $k + l + 2m - j \le 2k$ or when $j \ge 2m + l - k$. Since $l \ge k$, these non-zero derivatives occur only when $j = 2m$ and $l = k$. Then, we have:

$$K_{kl}^{m} = (-1)^l\,\delta_{kl}\,\frac{(-1)^{l+m}}{2^{2l}(l!)^2}\binom{l+m}{2m}\int_{-1}^{1}\left\{\begin{matrix}\left(x^2-1\right)^l \dfrac{d^{2m}}{dx^{2m}}\left[\left(1-x^2\right)^m\right] \\ \dfrac{d^{2l}}{dx^{2l}}\left[\left(1-x^2\right)^l\right]\end{matrix}\right\}dx$$

The factor $(-1)^l$ comes from switching the factor $(x^2 - 1)^k$ into $(1 - x^2)^l$.

(5) By knowing the general expression of the binomial theorem:

$$(x+y)^k = \sum_{j=0}^{k} \binom{k}{j} x^j y^{k-j} = \sum_{j=0}^{k} \binom{k}{j} x^{k-j} y^j =$$

Let us now expand the term $(1 - x^2)^k$ using this theorem:

$$\left(1-x^2\right)^k = \sum_{j=0}^{k} \binom{k}{j} (-1)^{k-j} x^{2(k-j)}$$

The only term that survives the differentiation 2^k times is the x^{2k} term, which after differentiation gives:

$$(-1)^k \binom{k}{0} 2k! = (-1)^k (2k)!$$

Then, the integration becomes:

$$K_{kl}^m = (-1)^l \delta_{kl} \frac{(-1)^l (-1)^{l+m}}{2^{2l} (l!)^2} (2l)! \binom{l+m}{2m} \int_{-1}^{1} \left(x^2-1\right)^l dx$$

$$K_{kl}^m = (-1)^l \delta_{kl} \frac{1}{2^{2l} (l!)^2} (2l)! \frac{(l+m)!}{(l-m)!} \int_{-1}^{1} \left(x^2-1\right)^l dx$$

(6) The integration of $(x^2 - 1)^l$ gives:

$$\int_{-1}^{1} \left(x^2-1\right)^l dx = (-1)^l \frac{2^{2l+1} (l!)^2}{(2l+1)!}$$

(7) Then, we have:

$$K_{kl}^m = (-1)^l \delta_{kl} \frac{1}{2^{2l} (l!)^2} (2l)! \frac{(l+m)!}{(l-m)!} (-1)^l \frac{2^{2l+1} (l!)^2}{(2l+1)!}$$

$$K_{kl}^m = \delta_{kl} \frac{2}{2l+1} \frac{(l+m)!}{(l-m)!}$$

The associated Legendre equation is used for the solution of the spherical harmonics for the particle in a sphere (chapter sixteen) and the angle part of the hydrogen atom (chapter seventeen).

13. Partial differential equation and separation of variables

As stated in the first section of this chapter, there are two types of differential equations: ordinary and partial. When a differential equation has two or more independent variables, then it is called partial differential equation.

When a partial differential can be reduced to a set of ordinary differential equations, the solutions of the partial differential equation are the product of the solutions of the ordinary equations. This is the method of separation of variables. Let us take the following example:

$$\frac{\partial f(x,y)}{\partial x}-t\frac{\partial^2 f(x,y)}{\partial y^2}=0$$

Where the function *f*(x,y) is a product in which the dependence on x and y is separated.

$$f(x,y)=X(x)Y(y)$$

Let us substitute the above equation in the partial differential equation.

$$\frac{\partial\left[X(x)Y(y)\right]}{\partial x}-t\frac{\partial^2\left[X(x)Y(y)\right]}{\partial y^2}=0$$

$$Y(y)X'(x)-tX(x)Y''(y)=0$$

$$\frac{Y''(y)}{Y(y)}=\frac{X'(x)}{tX(x)}$$

Both sides of the above equation are equal to the same constant –λ.

$$X'(x)=-\lambda tX(x)$$

$$Y''(y)=-\lambda Y(y)$$

The equations above are eigenequations and –λ is the eigenvalue. They have the following general solutions:

$$X(x)=Ae^{-\lambda tx}$$

$$Y(x)=B\sin\left(\sqrt{\lambda}x\right)+C\cos\left(\sqrt{\lambda}x\right)$$

$$f(x,y)=\left(Ae^{-\lambda tx}\right)\left(B\sin\left(\sqrt{\lambda}x\right)+C\cos\left(\sqrt{\lambda}x\right)\right)$$

14. Undimensionalization of a differential equation

The process of turning a dimensional differential equation into a dimensionless differential equation has some advantages. It simplifies the equation and it becomes independent of the system of units used or the type of coordinates. In the undimensionalization, physical quantities such as mass, length and time which have [M], [L] and [T] dimensions, respectively, become dimensionless numbers without units, i.e. they acquire the unit dimension [1].

Let us take the equation of motion of an object launched vertically from the surface of earth (of radius R) with a determined initial velocity, V_0, and initial height y = 0.

$$\frac{d^2 y}{dt^2}=-\frac{gR^2}{(y+R)^2}$$

Where t is the time and g is the gravitational acceleration.

Let us separate the variables (i.e., the independent variables) and parameters (i.e., the dependent variables) of the equation.

Variables/dimensions	Parameters/dimensions
y [L]	g [L/T^2]
t [T]	R [L]
	V_0 [L/T]

Let us express the time variable in terms of the parameters V_0 and R in order to obtain the dimensionless τ (dimensionless time).

$$\tau = \frac{t}{R/V_0} = \frac{tV_0}{R}, \quad \left[\frac{T\left(\frac{L}{T}\right)}{L}\right] = [1]$$

$$t = \frac{\tau R}{V_0}$$

Let us now undimensionalize the height, y, using the Y quantity. See below.

$$Y = \frac{y}{b} = \frac{y}{R}, \quad \left[\frac{L}{L}\right] = [1]$$

$$y = YR$$

Then, let us put t and y in terms of their dimensionless quantities, τ and Y, respectively in the differential equation of motion.

$$\frac{d^2(YR)}{d\left(\frac{R}{V_0}\tau\right)^2} = -\frac{gR^2}{(YR+R)^2}$$

$$\frac{d^2(YR)}{d\left(\frac{R}{V_0}\tau\right)^2} = -\frac{gR^2}{(YR+R)^2} = -\frac{gR^2}{(YR)^2+R^2+2YR^2} = -\frac{gR^2}{R^2(Y+1)^2}$$

$$\frac{d^2(YR)}{d\left(\frac{R^2}{(V_0)^2}\tau^2\right)} = -\frac{g}{(Y+1)^2}, \quad R = const., V_0 = const.$$

$$\frac{(V_0)^2}{R^2} \times R\frac{d^2Y}{d\tau^2} = \frac{V_0^2}{R}\frac{d^2Y}{d\tau^2}$$

$$\frac{V_0^2}{R}\frac{d^2Y}{d\tau^2} = -\frac{g}{(Y+1)^2}$$

$$\frac{d^2Y}{d\tau^2} = -\frac{gR/V_0^2}{(Y+1)^2}$$

$$\frac{d^2Y}{d\tau^2} = -\frac{\alpha}{(Y+1)^2}, \qquad \alpha = gR/V_0^2 = const.$$

Then, the dimensionless equation of motion becomes:

$$\frac{d^2Y}{d\tau^2} = -\frac{\alpha}{(Y+1)^2}$$

$$\alpha = gR/V_0^2, \quad Y = y/R, \quad \tau = V_0 t/R$$

15. Numerical solutions of first-order differential equations

In many cases of differential equations, it is possible to transform a higher order differential equation into a set of first order differential equations (see introduction section). It is also important to know the boundary conditions of this equation and the interval of the independent variable, e.g., variable x from 0 to 1 [0,1].

(i) The Euler's method

The Euler's method is used for first order differential equation of the type:

$$\frac{dy}{dx} = f(x, y)$$

where the left side of the equation is substituted by the numerical approximation of a derivative.

$$\frac{dy}{dx} = f'(x) = \frac{f(x+h) - f(x)}{h} + O(h)$$

Where O(h) is the numerical error of the equation.

By replacing $f(x+h)$ and $f(x)$ into y_{n+1} and y_n, respectively, and we consider the first differential equation at the point x_n.

$$\frac{y_{n+1} - y_n}{h} + O(h) = f(x_n, y_n)$$

Which gives:

$$y_{n+1} = y_n + hf(x_n, y_n) + O(h^2)$$

One can see that the numerical error decreases to $O(h^2)$, which is the local error. By taking n steps for integrating the above equation in recursive procedure, leads to a higher inaccuracy (a higher global error, O(h)). To tackle this problem is needed small h and large number of steps in the summation procedure.

In the Euler's algorithm, the recursive process begins at the boundary conditions, for example, for x = 0 and known y_0, and goes to another boundary at x_1, for example, x = 3. The value of x and number of steps, N, is given below. By using different values of h, the smallest h gives the smallest Δx (the smallest interval between x_n and x_{n+1}), and, as a consequence, the highest number of steps. *By slicing up smaller and smaller the intervals between x_n and x_{n+1}, and, as a consequence, the intervals between y_n and y_{n+1}, it guarantees a higher precision in the numeric calculation of the differential equation.*

$$Boundary: x_0, x_1$$

$$x_n = i * h, \quad i = x_0, N, \quad N = x_1/h$$

$$y_0 \Leftrightarrow x_0$$

$$y_{n+1} = y_n + hf(x_n, y_n) + nO(h^2)$$

$$nO(h^2) \approx O(h)$$

In this case, as the number of steps, N, increases, the error decreases. In addition, as the h decreases, the error decreases as well. Notice that Y_{n+1} is associated with X + h, i.e., X_{n+1}.

For example, consider the differential equation and its boundary condition below:

$$\frac{dy}{dx} = -xy, \quad y(0) = 1$$

Whose solution is:

$$\frac{dy}{y} = -xdx, \quad \int \frac{dy}{y} = -\int xdx$$

$$\ln y = -x^2/2, \quad y = e^{-x^2/2}$$

Let us use the equation below to find the numerical solution in the interval [0,3]

$$Boundary: x_0 = 0, y_0 = 0, x_1 = 3$$

$$x_n = i * h, \quad i = 0, N, \quad N = 3/h$$

$$y_{n+1} = y_n + h(-x_n \cdot y_n)$$

Table 4.1 shows the Euler's method for h = 0.5 and h = 0.05 for different values of Δx. The Δx gives the inverse relation with the number of steps. For h = 0.5, the Y_{n+1} reaches zero in a much shorter interval than that for h = 0.05. When the Y_{n+1} reaches zero, the algorithm becomes useless for the next values of X because they will give only zero values for Y_{n+1}.

Table 4.1: Euler's method for dy/dx = –xy according to different values of h (0.5 and 0.05) and Δx.

X_n	Y_n	$h*(-X_n*Y_n)$	X_{n+1}	Y_{n+1}	error
		h = 0.05 Δx = 1.0			
0	1	0	**1.0**	**1**	**0.3935**
1.0	1.0000	–0.0500	2.0	0.9500	0.8147
2.0	0.9500	–0.0950	3.0	0.8550	0.8439
3.0	0.8550	–0.1283	4.0	0.7268	0.7264
		h = 0.05 Δx = 05.0			
0	1	0	0.50	1	0.1175
0.50	1.0000	–0.0250	**1.00**	**0.9750**	**0.3685**
1.00	0.9750	–0.0488	1.50	0.9263	0.6016
1.50	0.9263	–0.0695	2.00	0.8568	0.7214
		h = 0.05 Δx = 0.25			
0	1	0	0.25	1	0.0308
0.25	1.0000	–0.0125	0.50	0.9875	0.1050
0.50	0.9875	–0.0247	0.75	0.9628	0.2080
0.75	0.9628	–0.0361	**1.00**	**0.9267**	**0.3202**
1.00	0.9267	–0.0463	1.25	0.8804	0.4225
		h = 0.5 Δx = 1.0			
0.0	1.0000	0.0000	**1.0**	**1.0000**	**0.3935**
1.0	1.0000	–0.5000	2.0	0.5000	0.3647
2.0	0.5000	–0.5000	**3.0**	**0.0000**	–0.0111
3.0	0.0000	0.0000	**4.0**	**0.0000**	–0.0003
		h = 0.5 Δx = 0.5			
0	1	0	0.50	1	0.1175
0.50	1.0000	–0.2500	**1.00**	**0.7500**	**0.1435**
1.00	0.7500	–0.3750	1.50	0.3750	0.0503
1.50	0.3750	–0.2813	2.00	0.0938	-0.0416
2.00	0.0938	–0.0938	**2.50**	**0.0000**	-0.0439
		h = 0.5 Δx = 0.25			
0	1	0	0.25	1	0.0308
0.25	1.0000	–0.1250	0.50	0.8750	–0.0075
0.50	0.8750	–0.2188	0.75	0.6563	–0.0986
0.75	0.6563	–0.2461	**1.00**	**0.4102**	**–0.1964**
1.00	0.4102	–0.2051	1.25	0.2051	–0.2528

Next, we introduce the code for the Euler´s method using the previous example.

```
!Program name: EULER
!Euler's method for differential equation dy/dx=-xy
!The exact solution: y=exp(-x^2/2)
!Interval X=0 to X=3 where y(0)=1
!y(n+1)=y(n)+h*(-xy)
!Y(n+1) <=> X+h
program euler
real :: x, y, h, exact, error
integer :: i
func(x,y)=-x*y
h=0.1
nstep=3./h ! 30 steps
y=1. !y(0)=1
do i=0,nstep
  x=i*h !It guarantees interval from 0 to 3
y=y+h*func(x,y)
!Y(n+1)=Y(n)+ h* f(x,y)
exact=exp((-(x+h)**2)/2)
  error=exact-y
  print *, i, x+h, y, exact, error
!x+h is associated with Yn+1. Then x+h = Xn+1
!for h=0.1 y(1)=0.6281 error=0.0216
!for h=0.1 y(3)= 0.00803 error=0.00308
!the error decreases as x increases
end do
Print *,
!smaller h and higher number of steps gives small error
h=0.05
!number of steps increases as h decreases
nstep=3./h !60 steps
y=1. !y(0)=1
do i=0,nstep
  x=i*h !It guarantees interval from 0 to 3
y=y+h*func(x,y)
!Y(n+1)=sum Y(n)+ h*sum f(x,y)
exact=exp((-(x+h)**2)/2)
  error=exact-y
  print *, i, x+h, y,exact, error
!for h=0.05 y(1)=0.6169 error=0.0104
!for h=0.05 y(3)=0.0094 error=0.0016
!the error decreases as x increases
!the error decreases as h decreases
end do
```

```
print *,
y=1.
h=0.1
do i=0,10 !10 steps
  x=i
  y=y+h*func(x,y)
!Y(n+1)=Y(n)+ h*f(x,y)
exact=exp((-(x+h)**2)/2)
  error=exact-y
  print *, i, x+h, y, exact, error
!for h=0.1 Y(1)=0.8999 error=-0.3529 (bad!)
!for h=0.1 Y(3)=0.5039 error=-0.4958 (bad!)
!the error decreases as x increases
end do
print *,
y=1.
h=0.05
do i=0,10 !10 steps
  x=i
  y=y+h*func(x,y)
!Y(n+1)=Y(n)+ h*f(x,y)
exact=exp((-(x+h)**2)/2)
  error=exact-y
  print *, i, x+h, y, exact, error
!for h=0.05 Y(1)=0.9499 error=-0.3737 (bad!)
!for h=0.05 Y(3)=0.7267 error=-0.7172(bad!)
!Smaller number of steps gives higher error!
!the error decreases as x increases
end do
end program euler
```

The Euler's method has low-order accuracy and it demands very low value of h and a huge number of steps. Then, integration with higher-order accuracy is preferable.

(ii) Taylor series

The Taylor series (see Chapter Two) is one order more accurate than Euler's method, but it is more useful when the function is known analytically. The Taylor series expansion for y_{n+1} around y_n is:

$$y_{n+1} = y(x_n + h) = y_n + hy'_n + \tfrac{1}{2}h^2 y''_n + O(h^3)$$

By knowing that for first differential equation, we have:

$$\frac{dy}{dx} = f(x, y), \qquad \frac{y_{n+1} - y_n}{h} + O(h) = f(x_n, y_n)$$

and:

$$y''_n = \frac{d}{dx} f(x_n, y_n), \qquad f = f(x_n, y_n)$$

Since f is a two-variable function, we use partial derivative:

$$y''_n = \frac{df}{dx} = \frac{\partial f}{\partial x} + \frac{\partial f}{\partial y}\frac{dy}{dx}$$

$$y_{n+1} = y_n + hf + \tfrac{1}{2}h^2\left[\frac{\partial f}{\partial x} + f\frac{\partial f}{\partial y}\right] + O(h^3)$$

Where f and its derivatives are evaluated at (x_n, y_n).

The same process of slicing up the intervals between x_n and x_{n+1}, and the intervals between y_n and y_{n+1} into small deltas, by using small h, is used in Taylor series algorithm.

$$Boundary: x_0, y_0, x_1$$

$$x_n = i * h, \qquad i = x_0, N, \qquad N = x_1/h$$

$$y_{n+1} = y_n + hf(x_n, y_n) + \tfrac{1}{2}h^2\left[\frac{\partial f}{\partial x} + f(x_n, y_n)\frac{\partial f}{\partial y}\right]$$

Let us use the Taylor series algorithm for the differential equation and its boundary condition below (the same used in the previous section):

$$\frac{dy}{dx} = -xy, \qquad y(0) = 1$$

```
!Program name: TAYLOR
!Taylor method for differential equation dy/dx=-xy
!The exact solution: y=exp(-x^2/2)
!Interval X=0 to X=3 where y(0)=1
!y(n+1)= y(n)+h*(-xy)+(h^2)/2*[-y+(x^2)*y)
!Y(n+1) <=> X+h
program taylor
real :: x, y, h, exact, error
integer :: i
h=0.1
nstep=3./h ! 30 steps
y=1. !y(0)=1
do i=0,nstep
  x=i*h !It guarantees interval from 0 to 3
y=y+h*(-x*y)+(0.5*(h**2))*(-y+(x**2)*y)
  exact=exp((-(x+h)**2)/2)
```

```
error=exact-y
  print *, i, x+h, y, exact, error
!x+h is associated with Yn+1. Then x+h = Xn+1
!for h=0.1 y(1)=0.60526 error=0.00163
!for h=0.1 y(3)= 0.01256 error=-0.00015
end do
Print *,
h=0.05
nstep=3./h !60 steps
y=1. !y(0)=1
do i=0,nstep
  x=i*h !It guarantees interval from 0 to 3
y=y+h*(-x*y)+(0.5*(h**2))*(-y+(x**2)*y)
  exact=exp((-(x+h)**2)/2)
error=exact-y
  print *, i, x+h, y,exact, error
!for h=0.05 y(1)=0.60621 error=0.000317
!for h=0.05 y(3)=0.011143 error=-0.000034
end do
stop
end
```

By comparing the results for h = 0.05 between Taylor and Euler methods, one ca see that Taylor method yields a more accurate result.

Euler's method for h = 0.05
y(1) = 0.6169 error = 0.0104
y(3) = 0.0094 error = 0.0016

Taylor method for h =0.05
y(1) = 0.60621 error = 0.000317
y(3) = 0.011143 error = –0.000034

(iii) Runge-Kutta method

The equation for second order Runge-Kutta algorithm has local error $O(h^3)$ but global error (in the recursive procedure) $O(h^2)$.

$$y_{n+1} = y_n + hf\left(x_n + \tfrac{1}{2}h, y_n + \tfrac{1}{2}k\right) + O(h^3)$$
$$k = hf(x_n, y_n)$$

The fourth-order Runge-Kutta algorithm is:

$$y_{n+1} = y_n + \tfrac{1}{6}\left(k_1 + 2k_2 + 2k_3 + k_4\right) + O(h^5)$$
$$k_1 = hf(x_n, y_n)$$
$$k_2 = hf(x_n + \tfrac{1}{2}h, y_n + \tfrac{1}{2}k_1)$$
$$k_3 = hf(x_n + \tfrac{1}{2}h, y_n + \tfrac{1}{2}k_2)$$
$$k_4 = hf(x_n + h, y_n + k_3)$$

Next, it follows the code for the second-order and fourth-order Runge-Kutta method using the same example in the previous sections. One can see that the error from second-order for y(3) is similar to Euler's method. On the other hand, there is nearly no detectable error (in single precision) when using fourth-order Runge-Kutta method. Then, fourth-order Runge-Kutta is the best method for the numerical differentiation besides the fact that there is no derivative operation in its equation as it happens in the Taylor series method.

```
!Program name: RK
!Runge-Kutta method for differential equation dy/dx=-xy
!The exact solution: y=exp(-x^2/2)
!Interval X=0 to X=3 where y(0)=1
!Y(n+1) <=> X+h
program Runge_Kutta
real :: x, y, h, exact, error
Real :: k1, k2, k3, k4
integer :: i
f(x,y)=-x*y
!fourth order Runge-Kutta
h=0.05
nstep=3./h !60 steps
y=1. !y(0)=1
do i=0,nstep
  x=h*i
  k1=h*f(x,y)
  k2=h*f(x+0.5*h,y+0.5*k1)
  k3=h*f(x+0.5*h,y+0.5*k2)
  k4=h*f(x+h,y+k3)
y=y+(1/6.*(k1+2.*k2+2.*k3+k4))
  exact=exp((-(x+h)**2)/2)
  error=exact-y
  print *, i, x+h, y,exact, error
!for h=0.05 y(1)=0.60653 error=-0.00000
!for h=0.05 y(3)=0.00111 error=-7.63E-08
end do
Print *,
!Second order RungeKutta
h=0.05
nstep=3./h !60 steps
y=1. !y(0)=1
do i=0,nstep
```

```
  x=i*h !It guarantees interval from 0 to 3
k=h*f(x,y)
  y=y+h*f(x+0.5*h,y+0.5*k)
exact=exp((-(x+h)**2)/2)
  error=exact-y
  print *, i, x+h, y,exact, error
!for h=0.05 y(1)=0.60137 error=--0.00515
!for h=0.05 y(3)=0.00871 error=-0.00240
end do
stop
end
```

Do-it-yourself experience

Use GNUPLOT to generate the graphs of the associated Legendre polynomials $P_{1,0}$; $P_{1,1}$; $P_{2,0}$; $P_{2,1}$ and $P_{2,2}$.

Use the following statements:

```
Set title "Associated Legendre polynomials" font ",14"
Set xtics font ",13"
Set ytics font ",13"
Set key font ",14"
Set xrange [-1:1]
P10(x)=x
P11(x)= -sqrt(1-x**2)
P21(x)=-3*x*sqrt(1-x**2)
P22(x)=3*(1-x**2)
P20(x)=0.5*(3*x**2-1)
Plot P10(x), P11(x), P20(x), P21(x), P22(x)
```

Exercises:

(1) Replace P, P' and P'' in the associated Legendre equation in order to obtain (see Section 11):

$$\left(1-x^2\right)p''-2x\left(m+1\right)p'+\left[l(l+1)-m(m+1)\right]p=0$$

(2) Use Frobenius' method to obtain the recurrence formula of the associated Legendre equation below (see section 11), assuming the solution in the series:

$$p=\sum_{n}^{\infty}a_j x^{r+n}$$

The indicial equation for the associated Legendre equation has solutions for r = 1 and r = 0. For r = 0, recurrence formula is:

$$a_{n+2}=a_n\left[\frac{n^2+(2m+1)n-l(l+1)+m(m+1)}{(n+1)(n+2)}\right]$$

(3) By knowing that x = cosθ, find the associated Legendre functions in polar angle for (a) l = 1, m = 1; (b) l = 1; m = –1; (c) l = 2, m = 1.

(4) Derive the polynomials of the associated Legendre functions $P_l^m(x)$ for (a) l = 2, m = 0; (b) l = 2; m= –1; (c) l = 2, m = 2.

(5) By knowing that the Schrödinger equation of the one-particle quantum harmonic oscillator is:

$$\left(-\frac{\hbar^2}{2m}\frac{d^2}{dx^2}+2\pi^2 m v^2 x^2\right)\psi(x)=E\psi(x)$$

And that it has two dimensional physical quantities (energy, E, and length, x), undimensionalize the above equation.

Tip: Table the unit (according to the international system of units) of the following physical quantities: Energy, angular frequency, length, mass and Planck's constant.

Part Two

Old Quantum Mechanics, Matrix Mechanics and Wave Mechanics

Absorption/Emission Spectroscopy and Spectral Lines

5

1. General overview of the network influence in quantum mechanics

The history of quantum mechanics can be divided into two parts. From 1850's to 1900's, Kirchhoff, Bunsen, Boltzmann, Lorentz, Planck, Einstein, Rydberg, Zeeman, Stark, Rayleigh, Thomson, Rutherford, among others, developed the first most important experiments and the first theoretical studies that led to the recognition of quantum states. During 1910's and 1920's, Bohr, Sommerfeld, Kramer, Born, Pauli, Heisenberg, Jordan, de Broglie, Dirac and Schrödinger developed the theoretical basis of the quantum mechanics.

The Fig. 5.1 shows the most prominent scientists which developed (or helped to develop) the quantum mechanics and the influence network among them along the time. The chronology starts from the top of figure and ends at the bottom. The most prominent characters of quantum mechanics history are Bohr, Heisenberg, Schrodinger and Dirac. Bohr, Heisenberg and Schrödinger developed the old quantum mechanics, matrix mechanics and wave mechanics, respectively. Dirac was the main responsible for the amalgamation of matrix and wave mechanics.

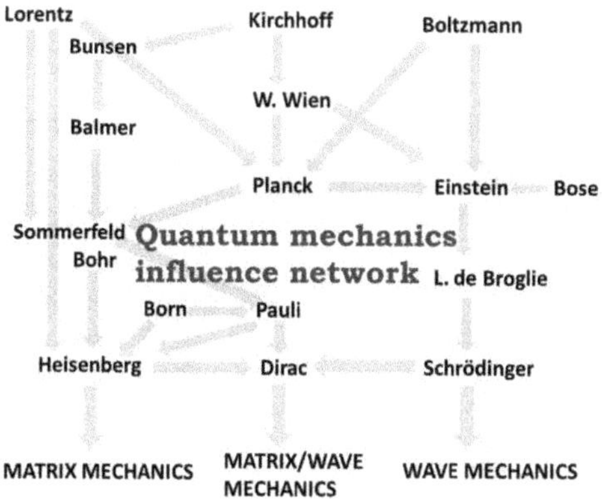

Fig. 5.1: Quantum mechanics network influence.

In the next chapters we will present the most important issues of the history and the mathematical development of the quantum mechanics which will partly follow the chronology of the Fig. 5.1.

2. The origin of the absorption/emission spectroscopy

Joseph Fraunhofer was a great expert in making homogenous glass in order to obtain homogeneous light (containing a single color) for the studies in optics science. He also observed that the light of a flame between red and yellow did not decompose in the prism, producing a homogeneous light. However, he pursued for the homogeneous light of every color which led him to develop an optical apparatus for this proposition (James 1981). In addition, he invented the modern spectroscope and objective telescope which enabled him to determine the refractive index of several types of glass (Fraunhofer 1817). He also detected the dark fixed lines of the solar spectrum (Fig. 5.2) and from other stars, having 574 different dark fixed lines (James 1981). Four decades later, Kirchhoff explained that the dark fixed lines were the atomic absorption lines (Kirchhoff and Bunsen 1862). Likewise Fraunhofer, Herschel succeeded in obtaining homogeneous light and described the spectra from several different flames produced by salts of several metals (Herschel 1822). Talbot was the first to use the light for chemical analysis in 1826 whose work was reprinted in 1861 (Talbot 1861) in which he also obtained the line spectrum of his studied substances. Talbot and Herschel were the first to observe that some metals of a salt have a particular line spectrum, then Talbot suggested that flame spectrum could be used for the identification of chemical substances. In the mid 1940s, Miller and Daniell were the first to analyze the spectra of various gases and they also found differences in the line spectrum of iodine and bromine as pure substances and a new possibility for the chemical analysis had definitely been born (James 1981).

In 1835, by applying the electromagnetic spark (magnetism inducing electricity) in which a voltaic battery is applied between two charcoal poles and a cup of mercury at the end of the wire, Wheatstone produced a light that when analyzed by

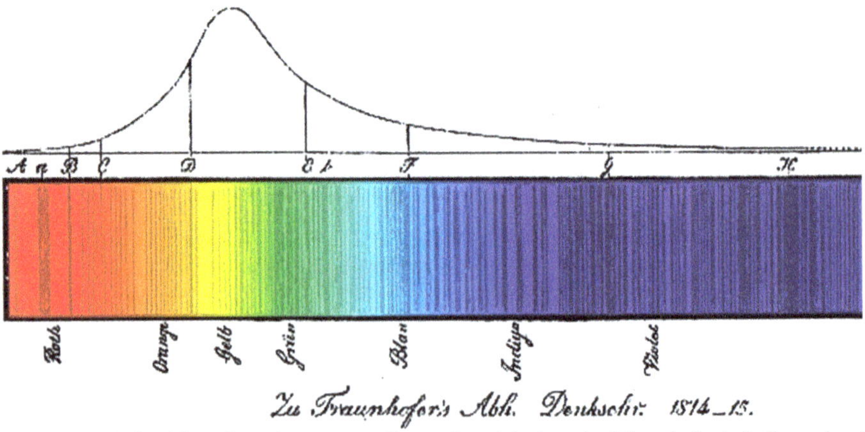

Fig. 5.2: Dark fixed lines from the sun according to the original work of Fraunhofer including colored scheme adaptation. Image from Twitter account of the HOPOS Journal in March, 10th, 2017. (see acknowledgment section).

a spectroscope from the surface of mercury have originated the line spectrum of the mercury (few definite lines of light separated by very wide dark intervals). He also obtained the line spectra of fluid volatile metals such as zinc, cadmium, bismuth, tin and lead by using the same electromagnetic spark equipment. His work had a reprint in 1861 (Wheatstone 1861). Later, he observed that the light from burning the metals was different from the light produced by voltaic spark, probably from the difference of temperature.

By investigating the emission spectra of some hydrocarbons, Swan obtained identical spectral lines for these compounds and observed that the lines were caused by individual atoms of carbon and hydrogen (Swan 1857). In 1858, Cartmell, one of Bunsen's student, observed that some substances when heated in the 'Bunsen burner' produced characteristic colors (Cartmell 1858).

Probably inspired by Brewster's work, Kirchhoff suggested Bunsen to use the light spectrum apparatus (prism spectroscope) including Bunsen burner (having low intrinsic luminosity and hot flame) for the chemical analysis (Fig. 5.3). Kirchhoff and Bunsen obtained the spectral lines of several metals and their salts and they concluded that within a wide range of conditions every chemical element has always unique spectral lines (Kirchhoff and Bunsen 1860). They also observed that the sodium spectral lines were similar to those from the sun (Kirchhoff and Bunsen 1862). Kirchhoff himself had already observed that the vapors of sodium emit the D-lines of the solar spectrum (yellow spectrum) and that the D-lines of the candle light were caused by the presence of sodium (Kirchhoff 1860). Kirchhoff also stated that the rays at a specific wavelength and temperature have the same ratio of emissive power (the amount of emitted radiation from the plate) and absorptive power (the amount of radiation incident on the plate) knows as the Kirchhoff's radiation law (Kirchhoff 1859).

3. The prism spectroscope and the grating spectroscope

The spectral lines can be obtained from prism spectroscope (as a result of a dispersion phenomenon) and from grating spectroscope (as a result of a diffraction phenomenon).

The Kirchhoff spectroscope (Fig. 5.3) uses a glass prism that refracts the light into different wavelengths (monochromatic beams) and different angles. The index of refraction depends on the wavelength and, as a consequence, each wavelength has a specific bend angle (the angle which the light is refracted) yielding the spectrum of thin lines of light. This phenomenon is called dispersion of the light (a wave that

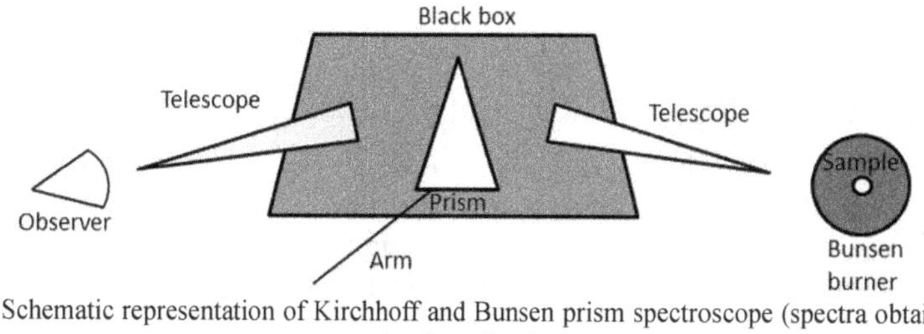

Fig. 5.3: Schematic representation of Kirchhoff and Bunsen prism spectroscope (spectra obtained from a prism by refraction).

spreads out as it passes through a dispersive medium). Due to the nature of the prism (a dispersive element), the light is spread out into monochromatic beams having specific angle according to corresponding wavelength of the monochromatic beam.

According to the Fig. 5.3, the telescope close to the light source (a sample in a Bunsen burner or a vapor lamp) has a collimating lens that aligns the thin strip of the light which then passes through a prism undergoing two refractions. Thereafter, the monochromatic beams pass through a second telescope which magnifies the final image.

The grating spectroscope, on the other hand, gets monochromatic beams of the light from diffraction phenomenon. The diffraction spreads out the light as it passes through narrow, parallel slits. The diffraction gratings have a periodic structure of narrowly spaced slits (transmissive or reflective gratings) that diffracts the light into several monochromatic beams. The diffraction depends linearly on the spacing of the grating and the wavelength of the monochromatic beam. The diffraction grating is less expensive (e.g., CD and DVD having finely-spaced tracks); easier to calibrate; and has a higher resolution than a dispersive element such a prism.

By supposing that the set of slits in the grating has the spacing d wider than the wavelength λ in order to its monochromatic light to be diffracted, then each point in the slit scatters the light in all directions. When the same monochromatic light of the same wavelength λ is diffracted in a slit below the former it causes constructive and destructive interference. The two rays will interfere constructively (producing bright beam) if the extra path distance (d.sin θ, where θ is the angle between the incident beam and the observer direction) is an integer of the wavelength.

$$d \sin \theta = m\lambda$$

Where m is an integer (m = 0, 1, 2, ...). The incident light made up of a spectrum of distinguished monochromatic waves, each single wave will have a specific θ angle for its brightest diffracted beam. Then, the wavelengths of each monochromatic light can be determined by measuring its specific angle of highest constructive interference.

4. Hydrogen spectral lines, Balmer and Rydberg series

Masson used voltaic spark with his own spectroscope to observe the emission of the air. Then, he replaced air by hydrogen gas and observed its emission for the first time (Masson 1851). Masson mentioned that the spark hydrogen emission is always dim, red light similar to that from thin air. Later, he observed that the hydrogen spectrum has four bright lines (red, two green and one blue) separated by dark bands (Masson 1855).

Subsequently, Angstrom reported a study of the spark spectra of hydrogen (besides other several elements) and depicted two spectra of the hydrogen containing four distinguished lines (Fig. 5.4). In his own words: “Remarkable, in the case of the hydrogen, are the strongly luminous and wide lines at the red end of the spectrum, which, moreover, besides a feeble line in the vicinity of the former, contains only two bright portions, one at the limit of blue and green, and the other in the extreme blue” (Angstrom 1855).

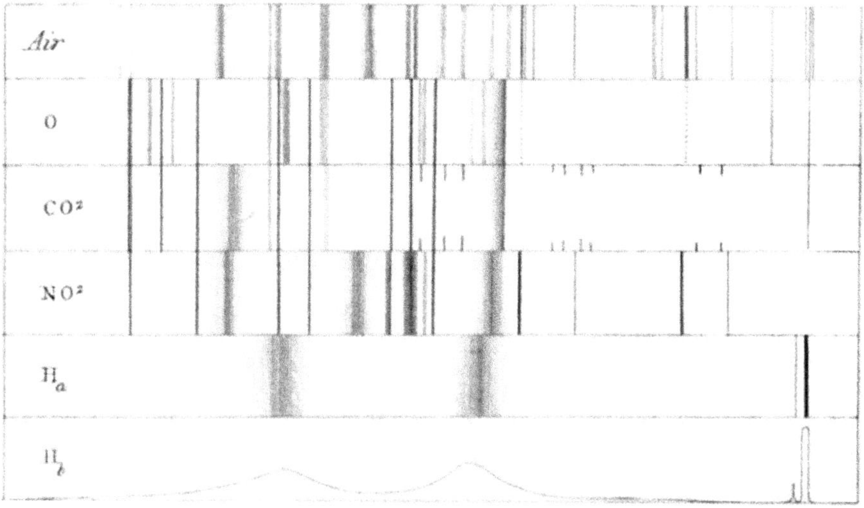

Fig. 5.4: Angstrom's spectral lines of air, O, CO_2, NO_2, and H (Angstrom 1855). The H_b represents the luminous intensity of the different parts of the hydrogen spectrum, in an approximate manner. See Acknowledgement section.

Based on the exact numbers of Angstrom's experimental work for the hydrogen spectrum (whose values are 6562.10; 4860.74; 4340.1; and 4101.2), Balmer observed that the four lines of the hydrogen spectrum have the following relation: 9/5; 4/3; 25/21; 9/8 to one another and they have a common factor h = 3645.6 mm/10^7 in the conversion to the corresponding wavelength. By multiplying both numerator and denominator of the second and fourth ratios (which does not change their results) gives: 9/5; 16/12; 25/21; 36/32. The numerators are 3^2, 4^2, 5^2 and 6^2, respectively, being represented by m^2. Then, the fractions are:

$$\frac{m^2}{(m^2 - n^2)}$$

Where n = 2, for this set of ratios (for the hydrogen spectrum). This is known as the Balmer series (Balmer 1885). Then, the wavelengths are given by:

$$\lambda = A\left(\frac{m^2}{m^2 - n^2}\right), \qquad n = 2$$

where A= 364.56 × 10^{-9}. This equation has a precision to the second or first decimal place.

The values of the Balmer lines (H-alfa, H-beta, H-gamma, and so on) are: 656.3 nm, 486.1 nm, 434.0 nm, 410.2 nm, 397.0 nm and 364.6 nm. The first four are the visible part of the hydrogen spectrum. The Balmer lines refer to the transitions from the second shell of the hydrogen atom (n = 2) to the subsequent levels (m = 3, 4, 5 and 6). See Chapter seven for more information.

The Balmer formula can be rewritten as:

$$\frac{1}{\lambda} = R\left(\frac{1}{4} - \frac{1}{n^2}\right)$$

Where R is the 4/A and it is called the Rydberg constant.

A few years later, Rydberg himself did a comprehensive study of the emission spectra of chemical elements (Rydberg 1890). Rydberg expressed a more general formula in terms of wavenumber (or frequency), not wavelength (as Balmer did).

$$\nu_n = \frac{R}{(n_1 + \mu_1)^2} - \frac{R}{(n_2 + \mu_2)^2}$$

Where ν_n is the frequency of the nth member of the series, and μ denotes the spectral terms. When $\mu_1 = 0$, $\mu_2 = 0$, $n_1 = 2$ (or n = 2) and n_2 = (or m =) 3,4,5,... Rydberg's formula reduces to Balmer's series. Rydberg was the first to distinguish between sharp series (S) and diffuse series (D). Other types of series discovered later were principal series (P) and fundamental series (F). They form the four chief series (S, P, D, F).

5. Zeeman effect and Lorentz model

Before Zeeman, other scientists tried to obtain the influence of magnetism on the spectral lines of a substance. Michael Faraday demonstrated that the plane of the polarized light rotates when placed in a magnetic field (Faraday 1846), but he failed "to detect any change in the lines of the spectrum of a flame when the flame was acted on by a powerful magnet", as announced by Maxwell (Arabatzis 1992).

In his doctoral research, Zeeman studied the reflection of the polarized light on the magnetized surface (known as the Kerr effect). Three years after finishing his doctoral thesis, he managed to circumvent the experimental difficulties to analyze the magnetic influence on the spectral lines of the light after some fruitless early attempts. As Zeeman said "In consequence of my measurements of Kerr's magneto-optical phenomena, the thought occurred to me whether the period of the light emitted by a flame might be altered when the flame is acted upon by the magnetic force" (Zeeman 1897).

The equipment used in Zeeman experiment was made up with Rühmkorff electromagnet (with magnetizing current), a Rowland grating and a Bunsen burner. The flame of the Bunsen heated a recipient (a tube closed at both ends of asbestos) containing impregnated common salt. This tube was placed horizontally between the poles of the electromagnet at the right angle of the magnetic force. A burner was placed between the poles of the electromagnet and was heated by the flame of the Bunsen burner. When the electromagnet was off, there were two narrow, sharp D-lines of the sodium spectrum. When the electromagnet was on the two D-lines were broadened. When replacing the Bunsen burner by a flame of oxyhydrogen, the D-lines became three or four times wider. The same result was obtained when replacing sodium by lithium salt (Arabatzis 1992). In Fig. 5.5(A), there is pictorial representation of a normal Zeeman effect with characteristic three lines.

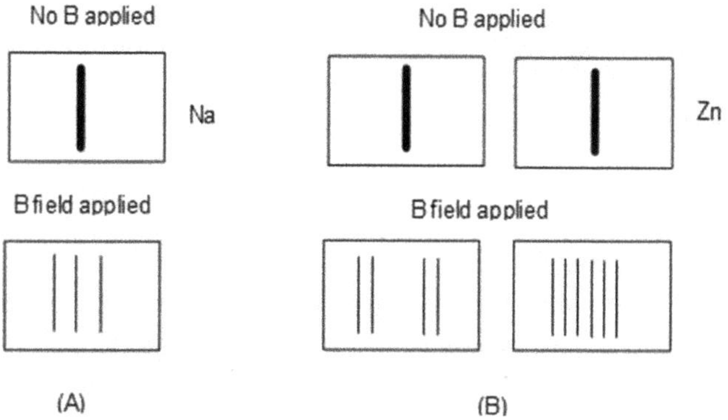

Fig. 5.5: Pictorial representation of the spectral lines of: (A) the normal Zeeman effect of zinc and; (B) the anomalous Zeeman effect in sodium.

The Zeeman experiment became known as Zeeman effect—the splitting of the spectral lines of light (from emission spectroscopy of a determined substance) under the influence of a static magnetic field. The spectral lines are split into three components in a magnetic field: one polarized parallel to the magnetic field and two perpendicular to it. This triplet is called the normal Zeeman effect which could be observed for zinc, copper, cadmium and others. The low resolving power of Zeeman's experiment did not allow him to observe duplets-triplets of sodium. He could only observe that the doublet of sodium (without magnetic field) was widened when the (weak) magnetic field was applied.

Lorentz had been working on electrodynamics since 1892 (according to Lorentz, the oscillations of the charged particles in the atom were the source of light) and right after the discovery of the Zeeman effect, Lorentz gave its classical theoretical interpretation. He explained that atoms contain charged particles called 'ions' (later called electrons) harmonically bound to a center. The frequencies of their vibrations correspond to the frequencies of the spectral lines of the analyzed substance. When a magnetic field is applied, the vibrating particles will experience a Lorentz force in addition to the harmonic force. The Lorentz force F describes the force acting on a particle of charge q moving with a velocity v in an electric field E and a magnetic field B experiences a force:

$$F = qE + qv \times B$$

Instead of one rotational frequency ω_0 of the electrons in the absence of a magnetic field, three frequencies appear when the magnetic field is applied. For a field H applied in z-direction, the equations of motion in x-direction and y-direction include the applied field H along with the harmonic force (–kx and –ky):

$$m\frac{d^2x}{dt^2} = -kx + \frac{eH}{c}\frac{dy}{dt}$$

$$m\frac{d^2y}{dt^2} = -ky - \frac{eH}{c}\frac{dx}{dt}$$

$$m\frac{d^2y}{dt^2} = -kz$$

The solutions of the equations give three rotational frequencies ω_0, ω_1, and ω_2 as a consequence of the splitting of the line spectra under the influence of the magnetic field (Kox 1997).

It is most probable that Lorentz's theory to elucidate the Zeeman effect was the basis for Planck's theory on the thermal black body, Slater's virtual oscillator model and the Heisenberg's matrix mechanics, that is, an electron is harmonically bound to the nucleus of the atom, following the equation of motion of a harmonic oscillator according to Hooke's law (F = –kr).

On the other hand, the Zeeman effect was also important to reinforce the quantized nature of the particles under the influence of the magnetic field since its result proves that the splitting of the spectrum is discrete—a quantization phenomenon—and not a continuum. The splitting occurs by the torque of magnetic field B on the magnetic dipole, $\mu_{orbital}$, which is associated with the orbital angular momentum, L.

$$\mu_{orbital} = -\frac{e}{2m_e}L$$

Where m_e is the electron mass.

One year after Zeeman's discovery, by applying a strong magnetic field, Preston discovered the so-called anomalous Zeeman effect (Preston 1898). He observed the two D lines of sodium (without B applied) were split into a quadruplet and a sextuplet (see Fig. 5.5(B)).

Zeeman effect and anomalous Zeeman effect played a very important role in spectroscopy in the early 20th century. Paschen, Landé, Sommerfeld among others interpreted theoretically their experimental results and they were one of the driving forces to the transition from old quantum theory to the modern quantum theory.

6. Stark effect

The Stark effect is similar to Zeeman effect where the applied field changes from magnetic to electric. Then, the Stark effect is the splitting of the spectral lines of light (from emission spectroscopy of a determined substance) under the influence of an electric field. No classical explanation could account for this effect. Bohr and coworkers were the first to rationalize theoretically to this phenomenon using the principle of correspondence (see Chapter seven). Afterwards, Schrödinger used a new theoretical approach to calculate the splitting of the spectral lines from the Stark effect (Schrödinger 1926).

7. Do-it-yourself activity: home-made spectroscope

Build your own grating spectroscope. The necessary material for your home-made spectroscope is: shoe box, cellophane tape, toilet paper roll, CD or DVD, aluminum foil, two razor blades, black marker, black paint, utility knife, ruler and glue.

(1) Assume that this box has a rectangular shape with four lateral parts (two smaller ones and two bigger ones), one bottom and one top parts.

(2) Paint in black color the inner part of shoe box to improve the resolution of the spectroscope;

(3) Place the CD/DVD outside the box against the edge of one smaller lateral part of the box and draw the inner circle of the CD/DVD in the box;

(4) Place the toilet paper roll over the circle you have just drawn and trace another circle. There will be two concentric circles (Fig. 5.6(A));

(5) Move the tube about 1.5 cm in the direction of the other side of the box (not to the top neither to the bottom) and draw another circle;

(6) Cut out the shape made up with the two bigger circles;

(7) Place the toilet paper roll halfway through the hole you have just made at nearly 60 degrees (see Fig. 5.6(A));

(8) Fix it with cellophane tape and seal around the tube and all lateral part of the box (where the tube is located) with aluminum foil to block out the light;

(9) Turn the box counter-clockwise to one of its bigger lateral parts and place the CD/DVD farthest apart from the fixed tube at the other side. Draw another inner circle of the CD/DVD in the box;

(10) Draw a rectangle passing by the circle having 1.5 cm wide and 5 cm tall and cut out this rectangle;

(11) Tape both razor blades together on the box so that their sharp edges almost touch each other in order to make a very narrow slit in the box. If the light is dim, the blades are too close and if the spectrum is blurry, the blades are too far away.

(12) Tape the CD/DVD inside the box opposite the razor blade slit (Fig. 5.6(B));

Obs.: steps 10 and 11 can be replaced by a single step where one makes a narrow slit with a knife (Fig. 5.6(A)).

(13) Tape the open, top part of the box and seal the whole box with aluminum tape to block out the light (Fig. 5.6(C));

(14) The sample spectra might be all commercial types of light sources (light bulb, fluorescent lamp, halogen lamp, neon lamp, metal-halide lamp, etc.) and

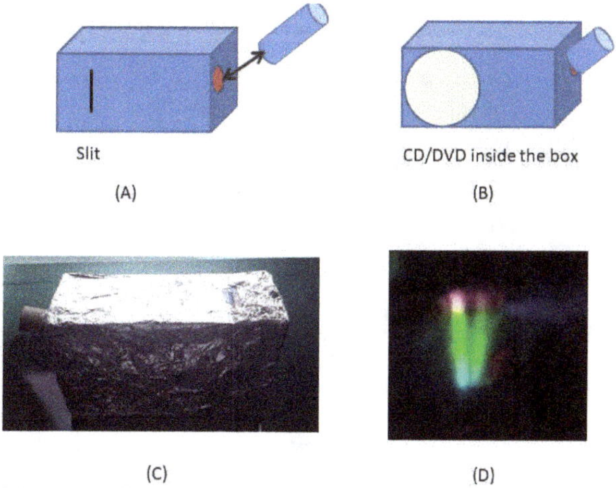

Fig. 5.6: (A) and (B) Schematic representations of the home made spectroscope; (C) photograph of the home-made spectroscope; (D) photograph of fluorescent light spectrum from the home made spectroscope.

sunlight. Figure 5.6(D) shows the spectrum from fluorescent lamp. Tip: put the light source close to the slit.

Acknowledgment section

We thank HOPOS The Journal of the International Society for the History of Philosophy of Science. We thank Francis & Taylor Group for the permission of publishing Fig. 2.3 from A. J. Angström (1855) XLVIII. Optical researches, The London, Edinburgh, and Dublin Philosophical Magazine and Journal of Science, 9: 60, 327–342, DOI: 10.1080/14786445508641880'.

References cited

Angstrom, A.J. 1855. Optical researches. Phil. Mag. 9: 327–342.
Arabatziz, T. 1992. The Discovery of the Zeeman effect: A case study of the interplay between theory and experiment. Stud. Hist. Phil. Sci. 23: 365–388.
Balmer, J.J. 1885. Notiz uber die spectrallinien des wassertoffs. Ann. Phys. Chem. 25: 80–87.
Cartmell, R. 1858. On a photochemical method of recognizing the non-volatile alkalies and alkaline earths. Phil. Mag. 16: 328–333.
Faraday, M. 1846. On the magnetization of light and the illumination of magnetic lines of force. Phil Trans. Roy. Soc. 136: 1–20.
Fraunhofer, J. 1817. Bestimmung des brechungs und des farben-zerstreuungs vermogens verschiedener glasarten, in bezug auf die vervokkkommnung achromatischer fernrohre. Ann. Phys. 56: 264–313.
Herschel, J.F.W. 1822. On the absorption of light. Trans. Roy. Soc. Edinb. 9: 446.
James, F.A.J.L. 1981. The early development of spectroscopy and astrophysics. PhD thesis. Department of History of Science and Technology, University of London.
Kirchhoff, G. 1859. Uber den Zusammenhang zwischen Emission und Absorption von Licht und Warme. Monatsber. Kon. Preuss. Akad. Wiss. Berlin 783–787.
Kirschhoff, G. 1860. Ueber die Fraunhofer'schen linien. Ann. Phys. 185: 148–150.
Kirchhoff, G. and Bunsen, R. 1860. Chemische analyse durch spectralbeobachtungen. Ann. Phys. 186: 161–189.
Kirchhoff, G. and Bunsen, R. 1862. Untersuchungen uber das sonnenspektrum und die spektren der chemischen elemente. Abhandlungen Kon. Preuss. Akad. Wiss. Berlin 74: 63–95.
Kox, A.J. 1997. The discovery of the electron: II. The Zeeman effect. Eur. J. Phys. 139–144.
Masson, M.A. 1851. Études de photométrie électrique. De la nature de l'étincelle életrique et de sa cause. Ann. Chim. et Phys. 31: 295–326.
Masson, M.A. 1855. Études de photométrie électrique. De l'influencce exercée par les pôles de l'étincelle électrique sur les raies brillantes du spectre. Ann. Chim. et Phys. 45: 385–454.
Preston, T. 1898. Radiation phenomena in a strong magnetic field. Scientific. Trans. Roy. Dublin Soc. 6: 385–389.
Rydberg, J.R. 1890. Recherches sur la constitution des spectres d'emission des éléments chimiques Kunglia vetenskapsakademiens handlingar 23: 1–177.
Schrödinger, E. 1926. Quantisierung als Eigenwertproblem. Ann. Phys. 385: 437–490.
Swan, W. 1857. On the prismatic spectra of the flames of compounds of carbon and hydrogen. Trans. Roy. Soc. Edinb. 21: 411–429.
Talbot, H.F. 1861. Researches on the spectra of artificial light from different sources. Experiments on coloured flames. Chem. News J. Phys. Sci. 3: 261–261.
Wheatstone, C. 1861. On the prismatic decomposition of the electric, voltaic and electro-magnetic sparks. Chem. News. J. Phys. Sci. 3: 198–201.
Zeeman, P. 1897. The effect of magnetisation on the nature of light emitted by a substance. Nature 55: 347.

Black-body Radiation, Einstein and Planck's Law

6

1. The black-body radiation and spectral radiance

In the mid 19th century, it was well understood that when any object is heated it would gain energy and it would emit light from red to white as the temperature increased. This is the behavior of the cavity radiation where the radiation is trapped inside a cavity of the black-body material with a small pinhole such as closed cylinders made of porcelain and platinum (Baggott 2011). A black-body radiation is an evacuated box containing thermal electromagnetic radiation in thermal equilibrium with the walls of the box. The radiation is emitted by an idealized opaque, non-reflective body (in a perfectly insulated enclosure)—the black-body box. It absorbs all light outside (by its outside walls that isolate the black-body from external light) and it emits all light spectrum from its inner, heated walls (a thermal emission) through the small hole (Fig. 6.1). The black-body will emit radiation (having all wavelengths present in a continuous spectra, unlike absorption and emission spectra) through a hole small enough to avoid disturbance upon the equilibrium. The radiation is characterized by its spectral density or spectral radiance that is the energy of radiation at a given frequency or wavelength.

According to Maxwell's electromagnetism, the energy U of electromagnetic waves in a cavity is:

$$U = \frac{1}{2}\iiint\left(\varepsilon_0 |E|^2 + \mu_0 |B|^2\right)d\tau$$

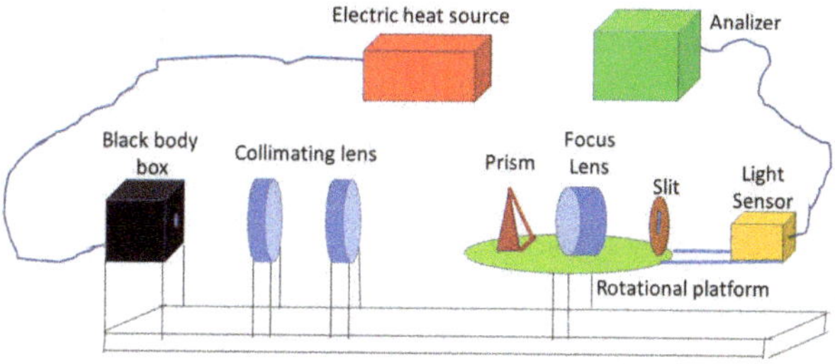

Fig. 6.1: Black-body radiation modern apparatus.

Where E and B are the electric and magnetic field vectors, ε_0 is the vacuum permittivity and μ_0 is the vacuum permeability. The electromagnetic cavity is an empty container for the electromagnetic fields containing photon's wave function inside.

In 1860, Kirchhoff made a theoretical study for designing the black-body experiment and its related laws. The black-body material consisted of two black screens S1 and S2 containing openings 1 and 2, respectively, whose dimensions are infinitely small compared to their separation distance and a body C with a black covering whose S1 forms one of its walls (Fig. 6.2).

Kirchhoff stated that: "all bodies emit rays, the quality and intensity of which depend on the nature and temperature of the bodies themselves." He proposed the experiment with the black-body material where its nature does not influence on the nature of the emitted radiation and its temperature is kept constant due to the covering impermeable to heat (like a perfectly reflecting surface). Since the temperature of the body is constant, the intensity of the incident rays (which are entirely absorbed) is equal to that of the emitted rays. Kirchhoff coined the term 'black body' as "infinitely small thicknesses, completely absorbs all incident rays, and neither reflect nor transmit any" which was used to investigate the radiating power of the bodies (Kirchhoff 1860). For every value of frequency ν there is a relation below:

$$\frac{E_\nu}{A_\nu} = J(\nu, \theta)$$

Where E_ν is the radiating power of the body (the amount of energy emitted by the body in the frequency ν from S1 to S2), A_ν is the power of absorption of the body for the frequency ν (from rays coming from S2 to S1), and J is the radiating power emitted by the black-body material (the recipient C in Fig. 6.2). The magnitudes of A and E depend only on the temperature θ and frequency ν. This ratio is independent of the nature of the body.

Kirchhoff stated: "A body placed within a covering whose temperature is the same as its own, is unaffected by radiation, and must therefore absorb as many rays as it emits. Hence, it has long been concluded that, at the same temperature, the ratio between the radiating and absorbing powers of all bodies is the same—it being, however, assumed that bodies only emit rays of one kind" (Kirchhoff 1860).

Rayleigh (John William Strutt—discoverer of argon element along with William Ramsay) used the term "complete radiation" to describe "the radiation from an ideally black body, which according to Stewart and Kirchhoff is a definite function of the absolute temperature, θ, and the wave-length, λ" in the expression $\theta^5\phi(\theta\lambda)d\lambda$, where ϕ is an arbitrary function of the single variable θλ (Rayleigh 1898, 1900).

In 1879, based on the measurements of physicists Dulong and Petit (Petit and Dulong 1819), Stefan stated empirically that the total energy density (the energy

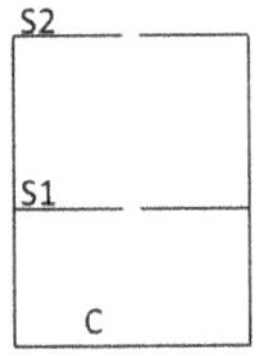

Fig. 6.2: Kirchhoff's designing of the black-body material.

radiated per unit surface area) from a black-body (or the radiated power density in Wm^{-2}), integrated over all frequencies, is proportional to the fourth power of the temperature:

$$E = \int_{0}^{\infty} Q(v,T) = \sigma T^4$$

Where $\sigma = 5.66 \times 10^{-8}\ Wm^{-2}K^{-4}$ (Stefan 1879).

However, Stefan's equation is not true for all situations. Boltzmann, a Stefan's former student, derived theoretically Stefan's equation. He applied the second law of thermodynamics to the radiation by treating it as a gas whose pressure was the radiation pressure of the electromagnetic theory and he obtained a general equation which became known as Stefan-Boltzmann law (Boltzmann 1884). Boltzmann used the electromagnetic theory of Maxwell for a black-body where

$$J(v,\theta) = \left(\frac{c}{8\pi}\right)\rho(v,\theta)$$

In the above equation, $\rho(v, \theta)$ is the energy density per unit volume for the frequency v. The energy density is the amount of energy stored per unit volume. Then, Boltzmann found (Pais 1979):

$$E(\theta) = V\int \rho(v,\theta)dv = \sigma V T^4$$

Where V is the volume of the cavity.

In a general form of the Stefan-Boltzmann law, the total thermal energy radiated by an object is given by

$$E = A\varepsilon\sigma T^4$$

Where A is the surface area of the object and ε is the emissivity (which is a unit for a black body).

In 1893, Wilhelm Wien developed the relationship between maximum wavelength of the black-body radiation and the absolute temperature, T, of the black-body radiation:

$$T\lambda_{max} = b$$

where b is a constant (Wien's constant, b = 2897.77 μm.K). This relationship shows that each radiation curve will peak at different wavelength (Wien 1893), known as Wien's displacement law.

In his paper entitled "on the energy distribution in the emission spectrum of a black-body", Wein was the first to propose the law of the distribution of the radiant energy (or spectral radiance), ρ, which was generally accepted at that time (Wien 1896).

$$\rho_\lambda = \frac{C}{\lambda^5}e^{-\frac{c}{\theta\lambda}}$$

Where C and c are constants, and ρ_λ is the spectral radiance. The above equation is known as Wien's energy distribution law.

One year later, by using detection in near-infrared range, Paschen proved the correctness of Wien's law (Pais 1979). However, three years later, when using a detector for larger extent of wavelength, Wien's law lost its experimental support (see next section).

Rayleigh developed the law of distribution of the radiant energy (or spectral radiance) in the expression

$$\theta\lambda^{-4}d\lambda$$

which belongs to the Rayleigh-Jeans expression (Rayleigh 1900, Jeans 1905a). In 1900, before Planck's law, Rayleigh applied Maxwell-Boltzmann partition energy to develop the relation:

$$\rho = cv^2T$$

Where c is a constant which Einstein later discovered yielding the equation of the spectral distribution, the energy per unit volume in a frequency ν (Pais 1979):

$$\rho(v,\theta) = \frac{8\pi v^2}{c^3}\frac{R}{N}T$$

Jeans stated that: "The law of radiation from a perfect radiation is of the form $\lambda^{-4}\, Tf(\lambda T)d\lambda$ so that Stefan's law and Wien's displacement law are accurately obeyed by the radiation from this ideal radiator" (Jeans 1905b).

For a perpendicular radiated energy, the radiated power per unit wavelength, $B_\lambda(T)$, i.e., the spectral radiance, is given by:

$$B_\lambda(T) = \frac{2\pi ckT}{\lambda^4}$$

Where k is the Boltzmann constant and $B_\lambda(T)$ is the Rayleigh-Jeans' energy distribution law. See in the next chapter the Rayleigh-Einstein-Jeans law.

The above equation is known as Rayleigh-Jeans law as a function of the wavelength. It produces good agreement in the low frequency (large wavelength) limit, but for high frequencies leads to the ultraviolet catastrophe.

2. The black-body experiment

There was a long period (of 40 years) from Kirchhoff theorem and the proper experiment on the black-body radiation due to the difficulty in building a perfect black-body and a proper device to detect the radiation to a large extent of the wavelength.

The black-body emits a light spectrum whose spectral radiance is dependent upon the temperature. The spectral radiance is the radiance (radiant flux emitted, reflected, transmitted or received by a given surface, per unit solid angle, in steradian unit, per unit projected area) of a surface per unit wavelength (or frequency). The unit of spectral radiance is watt per steradian per square meter, per meter ($W.sr^{-1}.m^{-2}.m^{-1}$). At room temperature, the black-body emits only infrared radiation. As the temperature increases, the black-body starts to emit visible light changing its color from red to blue. When it is white, it is emitting ultraviolet radiation. At each temperature, there is a distinguished energy distribution among different wave lengths.

The Fig. 6.3(A) and (B) depict the spectral radiance obtained from the author's own data using Planck's law. At each temperature, there is a specific spectral radiance curve. As the temperature increases, the maximum intensity increases and the wavelength of the maximum radiance, λ_{max}, tends to decrease. At lower temperatures (from 500 to 1000 K in Fig. 6.3(A)), the range of the higher radiance (10^5 to 10^9 W $sr^{-1}m^{-2}\mu m^{-1}$) oscillates from 1500 to 8000 nm. At higher temperatures (from 3000 to 5000 K in Fig. 6.3(B)), the range of higher radiance (10^9 to 10^{13} W $sr^{-1}m^{-2}\mu m^{-1}$) oscillates from 250 to 2000 nm. Then, at higher temperatures, the material emits visible light (380–740 nm).

The experimental investigations of the black-body by Lummer, Pringsheim, Rubens and Kurlbaum at Physikalisch-Technische Reichsanstalt laboratory in Berlin led to the full spectrum of the cavity radiation since they detected the radiation in far infrared (Pais 1979). Their results confirmed that the energy distribution law in the normal spectrum from Wien (Wien 1896) was valid only for short wavelengths and low temperatures. As to long wavelengths only the Rayleigh-Jeans curve adequately described it (Fig. 6.4). Planck communicated this experimental finding in his second 1900 paper (Planck 1900a).

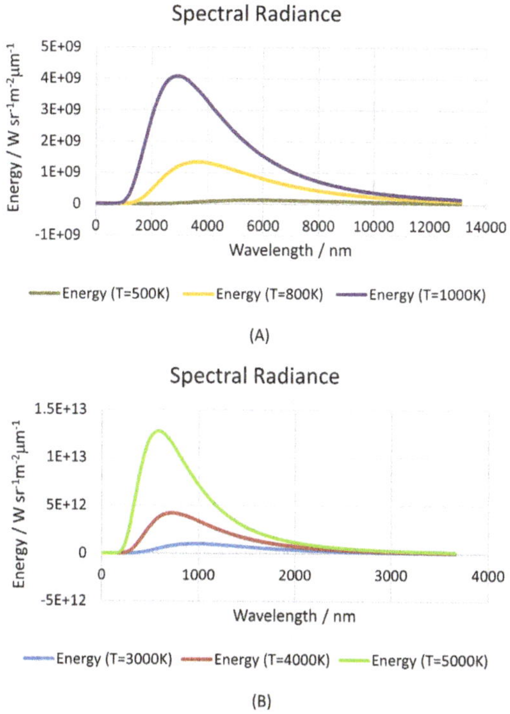

Fig. 6.3: Spectral radiance versus wavelength at (A) 500 to 1000 K and (B) 3000 to 5000 K from the author's own data.

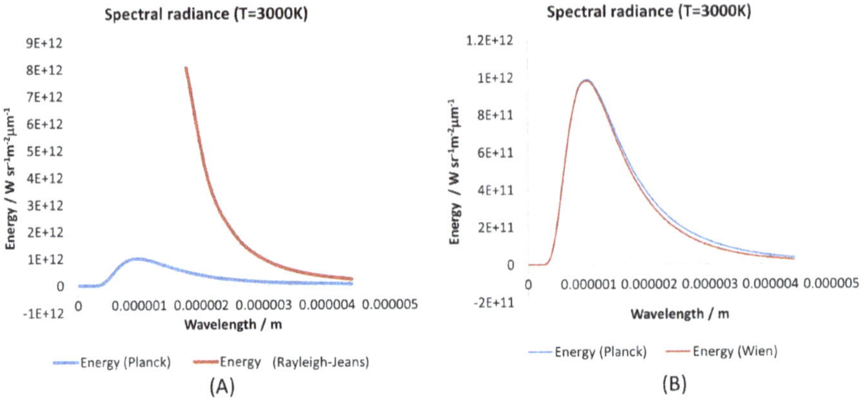

Fig. 6.4: Spectral radiance at 3000 K from (a) Planck's law and Rayleigh-Jeans' law and (b) Planck's law and Wien´s law from author's own data.

Important to add that it was Planck who included fundamental constants to the Wein's distribution law:

$$E = \frac{2hc^2}{\lambda^5} e^{-\frac{hc}{\lambda k \theta}}$$

3. Boltzmann and Planck's law

Before Planck began to work with black-body radiation he had devoted his career to the study of the second law of classical thermodynamics. In the study of the black-body radiation, he tried to find the basis for his idea of irreversibility for conservative system of the electromagnetic radiation in an enclosure with reflecting walls interacting with a collection of charged harmonic oscillators (a model to represent the experiment of the black box radiation). In another words, Planck aimed to prove the apparently irreversible alteration of the form of an electromagnetic wave from incident plane wave to outgoing spherical wave which approached the equilibrium state according to the second law of thermodynamics—the entropy increases monotonically in time (Planck 1897). The harmonic oscillators were chosen because of its realistic model of matter and Kirchhoff's assertion that the equilibrium radiation distribution was independent of the system with which the radiation interacted.

A harmonic oscillator is a system that is displaced from its equilibrium position experiencing a restoring force proportional to its displacement having sinusoidal oscillations about the equilibrium point. Simple mechanics examples are pendulums and masses connected to springs. The light has two sinusoidal oscillating fields (electric and magnetic fields) and it is a two-coupled harmonic oscillators. In Section 4 of the Chapter four we presented the solution for the classical harmonic oscillator. As we observed in Chapter five, it is probable that the Lorentz's theory (an electron is harmonically bound to the nucleus of the atom, following the equation of motion of a harmonic oscillator) used to elucidate the Zeeman effect was the basis for Planck's theory on the thermal black body.

Important to add that in the Section 6 of this chapter we will see another model to represent the black body radiation based on the equipartition law establishing that the energy of the "ether" is in equilibrium with that of the matter.

Planck tried to obtain a suitable relationship between the energy and the entropy of the oscillator where he devised the equation (Planck 1899):

$$S = -\frac{u}{\beta v}\ln\frac{u}{aev}$$

$$a = \alpha c^3/8\pi = 6.885\times10^{-27}\,erg\,\text{sec}$$

$$\beta = 0.4818\times10^{-10}\ \text{sec K}$$

Where a and β are constants obtained numerically by Planck (the same used in Wien distribution law), e is the base of the natural logarithms and u is the energy. Planck discussed his result saying that: "I believe that it must therefore be concluded that the definition given for the entropy of radiation, and also the Wien distribution law for radiation that goes with it, are necessary consequences of applying the principle of entropy increase to the electromagnetic theory of radiation" (Klein 1966).

Consider the equation of motion of a linear oscillator with a mass m and charge e interacting with a monochromatic periodic electric field of frequency ω whose energy is:

$$E = \frac{e^2F^2}{2m}\frac{1}{4\pi(v-\omega)^2+\gamma^2}$$

Where F/2 is the electric field energy density (Planck 1900b). Let F/2 be:

$$\frac{F}{2} = \frac{4\pi\rho(\omega,T)d\omega}{3}$$

Integrating over ω, we have:

$$U = \frac{4\pi e^2}{3m}\int\frac{\rho(\omega,T)d\omega}{4\pi(v-\omega)^2+\gamma^2}$$

Since γ is very small, the maximum response of the oscillator occurs when ω = ν. Extending the integration from −∞ to +∞, we have the spectral distribution equation related to the average energy U(ν,T):

$$\rho(v,T) = \frac{8\pi v^2}{c^3}U(v,T)$$

This average energy could be determined from dependence of the entropy S of the oscillator on its energy S(U). The spectral radiance (or distribution) ρ(ν,T) can be obtained from U(ν,T). Planck integrated the equation T(U)dS = dU to obtain the entropy of the linear oscillator:

$$S = k\left[\left(1+\frac{U}{hv}\right)\ln\left(1+\frac{U}{hv}\right)-\frac{U}{hv}\ln\frac{U}{hv}\right]$$

In an outstanding work of Boltzmann in 1884, he proved that Clausius entropy is equivalent to the statistical entropy (so-called permutability measure, Ω) times a constant at thermal equilibrium (Boltzmann 1877).

$$\int \frac{dQ}{T} = \frac{2}{3}\Omega$$

Besides the Wien's law, the Boltzmann's work on the entropy had a great influence on Planck's investigation of the spectral radiance. Important to mention that so far, Planck had neglected and criticized Boltzmann's statistical entropy but he had not found other way except for accepting Boltzmann's advice on the importance of the statistical entropy to produce a monotonic approach to equilibrium (Klein 1966). He tried to solve the problem of the equilibrium between matter and radiation without success for six years. Then, in an act of desperation, Planck calculated the thermodynamic probability of a state in which a certain energy was shared among the many oscillators of the same frequency.

Planck chose the inverse of the second derivative of the entropy, S, of the oscillator coupled with the field with respect to its energy, U. He found that a deep connection between entropy and energy was the pathway to find the expression for the spectrum radiance (Planck 1900c).

$$\frac{\partial^2 S}{\partial U^2} = -\frac{\alpha}{U}$$

Planck observed that the energy distribution law is according to the entropy of a linear resonator (device that produces waves of specific frequencies) which interacts with the radiation as a function of the vibrational energy U and stated that Wien's law follows the above equation where entropy is a function of (U/ν), where ν is the frequency of the resonator (Planck 1900c).

After the experimental results showing that Wien's law failed to predict spectrum radiance for long wavelengths, Planck constructed an arbitrary expression to satisfy thermodynamic and electromagnetic theories from Wien's thermodynamic derivation (Planck 1900c), to yield the expression below which could encompass all experimental observations (Planck 1900a).

$$\frac{\partial^2 S}{\partial \varepsilon^2} = -\frac{\alpha}{U(\beta + U)}$$

The integration of the above equation yields (Barraca 2005):

$$\frac{1}{T} = \frac{\partial S}{\partial U} = \left(\frac{\alpha}{\beta}\right) \ln\left[\left(\frac{\beta + U}{U}\right)\right]$$

This equation, according to Planck, "lead to S as a logarithmic function of U—which is suggested from probability considerations—and which moreover for small values of U reduces to Wien's expression" (Planck 1900a). Henceforth, Planck obtained the spectral radiance formula, ρ, with two constants C and c:

$$\rho(\lambda,T)=\frac{C\lambda^{-5}}{e^{c/\lambda T}-1}$$

Later, Planck proposed that the entropy of the total amount of identical resonators (oscillators), in the same stationary radiation field, S_N, is proportional to the logarithm of its probability W and an arbitrary additive constant (Planck 1901). Consider a large number N of linear oscillators having frequency ν.

$$W=\frac{(N-1+P)!}{P!(N-1)!}$$

$$S_N=k\log W+const.$$

$$S_N=NS$$

$$U_N=NU$$

Where W is the number of P indistinguishable energy elements which can be distributed over N distinguishable oscillators; S is the average entropy of a single resonator; the total entropy S_N depends on the disorder with which the total energy U_N is distributed among the individual resonators; and U is the constant energy of a single resonator (Planck 1901). U_N is made up out of finite energy elements ε:

$$U_N=P\varepsilon$$

Planck added that for finding probability W so that N resonators have the vibrational energy U_N, is necessary to interpret it not as continuous quantity, but as discrete quantities composed of finite equal parts having energy element ε, where $U_N = P\varepsilon$ and P is an integral number.

Planck used the Boltzmann's idea to obtain the distribution of the P energy elements among the N resonators. By inserting W equation in $S_N = k\ln W$, using $P/N = U/\varepsilon$, $S_N = NS$ and applying the Stirling approximation he obtained the equation:

$$S_N=kN\left[\left(1+\frac{U}{\varepsilon}\right)\ln\left(1+\frac{U}{\varepsilon}\right)-\frac{U}{\varepsilon}\ln\frac{U}{\varepsilon}\right]$$

$$S=k\left[\left(1+\frac{U}{\varepsilon}\right)\ln\left(1+\frac{U}{\varepsilon}\right)-\frac{U}{\varepsilon}\ln\frac{U}{\varepsilon}\right]$$

By examining the Wien's displacement law in the form of Thiesen's work about the dependence of entropy of the resonator on its energy $\{\rho(\nu,\theta)=\nu^3 f(\nu/\theta)\}$, Planck obtained the entropy as a function of U/ν, where ν is the frequency of the resonator.

In a comparison among the previous equations of the entropy he found the expression that gives birth to the quantum theory:

$$\varepsilon=h\nu$$

And the spectral radiance in terms of frequency:

$$\rho(\nu,T)=\frac{8\pi h\nu^3}{c^3}\frac{1}{e^{h\nu/k\theta}-1} \quad or \quad \rho(\nu,T)=\frac{2h\nu^3}{c^2}\frac{1}{e^{h\nu/k\theta}-1}$$

Where h and k are universal constants calculated by Planck with the aid of some available measurements:

$$h=6.55\cdot 10^{-27}\,erg.\sec$$

$$k=1.346\cdot 10^{-16}\,erg/\deg$$

In terms of wavelength instead of frequency, the spectral radiance becomes:

$$\rho(\lambda,T)=\frac{8\pi ch}{\lambda^5}\frac{1}{e^{hc/k\lambda\theta}-1} \quad or \quad \rho(\lambda,T)=\frac{2\pi c^2}{\lambda^5}\frac{1}{e^{hc/k\lambda\theta}-1}$$

Although the success of Planck's discovery, the constant h still lacked fundamental significance until 1905, according to Jeans (Pais 1979). In 1906, Einstein gave the correct quantum definition to the Planck's law: "The energy of Planck oscillator can only take on values which are integral multiples of hν; in emission and absorption the energy of a Planck oscillator changes by jumps which are multiples of hν" (Pais 1979).

Twelve years after his famous paper, Planck postulated that Planck's constant, h, was the smallest phase space cell for the oscillators in the black-body radiation (Planck 1912). Similar result was reached by Sommerfeld in 1916 (see Chapter seven).

$$\iint dq\,dp=h$$

Where q and p are the coordinate and momentum of a one-dimensional harmonic oscillator.

4. Planck's pathway from classical towards statistical thermodynamics

Before Planck's law in 1901, there was a debate between the continuous energy/matter model and the statistical-mechanical model (composed of discrete matter and energy—an atomist view). Until early 1900, Planck opposed the atomist and probabilistic views in physics. Planck chose the cavity radiation to prove its classical view of the thermodynamics since it seemed to have no connection with atoms and molecules. Important to add that his doctoral thesis was on the second law of the classical thermodynamics—in which the entropy of a gas always increases until its equilibrium (Baggott 2011).

Boltzmann, the most prominent developer of the statistical mechanics, stated that the maximum entropy is related to the highest number of possible permutations according to the total energy of the gas. Boltzmann parceled up the continuous energy into 'energy elements', nε (n = 1,2,3...), so that he could count the number of molecules having each energy element in order to obtain the number of different possible permutations. Boltzmann's work played a decisive role in Planck´s law and the development of the quantum mechanics.

In 1900, Planck changed his mind with respect to the probabilistic view of the thermodynamics. In his second paper in 1900, Planck devised the general equation for the spectral radiance without any mention to the statistical mechanics (Planck 1900a). However, in his next paper, he surrendered to the Boltzmann's probabilistic arguments in order to find the correct expression for the spectral radiance (Planck 1901). He said: "I busied myself, from then on, that is, from the day of its establishment, with the task of elucidating a true physical character for the [new distribution law], and this problem led me automatically to a consideration of the connection between entropy and probability, that is, Boltzmann's trend of ideas". Planck took some weeks of the most strenuous work in order to convince himself that the mechanical statistics was the unique way to find his energy distribution law (Baggott 2011).

Planck used his expertise in entropy and the second law to derive the equation of the spectral radiance based on an expression for the entropy of an individual oscillator in terms of its internal energy and its frequency of oscillation. However, the mathematics led him to the direction he was avoiding: the probabilistic interpretation of the entropy. Thereafter, Planck became an enthusiast of the atomist view where "the energy is composed of a very definite number of equal finite packages" (Baggott 2011). The statistics of large collections of simple harmonic oscillators with equiprobable states in phase space was used until 1913 with the arrival of the Bohr's model in which physicists began to focus on individual atoms to explain their spectral lines with or without external (electric or magnetic) fields (see Chapter seven).

5. Einstein and Planck's constant

In his first paper on quantum theory in 1905, Einstein recognized that Planck's equation for the thermal radiation was in accordance with the experiment but he opposed the theory. Einstein wrote: "while we conceive of the state of a body as being completely determined by the positions and velocities of a very large but finite number of atoms and electrons, we use continuous spatial functions to determine the electromagnetic state of a space, so that a finite number of quantities cannot be considered as sufficient for the complete description of the electromagnetic state of a space. (…) In the following, we shall consider 'black-body radiation' in connection with experience without basing it on any model for the production and propagation of radiation" (Einstein 1905). It is important to emphasize Pais' sentence: "In 1900 Planck has discovered the black-body radiation law without using light-quanta. In 1905 Einstein discovered light-quanta without using Planck's law" (Pais 1979). Einstein's main criticisms to the Planck's law were that the U is the equilibrium energy of a one-dimensional harmonic oscillator, and the equation of the spectral radiance obtained by Einstein (see below) using Planck's law and the equipartition law (the partition of energy) of the classical statistical mechanics $U = k_bT$ is in disagreement with the experiment (Pais 1979).

$$\rho(\nu,\theta)=\frac{8\pi\nu^2}{c^3}\frac{R}{N}T=\frac{8\pi\nu^2}{c^3}k_bT$$

Where R is the universal gas constant and N is the number of real molecules in one gram-equivalent. Einstein said that the equipartition law was already known at that time for over three decades. Planck's omission on the use of equation $U = k_bT$ in its spectral radiance equation is a consequence of his negative attitude towards Boltzmann's statistical mechanics. However, due to the failure of the Rayleigh-Einstein-Jeans law (see below) to high frequencies, the second criticism of Einstein is only consistent with the lack of a theoretical basis in Planck's law.

Einstein recognized that Rayleigh-Jeans law is theoretically correct but in disagreement with the experiment of the black-body thermal radiation (Pais 1979). Actually, Einstein gave important contribution to Rayleigh-Jeans law in the same year when Rayleigh and Jeans found the value for the constant c. Then, Einstein found that Rayleigh-Jeans law should be written as:

$$\rho(\nu,\theta)=c\nu^2T, \quad \textit{Rayleigh}$$

$$\rho(\nu,\theta)=\frac{8\pi\nu^2}{c^3}\frac{R}{N}T, \quad \textit{Einstein}$$

Such contribution gave the proper name to that law: Rayleigh-Einstein-Jeans law (Pais 1979). However, the failure of the Rayleigh-Einstein-Jeans law propelled the success of Planck's law. Rayleigh said that "we must admit the failure of the law of equipartition in these extreme cases (high frequencies)". Jeans had a different view: "the equipartition law is correct but the supposition that the energy of the ether is in equilibrium with that of matter is utterly erroneous in the case of ether vibrations of short wavelength under experimental conditions" (Pais 1979).

In the early 1900's Einstein recognized that Planck's law agreed with the experiment but denied the existence of a theoretical basis in Planck's law. Einstein realized that Planck used the ρ-U relation in Planck's law from the classical mechanics and electrodynamics, but by introducing the quantization of the energy he came to an orthodox consequence of the classical theory. Einstein himself followed a different path that is the quantization of ρ known as the light-quantum theory.

6. Einstein and the photoelectric effect

Elster and Geitel developed the first practical photoelectric cells (or photovoltaic cell) which converts light energy into electrical energy containing two electrodes in an evacuated glass (emitter and collector) where the incident light reaches the emitter metal plate and emits electrons to the collector electrode.

Heinrich Hertz was the first to observe the photoelectric phenomenon accidentally during the investigations of the electromagnetic nature of light in 1887 (Hertz 1887). He was studying the spark discharges generated by potential differences between two metal surfaces. His apparatus consisted of a spark generator (RLC circuit), a spark gap and a receiver (induced sparks and tuner) to generate and detect electromagnetic waves. In a second experiment, he increased the distance between the plates so that no spark could be generated by the spark generator and after he illuminated the metal surfaces with a nearby electric arc lamp which gave rise to the sparks.

In 1888, Hallwachs used a different apparatus containing an ultraviolet source, electroscope and a zinc plate (connected to the electroscope) between them which absorbed the ultraviolet light from a ultraviolet source removing electrons from the plate which is confirmed when the plates of the electroscope are discharged to provide electrons to the discharged zinc plate (Hallwachs 1888). From 1888 to 1890, Stoletov invented a new experimental apparatus for the photoelectric effect and discovered a direct proportionality between the intensity of light acting on a metallic plate and the photocurrent induced by this radiation (Stoletov 1888). Thomsom was the first to state that the emitted particles from the photoeffect are electrons, i.e., photoelectrons (Thomson 1897, 1899) from his investigation of the ultraviolet light in Crookes tubes.

In 1905, Einstein proposed a simple picture for the photoelectric effect. A light quantum transfers all its energy to a single electron which is independent of the presence of other light-quanta. He also noticed that the ejected electron losses some energy before leaving the surface of the metal. The maximum kinetic energy of the ejected electron is given by:

$$E = h\nu - P$$

Where ν is the frequency of the incident monochromatic radiation and P is the work function (the energy needed to escape the metal surface).

If the energy of the photon is less than the work to remove the single electron from the surface metal, no photoelectron is observed. If the energy of the photon is greater than the work to remove the single electron, then it will be ejected with a maximum kinetic energy E.

$$E = h\nu - h\nu_0$$

$$P = h\nu_0$$

$$E = h(\nu - \nu_0)$$

Then, Einstein proposed that the maximum kinetic energy of the photoelectron is a linear function of the frequency of the incident radiation (Einstein 1905).

In 1914–1915 Millikan used visible light and various alkaline metals as targets linear plots of the retarding potential (V) versus the frequency where the slope is h and the intersection with the frequency axis is P/ν_0, as predicted by Einstein (Millikan 1916).

After this achievement, the Einstein's theory of the failure of equipartition of the black-body radiation (demonstrated in his 1905 paper) prevailed over Planck's non-equilibrium model (the irreversibility for conservative system).

7. Einstein's model of equipartition and the average energy

According to Einstein's model of equipartition to the black-body radiation (Pais 1979), the entity executing simple harmonic oscillations can only have discrete total energies ($E = nh\nu$, $n = 0,1,2,3,\ldots$). This is the quantized energy. The average energy is given by:

$$\bar{\varepsilon} = \frac{\sum_{n=0}^{\infty} \varepsilon.P(\varepsilon)}{\sum_{n=0}^{\infty} P(\varepsilon)} = \frac{\sum_{n=0}^{\infty} A\varepsilon.e^{-\varepsilon/kT}}{\sum_{n=0}^{\infty} Ae^{-\varepsilon/kT}} = \frac{\sum_{n=0}^{\infty} nh\nu.e^{-\alpha nh\nu}}{\sum_{n=0}^{\infty} e^{-\alpha nh\nu}}, \qquad \alpha = \frac{1}{kT}$$

The average energy can also be written as:

$$\bar{\varepsilon} = -\frac{d}{d\alpha}\ln\sum_{n=o}^{\infty}\exp(-\alpha nh\nu)$$

$$\sum_{n=o}^{\infty}\exp(-\alpha nh\nu) = \left(1-\exp(-\alpha h\nu)\right)^{-1}$$

$$\bar{\varepsilon} = -\frac{d}{d\alpha}\ln\left(1-e^{(-\alpha h\nu)}\right)^{-1}$$

$$\bar{\varepsilon} = -\left(1-e^{(-\alpha h\nu)}\right)\ (-1)\left(1-e^{(-\alpha h\nu)}\right)^{-2}\left[-(-h\nu)e^{-\alpha h\nu}\right]$$

Let us work on the last equation:

$$y = \left(1-e^{-\alpha h\nu}\right)^{-1}, \qquad u = 1-e^{-\alpha h\nu}$$

$$\frac{d\ln y}{d\alpha} = \frac{d\ln y}{du}\frac{du}{d\alpha}$$

$$\frac{d\ln y}{du} = \frac{d\ln u^{-1}}{du} = (-1)\frac{1}{\left(1-e^{-\alpha h\nu}\right)}$$

$$\frac{d\ln y}{du} = (-1)\left(1-e^{-\alpha h\nu}\right)^{-1}$$

$$\frac{d\ln y}{du} = -\left(1-e^{-\alpha h\nu}\right)\left(1-e^{-\alpha h\nu}\right)^{-2}$$

$$\frac{d}{d\alpha}e^{u'} = \frac{de^{u'}}{du'}\frac{du'}{d\alpha} = e^{u'}\frac{du'}{d\alpha}$$

$$u' = -\alpha h\nu$$

$$\frac{du}{d\alpha} = -(-h\nu)e^{-\alpha h\nu}$$

Then, we have:

$$\bar{\varepsilon} = \frac{h\nu e^{-\alpha h\nu}}{1-e^{-\alpha h\nu}} = \frac{h\nu}{e^{\alpha h\nu}-1} = \frac{h\nu}{e^{h\nu/kT}-1}$$

Returning to the Planck's law

$$\rho_T(\nu) = \frac{8\pi\nu^2}{c^3}\frac{h\nu}{e^{h\nu/kT}-1}$$

We see that it can rewritten as:

$$\rho_T(\nu) = \frac{8\pi\nu^2}{c^3}\bar{\varepsilon}$$

8. Einstein's theory of interaction between matter and radiation

In his 1917's paper, Einstein started with the Bohr's assumptions (see Chapter seven) that a molecule can only exist in a discrete set of states with energies ε_1, ε_2, etc., apart from its orientation and translatory motion. Einstein stated that the frequency of the nth quantum state (W_n) of a gas at temperature T is given from a Boltzmann derivation as:

$$W_n = p_n e^{\frac{-\varepsilon_n}{kT}}$$

Where p_n is the statistical weight of this state (Einstein 1917).

Einstein assumed two possible quantum states Z_n and Z_m of the molecule where $\varepsilon_m > \varepsilon_n$, and he assumed that a transition from Z_n into Z_m is possible with absorption and emission energies of same value as: $\varepsilon_m - \varepsilon_n$, and then, he stated that the probability for this emission within the time interval dt is given by:

$$dW = A_m^n \, dt$$

Where A_m^n is a constant, and the probability within the time interval dt for this absorption is given by:

$$dW = B_m^n \rho dt$$

Where A_m^n and B_m^n are constants.

9. Do-it-yourself activity: Plot spectral radiance vs wavelength

Firstly, consider the following constants in the international system of units.
Planck constant (h): $6.626*10^{-34}$ J.s
Light speed (c): 299 792 458 m s^{-1}
Natural exponent (e): 2.718
Boltzmann constant (k): $1.38*10^{-23}$ J K^{-1}
Sample temperature (θ): 3000 K
Secondly, use the following equation:

$$E = \frac{2\pi c^2}{\lambda^5}\frac{1}{e^{hc/k\lambda\theta} - 1}$$

Choose one spreadsheet program. In the first line, write "Lambda" (first column), "E first term" (second column), "power of e" (third column), "E second term" (fourth column), "E total" (fifth column).

In the second line of each column write the following equations: = 1*10^–9 (first column – A2); = (2*6.262*10^–34*(299792458^2))/(A2^5), (second column – B2); = (6.626*10^–34*299792458)/(A2*3000*1.38*10^–23) (third column – C2); = 1/((2.718^C2)–1), (fourth column – D2); = B2*D2.

In the third line of the first column (Lambda), write the equation: = A2*1.2, and drag it to the subsequent lines underneath until the value 9.7368E-05. Drag from the third line in the first column to the other columns in the third line. Drag each column in downward direction.

Paste the values of the first and fifth columns side by side and draw the corresponding plot.

Exercises

(1) Give the algorithm and Fortran source code (see Chapter one) of the Planck's law according to the last section (do-it-yourself experiment).

(2) According to the classical statistical mechanics, the average energy $< \varepsilon >$ for systems in thermal equilibrium energy is:

$$\langle \varepsilon \rangle = \frac{E_{total}}{N_{total}} = \frac{\iint \varepsilon e^{-\varepsilon/k_B T} dpdx}{\iint e^{-\varepsilon/k_B T} dpdx}$$

Where p and x are linear momentum and position, respectively. For a simple harmonic oscillator whose energy, ε, is:

$$\varepsilon = \frac{p^2}{2m} + \frac{1}{2} kx^2$$

Obtain its average energy.
Solution: $\langle \varepsilon \rangle = k_B T$

References cited

Baggott, J. 2011. The Quantum Story—A History in 40 Moments. Oxford University Press. New York.

Baracca, A. 2005. 1905, annus mirabilis: The roots of the 20th-century revolution in physics and the take-off of the quantum theory. L.L.U.L.L. 28: 295–382.

Boltzmann, L. 1877. Ableitung des Stefan'schen gesetzes, betreffend die abhangigkeit der warmestrahlung von der temperatur aus der electromagnetischen lichttheorie. Ann. Phys. Chem. 258: 291–294.

Boltzmann, L. 1887. Uber die beziehung zwischen dem zweiten hauptsatze der mechanischen warmetheorie und der wahrscheinlichkeinlichkeitsrechunung respektive den satzen uber das warmegleichgewicht. Wiener Berichte 76: 373–435.

Einstein, A. 1905. On a heuristic point of view concerning the production and transformation of light. Ann. Phys. 17: 132–148.

Einstein, A. 1917. Zur quantentheorie der Strahlung. Phys. Z. 18: 121–128.

Hallwachs, Wilhelm. 1888. Ueber den Einfluss des Lichtes auf electrostatisch geladene Körper. Ann. Phys. 269: 301–312.

Hertz, H. 1887. Ueber einen Einfluss des ultravioletten Lichtes auf die electrische Entladung. Ann. Phys. 267: 983–1000.

Jeans, J.H. 1905a. Letters to the editor: The dynamical theory of gases and of radiation. Nature 72: 101–102.

Jeans, J.H. 1905b. On the laws of radiation. Proc. R. Soc. Lond. A 76: 545–552.

Kirchhoff, G. 1860. I. On the relation between the radiating and absorbing powers of different bodies for light and heat. Phil. Mag. 130: 1–21.

Klein, M.J. 1966. Thermodynamics and quanta in Planck's work, Phys. Today, 294–302.

Millikan, R.A. 1916. A direct photoelectric determination of Planck's "h". Phys. Rev. 7: 355–388.

Pais, A. 1979. Einstein and the quantum theory. Rev. Mod. Phys. 51: 863–914.

Planck, M. 1897. S.-B Preuss. Akad. Wiss. 57, papers I, 493.

Planck, M. 1899. Uber irreversible Strahlungsvorgänge. S.-B Preuss. Akad. Wiss. 440–480.

Planck, M. 1900a. Uber eine verbesserung der Wien´schen spektralgleichung, Verhandl. Deutsch. Phys. Gesellschaft 2: 202–204.

Planck, M. 1900b. Ueber irreversible Strahlungsvorgänge. Ann Phys. 306: 69–122.

Planck, M. 1900c. Entropie und Temperatur strahlender Warme. Ann. Phys. 306: 719-737.

Planck, M. 1901. Ueber das gesetz der energieverteilung im normalspectrum. Ann. Phys. 4: 553–563.

Planck, M. 1912. La loi du rayonnement noir et l'hypothèse des quantités. La théorie dy rayonnement et les quanta. 1911 Solvay congress, Brussels.

Petit, A.-T. and Dulong, P.-L. 1819. Recherches sur quelques points importants de la théorie de la chaleur. Ann. Chim. Phys. 10: 395–413.

Rayleigh, F.R.S. 1898. Note on the pressure of radiation, showing an apparent failure of the usual electromagnetic equations. Phil. Mag. 45: 522–525.

Rayleigh, F.R.S. 1900. LIII. Remarks upon the law of complete radiation. Phil. Mag. 49: 539–540.

Rayleigh, F.R.S. 1900. The dynamical theory of gases and of radiation. Nature 72: 54–55.

Stefan, J. 1879. Uber die beziehung zwischen der warmestrahlung un der temperatur. Proc. Imperial Phil. Acad. Vienna: Math. Sci. Class 79: 391–428.

Stoletov, M.A. 1888. On a kind of electrical current produced by ultraviolet rays. Phil. Mag. 26: 317–319.

Thomson, J.J. 1897. Cathode rays. Phil. Mag. 44: 293–316.

Thomson, J.J. 1899. On the masses of the ions in gases at low pressures. Phil. Mag. 48: 547–567.

Wien, W. 1893. Eine neue Beziehung der Strahlung schwarzer Körper zum zweiten Hauptsatz der Wärmetheorie. Ann. Phys. Chem. 49: S. 633–641.

Wien, W. 1896. Ueber die energievertheilung in emissionsspectrum eines schwarzen körpers. Ann. Phys. 294: 662–669.

Bohr, Sommerfeld and Old Quantum Mechanics

7

1. Thomson and Rutherford models of the atom

Before Thomson's experiment on cathode rays (later known as electrons), they were thought as "they are due to some process in the ether to which (...) no phenomenon hitherto observed is analogous" (Thomson 1897).

In 1897, Thomson firstly improved Perrin's experiment where he showed that the magnetic forces deflect the cathode rays and the negative electrification follows the same path as the rays. Later, he used the Crookes tube (Fig. 7.1(A)) to prove the existence of the electrons (Thomson 1897). The Crookes tube is an electrical discharge tube emitting cathode rays (electrons) from the cathode to the anode after ionization of residual air by high voltage (See Fig. 7.1(A)). Thomson proved that the cathode rays are "carriers of the negative electricity" and they are bodies (or corpuscles), which were later called electrons. Thomson also showed that the electrons are much smaller that the lightest atom. Later, in 1904 he created a model for the atom, known as the plum pudding (Thomson 1904). In Thomson's model, the electrons were similar to raising 'embedded' in a plum pudding (Fig. 7.1(B)). Thomson stated in his 1904 paper: "The view that the atoms of the elements consist of a number of negatively electrified corpuscles enclosed in a sphere of uniform positive electrification, suggests, among other interesting mathematical problems, the one discussed in this paper, that of the motion of a ring of n negatively electrified particles placed inside a uniformly electrified sphere" (Thomson 1904). At that time, neither protons nor neutrons were discovered.

Prior to the Thomson's experiment on cathode rays, Rutherford has worked with Thomson in the study of the electric conductivity of gases when exposed to X-rays (Röntgen rays). These gases are insulators but when exposed to X-rays, they have conducting properties (Thomson and Rutherford 1896). In this work, they discovered the negatively charge corpuscles later presented by Thomson in 1897.

After his work with Thomson, Rutherford started to work with radioactivity where he coined the terms alpha and beta rays (Rutherford 1899). In 1902, during a period that atoms were thought to be indestructible, Rutherford and Soddy, stated that radioactive atoms break down to another atom (Rutherford and Soddy 1902). In 1906, Rutherford invited Geiger to work with him and they designed alpha particles counting device with two electrodes in a glass tube. In 1908, Gieger constructed a long glass tube with a source of alpha particles from radium atoms at one extremity and a phosphorescent screen at the opposite end with a spectroscope to count the scintillations and measure the spread of alpha particles. In the middle, there was a plate with a gold foil which scattered the alpha particles, but the deflection was not high enough. In 1909, Geiger and his student Marsden, under the direction of Rutherford, improved previous Geiger's experiment where they proved that alpha particles can be deflected by more than 90 degrees. They observed that metals such as gold—with higher atomic mass—reflected more alpha particles than aluminum—with lighter atomic mass (Geiger and Marsden 1909). However, they observed that only a small fraction of alpha particles (1 over 20,000 particles) deflected 90 degrees in passing through a 0.4 mm gold-foil. The scheme of Geiger and Marsden experiment is simplified in Fig. 7.1(C), although is not identical to the scheme presented in their paper. This result is incompatible with Thomson's model of the atom. Rutherford deduced that the atom has a heavy central mass with a small volume surrounded by light masses of opposite charges.

Rutherford had already observed that radioactive alpha particles are made up of two positively charged particles (i.e., the helium nucleus), then in 1911, he developed another atomic model (Rutherford 1911) to account for the fact that most of alpha particles passed through the metal leaf with no deflection. Then, Rutherford's experiment showed that the nucleus was small and dense (Fig. 7.1(D)). The Rutherford's atomic model in Bohr's words is: "the atoms consist of a positively charged nucleus surrounded by a system of electrons kept together by attractive forces from the nucleus; the total negative charge of the electrons is equal to the positive charge of the nucleus. Further, the nucleus is assumed to be the seat of the essential part of the mass of the atom, and to have linear dimensions exceedingly small compared with the linear dimensions of the whole atom" (Bohr 1913a).

The problem of the Rutherford's atomic model was to conciliate with the Larmor formula (classical electrodynamics) that gives the total power, P, radiated by non-relativistic point charge as it accelerates. For example, an alternate current

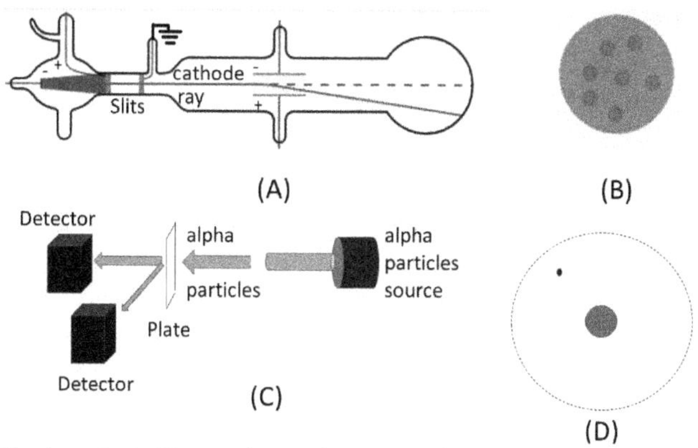

Fig. 7.1: (A) Crookes tube in Thomson's experiment; (B) Thomson's atomic model; (C) Rutherford's experiment; (D) Rutherford's atomic model.

source connected with two pieces of wire placed according to antenna design criterion (a broadcasting dipole antenna) generates accelerated electrons which emit electromagnetic waves.

$$P = \frac{2}{3}\frac{m_e r_e a^2}{c}$$

Where m_e and r_e are the mass and classical radius of the electron; c is the speed of light and a is the acceleration. As a consequence, an electron orbiting around a nucleus should lose energy and collapse into the nucleus since the electron is under the influence of the electric force yielding a centripetal acceleration of the orbiting electron.

Other problems in Rutherford´s atomic model were: (1) the fact that it could not explain the spectral lines of hydrogen and other atoms; (2) the lack of natural length in the theory.

2. Bohr model of the atom

After his Ph.D. in Copenhagen in 1911, Bohr went to Cambridge to work as postdoctoral fellow with Thomson and also spent some time with Rutherford in Manchester. Both physicists had a great influence in the early studies of Bohr's model of the atom.

Bohr knew about the limitation of the Rutherford's atomic model, and he also was aware of the Thomson's model is disagreement with Rutherford's experiment on large angle scattering of the alfa rays. In his first paper in 1913, he wrote: "The principal difference between the atom-models proposed by Thomson and Rutherford consists in the circumstance that the force acting on the electrons in the atom-model of Thomson allow of certain configurations and motions of the electrons for which the system is in a stable equilibrium; such configurations, however, apparently do not exist for the second atom-model" (Bohr 1913a).

Bohr resolved the problem of Rutherford's model with his postulate of stationary states where electrons could move in orbit without losing energy. He observed "the inadequacy of the classical electrodynamics in describing the behavior of systems of atomic size. Whatever the alteration in the laws of motion of the electrons may be, it seems necessary to introduce in the laws in question a quantity foreign to the classical electrodynamics, i.e., Planck's constant (...). By the introduction of this quantity the question of the stable configuration of the electrons in the atoms is essentially changed" (Bohr 1913a). Then, Planck's constant was the missing piece in the Rutherford's atomic model by which one can change the classical electrodynamic laws of the electron in the atom into a new quantum electrodynamic law. That is the Bohr's atomic model.

For the case of the hydrogen atom, where the charge of nucleus is equal to the charge of one electron, the total energy radiated by the formation of one of the stationary states is:

$$E = \frac{2\pi^2 me^4}{h^2 n^2}$$

Where m is the mass of the electron, e is the charge of the electron and n is a non-negative integer number and h is Planck constant. The amount of energy emitted after electronic excitation is:

$$E_{n_2} - E_{n_1} = \frac{2\pi^2 me^4}{h^2}\left(\frac{1}{n_1^2} - \frac{1}{n_2^2}\right)$$

Where n_1 is the principal quantum number of lower energy level and n_2 is the principal quantum number of upper energy level.
From photoelectric effect, by knowing that:

$$E_{n_2} - E_{n_1} = h\nu$$

We have the law connecting the lines of the spectrum of hydrogen.

$$\nu = \frac{2\pi^2 me^4}{h^3}\left(\frac{1}{n_1^2} - \frac{1}{n_2^2}\right)$$

Bohr found the expression of Balmer series in his atomic model.
Alternatively, the above equation can be written as:

$$\nu = Z^2 R_\infty\left(\frac{1}{n_1^2} - \frac{1}{n_2^2}\right)$$

Where Z is the atomic number of the hydrogen-like atom and R_∞ is the Rydberg constant (1.09677×10^7 m^{-1}).

In 1915, Bohr has made a more general formula including the mass of the nucleus, M, and the correction from the theory of relativity (Bohr 1915a):

$$\nu = \frac{2\pi^2 mMe^4}{h^3(m+M)}\left(\frac{1}{n_1^2} - \frac{1}{n_2^2}\right)\left[1 + \frac{\pi^2 e^4}{c^2 h^2}\left(\frac{1}{n_1^2} + \frac{1}{n_2^2}\right)\right]$$

In his second 1913 paper, Bohr gave the condition of dynamical equilibrium as the equivalence between the attractive electrostatic force between proton and electron of the hydrogen atom and the centripetal force exerted on the electron on a circular orbit around the nucleus Z = 1 (Bohr 1913b).

$$F_{centripetal} = F_{electrostatic(e-Z)}$$

$$F_{centripetal} = \frac{2T}{a_0}, \qquad F_{electrostatic} = K\frac{(Ze)e}{a_0^2}$$

$$Z = 1, \qquad K = \frac{1}{4\pi\varepsilon_0}$$

$$K\frac{e^2}{a_0^2} = \frac{mv^2}{a_0}$$

where a_0 is the Bohr radius and v is the linear velocity of the electron. In classical physics, the angular momentum is given by the product of the moment of inertia, I (needed torque to yield angular acceleration), and angular velocity, ω :

$$L = I\omega, \qquad I = mr^2, \qquad \omega = v/r, \qquad L = mvr$$

Bohr also stated that the angular momentum of an electron in a circular orbit around the nucleus of an atom is constrained to discrete values according to the quantum number, n.

$$L = \frac{nh}{2\pi}$$

Bohr introduced the condition of universal constancy of the angular momentum of the electron where there is an equivalence between both angular momentum equations.

$$mva_0 = \frac{nh}{2\pi}, \qquad n = 1$$

$$mva_0 = \frac{h}{2\pi}$$

Then, the square velocity of the electron is obtained from the previous equation.

$$v^2 = \frac{h^2}{4\pi^2 m^2 a_0^2}$$

And it is replaced in the relation of dynamical stability.

$$\frac{m}{a_0}\frac{h^2}{4\pi^2 m^2 a_0^2} = K\frac{e^2}{a_0^2}$$

$$\frac{1}{a_0}\frac{h^2}{4\pi^2 m} = Ke^2$$

$$a_0 = K^{-1}\frac{\hbar^2}{me^2} = \frac{4\pi\varepsilon_0\hbar^2}{me^2}$$

The a_0 is the Bohr radius—the most probable distance between the nucleus and the electron in a hydrogen atom in its ground state—and is 0.5292Å.

In his theory's review, Bohr emphasized the two postulates of his theory: "1. *Among the conceivably possible states of motion in an atomic system there exist a number of so-called stationary states which, in spite of the fact the motion of the particles in these states obeys the laws of classical mechanics to a considerable extent, possess a peculiar, mechanically unexplainable stability, of such a sort that every permanent change in the motion of the system must consist in a complete transition from one stationary state to another.* 2. *While in contradiction to the classical electromagnetic theory no radiation takes place from the atom in the stationary states themselves, a process of transition between two stationary states can be accompanied by the emission of electromagnetic radiation, which will have the same properties as that which would be sent out according to the classical theory from an electrified particle executing a harmonic vibration with constant frequency. This frequency v has, however, no simple relation to the motion of the particles of the atom but is given by the relation hv = E′–E′′, where h is Planck's constant, and E′ and E′′ are the values of the energy of the atom in the two stationary states that from the initial and final state of the radiation process*" (Bohr 1923).

Another Bohr's postulate appeared in his 1915's paper: "That the various possible stationary states of a system consisting of an electron rotating round a positive nucleus are determined by the relation:

$$T = \tfrac{1}{2} nh\omega$$

Where T is the mean value of the kinetic energy of the system, ω the frequency of rotation, and n a whole number" (Bohr 1915b).
In his 1915's paper Bohr used the equation of frequency of emission:

$$\nu = Z^2 \frac{2\pi^2 mMe^4}{h^3 (m+M)} \left(\frac{1}{n_1^2} - \frac{1}{n_2^2} \right)$$

Where M is the mass of the nucleus and he stated that the energy necessary to remove the electron to infinite distance from the nucleus was:

$$W_n = Z^2 \frac{2\pi^2 mMe^4}{h^2 (m+M)} \frac{1}{n^2}$$

In order to explain the Stark effect on hydrogen atom, he obtained the equation (Bohr 1915b):

$$\nu = Z^2 \frac{2\pi^2 mMe^4}{h^3} \left(\frac{1}{n_1^2} - \frac{1}{n_2^2} \right) \pm E \frac{3h}{8\pi^2 Zem} \left(n_2^2 - n_1^2 \right)$$

Where he found a good agreement with experimental results.

The results obtained in 1920's proved that the maximum angular momentum of the nth quantum state is different from that indicated by Bohr and, except for the hydrogen atom, the spectral lines of other atoms were not predicted by Bohr's model. Further refinement was made by Sommerfeld to predict the spectral lines of hydrogen and other atoms (Bohr 1923).

The early 1910's Weiss' work on some magnetic properties (e.g., the magnetic susceptibility) of certain metals was important to give further support to the Bohr's atomic model. Weiss observed the existence of a fundamental atomic unit of magnetism (Weiss 1911). He inferred that a molecule could have either no magnetic moment or an integral number of magnetons. If an orbiting electron generates a magnetic dipole moment (see Chapter twelve), the existence of a fundamental unit of magnetism implies that the electron's angular momenta (under atomic orbits) are fixed which is in accord with Bohr's hypothesis. Weiss reported that Einstein and collaborators observed that this unit should contain the Planck's constant. This unit was later known as Bohr magneton represented by $-q_e\hbar/2\ m_0c$ where q_e is the electron's charge, m_0 is the electron's rest mass and c is the speed of light (see Chapter twelve).

In 1921, Bohr developed the 'aufbauprinzip' (also known as the aufbau principle or the construction principle or the building-up principle) in order to explain the periodic occurrence of chemical properties in the periodic table. In the construction principle, the bigger atoms are built from hydrogenic structure by adding further electrons one by one assuming that the latest addition did not change the quantum numbers already assigned to the earlier electrons. Every newly added electron seeks

Table 7.1: Electronic configurations of the noble gases according to Bohr's paper (Bohr 1921).

Helium	(2_1)	Krypton	$(2_1\ 8_2\ 18_3\ 8_2)$
Neon	$(2_1\ 8_2)$	Xenon	$(2_1\ 8_2\ 18_3\ 18_3\ 8_2)$
Argon	$(2_1\ 8_2\ 8_2)$	Niton	$(2_1\ 8_2\ 18_3\ 32_4\ 18_3\ 8_2)$

the state of the lowest energy which is located in the orbit having the lowest energy and not fully occupied (Bohr 1921). It is the first successful interpretation of the electronic configuration of the elements according to their increasing atomic number and their arrangement in the periodic table (Table 7.1). Bohr attempted to give a detailed picture of the electronic configuration of the atoms. For example, in the neon atom, there are 2 electrons in 1_1 orbit (where in n_k notation n is the principal quantum number and k is the azimuthal quantum number), surrounded by a shell of 4 electrons in 2_1 orbits which, in turn, is surrounded by more 4 electrons in 2_2 orbits having a disturbed tetrahedral symmetry (MacKinnon 1977).

Important to emphasize that after Bohr's work, Stoner and Pauli gave further and important contributions for the building-up principle (see Chapter twelve) which ultimately led to the Pauli's exclusion principle (Stoner 1924, Pauli 1925).

From 1917 to 1920, Bohr developed the concept of the "correspondence principle" which is the expectation or necessary connection between the classical mechanics and quantum mechanics in the limit of large quantum numbers and low frequencies (Waerden 1967). The correspondence principle has been an important parameter for the study of quantum mechanics from 1918 to 1925.

An important extension of the correspondence principle was given by Van Vleck in 1924. As Waerden said: "Van Vleck's idea is: if we want to estimate the absorption by means of the principle of correspondence, we have to compare the absorption, computed classically, with the difference between absorption and induced emission, computed from Einstein's formulae. In the limit of the high quantum numbers, this difference must become equal to the classical absorption" (Waerden 1967).

Van Vleck used the Fourier series to represent the multiple periodic motion of the electron in an atom:

$$x = \sum_{\tau_1\tau_2\tau_3} X(\tau_1,\tau_2,\tau_3)\cos\left[2\pi(\tau_1\omega_1 + \tau_2\omega_2 + \tau_3\omega_3)t + \gamma^{(x)}_{\tau_1\tau_2\tau_3}\right]$$

and there are similar expressions for y and z coordinates, where ω_1, ω_2, and ω_3 are the intrinsic orbital frequencies and the summation is extended to all possible positive and negative integral values of the integers τ_1, τ_2, and τ_3. He stated that the classical electron radiates simultaneously all the combination overtones in the multiple of Fourier expansion rather than just one harmonic vibration component $\tau_1\omega_1 + \tau_2\omega_2 + \tau_3\omega_3$.

Van Vleck also formulated the quantum expression for the amount of energy emitted in a time interval Δt by the transitions from state r to state s:

$$\Delta E_{r\to s} = h\nu_{rs} N_r \left[A_{r\to s} + B_{r\to s}\rho(\nu_{rs})\right]\Delta t$$

$$N_r = \frac{Ne^{-W(r)/kT}}{\sum_r e^{-W(r)/kT}}$$

Where N_r is the number of atoms in the state r, N is the total number of atoms and W(r) is the energy of the stationary state r. He compared the expression above with the corresponding classical expression:

$$\Delta E = \left(16\pi^4 e^2/3c^3\right)\nu^4\left[D\left(\tau_1,\tau_2,\tau_3\right)\right]^2 \Delta t$$

Where $\left(2e^2/3c^3\right)\dot{v}^2$ is the classical rate of radiation from an electron having a vector acceleration $\dot{v}$.

Then, he arrived at the expression for the probability coefficient $A_{r\to s}$ for high quantum numbers (van Vleck 1924a).

3. Bohr, Kramer and Slater's quantum theory of radiation

In 1918, Bohr used Einstein's considerations (Bohr 1918) of probability law (dW) of the transitions of the electron from a stationary state to another within the atom under the influence of light (Einstein 1917). The equation below is the probability law for the process to occur during the time dt for the transition from state n to state m by absorbing radiative energy $\varepsilon_m - \varepsilon_n$.

$$dW = B_n^m \rho dt$$

Where ρ is the radiation density for the frequency ν (see Chapter six).

In 1924, in collaboration with Kramers and Slater, Bohr continued his study of the principles of the interaction between light and matter, which became known as Bohr-Kramers-Slater theory (Bohr et al. 1924). Bohr, Kramers and Slater made an extensive use of the *virtual oscillators* (which came from Slater's ideas) as a basis for treating the interactions between matter and light where a collection of atoms is replaced by a collection of simple harmonic oscillators. The idea of virtual oscillators was not treated as a model of an individual atom, but simply to treat the interaction between radiation and matter. Bohr described the oscillators as having a virtual existence and he did not commit himself as to link the virtual oscillator model and the orbital model.

The initial Slater's idea of virtual oscillator was: "Of course, the quanta can't travel in a straight line with the speed of light (from de Broglie's theorem); but it seems possible to suppose that there is an electromagnetic field, produced not by the actual motion of the electrons, but with motions with the frequency of possible emission lines and amplitudes determined by the correspondence principle, the function of this field being to determine the motion of the quanta. If this motion is determined by the condition that Poynting's theorem shall hold over an average take over a long period of time, definite patterns are described, and the probability of moving along the paths is such, for example, as to account for interference, many quanta being led to the bright spots in the field" (Hendry 1981). As said: "On receiving Slater's idea, Bohr apparently reacted warmly to the extension of the oscillator approach to include some kind of wave field but rejected outright the introduction of light-quanta" (Hendry 1981). Then, Bohr and Kramers changed Slater's initial idea into: "An atom may, in fact, be supposed to communicate with other atoms all the time it is in a stationary state, by means of a virtual field of radiation, originating from the oscillators having the frequencies of possible quantum transitions, and the function of which was to provide for stationary states conservation of energy and momentum by determining the probabilities of quantum transitions" (Hendry 1981). Slater opposed to the new interpretation of his idea (mainly the refutation of light-quanta) and left Copenhagen prematurely. Sommerfeld and Compton (from the Compton effect) insisted in the necessity of the association between light-quanta and energy-momentum conservation.

Bohr stated that the quantum theory must in a certain sense be a natural generalization of the classical electrodynamics. As he said: "This is evident from the condition that in the limit, where we consider processes which depend on the statistical behavior of a large number of atoms, and which involve stationary states where the difference between neighboring stationary states is comparatively little, the classical theory leads to conclusions in agreement with the experiments" (Bohr et al. 1924). This is the case of the Zeeman effect where Lorentz could explain it classically. Bohr also added that "In the case of emission and absorption of spectral lines, this connection between the two theories has led to the establishment of the *correspondence principle,* which postulates a general conjugation of each of the various possible transitions between stationary states with one of the harmonic oscillation components in which the electrical moment of the atom, considered as a function of the time, can be resolved. This principle has afforded a basis for an estimation of probabilities of transition, and thereby for bringing the problem of intensities and polarization of spectral lines (see Chapter eight) in close connection with the motion of the electrons in the atom" (Bohr et al. 1924). With the aid of the *correspondence principle* it was possible to give a quantum explanation of the Zeeman effect for the hydrogen lines whose results were similar to those from the classical explanation given by Lorentz, although the essence of both assumptions were different (Bohr 1923). According to Bohr,it was important to give a quantum explanation to the Zeeman effect since the classical electrodynamical theory could not explain several phenomena which are contradictory to the classical electrodynamical theory.

As to the Stark effect, Kramers in his Ph.D thesis calculated the intensities of the lines of the fine structure of the hydrogen atom from Stark effect (Kramers 1919). As Bohr said: "the position of the lines corresponds to the frequencies calculated for the different transitions, and the lengths of the lines are proportional to the probabilities as calculated on the basis of the correspondence principle (…) the Stark effect reflects down to the smallest details the action of the electric field on the orbit of the electron in the hydrogen atom" (Bohr 1923). In Bohr's paper (Bohr 1923) there are two figures (Figs. 6 and 7) showing the experimental and the theoretical splitting of the spectral lines of the hydrogen atom and it is important to add that their similarities are remarkable.

In Kramers' 1924-paper he stated: "It is well known that a consistent description of the phenomena of dispersion, reflection and scattering of electromagnetic waves by material media can be given on the fundamental assumption that an atom, when exposed to radiation, becomes a source of secondary spherical wavelets, which are coherent with incident waves" (Kramers 1924). He imagined that the incident radiation was a plane monochromatic train of polarized harmonic waves (which corresponds to coherent and incoherent scattering) where primary wavelets had frequency ν and amplitude E (intensity of the incident electromagnetic plane wave) and the secondary wavelets (from a varying electrical doublet, i.e., a varying electric charged dipole) had frequency ν and amplitude P (the induced polarization) which is proportional to the amplitude E of the incident waves whose phase difference between them is ϕ. Their electric vectors are given below and the relation between E and P related to an atom with electron of charge -e and mass m isotropically bound to a position of equilibrium with natural frequency ν_1 is given below:

$$\vec{\Im} = E\vec{v}\cos(2\pi\nu t)$$

$$\vec{\wp} = P\vec{w}\cos(2\pi\nu t)$$

$$P = E\frac{e^2}{m}\frac{I}{4\pi^2(\nu_1^2 - \nu^2)}$$

Then, Kramers arrived at an equation for P that worked but never had been given a theoretical justification.

$$P = E\sum_i A_i^a \tau_i^a \frac{e^2}{m}\frac{I}{4\pi^2(\nu_i^{a2} - \nu^2)} - E\sum_j A_j^e \tau_j^e \frac{e^2}{m}\frac{I}{4\pi^2(\nu_j^{e2} - \nu^2)}$$

Where A is the probability of the isolated atom performing in unit time one transition and τ is the decay characterizing the same transition.

4. The states of the system and the Sommerfeld's ellipses

The (linear) momentum, p, of a point mass, m, moving at a velocity, v, is:

$$p = m\cdot v$$

$$p = m\cdot\frac{dq}{dt}$$

Where q is the position coordinates. It is also known from the classical mechanics that:

$$\frac{dp}{dt} = F = -\frac{\partial E_{pot}}{\partial q}$$

Where E_{pot} is the potential energy (a function of the position coordinates). The kinetic energy is given by:

$$E_{kin} = \frac{m}{2}\left(\frac{dq}{dt}\right)^2 = \frac{1}{2m}p^2$$

The Hamiltonian function, H, is the total energy of the point mass:

$$H(q,p) = E_{kin} + E_{pot}$$

The momentum p and position coordinates q determine the state of the system.

As we mentioned before, the idea of the electron moving as a harmonic oscillator where the nucleus of the atom is the rest point came from Lorentz's theory, and,

Planck, Sommerfeld and Heisenberg used this in their theories as a starting point or inspiration. Classically, the oscillator is a spring attached to a point mass m at rest position whereby the point mass moves in one direction from the rest position. Due to the spring (which in the case of the atom is the electric field between proton of the nucleus and the electron), the mass experiences a restoring force. The movement of the point mass is described by a sinusoidal wave:

$$x = q = a\sin 2\pi\nu t$$

Where ν is the oscillation frequency, t is time and a is the amplitude of the oscillation. The momentum p of the harmonic oscillator is:

$$p = m\frac{dq}{dt} = 2\pi\nu ma\cos 2\pi\nu t$$

Let us eliminate the variable time, t, from the expressions above by squaring the last two equations and using a trigonometric identity:

$$q^2 = a^2\sin^2 x, \quad x = 2\pi\nu t$$

$$p^2 = b^2\cos^2 x, \quad b = 2\pi\nu ma$$

$$\sin^2 x + \cos^2 x = 1$$

$$\sin^2 x = \frac{q^2}{a^2}, \quad \cos^2 x = \frac{p^2}{b^2}$$

Then, we have:

$$\frac{q^2}{a^2} + \frac{p^2}{b^2} = 1, \quad b = 2\pi\nu ma$$

The above equation is an ellipse in the p,q-place where b is the semi-minor axis and the ratio b/a is a constant.

$$\frac{b}{a} = 2\pi\nu m$$

The area of the ellipse (abπ) is equivalent to the ratio E/ν, where E is the constant vibrational energy. For example, in t = 0 in the rest position (where $E_{pot} = 0$), the kinetic energy is equal to E and we have:

$$E_{kin} = \frac{m}{2}\left(\frac{dq}{dt}\right)^2$$

$$p = \frac{dq}{dt} = 2\pi\nu a\cos 2\pi\nu t$$

$$E_{kin} = \frac{m}{2}(2\pi\nu a)^2\cos^2 2\pi\nu t$$

$$t = 0, \quad \cos^2 0 = 1$$

$$E_{kin} = E = (2\pi\nu ma)(\pi a)\nu = b\pi a\nu$$

$$\frac{E}{\nu} = ab\pi$$

By changing the value of E, one can obtain similar ellipses in the plane (p,q). The area difference between two consecutive ellipses is the Planck constant.

$$\frac{\Delta E}{\nu} = h, \qquad \Delta E = h\nu$$

In the classical theory, all points of the state plane of the ellipses are equal. In quantum theory, those states are distinguished from each ellipse they belong to. They are the stationary states of the oscillator, i.e., the charged mass point without radiation within a specific ellipse. When the charged mass point moves from one ellipse to a smaller ellipse, it emits energy hν.

Sommerfeld's work improved Bohr's atomic theory by including a second quantum number (subordinate quantum number) and changing the perspective from a simply periodic to a doubly periodic system ruled by a (p,q) plane state. Each elliptical orbit is designated by a symbol n_k (or simply k) where n is the principal quantum number and k is the subordinate quantum number. All orbits with the same principal quantum number have the same major axis, a, and all orbits with the same subordinate quantum number will have the same semi-minor axis, b (Bohr 1923).

5. Sommerfeld extension of Bohr's atomic theory

Sommerfeld was doctoral advisor of Heisenberg and Pauli. He introduced them to Bohr and Born who also worked with them. Besides his great contribution to the old quantum mechanics, he wrote the famous book of spectroscopy: 'Atombau und Spectrallinien' (atomic structure and spectral lines) which influenced several famous physicists.

In 1916, Sommerfeld wrote two lengthy papers with the same title about the quantum theory of the spectral lines which became an extension of the Bohr's model (Sommerfeld 1916a, 1916b).

Sommerfeld adapted Bohr's quantization condition to Planck's phase space ideas and he developed the phase-space quantization rule which restricts the values of the integral of the conjugate momentum p of some generalized coordinate q over one period of motion to integer multiples of h. In addition, Sommerfeld generalized Bohr's model by allowing elliptical orbits as well as circular orbits. He recovered Bohr's quantum number n as the sum of two quantum numbers r and k.

Sommerfeld showed his quantization condition from Planck's quantization condition (Sommerfeld 1916a):

$$Planck: \iint dpdq = h$$

$$\iint dpdq = \int p_n dq - \int p_{n-1} dq = h$$

$$n = 1, 2, 3, \ldots$$

$$\int p_1 dq - \int p_0 dq = h$$

$$\int p_2 dq - \int p_1 dq = h$$

$$\int p_3 dq - \int p_2 dq = h$$

...

$$\int (p_n - p_0) dq = nh$$

$$as : \int p_0 dq = 0$$

$$then : \int p_n dq = nh$$

Born commented on Sommerfeld's work that: "He investigated the elliptic motion of the electron and found that there are actually three quantum numbers connected with it which, however, for an 'undisturbed' atom combine to one number, that appearing in Balmer's formula. But by exposing the atoms to perturbations the effect of these three numbers can be separated. Thus, the splitting of the spectral lines by magnetic and electric fields, the Zeeman- and Stark-effect, were explained. In the magnetic case the different energy levels correspond to precessional motions of the plane of the elliptic orbit at different inclinations to the direction of the field" (Born 1952).

In Sommerfeld's work, it was used two-dimensional Keplerian motion in polar coordinates with azimuthal (φ) and radial (r) coordinates associated with the corresponding momenta p_φ and p_r in two integrations over a complete elliptical orbit (Eckert 2014).

$$\oint p_\varphi d\varphi = kh$$

$$\oint p_r dr = rh$$

Where k and r are the azimuthal and the radial quantum numbers (non-negative integers), respectively. The sum of both quantum numbers gives Bohr's principal quantum number, n.

$$n = r + k$$

Sommerfeld quantized separately the linear and angular momenta associated with the quantum numbers k and r. The quantum numbers r and k determined respectively the size and shape of the orbit, respectively.

Bohr proposed that the electron in an orbit could only exist in certain well-defined, stable orbits which satisfied the integral:

$$\oint pdq = nh, \quad n \in \mathbb{N}$$

Which became known as Bohr-Sommerfeld quantization rule.

In Bohr-Sommerfeld model, the angular momentum, L, of the electron is:

$$L = \frac{kh}{2\pi} = k\hbar$$

$$where : k = 1, 2, 3, ..., (n/k)$$

Where n is the length of the major axis and k the length of the minor axis of the ellipse. When n = k, the orbit is circular.

Each stationary state corresponds to a n_k-orbit for the electron, where n is the principal quantum number and k is the azimuthal quantum number. Hydrogen and helium have one or two 1_1 orbits, respectively. Lithium and berilium have two 1_1 orbits, plus one or two 2_1 orbits, respectively. And so on. One electron in each orbit, each chemical element is built up by adding an outer electron to the electronic configuration of the previous element according to the atomic number.

The first observation of doublets, quartets and sextuplets in 1920's indicated that the multiplicity in the energy levels were not correctly accounted for the n and k alone. Then, Sommerfeld created a third quantum number j (called inner quantum number) which corresponds to energy-sublevels that n_k orbits split to give rise to the multiplets (Sommerfeld 1920). The third quantum number lacked clear geometrical or physical meaning, but it was a useful empirical rule for doublets and triplets: the doublets originate from j = k, k-1 and triplets come from j = k, k-1, k-2.

One year later Landé gave a physical meaning to the inner atomic number j, called atomic core model (Landé 1921) where the atomic core of any atom (core electrons and nucleus) has a non-zero angular momentum denoted by a core quantum number r (or s) and it has a corresponding vector R (or S in modern notation). The orbital angular momentum of the valence electrons is denoted by the azimuthal quantum number k (or l) and it has a corresponding vector K (or L in modern notation). The sum or R and K (or S and L) gives the total angular momentum vector J. Landé identified the total angular momentum with Sommerfeld's inner quantum number j. The atomic core angular momentum was later substituted by the electron's spin angular momentum (see Chapter twelve).

6. The limitations of Bohr and Sommerfeld's atomic model

As van Vleck has appointed in 1924: "The study of helium, the simplest atom except hydrogen, should be a key to a generalized Bohr theory of atomic structure. However, no satisfactory model of normal helium has yet been devised, for the models of Bohr, Langmuir, Franck and Reiche, and Landé all give the wrong ionization potentials if the non-radiating orbits are determined by Sommerfeld quantum conditions" (van Vleck 1924b).

In the same year, Heisenberg pointed out that: "Since the empirical material on the anomalous Zeeman effects was systematically organized by Landé according to the previous quantum theoretical principles and put into formulas, it has become increasingly clear that an explanation of the phenomena of the anomalous Zeeman effect must bring about profound changes in our quantum theoretical ideas (that is, the old quantum theory). This is shown particularly impressively in the failure of the structural principle with respect to the statistical weights of atomic residue and electrons (Heisenberg 1924).

Then, the helium atom was responsible for the turning point in Physics from the Bohr and Sommerfeld's atomic model (later known as the old quantum theory) into modern quantum theories of Heisenberg and Schrödinger.

Other limitation of Bohr's model was the refutation of light-quanta. As Pauli wrote to Kramers: "[The ideas of BKS] thus move in completely wrong direction: it is not the energy concept that is to be modified but the concepts of motion and force. One can indeed define no fixed path for the light-quanta where interference phenomena are present, but nor can one define any such paths for the electrons in atoms; and to doubt the existence of light-quanta on the grounds of interference phenomena is just as little justified, therefore, as to doubt the existence of the electron would be" (Hendry 1981). The Compton effect (the scattering of a photon by an electron, increasing its wavelength) proved the existence of the light-quanta, i.e., that the light cannot be described only as wave (Compton 1923), which Bohr and Kramers rejected till the end of the old quantum theory.

7. Do-it-yourself activity: find the transitions of visible spectrum of the hydrogen atom

By knowing the values of the Balmer lines (656.3 nm, 486.1 nm, 434.0 nm and 410.2 nm), *(1) find the corresponding frequencies in wavenumber for the Balmer lines.*

After that, convert the frequency unit in wavenumber (m^{-1}) into energy unit (eV) by using the relation below.

Important to know that the wavenumber, κ, is analogous to the frequency and it informs how many wavelengths fit into a unit of distance. The wavenumber has an inverse relation with wavelength, λ, that is, $\lambda = 1/\kappa$.

$$\nu = \frac{c}{\lambda} = \frac{c}{1/\kappa} = c\kappa$$

$$E_{photon} = h\nu = hc\kappa$$

$$hc = 1.99 \times 10^{-25} J \cdot m = 1.23984 \times 10^{-6} eV \cdot m$$

$$E_{photon} = 1.23984 \times 10^{-6} \left[eV \cdot m\right] \times \kappa \left[m^{-1}\right]$$

$$E_{photon} = \frac{1}{806554} \times \kappa \left[eV\right]$$

$$For: \kappa = 806554 m^{-1} \Rightarrow E_{photon} = 1 eV$$

Then, we use the relation below:

$$E_{photon} = \frac{1}{806554} \times \kappa$$

The energy levels for the hydrogen atom are: n = 2 (E = –3.4 eV), n = 3 (E = –1.51 eV), n = 4 (E = –0.850 eV), n = 5 (E = –0.544 eV), n = 6 (–0.378 eV). *(2) Obtain the energy differences according to the transitions in the Balmer series (from n = 2 to n = 3, 4, 5 and 6)*

Then, use the relation above (whose input is the frequency associated with the transitions of the Balmer lines) in order to *(3) find the same values of energy difference* (or the energy of the emitted photon from the transition in the hydrogen atom). Finally, *(4) establish the relation between the energy (and frequency) of the emitted photon and its corresponding transition.*

Answers: Values of the frequencies of the emitted photons and their corresponding transitions: 1.524×10^6 m^{-1} (n = 2 => 3), 2.057×10^6 m^{-1} (n = 2 => 4), 2.304×10^6 m^{-1} (n = 2 => 5) and 2.437×10^6 m^{-1} (n = 2 => 6).

References cited

Bohr, N. 1913a. On the constitution of atoms and molecules. Phil. Mag. S.6(26): 1–23.

Bohr, N. 1913b. On the constitution of atoms and molecules. Part II—systems containing only a single nucleus. Phil. Mag. S. 6(26): 476–502.

Bohr, N. 1915a. On the series spectrum of hydrogen and the structure of the atom. Phil. Mag. 6(29): 332–335.

Bohr, N. 1915b. On the quantum theory of radiation and the structure of the atom. Phil. Mag. 6(30): 394–415.

Bohr, N. 1918. On the quantum theory of line-spectra. Kgl. Danske Vid. Selsk. Skr., nat.-math. Afd., 8. Raekke IV. 1. Part I.

Bohr, N. 1921. Atomic structure. Nature 107: 104–107.

Bohr, N. 1923. The structure of the atom. Nature 112: 29–44.

Bohr, N., Kramers, H.A. and Slater, J.C. 1924. The quantum theory of radiation. Phil. Mag. S. 6, 47: 785–802.

Born, M. 1952. Arnold Johannes Wilhelm Sommerfeld. 1868–1951. Obituary notices of fellows of the royal society 8: 274–296.

Compton, A.H. 1923. A quantum theory of the scattering of x-rays by light elements. Phys. Ver. 21: 483–502.

Eckert, M. 2014. How Sommerfeld extended Bohr's model of the atom (1913–1916). Eur. Phys. J. 39: 141–156.

Einstein, A. 1917. Zur quantentheorie der Strahlung. Phys. Z. 18: 121–128

Geiger, H. 1908. On the scattering of the alpha and beta particles by matter. Proc. Roy. Soc. A 81: 174–177.

Geiger, H. and Marsden, E. 1909. On a diffuse reflection of the alpha particles. Proc. Roy. Soc. A: 82: 495–500.

Heisenberg, W. 1924. Über eine abänderung der formalin regeln der quantentheorie beim problem der anomalen Zeeman-effekte. Z. Phys. 26: 291–307.

Hendy, J. 1981. Bohr-Kramers-Slates: A virtual theory of virtual oscillators and its role in the history of quantum mechanics. Centaurus 25: 189–221.

Kramers, H.A. 1919. Intensities of spectral lines. On the application of the quantum theory to the problem of relative intensities of the components of the fine structure and of the Stark effect of the lines of the hydrogen spectrum. Roy. Danish Academy.

Kramers, H.A. 1924. The law of dispersion and Bohr's theory of spectra. Nature 113: 673–674.

Landé, A. 1921. Uber den anomalen Zeemaneffekt. Z. Phys. 5: 231–241.

MacKinnon, E. 1977. Heisenberg, models and the rise of the matrix mechanics. Hist. Stud. Phys. Sci. 8: 137–188.

Pauli, W. 1925. Über den Zusammenhang des Abschlusses der Elektronengruppen im Atom mit der Komplexstruktur der Spektren. Z. Phys. 31: 765–783.

Rutherford, E. 1899. Uranium radiation and the electrical conductions produced by it. Phil. Mag. 47: 109.

Rutherford, E. and Soddy, F. 1902. The cause and nature of radioactivity part I. Phil. Mag. 6(4): 370–396.

Rutherford, E. 1911. The scattering of α and β particles by matter and the structure of the atom. Phil. Mag. 6(125): 669–688.

Sommerfeld, A. 1916a. Zur quantentheorie der spektrallinien. Ann. Phys. 51: 1–94.

Sommerfeld, A. 1916b. Zur quantentheorie der spektrallinien (part II). Ann. Phys. 51: 125–167.

Sommerfeld, A. 1920. Allgemeine spektroskopische gesetze, insbesondere ein magnetooptischer zerlegungssatz. Ann. Phys. 368: 221–263.

Stoner, E.C. 1924. The distribution of electrons among atomic levels. Phil. Mag. 48: 719–736.

Thomson, J.J. and Rutherford, E. 1896. On the passage of electricity through gases exposed to Röntgen rays. Phil. Mag. 5(42): 392–407.

Thomson, J.J. 1897. Cathode rays. Phil. Mag. 5(44): 293–316.

Thomson, J.J. 1904. On the structure of the atom: An investigation of the stability and periods of oscillation of a number of corpuscles arranged at equal intervals around the circumference of a circle; with application of the results to the theory of atomic structure. Phil. Mag. 6(39): 237–265.

Van Vleck, J.H. 1924a. The absorption of radiation by multiply periodic orbits, and its relations to the correspondence principle and the Rayleigh-Jeans Law. Part I: Some extensions of the correspondence principle. Phys. Rev. 24: 330–347.

Van Vleck, J.H. 1924b. The dilemma of the helium atom. Phys. Rev. 19: 419–420.

Waerden, B.L. van der. 1967. Sources of quantum mechanics. Dover Publications Inc. Mineola, New York.

Weiss, P. 1911. Uber die rationalen Verhältnisse der magnetischen momente der molekule und das magneton. Phys. Z. 12: 935–952.

Heisenberg's Matrix Quantum Mechanics

8

1. Heisenberg before 1925's famous paper: the turning point from old to modern quantum mechanics

Heisenberg was Sommerfeld's doctorate student and submitted his dissertation under the title "On the stability and turbulence of liquid currents" in 1923. He was only 21 years old and had "complete command of the mathematical apparatus and daring physical insight" as his mentor affirmed. The problem proposed by Sommerfeld (which was a very difficult one) was to determine the transition from laminar fluid to a turbulent fluid. Heisenberg's results were only confirmed twenty-five years later.

In 1922, Heisenberg wrote to Sommerfeld: "I am now working with Born to improve and refine the Born-Pauli method (of perturbation theory); with its help, for instance, one can prove that the quantum theory demands phase relations between the electrons of an atomic system" (Rechenberg 2000). Then, Heisenberg was simultaneously working with Sommerfeld in Munich and Born in Gottingen during his doctorate course.

Due to Heisenberg's lack of interest in experimental physics, Heisenberg had failed to derive the resolving power of telescope, interferometer and microscope during the final oral of his doctorate course. The final grade was based on his dissertation and final oral. He had the lowest passing grade from Wien (an experimental physicist) and highest passing grade from Sommerfeld (a theoretical physicist). Heisenberg was shocked for receiving the lowest passing grade in his doctorate. In the same night, he moved from Munich to Gottingen to work with Born after completing his doctorate.

Born (a former PhD student of David Hilbert) and Heisenberg tried to establish phase relations between the two orbiting electrons of the helium atom (Born and Heisenberg 1923a). In this work, they used the Hamiltonian function (see Chapter seven), perturbation theory (see Chapter eighteen) and relations similar to those given below:

$$H(q,p) = E_{kin}(p) + E_{pot}(q),$$

$$Then: \frac{\partial H}{\partial q} = \frac{\partial E_{pot}}{\partial q}, \frac{\partial H}{\partial p} = \frac{\partial E_{kin}}{\partial q} = \frac{p}{m}$$

$$And: \frac{dq}{dt} = \frac{\partial H}{\partial p}, \frac{dp}{dt} = -\frac{\partial H}{\partial q}$$

In a subsequent paper (The electron trajectories in the excited helium atom), Born and Heisenberg adapted the Bohr's core model to calculate allowed energy levels of the excited helium (Born and Heisenberg 1923b). They expressed the Hamiltonian function of the helium atom as:

$$H = \frac{1}{2m}\left(p_{r1}^2 + \frac{1}{r_1^2}\left[\frac{1}{\sin^2\theta_1}p_{\varphi 1}^2 + p_{\theta 1}^2\right]\right) + \frac{1}{2m}\left(p_{r2}^2 + \frac{1}{r_2^2}\left[\frac{1}{\sin^2\theta_2}p_{\varphi 2}^2 + p_{\theta 2}^2\right]\right)$$

$$-\frac{e^2Z}{r_1} - \frac{e^2Z}{r_2} + \frac{e^2}{\sqrt{r_1^2 + r_2^2 - 2r_1r_2\left[\cos\theta_1\cos\theta_2 + \sin\theta_1\sin\theta_2\cos(\varphi_1 - \varphi_2)\right]}}$$

Where r_1, θ_1, φ_1 and r_2, θ_2, φ_2 are polar coordinates of the electrons 1 and 2.

Their formulation placed the excited electron in an outside orbit above the core (the nucleus and the unexcited electron). Though they did not succeed to obtain results similar to the experimental ones, they obtained the correct general form of the spectral lines (Born and Heisenberg 1923b).

Born himself had already adventured on molecular theories when he wrote in 1922 a paper entitled "On the model of the hydrogen molecule" also using the perturbation theory (Rechenberg 2000). In 1924, Born and Heisenberg studied the quantum theory of molecules (Born and Heisenberg 1924) where they used the perturbation theory for the Hamiltonian function of a molecule in which they expanded the energy of the states as (Born and Heisenberg 1924):

$$H = H_0 + \lambda^2 H_2 + \lambda^3 H_3 + \lambda^4 H_4, \lambda = \sqrt{m/M}$$

Where m and M are the mass of the electron and nuclei, respectively. Important to note that Born had already used the perturbation theory for the Hamiltonian function of the atom (Born 1924). In this Born-Heisenberg approach, they developed the steps for the appropriate treatment of the quantum theory for molecules without the need of a descriptive picture of the molecule:

(a) The molecule is treated as a rigid rotator;
(b) The vibrations of the nuclei are included and added to the rotational motions;
(c) The interactions between rotations and nuclear vibrations are added;
(d) The inclusion of electronic and nuclear angular momentum to the previous results;
(e) The full treatment of the nuclear and electronic structure of molecules.

Still in 1924, Heisenberg completed his habilitation thesis working on anomalous Zeeman effect (Heisenberg 1924) under Born. The Zeeman effect (the splitting of each spectral line into three lines under magnetic field for singlet atoms where spin does not influence the splitting) had been explained by Sommerfeld, but the anomalous Zeeman effect (the splitting of each spectral line into three or more lines of an open-shell atom, doublet or triplet, for example, under magnetic field where the spin takes its part) is a more complex problem.

Despite the sad outcome of Heisenberg's doctoral grade, Sommerfeld was very important for presenting Heisenberg his atomic model and for introducing him to

Bohr. In late 1920's Sommerfeld assigned the problem of the anomalous Zeeman effect to Heisenberg. They also worked together using the correspondence principle for the study of X-ray doublets (Sommerfeld and Heisenberg 1922). In Sommerfeld's model (later known as Heisenberg's core model), there is an optical electron (which is an old term for the valence electron) in an outside orbit and the Z-1 core electrons in inner orbits (see Chapter twelve). For Sommerfeld, the optical electron can have a magnetic interaction with the core to produce the splitting of the energy levels. Heisenberg adapted this model by assigning the core (the core electrons and the nucleus) an average angular momentum of ½ and the optical electron the value n-½ (Heisenberg 1922). Depending on whether the two angular momenta are parallel or anti-parallel would determine the exchange energy between them. Then, Heisenberg developed formulas for the anomalous Zeeman effect (MacKinnon 1977). In his subsequent work in collaboration with Landé Heisenberg developed the second core model on the anomalous Zeeman effect based on neon ion and atom (Landé and Heisenberg 1924). At that time, Heisenberg had only modified the way of the coupling of the core and optical electrons preserving Landé g-values. See more details about Heisenberg's core model in Chapter twelve.

Bohr gave a series of lectures on quantum atomic physics in June 1922 which had awakened in Heisenberg a great interest on his atomic model. After becoming "privatdozent" in 1924, Heisenberg went to Copenhagen to research with Bohr from September 1924 to May 1925. Heisenberg worked with Kramers (former Bohr's doctorate student) on the problem of dispersion, i.e., the scattering of the photon by an atomic electron from where the incident radiation has wavelength larger than the atoms. Ladenburg had previously treated the dispersion electrons as oscillators (an extension of Lorentz model) and the electrons disturbed by electromagnetic radiation return to equilibrium positions through damped oscillations (Ladenburg 1921).

Kramers and Heisenberg presented a systematic treatment of the interaction between matter and radiation (Kramers and Heisenberg 1925). They continued previous Kramers' work (Kramers 1924) using his equation (see Chapter seven) for the problem of dispersion:

$$P = E\sum_i A_i^a \tau_i^a \frac{e^2}{m} \frac{I}{4\pi^2\left(\nu_i^{a2} - \nu^2\right)} - E\sum_j A_j^e \tau_j^e \frac{e^2}{m} \frac{I}{4\pi^2\left(\nu_j^{e2} - \nu^2\right)}$$

Heisenberg found that this equation fits Slater's virtual oscillator theory where an individual atom is treated as a set of virtual oscillators. The virtual oscillator model dispenses the notion of stationary states and uses systematic perturbations of the electron's motions rather than orbits. In the virtual state model the incoming radiation excites the outer electron (the optical electron) to a virtual state from which it decays. Heisenberg found that this approach worked better than Bohr model. They showed that the transition frequencies are not constant although the frequencies (derived from Fourier transform) of sharp classical orbits are equally spaced. As they stated: "The electric moment of the system as a function of these variables is assumed by the following multiple Fourier series" (Kramers and Heisenberg 1925). Important to note that the Lorentz model for the electron is a charged particle on a 'spring' (a dipole oscillator with a simple harmonic oscillator) its position, x(n,t), where the stationary state is labeled by n, can be expressed as a Fourier series (see Chapter two) in the frequency of the oscillator.

$$x(n,t) = \sum_{\alpha=-\infty}^{\infty} a_{\alpha} e^{i\omega(n)t}$$

The x coordinate can be generalized to the real coordinate q (since x is usually restricted to x direction), the angular frequency can be changed into the linear frequency, and the n quantum state can be neglected. Then, the Fourier series can be written as:

$$q(t) = \sum_{r=-\infty}^{\infty} q_r e^{r(2\pi i v t)}$$

$$\omega = 2\pi v$$

Which can also be represented as:

$$q(t) = a_0 + a_1 \cos(2\pi vt) + a_2 \cos 2(2\pi vt) + a_3 \cos 3(2\pi vt) + ...$$
$$+b_1 \sin(2\pi vt) + b_2 \sin 2(2\pi vt) + b_3 \sin 3(2\pi vt) + ...$$

They gave rise to the formulation later known as the Kramers-Heisenberg dispersion formula (or Kramers-Heisenberg scattering cross-section) based on the correspondence principle applied to the classical dispersion formula for light. Later on, Dirac developed its quantum mechanical derivation which could be used for the cases where the classical analogies are obscure or non-existent (Dirac 1927).

The turning point for Heisenberg to create the matrix quantum mechanics was when he studied the problem of the emission of polarized light by fluorescent sodium and mercury vapors under a magnetic field (Hanle effect). The Hanle effect is the reduction in the polarization of the light from gaseous atoms (as a source of light) when it excites the gaseous atoms of the same substance in another compartment (resonance tube) under a magnetic field in a particular direction. The states that are degenerate at zero magnetic field are split due to the magnetic field (similarly to the Zeeman effect). As stated by Wood and Ellett: "The polarization is normally rather feeble with the bulb at room temperature, on account of the strong secondary resonance of the vapor surrounding the region traversed by the primary beam (...) the best results are obtained with the stem of the bulb cooled in a mixture of ice and salt (...) with the earth's field neutralized we have obtained as high as 90 percent of polarization" (Wood and Ellett 1923). Two main observations by Woor and Ellett were: (1) as the magnetic field increases the polarization percentage decreases; (2) The polarization of the light can be destroyed by a magnetic field in a certain direction (Wood and Ellett 1923). A typical experimental arrangement for the polarization of resonant fluorescent radiation consists of a source of radiation, a focusing lens, a polarizing prism where the polarized light reaches the resonance tube and the light emitted from the resonance tube at right angle is focused to another lens passing through another polarizing prism and a photocell which measures the intensity of the polarization.

To Bohr's mind, the Slater's model did not contradict his model because they could be related through the correspondence principle (Bohr 1924). Bohr assumed that the virtual oscillator model had common features with the core model (a specialized version of Bohr's general model). Heisenberg did not agree with the above-mentioned statement for the problem of the polarization of the fluorescent light. To him, the virtual oscillator model was the correct one because it dispenses the notion of stationary states and uses systematic perturbations of the electron motions. From the case of the resonance fluorescence on, Heisenberg used only the virtual oscillator model as a basic tool (MacKinnon 1977).

Pauli, a Heisenberg's colleague from Göttingen, probably somehow influenced this change in Heisenberg's position towards Bohr's model because he was a strong critic of Bohr's model. Pauli criticized the idea of an electrical field at a point and the way physicists accepted the electronic orbits as a reality of the atoms (MacKinnon 1977). Pauli had shown that was possible to explain the anomalous Zeeman effect without the use of Bohr's core model. *He stated that any closed shell must have a net angular momentum of zero and that the magnetic effects (the ones associated with the polarization of fluorescence light) attributed to the core must be related only to the optical electron.* In this paper Pauli was postulating a new fourth degree of the electron which admits only two values instead of a continuous distribution associated with the classical degrees of freedom (Pauli 1924).

2. The birth of modern quantum mechanics: Heisenberg's 1925's paper

Heisenberg's matrix mechanics disrupts with Borh's atomic model and provides the matrix elements that represent the dynamical variables which determine the intensities and the frequencies of radiation emitted.

Heisenberg began 1925's paper by disrupting with his own previous attachments with the old quantum theory and following Pauli's advice: "discard all hope of observing hitherto unobservable quantities, such as position and period of the electron, and to try to establish a theoretical quantum mechanics, analogous to classical mechanics, but in which only relations between observable quantities occur" (Aitchison et al. 2004). Heisenberg solved the problem in two different ways, (1) using the virtual oscillator model; and (2) treating the harmonic potential term as a perturbation to the harmonic oscillator, yielding the same results. Then, he realized that he could dispense with the interpretation of the virtual oscillations. Heisenberg was isolated in a small island (Helgoland) in North Sea to recover from a hay fever when he wrote this paper.

In Heisenberg's work (Heisenberg 1925), he stated that in the atomic range, classical mechanics is no longer valid and one of the main conditions was to fulfill the correspondence principle (for large quantum numbers, the new results should converge to those obtained in classical mechanics). *Heisenberg replaced the motion of the electron (a non-observable property) into the transition of the electron (an observable property).* He stated: "It is necessary to bear in mind that in quantum theory it has not been possible to associate the electron with a point in space, considered as a function of time, by means of observable quantities. However, even in quantum theory it is possible to ascribe to an electron the emission of radiation" (Heisenberg 1925).

The motion of the electron is represented in a Fourier series:

$$x(n,t)=\sum_{\alpha=-\infty}^{\infty} a_\alpha e^{i\omega(n)t}$$

Where n is the state of the electron and ω(n) is the fundamental frequency. In some texts, a_α, is replaced by $X_\alpha(n)$, and x(n,t) is replaced by q(n,t).

In order to characterize this radiation (the emission of radiation), it is needed the frequencies as a function of two variables (n, α) where n → n – α is the transition during the emission of radiation.

$$\nu(n,n-\alpha)=\frac{1}{h}\left[E(n)-E(n-\alpha)\right]$$

In Heisenberg's work, the motion of the electron became the transition of the electron from state n to m:

$$x_{nm}=a(n,m)e^{i\omega(n,m)t}$$

$$or: x(t)=X(n,n-\alpha)e^{i\omega(n,n-\alpha)t}$$

Where the transition amplitude a(n,m) or X(n, n – α) depends on two discrete variables. *The ensemble of all quantities x(t) can be represented in a matrix.*

Next, Heisenberg wanted to know the quantum meaning for $|x(t)|^2$. In classical theory, $|x(t)|^2$ is represented via Fourier series.

$$|x(t)|^2=\sum_{\alpha=-\infty}^{\infty} a_\alpha(n)a_{\beta-\alpha}(n)e^{i(\alpha+\beta-\alpha)\omega(n)t}$$

$$Then: |x(t)|^2=\sum_{\beta=-\infty}^{\infty} b_\beta(n)e^{i\beta\omega(n)t}$$

$$where: b_\beta(n)=\sum_{\alpha=-\infty}^{\infty} a_\alpha(n)a_{\beta-\alpha}$$

The square of *x(t)* in quantum theory depends only on the transitions, then the quantity $b_\beta(n)\exp\left[i\beta\omega(n)t\right]$ is expressed as: $b(n,n-\beta)\exp\left[i\omega(n,n-\beta)t\right]$ and then, we have:

$$b(n,n-\beta)e^{i\omega(n,n-\beta)t}=\sum_\alpha a(n,n-\alpha)a(n-\alpha,n-\beta)e^{i\omega(n,n-\beta)t}$$

This equation can be written as:

$$b(n,n-\beta)e^{i\omega(n,n-\beta)t}=\sum_\alpha a(n,n-\alpha)a(n-\alpha,n-\beta)e^{i\omega(n,n-\beta)t}$$

Then, we have:

$$b(n,n-\beta)=\sum_\alpha a(n,n-\alpha)a(n-\alpha,n-\beta)$$

This equation indicates that the term b is computed by summing up all the products of some group of terms a(n, n – α) by some group of terms a(n – α, n – β). In classical theory, the energy emitted per unit time in a transition is:

$$-\left(\frac{dE}{dt}\right)_\alpha=\frac{e^2}{3\pi\varepsilon_0 c^3}\left[\alpha\omega(n)\right]^4|a_\alpha(n)|^2$$

In Heisenberg's work, the energy derivative is replaced by the product of the transition probability per unit time, P(n, n – α):

$$P(n,n-\alpha)=\frac{e^2}{3\pi\varepsilon_0\hbar c^3}\left[\omega(n,n-\alpha)\right]^3\left|a(n,n-\alpha)\right|^2$$

The above equation refers to only one specific transition.

On the quantum theory of Heisenberg, q(t) or X(t) is expressed in terms of the amplitudes and frequencies of the spectral lines related to q(t). The frequencies, ν(nm), are the differences of the Ritz term series: T_1, T_2, T_3,..., where **T** is the diagonal matrix:

$$\mathbf{T}=\begin{pmatrix} T_1 & 0 & 0 & 0 \\ 0 & T_2 & \cdots & 0 \\ \vdots & \vdots & \ddots & 0 \\ 0 & 0 & 0 & T_n \end{pmatrix}$$

$$\nu(nm)=T_n-T_m$$

$$\nu(mn)=-\nu(nm)$$

$$\nu(nn)=0$$

Just as Fourier series represent a real coordinate q (sometimes named X), the square of modulus of q is:

$$q(mn)=q^*(nm)$$

$$then: q(nm)q(mn)=\left|q(nm)\right|^2$$

Where we have:

$$q=\sum_r q_r\exp(2\pi i r\nu t)$$

$$q^2=\left(\sum_\sigma q_\sigma\exp(2\pi i\sigma\nu t)\right)\left(\sum_{\sigma'} q_{\sigma'}\exp(2\pi i\sigma'\nu t)\right)$$

$$r=\sigma+\sigma'$$

$$q^2=\sum_r\sum_\sigma q_\sigma q_{r-\sigma}\exp(2\pi i r\nu t)$$

$$q^2=\sum_r Q_r\exp(2\pi i r\nu t)$$

$$where: Q_r=\sum_\sigma q_\sigma q_{r-\sigma}$$

3. Born and Jordan's contribution to quantum mechanics

Before publishing this work, Heisenberg delivered the manuscript to Born who realized that the last two equations were a matrix multiplication which Heisenberg did not know it at that time. Six months after the publication of Heisenberg's work

(Heisenberg 1925), Born and Jordan wrote a paper where they established the matrix form for all possible transitions (Born and Jordan 1925):

$$\mathbf{X}=\left(a(n,m)e^{i2\pi\nu(n,m)t}\right)$$

Then, we have:

$$\mathbf{X}=\begin{pmatrix} 0 & a(0,1)e^{2\pi i\nu(0,1)t} & a(0,2)e^{2\pi i\nu(0,2)t} & \cdots \\ a(1,0)e^{2\pi i\nu(1,0)t} & 0 & a(1,2)e^{2\pi i\nu(1,2)t} & \cdots \\ a(2,0)e^{2\pi i\nu(2,0)t} & a(2,1)e^{2\pi i\nu(2,1)t} & 0 & \cdots \\ \vdots & \vdots & \vdots & \ddots \end{pmatrix}$$

$$where: \nu(n,n)=0$$

The off-diagonal elements of the matrix correspond to transitions. Let us derive the matrix elements x(n,m) with respect to time.

$$\frac{dx(n,m)}{dt}=2\pi i\nu(n,m)a(n,m)e^{i2\pi\nu(n,m)t}$$

$$\frac{dx(n,m)}{dt}=2\pi i\nu(n,m)x(n,m)$$

$$where: x(n,m)=a(n,m)e^{i2\pi\nu(n,m)t}$$

Since:

$$E(n,m)=h\nu(n,m)\therefore \nu(n,m)=\frac{E(n,m)}{h}$$

Then,

$$\frac{dx(n,m)}{dt}=2\pi i\frac{E(n,m)}{h}x(n,m)=\frac{i}{\hbar}E(n,m)x(n,m)$$

Then, we come to a new matrix $\dot{\text{X}}$, that is *called the quantum mechanical equation of motion* or *Heisenberg equation of motion*:

$$\dot{\mathbf{X}}=\frac{i}{\hbar}(\mathbf{EX}-\mathbf{XE})$$

$$where: \mathbf{E}_n x(n,m)=(\mathbf{EX})(n,m)$$

$$and: \mathbf{E}_m x(n,m)=(\mathbf{XE})(n,m)$$

The last two relations are true because E is a diagonal matrix.
The Heisenberg equation of motion can be written in the form:

$$\frac{dA}{dt}=\frac{i}{\hbar}[\mathbf{H},\mathbf{A}]=\frac{i}{\hbar}(\mathbf{HA}-\mathbf{AH})$$

Where A is the observable and the Hamiltonian H is equivalent to E. In Heisenberg equation of motion, the state vectors are time-independent while the operators (and observables) are time-dependent.

The matrix **X** is Hermetian because ν(n, m) = −ν(m, n) and a(m, n) = a*(n, m), then x(n, m) = x*(m, n). When transposing a Hermetian matrix, each component becomes its complex conjugate. Then, the elements of a matrix multiplication $\mathbf{XX}^{\dagger}$ the elements of the product matrix are:

$$\mathrm{a}(n,m)\,a(m,n)=\mathrm{a}(n,m)\,\mathrm{a}^{*}(n,m)=\left|a(n,m)\right|^{2}$$

Where $|a(n, m)|^2$ is the probability for the transitions between n and m states. Later, Born discovered the connection between $|\Psi(x)|^2$ and position probability via a quantum electrodynamic argument.

Born said that "And one morning about 10 July 1925, I suddenly saw the light: Heisenberg's symbolic multiplication was nothing but the matrix calculus well known to me since my student days from the lectures of Rosanes in Breslau (...). I meant that the two matrix products **PQ** and **QP** are not identical. I was familiar with the fact that matrix multiplication is not commutative" (Fedak and Prentis 2009). Born and Jordan introduced the position and momentum matrices in a similar way as described previously for matrix **X**:

$$\mathbf{Q}=\left(q(n,m)\,e^{i2\pi\nu(n,m)t}\right)$$

$$\mathbf{P}=\left(p(n,m)\,e^{i2\pi\nu(n,m)t}\right)$$

As Fedak and Prentis highlighted: "For Born and Jordan, **Q** and **P** do not specify the position and (linear) momentum of an electron in an atom. Heisenberg stressed that quantum theory should focus only on the observable properties, namely the frequency and intensity of the atomic radiation and not the position and period of the electron. The quantities **Q** and **P** represent position and momentum in the sense that **Q** and **P** satisfy matrix equations of motion that are identical in the form to those satisfied by the position and momentum of classical mechanics (...). In Heisenberg-Born-Jordan atom there is no longer orbit, but there is some sort of periodic 'quantum motion' of the electron characterized by the set of frequencies ν(n, m) and amplitudes q(n, m). Physicists believed that something inside the atom must vibrate with the right frequencies even though they could not visualize what the quantum oscillations looked like" (Fedak and Prentis 2009).

From Hamiltonian matrix, one gets the derivatives of **Q** and **P** with respect to **P** and **Q**, respectively.

$$\dot{\mathbf{Q}}=\frac{\partial\mathbf{H}}{\partial\mathbf{P}}=\frac{1}{m}\mathbf{P}=\frac{2\pi i}{h}[\mathbf{H},\mathbf{Q}]$$

$$\dot{\mathbf{P}}=-\frac{\partial\mathbf{H}}{\partial\mathbf{Q}}=-\frac{\partial\mathbf{U}}{\partial\mathbf{Q}}=\frac{2\pi i}{h}[\mathbf{H},\mathbf{P}]$$

And they stated that the diagonal elements H(n, n) of **H** are the energies of the various states of the system. In matrix form, energy of the states of the electron are naturally quantized.

The commutator of **P** and **Q** is:

$$[\mathbf{P},\mathbf{Q}]=(\mathbf{PQ}-\mathbf{QP})$$

The differentiation of the commutator with respect to time is:

$$\frac{d[\mathbf{P},\mathbf{Q}]}{dt} = \left[\dot{\mathbf{P}},\mathbf{Q}\right] + \left[\mathbf{P},\dot{\mathbf{Q}}\right]$$

$$where: \dot{\mathbf{P}} = \frac{d\mathbf{P}}{dt} = m\frac{d\mathbf{V}}{dt}$$

$$and: \dot{\mathbf{Q}} = \frac{d\mathbf{Q}}{dt} = \mathbf{V}$$

$$then: \dot{\mathbf{P}} = m\ddot{\mathbf{Q}} = f(Q)$$

where V is the velocity matrix. The last relation above is the Newton's law in matrix form. Let us replace the Newton's law in the time derivative of the commutator.

$$\frac{d[\mathbf{P},\mathbf{Q}]}{dt} = [f(\mathbf{Q}),\mathbf{Q}] + \left[\mathbf{P},\frac{\mathbf{P}}{m}\right]$$

$$where: \dot{\mathbf{Q}} = \mathbf{V}$$

$$and: \mathbf{P} = m\mathbf{V}$$

$$then: \dot{\mathbf{Q}} = \frac{\mathbf{P}}{m}$$

Which yields:

$$\frac{d[\mathbf{P},\mathbf{Q}]}{dt} = f(\mathbf{Q})\mathbf{Q} - \mathbf{Q}f(\mathbf{Q}) + \frac{\mathbf{P}^2}{m} - \frac{\mathbf{P}^2}{m} = 0$$

Where $f(\mathbf{Q})$ commutes with **Q**. Consequently, the commutator [**P**,**Q**] is a constant matrix, i.e., it does not vary with time. In addition, it is a diagonal matrix.
The Hamiltonian matrix is given by:

$$\mathbf{H} = \frac{1}{2m}\mathbf{P}^2 + \mathbf{V}(\mathbf{Q})$$

Where **V**(**Q**) and **P** are the potential energy and the momentum matrices. Since the total energy of the system is constant, then $\dot{H} = 0$ and the Hamiltonian matrix is constant with time. As a consequence, it must be a diagonal matrix like the commutator [**P**,**Q**]. Therefore, the diagonal elements of the Hamiltonian matrix, **H**(i,i) are the constant energy of the ith system state, i.e., **H**(i,i) = **E**(i). According to the equation:

$$\dot{\mathbf{X}} = \frac{i}{\hbar}(\mathbf{EX} - \mathbf{XE})$$

We have that:

$$\dot{\mathbf{Q}} = \frac{i}{\hbar}(\mathbf{HQ} - \mathbf{XQ})$$

$$where: \mathbf{X} = \mathbf{Q}$$

$$as: \mathbf{P} = m\dot{\mathbf{Q}}$$

$$then: \mathbf{P} = \frac{i}{\hbar}m[\mathbf{H},\mathbf{Q}]$$

Since [**V**(**Q**),**Q**] = 0, we have:

$$\mathbf{P} = \frac{i}{\hbar}m(\mathbf{HQ} - \mathbf{QH})$$

$$\mathbf{P} = \frac{i}{\hbar}m\left(\left(\frac{1}{2m}\mathbf{P}^2 + \mathbf{V(Q)}\right)\mathbf{Q} - \mathbf{Q}\left(\frac{1}{2m}\mathbf{P}^2 + \mathbf{V(Q)}\right)\right)$$

$$\mathbf{P} = \frac{i}{\hbar}m\left(\frac{1}{2m}\mathbf{P}^2\mathbf{Q} + \mathbf{V(Q)Q} - \mathbf{Q}\frac{1}{2m}\mathbf{P}^2 - \mathbf{QV(Q)}\right)$$

$$\mathbf{P} = \frac{i}{\hbar}m\left(\frac{1}{2m}\mathbf{P}^2\mathbf{Q} - \mathbf{Q}\frac{1}{2m}\mathbf{P}^2\right) + [V(\mathbf{Q}),\mathbf{Q}]$$

$$[\mathbf{V(Q)},\mathbf{Q}] = 0$$

$$\mathbf{P} = \frac{i}{\hbar}m\left(\frac{1}{2m}\mathbf{P}^2\mathbf{Q} - Q\frac{1}{2m}\mathbf{P}^2\right) = \frac{\pi i}{h}\left[\mathbf{P}^2,\mathbf{Q}\right]$$

The commutator [$\mathbf{P}^2$,**Q**] is:

$$\left[\mathbf{P}^2,\mathbf{Q}\right] = \mathbf{P}^2\mathbf{Q} - \mathbf{QP}^2 - \mathbf{PQP} + \mathbf{PQP}$$

$$\left[\mathbf{P}^2,\mathbf{Q}\right] = \mathbf{P}(\mathbf{PQ}) - (\mathbf{QP})\mathbf{P} - \mathbf{P}(\mathbf{QP}) + (\mathbf{PQ})\mathbf{P}$$

$$\left[\mathbf{P}^2,\mathbf{Q}\right] = \mathbf{P}(\mathbf{PQ}) - \mathbf{P}(\mathbf{QP}) + (\mathbf{PQ})\mathbf{P} - (\mathbf{QP})\mathbf{P}$$

$$\left[\mathbf{P}^2,\mathbf{Q}\right] = \mathbf{P}[\mathbf{P},\mathbf{Q}] + [\mathbf{P},\mathbf{Q}]\mathbf{P}$$

The equation for the momentum matrix can be simplified to:

$$\mathbf{P} = \frac{\pi i}{h}\left[\mathbf{P}^2,\mathbf{Q}\right]$$

$$\mathbf{P} = \frac{\pi i}{h}(\mathbf{P}[\mathbf{P},\mathbf{Q}] + [\mathbf{P},\mathbf{Q}]\mathbf{P})$$

Since the commutator [**P**,**Q**] is a diagonal matrix, let us replace [**P**,**Q**] by the diagonal matrix **D**:

$$\mathbf{P} = \frac{\pi i}{h}(\mathbf{PD} + \mathbf{DP})$$

$$\mathbf{PD} + \mathbf{DP} = \frac{h}{\pi i}\mathbf{P}$$

In terms of the components of the above equation, we have:

$$p(n,m)d_m + d_n p(n,m) = \frac{h}{\pi i}p(n,m)$$

Since p(n, m) ≠ 0 for n ≠ m, we have:

$$(d_m + d_n)\,p(n,m) = \frac{h}{\pi i}p(n,m)$$

$$d_m + d_n = \frac{h}{\pi i}$$

$$as: d_m = d_n$$

$$then: d_m = d_n = \frac{h}{2\pi i} = \frac{\hbar}{i}$$

Then, the commutator [P,Q] or the diagonal matrix D is:

$$\mathbf{D} = [\mathbf{P}, \mathbf{Q}] = \frac{\hbar}{i}1$$

Then, we come to the quantum condition:

$$[\mathbf{P}, \mathbf{Q}] = \mathbf{PQ\text{-}QP} = \frac{\hbar}{i}\mathbf{I} = -i\hbar\mathbf{I}$$

$$where: \frac{1}{i} = \frac{1}{i} \cdot \frac{i}{i} = \frac{i}{i^2} = \frac{i}{-1} = -i$$

and I is the identity matrix.
On the other hand:

$$[\mathbf{Q}, \mathbf{P}] = i\hbar\mathbf{I}$$

4. Born, Heisenberg and Jordan (third work of matrix mechanics)

The canonical transformation is used in classical mechanics to change variables **q** and **p** into **Q** and **P**, respectively. In the third work of Matrix mechanics (Born et al. 1925), it was used the canonical transformation of the variables **p**, **q** to yield new variables **P**, **Q** where the following relation must be respected:

$$\mathbf{pq} - \mathbf{qp} = \mathbf{PQ} - \mathbf{QP} = \frac{h}{2\pi i}$$

The applied canonical transformation in matrix mechanics is:

$$\mathbf{P} = \mathbf{SpS}^{-1}$$

$$\mathbf{Q} = \mathbf{SqS}^{-1}$$

$$where: \mathbf{S}\tilde{\mathbf{S}}^{*} = 1, \quad \tilde{\mathbf{S}}(nm) = \mathbf{S}(mn)$$

Where S is an arbitrary Hermite transformation matrix and the above equation represents the most general canonical transformation.

Now, we seek for a transformation matrix **S** that gives a new Hamiltonian matrix **H(P,Q)** from **H**(p,q):

$$\mathrm{H(P,Q)} = \mathrm{SH(p,q)S}^{-1}$$

If the new **H(P,Q)** is a diagonal matrix **E**:

$$\mathbf{SH(p,q)S}^{-1} = \mathbf{E}$$

Then, we obtain the energy of the system. The above equation is analogue to the Hamilton-Jacobi equation (see Chapter ten). As a consequence, we have:

$$E(nm) = \begin{cases} E(n), n = m \\ 0, n \neq m \end{cases}$$

$$\nu(nm) = \frac{E_n - E_m}{h}$$

On multiplying behind the above equation by **S** we obtain the relation:

$$\mathbf{SH(p,q)S^{-1}S = ES}, \quad \mathbf{S^{-1}S = I}$$

$$then: \mathbf{SH(p,q) = ES}$$

Now, we see one application of the matrix mechanics for the case of the Planck's oscillator where the Hamiltonian function is:

$$H = \frac{1}{2m_0} p^2 + \frac{1}{2} kq^2$$

From the Hamiltonian equations:

$$F = -kq = ma$$

$$p = m\frac{dq}{dt} = m\dot{q} \therefore \dot{q} = \frac{p}{m}$$

$$a = \ddot{q} = \frac{d}{dt}\dot{q} = \frac{1}{m}\frac{dp}{dt} = \frac{\dot{p}}{m}$$

$$F = ma = \frac{dp}{dt} = \dot{p}$$

$$\frac{dp}{dt} = \dot{p} = -kq$$

$$\ddot{q} = \frac{\dot{p}}{m} = -\frac{kq}{m}$$

$$\ddot{q} + \frac{k}{m}q = 0$$

$$\omega = 2\pi\nu = \sqrt{\frac{k}{m}} \therefore \frac{k}{m} = (2\pi\nu)^2$$

We obtain the equation:

$$\ddot{q} + (2\pi\nu)^2 q = 0$$

And the Hamiltonian function can be written as:

$$H = \frac{1}{2m_0}\left(m_0\dot{q}\right)^2 + \frac{1}{2}(2\pi\nu)^2 m_0 q^2$$

$$H = \frac{m_0}{2}\dot{q}^2 + \frac{1}{2}(2\pi\nu)^2 m_0 q^2$$

Note that the linear momentum in the matrix mechanics is depicted below. The proof for this equation is at the beginning of the Section 3.

$$p = m_0\dot{q} = \left(m_0 2\pi i\nu(nm)q(nm)\right)$$

Then, from the quantum condition:

$$qp - pq = \frac{ih}{2\pi} I$$

We obtain its diagonal elements as:

$$\sum_k \begin{Bmatrix} q(nk)m_0 2\pi i v(kn)q(kn)- \\ -m_0 2\pi i v(nk)q(nk)q(nk) \end{Bmatrix} = \frac{ih}{2\pi}$$

$$where: v(kn) = -v(nk)$$

$$and: q(nk)q(kn) = |q(kn)|^2$$

$$-\sum_k 4m_0 \pi i v(nk)|q(kn)|^2 = \frac{ih}{2\pi}$$

$$then: \sum_k v(nk)|q(kn)|^2 = -\frac{h}{8\pi^2 m_0}$$

The Hamiltonian matrix **H** becomes:

$$\mathbf{H} = \left(2\pi^2 m_0 \sum_k \begin{Bmatrix} -v(nk)q(nk)v(km)q(km)+ \\ v_0^2 q(nk)q(km) \end{Bmatrix} \right)$$

The energy **E**(n) of the nth stationary state is **H**(nn) which are diagonal elements of **H** (a diagonal matrix).

$$\mathbf{E}(n) = \mathbf{H}(nn) = 2\pi^2 m_0 \sum_k \left\{ v_0^2 + v^2(nk) \right\} |q(nk)|^2$$

Further development using the above equation and the equations below:

$$\ddot{q} + (2\pi v)^2 q = 0$$

$$\sum_k v(nk)|q(kn)|^2 = -\frac{h}{8\pi^2 m_0}$$

Yields the solution (Birtwistle 1928):

$$\mathbf{E}_n = \mathbf{H}(nn) = hv_0\left(n + \frac{1}{2}\right)$$

In Schrödinger's words about the matrix mechanics: "Heisenberg's theory connects the solution of a problem in quantum mechanics with the solution of a system of an infinite number of algebraic equations, in which the unknown—infinite matrices—are allied to the classical position and momentum coordinates of the mechanical system, and functions of these, and obey peculiar calculating rules" (Schrödinger 1926).

Important to add that Pauli also gave an important contribution to the matrix mechanics with the study of the hydrogen spectrum where the influence of external electric and magnetic fields were analyzed from the point of view the matrix mechanics (Pauli 1926). He found the correct energy spectrum of the hydrogen atom as well as the correct Stark effect corrections to the energy. Curiously, in this work Pauli used a new commutator, the **[E,x]** commutator:

$$\mathbf{EX} - \mathbf{XE} = \frac{\hbar}{i}\dot{\mathbf{X}}$$

$$\dot{\mathbf{X}} = \frac{d\mathbf{X}}{dt}, \quad \mathbf{H}(p,q) = \mathbf{E}(diagonalmatrix)$$

5. Dirac's complementation to the matrix mechanics

From the classical dynamics when x and y are functions of the coordinates and linear momentum, respectively, the expression:

$$\sum_{1}^{k}\left(\frac{\partial x}{\partial q_k}\frac{\partial y}{\partial p_k} - \frac{\partial x}{\partial p_k}\frac{\partial y}{\partial q_k}\right) = [xy]$$

Where [xy] is the Poisson bracket, gives the following relations:

$$[p_r p_s] = 0, [q_r q_s] = 0, [q_r p_r] = \begin{cases} 1, r = s \\ 0, r \neq s \end{cases}$$

After some derivation of the Poisson bracket relations, Dirac found the commutation relation for the two quantum magnitudes x and y:

$$xy - yx = \frac{i\hbar}{2\pi}[xy]$$

Then, Dirac arrived at the relations below known as Dirac-Heisenberg rules for q's and p's (Dirac 1925).

$$p_r p_s - p_s p_r = 0$$

$$q_r q_s - q_s q_r = 0$$

$$q_r p_s - p_s q_r = \begin{cases} \frac{i\hbar}{2\pi}, r = s \\ 0, r \neq s \end{cases}$$

6. Vector states and expectation value

An expectation value ⟨A⟩ of the observable A is the mean value of a series of the specific measurements of the quantum state vector. The quantum state vector provides a probability distribution for the outcomes of each measurement of a particle. There is no state which is simultaneously an eigenstate for all observables.

The state vector depends on the observable to be measured. For example, for the energy of the electron in hydrogen atom, the state vector is identified by four quantum numbers (n, l, m, s). When the spin of the electron is measured in any direction, there are two possible results: up (α) and down (β) which are represented by a two-dimensional complex vector (α, β).

Usually, a system is made up with a superposition of multiple different eigenstates which have a quantum uncertainty for a given observable and the vector state, ε, is represented by a linear combination of eigenstates $c_i e_i$. Each $c_i e_i$ describes the possibility of measurement related to the observable **A** to give the eigenvalue a_i.

$$\varepsilon = \sum_i c_i e_i$$

If ε and e_i are normalized, the product $c_i^* c_i = |c_i^2|$ is the probability that the measurement gives the eigenvalue a_i, as stated by Born.

$$\text{Probability density}: c_i^* c_i = \left|c_i^2\right|$$

$$\sum_i c_i^* c_i = 1$$

The measured mean value or the expectation value is:

$$\langle \mathbf{A} \rangle = \sum_i c_i^* c_i a_i = \varepsilon^\dagger \mathbf{A} \varepsilon$$

While the state of a particle is described by a vector, the state of an ensemble of particles is best described by density matrix, **D**.

$$\mathbf{D} = \varepsilon\varepsilon^\dagger = \sum_i c_i e_i \sum_j c_j^* e_j^T$$

The trace of the density matrix is:

$$trace(\mathbf{D}) = \sum_i \left|c_i^2\right| = 1$$

The square of the density matrix equals the density matrix:

$$\mathbf{D}^2 = \varepsilon\varepsilon^\dagger\varepsilon\varepsilon^\dagger \;\therefore\; \varepsilon^\dagger\varepsilon = 1$$

$$\mathbf{D}^2 = \varepsilon\varepsilon^\dagger = \mathbf{D}$$

The expectation value can be given from the trace of the matrix product **DA**.

$$\langle \mathbf{A} \rangle = trace(\mathbf{DA})$$

7. Matrix mechanics solution to the one-particle harmonic oscillator

The wave mechanics solution to the problem of one particle harmonic oscillator was already given in Chapter four and partially in Section 4. Here we focus on its solution using another approach of the quantum matrix mechanics.

One particle oscillator is under the restoring force from Hooke's law which is equivalent to Newton's law of the force.

$$m\frac{d^2x}{dt^2} = -kx$$

Where the particle of mass m attached to a spring with elastic constant k moves in the x-axis. The solution for this problem was given in Chapter 4 and the equation of motion of this problem is given below

$$\frac{d^2x}{dt^2} = -\frac{k}{m}x$$

$$solution: x(t) = x_m \cos(\omega t) + x_m \sin(\omega t)$$

$$where: \omega = \sqrt{\frac{k}{m}}$$

And ω is the angular frequency. The Hamiltonian function for this system is:

$$F = -\frac{dE_p(x)}{dx}, \qquad E_p = -\int F dx$$

$$E_p = -\int(-kx)\,dx = \frac{1}{2}kx^2$$

$$\omega = \sqrt{\frac{k}{m}}, \qquad k = \omega^2 m$$

$$H = \frac{m}{2}\dot{x}^2 + \frac{m}{2}\omega_0^2 x^2$$

$$p = m\dot{x}$$

$$H = \frac{1}{2m}p^2 + \frac{m}{2}\omega_0^2 x^2$$

The above equation can be modified to:

$$H = \hbar\omega_0\left(\frac{1}{2m\hbar\omega_0}p^2 + \frac{m}{2\hbar}\omega_0 x^2\right)$$

In matrix form: $\mathbf{H} = \hbar\omega_0\left(\frac{1}{2m\hbar\omega_0}\mathbf{P}^2 + \frac{m\omega_0}{2\hbar}\mathbf{X}^2\right)$

where, for simplicity, we state that: $\widetilde{\mathbf{P}} = \sqrt{\frac{1}{2m\hbar\omega_0}}\mathbf{P}, \qquad \mathbf{P} = \sqrt{2m\hbar\omega_0}\,\widetilde{\mathbf{P}}$

$$and: \widetilde{\mathbf{X}} = \sqrt{\frac{m\omega_0}{2\hbar}}\mathbf{X}, \qquad \mathbf{X} = \sqrt{\frac{2\hbar}{m\omega_0}}\widetilde{\mathbf{X}}$$

$$\mathbf{H} = \hbar\omega_0\left(\widetilde{\mathbf{P}}^2 + \widetilde{\mathbf{X}}^2\right)$$

Let us define the matrix **A** and its transpose:

$$\mathbf{A} = \widetilde{\mathbf{X}} + i\widetilde{\mathbf{P}}$$

$$\mathbf{A}^\dagger = \widetilde{\mathbf{X}} - i\widetilde{\mathbf{P}}$$

And the commutative relation between them is:

$$\left[\mathbf{A},\mathbf{A}^\dagger\right]=\mathbf{A}\mathbf{A}^\dagger-\mathbf{A}^\dagger\mathbf{A}$$

$$\left[\mathbf{A},\mathbf{A}^\dagger\right]=\left(\tilde{\mathbf{X}}+i\tilde{\mathbf{P}}\right)\left(\tilde{\mathbf{X}}-i\tilde{\mathbf{P}}\right)-\left(\tilde{\mathbf{X}}-i\tilde{\mathbf{P}}\right)\left(\tilde{\mathbf{X}}+i\tilde{\mathbf{P}}\right)$$

$$\left(\tilde{\mathbf{X}}+i\tilde{\mathbf{P}}\right)\left(\tilde{\mathbf{X}}-i\tilde{\mathbf{P}}\right)=\tilde{\mathbf{X}}^2+\tilde{\mathbf{P}}^2-i\left(\tilde{\mathbf{X}}\tilde{\mathbf{P}}-\tilde{\mathbf{P}}\tilde{\mathbf{X}}\right)$$

$$\left(\tilde{\mathbf{X}}-i\tilde{\mathbf{P}}\right)\left(\tilde{\mathbf{X}}+i\tilde{\mathbf{P}}\right)=\tilde{\mathbf{X}}^2+\tilde{\mathbf{P}}^2-i\left(\tilde{\mathbf{P}}\tilde{\mathbf{X}}-\tilde{\mathbf{X}}\tilde{\mathbf{P}}\right)$$

$$\left[\mathbf{A},\mathbf{A}^\dagger\right]=-i\left(\tilde{\mathbf{X}}\tilde{\mathbf{P}}-\tilde{\mathbf{P}}\tilde{\mathbf{X}}\right)+i\left(\tilde{\mathbf{P}}\tilde{\mathbf{X}}-\tilde{\mathbf{X}}\tilde{\mathbf{P}}\right)$$

$$\left[\mathbf{A},\mathbf{A}^\dagger\right]=-i\left[\left(\sqrt{\frac{m\omega_0}{2\hbar}}\mathbf{X}\right)\left(\sqrt{\frac{1}{2m\hbar\omega_0}}\mathbf{P}\right)-\left(\sqrt{\frac{1}{2m\hbar\omega_0}}\mathbf{P}\right)\left(\sqrt{\frac{m\omega_0}{2\hbar}}\mathbf{X}\right)\right]+$$

$$+i\left[\left(\sqrt{\frac{1}{2m\hbar\omega_0}}\mathbf{P}\right)\left(\sqrt{\frac{m\omega_0}{2\hbar}}\mathbf{X}\right)-\left(\sqrt{\frac{m\omega_0}{2\hbar}}\mathbf{X}\right)\left(\sqrt{\frac{1}{2m\hbar\omega_0}}\mathbf{P}\right)\right]$$

$$\left[\mathbf{A},\mathbf{A}^\dagger\right]=-i\left[\left(\sqrt{\frac{m\omega_0}{4\hbar^2m\omega_0}}\mathbf{XP}\right)-\left(\sqrt{\frac{m\omega_0}{4\hbar^2m\omega_0}}\mathbf{PX}\right)\right]+$$

$$+i\left[\left(\sqrt{\frac{m\omega_0}{4\hbar^2m\omega_0}}\mathbf{PX}\right)\left(\sqrt{\frac{m\omega_0}{4\hbar^2m\omega_0}}\mathbf{XP}\right)\right]$$

$$\left[\mathbf{A},\mathbf{A}^\dagger\right]=-i\left[\sqrt{\frac{m\omega_0}{4\hbar^2m\omega_0}}(\mathbf{XP}-\mathbf{PX})\right]+i\left[\sqrt{\frac{m\omega_0}{4\hbar^2m\omega_0}}(\mathbf{PX}-\mathbf{XP})\right]$$

$$\left[\mathbf{A},\mathbf{A}^\dagger\right]=-\frac{i}{2\hbar}(\mathbf{XP}-\mathbf{PX})+\frac{i}{2\hbar}(\mathbf{PX}-\mathbf{XP})$$

$$where:\mathbf{XP}-\mathbf{PX}=i\hbar I$$

$$And:\mathbf{PX}-\mathbf{XP}=-i\hbar I$$

$$Then:\left[\mathbf{A},\mathbf{A}^\dagger\right]=-\frac{i}{2\hbar}(i\hbar\mathbf{I})+\frac{i}{2\hbar}(-i\hbar\mathbf{I})=\mathbf{I}$$

As we have seen in the derivation of the above commutative relation, we have:

$$\mathbf{A}^\dagger\mathbf{A}=\tilde{\mathbf{X}}^2+\tilde{\mathbf{P}}^2-i\left(\tilde{\mathbf{P}}\tilde{\mathbf{X}}-\tilde{\mathbf{X}}\tilde{\mathbf{P}}\right)=\tilde{\mathbf{X}}^2+\tilde{\mathbf{P}}^2-\frac{1}{2}\mathbf{I}$$

$$Then:\tilde{\mathbf{X}}^2+\tilde{\mathbf{P}}^2=\mathbf{A}^\dagger\mathbf{A}+\frac{1}{2}\mathbf{I}$$

Let us substitute the above equation in the Hamiltonian matrix equation:

$$\mathbf{H}=\hbar\omega_0\left(\tilde{\mathbf{P}}^2+\tilde{\mathbf{X}}^2\right)=\hbar\omega_0\left(\mathbf{A}^\dagger\mathbf{A}+\frac{1}{2}\mathbf{I}\right)$$

The commutation relation between H and A is:

$$[\mathbf{H},\mathbf{A}]=\left[\hbar\omega_0\left(\mathbf{A}^{\dagger}\mathbf{A}+\frac{1}{2}\mathbf{I}\right)\mathbf{A}-\mathbf{A}\hbar\omega_0\left(\mathbf{A}^{\dagger}\mathbf{A}+\frac{1}{2}\mathbf{I}\right)\right]$$

$$[\mathbf{H},\mathbf{A}]=\left[\hbar\omega_0\left(\mathbf{A}^{\dagger}\mathbf{A}\mathbf{A}+\frac{1}{2}\mathbf{A}\right)-\hbar\omega_0\left(\mathbf{A}\mathbf{A}^{\dagger}\mathbf{A}+\frac{1}{2}\mathbf{A}\right)\right]$$

$$[\mathbf{H},\mathbf{A}]=\left[\hbar\omega_0\mathbf{A}^{\dagger}\mathbf{A},\mathbf{A}\right]=\hbar\omega_0\mathbf{A}^{\dagger}\mathbf{A}\mathbf{A}-\hbar\omega_0\mathbf{A}\mathbf{A}^{\dagger}\mathbf{A}$$

$$[\mathbf{H},\mathbf{A}]=\hbar\omega_0\left[\mathbf{A}^{\dagger},\mathbf{A}\right]\mathbf{A}$$

$$As:\left[\mathbf{A},\mathbf{A}^{\dagger}\right]=\mathbf{I}$$

$$Then:\left[\mathbf{A}^{\dagger},\mathbf{A}\right]=-\mathbf{I}$$

$$[\mathbf{H},\mathbf{A}]=-\hbar\omega_0\mathbf{A}$$

In a similar algebraic procedure, we find that the commutation relation between **H** and $\mathbf{A}^{\mathrm{T}}$ is:

$$\left[\mathbf{H},\mathbf{A}^{\dagger}\right]=\hbar\omega_0\mathbf{A}^{\dagger}$$

Since we have to find the eigenvalues, λ, and the eigenvectors, **e**, for this problem, we use the following equation:

$$\mathbf{He}=\lambda\mathbf{e}$$

Let us multiply $[\mathbf{H},\mathbf{A}]=-\hbar\omega_0\mathbf{A}$ by the eigenvector **e** from the right:

$$[\mathbf{H},\mathbf{A}]\mathbf{e}=-\hbar\omega_0\mathbf{Ae}$$

$$\mathbf{HAe}-\mathbf{AHe}=-\hbar\omega_0\mathbf{Ae}$$

$$\mathbf{HAe}=\mathbf{AHe}-\hbar\omega_0\mathbf{Ae}$$

And now, since $\mathbf{He}=\lambda\mathbf{e}$ we have:

$$\mathbf{HAe}=\mathbf{AHe}-\hbar\omega_0\mathbf{Ae}$$

$$\mathbf{HAe}=\mathbf{A}\lambda\mathbf{e}-\hbar\omega_0\mathbf{Ae}$$

$$\mathbf{HAe}=(\lambda-\hbar\omega_0)\,\mathbf{Ae}$$

Let us multiply $[\mathbf{H},\mathbf{A}]\mathbf{e}=-\hbar\omega_0\mathbf{Ae}$ by **Ae** from the right:

$$(\mathbf{HA}-\mathbf{AH})\,\mathbf{Ae}=-\hbar\omega_0\mathbf{AAe}$$

$$\mathbf{HA}^2\mathbf{e}-\mathbf{AHAe}=-\hbar\omega_0\mathbf{A}^2\mathbf{e}$$

$$\mathbf{HA}^2\mathbf{e}=\mathbf{AHAe}-\hbar\omega_0\mathbf{A}^2\mathbf{e}$$

Let us multiply $\mathbf{HAe} = (\lambda - \hbar\omega_0)\mathbf{Ae}$ by $\mathbf{A}$ from the left:

$$\mathbf{AHAe} = \mathbf{A}(\lambda - \hbar\omega_0)\mathbf{Ae}$$

$$\mathbf{AHAe} = (\lambda - \hbar\omega_0)\mathbf{A}^2\mathbf{e}$$

Let us replace the above equation in $\mathbf{HA}^2\mathbf{e} = \mathbf{AHAe} - \hbar\omega_0\mathbf{A}^2\mathbf{e}$, then we have:

$$\mathbf{HA}^2\mathbf{e} = \mathbf{AHAe} - \hbar\omega_0\mathbf{A}^2\mathbf{e}$$

$$\mathbf{HA}^2\mathbf{e} = (\lambda - \hbar\omega_0)\mathbf{A}^2\mathbf{e} - \hbar\omega_0\mathbf{A}^2\mathbf{e}$$

$$\mathbf{HA}^2\mathbf{e} = (\lambda - 2\hbar\omega_0)\mathbf{A}^2\mathbf{e}$$

From the last operations above, we have three eigenvalue equations:

$$\mathbf{He} = \lambda\mathbf{e}$$

$$\mathbf{HAe} = (\lambda - \hbar\omega_0)\mathbf{Ae}$$

$$\mathbf{HA}^2\mathbf{e} = (\lambda - 2\hbar\omega_0)\mathbf{A}^2\mathbf{e}$$

Each product of the $\mathbf{He} = \lambda\mathbf{e}$ by A gives a $\hbar\omega_0$ reduction of the eigenvalue of **H**, but it can never be negative (there is no state with energy lower than zero) and it has to end at some point before becoming negative. *The matrix A is called annihilation operator.* At this point, we establish that we will have the eigenvector e_0 where $\mathbf{Ae}_0 = 0$. In this case, we have:

$$As: \mathbf{H} = \hbar\omega_0\left(\mathbf{A}^\dagger\mathbf{A} + \frac{1}{2}\mathbf{I}\right)$$

$$\mathbf{He}_0 = \left(\hbar\omega_0\mathbf{A}^\dagger\mathbf{A} + \frac{1}{2}\hbar\omega_0\mathbf{I}\right)e_0 = \hbar\omega_0\mathbf{A}^\dagger\mathbf{Ae}_0 + \frac{1}{2}\hbar\omega_0\mathbf{e}_0$$

$$where: \mathbf{Ae}_0 = 0$$

$$Then: \mathbf{He}_0 = \frac{1}{2}\hbar\omega_0\mathbf{e}_0$$

By multiplying $[\mathbf{H}, \mathbf{A}^\dagger] = \hbar\omega_0\mathbf{A}^\dagger$ with e_0 from the right, we have:

$$[\mathbf{H}, \mathbf{A}^\dagger] = \hbar\omega_0\mathbf{A}^\dagger$$

$$\mathbf{HA}^\dagger\mathbf{e}_0 - \mathbf{A}^\dagger\mathbf{He}_0 = \hbar\omega_0\mathbf{A}^\dagger\mathbf{e}_0$$

$$where: \mathbf{He}_0 = \frac{1}{2}\hbar\omega_0\mathbf{e}_0$$

$$then: \mathbf{HA}^\dagger\mathbf{e}_0 - \frac{1}{2}\hbar\omega_0\mathbf{e}_0 = \hbar\omega_0\mathbf{A}^\dagger\mathbf{e}_0$$

$$\mathbf{HA}^\dagger\mathbf{e}_0 = \hbar\omega_0\left(1 + \frac{1}{2}\right)\mathbf{A}^\dagger\mathbf{e}_0$$

If we multiply $\left[\mathbf{H},\mathbf{A}^{\dagger}\right]=\hbar\omega_0\mathbf{A}^{\dagger}$ with the new eigenvector $A^{\dagger}e_0$ from the right, gives:

$$\mathbf{HA}^{\dagger}\mathbf{e}_0\mathbf{A}^{\dagger}-\mathbf{A}^{\dagger}\mathbf{He}_0\mathbf{A}^{\dagger}=\hbar\omega_0\mathbf{A}^{\dagger}\mathbf{e}_0\mathbf{A}^{\dagger}$$

$$\mathbf{H}\left(\mathbf{A}^{\dagger}\right)^2\mathbf{e}_0-\mathbf{A}^{\dagger}\mathbf{HA}^{\dagger}\mathbf{e}_0=\hbar\omega_0\left(\mathbf{A}^{\dagger}\right)^2\mathbf{e}_0$$

$$where:\mathbf{HA}^{\dagger}\mathbf{e}_0=\hbar\omega_0\left(1+\frac{1}{2}\right)\mathbf{A}^{\dagger}\mathbf{e}_0$$

$$then:\mathbf{H}\left(\mathbf{A}^{\dagger}\right)^2\mathbf{e}_0-\mathbf{A}^{\dagger}\hbar\omega_0\left(1+\frac{1}{2}\right)\mathbf{A}^{\dagger}\mathbf{e}_0=\hbar\omega_0\left(\mathbf{A}^{\dagger}\right)^2\mathbf{e}_0$$

$$\mathbf{H}\left(\mathbf{A}^{\dagger}\right)^2\mathbf{e}_0-\hbar\omega_0\left(1+\frac{1}{2}\right)\left(\mathbf{A}^{\dagger}\right)^2e_0=\hbar\omega_0\left(\mathbf{A}^{\dagger}\right)^2e_0$$

$$\mathbf{H}\left(\mathbf{A}^{\dagger}\right)^2\mathbf{e}_0=\hbar\omega_0\left(2+\frac{1}{2}\right)\left(\mathbf{A}^{\dagger}\right)^2e_0$$

From the last operations above, we have three eigenvalue equations:

$$\mathbf{He}_0=\frac{1}{2}\hbar\omega_0\mathbf{e}_0$$

$$\mathbf{HA}^{\dagger}\mathbf{e}_0=\hbar\omega_0\left(1+\frac{1}{2}\right)\mathbf{A}^{\dagger}\mathbf{e}_0$$

$$\mathbf{H}\left(\mathbf{A}^{\dagger}\right)^2\mathbf{e}_0=\hbar\omega_0\left(2+\frac{1}{2}\right)\left(\mathbf{A}^{\dagger}\right)^2\mathbf{e}_0$$

Then, we see that each product of the $\mathbf{He}=\lambda\mathbf{e}$ by $\mathbf{A}^{\dagger}$ gives a $\hbar\omega_0$ increment of the eigenvalue of **H**. *The matrix* $\mathbf{A}^{\dagger}$ *is called creation operator. The annihilation and creation operators are collectively known as ladder operators.*
As a general result from the operations above, we have:

$$\mathbf{He}_j=\hbar\omega_0\left(j+\frac{1}{2}\right)\mathbf{e}_j$$

$$where:\mathbf{e}_j=\left(\mathbf{A}^{\dagger}\right)^j\mathbf{e}_0$$

$$and:j=0,1,2,3,\ldots$$

Then, the energy of the one-particle harmonic oscillator is:

$$E_j=\hbar\omega_0\left(j+\frac{1}{2}\right),j=0,1,2,\ldots$$

$$for:j=0:E_0=\frac{1}{2}\hbar\omega_0$$

In Chapter fourteen, we see an alternative solution for the one-particle harmonic oscillator using the wave quantum mechanics.

8. Classical angular momentum: definition

The classical definition of the angular momentum, **L**, is the vector cross product of the Cartesian coordinates $(x, y, z) \equiv \mathbf{r}$ (the position vector) and their conjugate momenta $(p_x, p_y, p_z) = \mathbf{p}$. It is the rotational equivalent of the linear momentum. The angular momentum is orthogonal to the plane (x,y) containing the vectors r and p (See Fig. 8.1).

$$\mathbf{L} = \mathbf{r} \times \mathbf{p}$$

The angular momentum vector points in a direction that is perpendicular to the plane containing **r** and **p**, and **L** has a magnitude:

$$L = rp \sin \alpha$$

Where α is the angle between the vectors **r** and **p**.

Both linear momentum and angular momentum are conserved quantities, i.e., their total values remain constant in a closed system. The gyroscope is an example of conservation of the angular momentum. It remains upright while spinning because of the conservation of its angular momentum.

In Cartesian coordinates, the components of the angular momentum are:

$$L = \begin{bmatrix} y & z \\ p_y & p_z \end{bmatrix}\mathbf{i} - \begin{bmatrix} x & z \\ p_x & p_z \end{bmatrix}\mathbf{j} + \begin{bmatrix} x & y \\ p_x & p_y \end{bmatrix}\mathbf{k}$$

$$\mathbf{L}_x = \mathbf{y}\mathbf{p}_z - \mathbf{z}\mathbf{p}_y$$

$$\mathbf{L}_y = \mathbf{z}\mathbf{p}_x - \mathbf{x}\mathbf{p}_z$$

$$\mathbf{L}_z = \mathbf{x}\mathbf{p}_y - \mathbf{y}\mathbf{p}_x$$

These components are obtained from the determinant form of the vector product:

$$\mathbf{L} = \mathbf{r} \times \mathbf{p}$$

$$\mathbf{L} = \begin{vmatrix} \mathbf{i} & \mathbf{j} & \mathbf{k} \\ x & y & z \\ p_x & p_y & p_z \end{vmatrix}$$

The classic angular momentum can be measured by means the moment of inertia, I, which is a measure of the resistance of an object to be accelerated in its

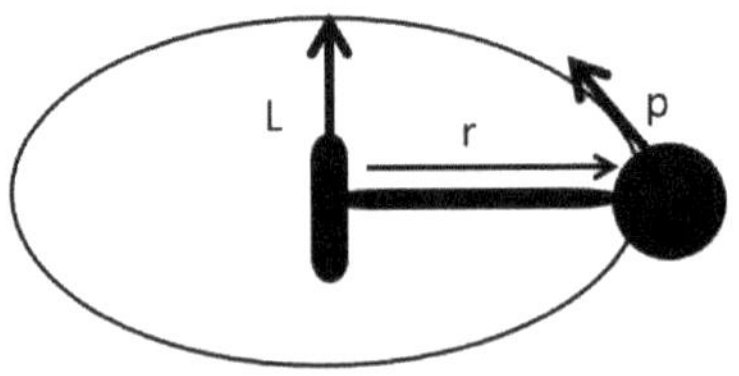

Fig. 8.1: Schematic representation of the vector of the classic angular momentum.

circular motion. The angular momentum is the scalar product of moment of inertia and angular velocity.

$$L = I \cdot \omega$$

The relation between the angular momentum and the kinetic energy, T, is given below:

$$T = \frac{mv^2}{2} = \frac{m(r\omega)^2}{2} = \frac{mr^2\omega^2}{2}$$

$$I = mr^2$$

$$then: T = \frac{I\omega^2}{2}$$

$$L = I\omega$$

$$multiply: I/I$$

$$T = \frac{I\omega^2}{2}\frac{I}{I}$$

$$then: T = \frac{I^2\omega^2}{2I} = \frac{L^2}{2I}$$

9. Quantum angular momentum

The quantum angular momentum components are obtained by replacing the classical operators of position and linear momentum by their corresponding quantum counterparts (see Chapter ten for the derivation of the quantum linear momentum operator).

$$\hat{p}_x = -i\hbar(\partial/\partial_x), \hat{p}_y = -i\hbar(\partial/\partial_y), \hat{p}_z = -i\hbar(\partial/\partial_z)$$

$$\hat{L}_x = -i\hbar(y\partial/\partial_z - z\partial/\partial_y)$$

$$\hat{L}_y = -i\hbar(z\partial/\partial_x - x\partial/\partial_z)$$

$$\hat{L}_z = -i\hbar(x\partial/\partial_y - y\partial/\partial_x)$$

The square angular momentum, $\mathbf{L}^2$, is another important operator in quantum mechanics.

$$\mathbf{L}^2 = \mathbf{L}_x^2 + \mathbf{L}_y^2 + \mathbf{L}_z^2$$

The commutative properties of the quantum angular momentum are:

$$[\mathbf{L}^2, \mathbf{L}_x] = 0, \quad [\mathbf{L}^2, \mathbf{L}_y] = 0, \quad [\mathbf{L}^2, \mathbf{L}_z] = 0$$

$$[\mathbf{L}^2, \mathbf{L}] = 0$$

$$[\mathbf{L}_x, \mathbf{L}_y] = i\hbar\mathbf{L}_z, \quad [\mathbf{L}_y, \mathbf{L}_z] = i\hbar\mathbf{L}_x, \quad [\mathbf{L}_z, \mathbf{L}_x] = i\hbar\mathbf{L}_y$$

Let us firstly derive the commutative relation between $\mathbf{L_x}$ and $\mathbf{L_y}$ is:

$$[\mathbf{L}_x, \mathbf{L}_y] = \mathbf{L}_x\mathbf{L}_y - \mathbf{L}_y\mathbf{L}_x$$

$$[\mathbf{L}_x, \mathbf{L}_y] = (\mathbf{yp}_z - \mathbf{zp}_y)(\mathbf{zp}_x - \mathbf{xp}_z) - (\mathbf{zp}_x - \mathbf{xp}_z)(\mathbf{yp}_z - \mathbf{zp}_y)$$

$$[\mathbf{L}_x, \mathbf{L}_y] = \mathbf{yp}_z\mathbf{zp}_x - \mathbf{yp}_z\mathbf{xp}_z - \mathbf{zp}_y\mathbf{zp}_x + \mathbf{zp}_y\mathbf{xp}_z$$

$$-\mathbf{zp}_x\mathbf{yp}_z + \mathbf{zp}_x\mathbf{zp}_y + \mathbf{xp}_z\mathbf{yp}_z - \mathbf{xp}_z\mathbf{zp}_y$$

Since:

$$\mathbf{yp}_z\mathbf{xp}_z = \mathbf{xp}_z\mathbf{yp}_z$$

$$\textit{and}: \mathbf{zp}_x\mathbf{zp}_y = \mathbf{zp}_y\mathbf{zp}_z$$

We have:

$$[\mathbf{L}_x, \mathbf{L}_y] = (\mathbf{yp}_z\mathbf{zp}_x - \mathbf{zp}_x\mathbf{yp}_z) + (\mathbf{zp}_y\mathbf{xp}_z - \mathbf{xp}_z\mathbf{zp}_y)$$

$$[\mathbf{L}_x, \mathbf{L}_y] = [\mathbf{yp}_z, \mathbf{zp}_x] + [\mathbf{zp}_y, \mathbf{xp}_z]$$

From the commutator's identity (see Chapter three):

$$[\mathbf{AB}, \mathbf{CD}] = \mathbf{A}[\mathbf{B}, \mathbf{C}]\mathbf{D} + [\mathbf{A}, \mathbf{C}]\mathbf{BD} + \mathbf{CA}[\mathbf{B}, \mathbf{D}] + \mathbf{C}[\mathbf{A}, \mathbf{D}]\mathbf{B}$$

we have:

$$[\mathbf{yp}_z, \mathbf{zp}_x] = \mathbf{y}[\mathbf{p}_z, \mathbf{z}]\mathbf{p}_x + [\mathbf{y}, \mathbf{z}]\mathbf{p}_z\mathbf{p}_x + \mathbf{zy}[\mathbf{p}_z, \mathbf{p}_x] + \mathbf{z}[\mathbf{y}, \mathbf{p}_x]\mathbf{p}_z$$

$$\textit{where}: [\mathbf{y}, \mathbf{z}] = 0, [\mathbf{p}_z, \mathbf{p}_x] = 0, [\mathbf{y}, \mathbf{p}_x] = 0$$

$$\textit{then}: [\mathbf{yp}_z, \mathbf{zp}_x] = \mathbf{y}[\mathbf{p}_z, \mathbf{z}]\mathbf{p}_x$$

$$[\mathbf{zp}_y, \mathbf{xp}_z] = \mathbf{z}[\mathbf{p}_y, \mathbf{x}]\mathbf{p}_z + [\mathbf{z}, \mathbf{x}]\mathbf{p}_y\mathbf{p}_z + \mathbf{xz}[\mathbf{p}_y, \mathbf{p}_z] + \mathbf{x}[\mathbf{z}, \mathbf{p}_z]\mathbf{p}_y$$

$$\textit{where}: [\mathbf{p}_y, \mathbf{x}] = 0, \quad [\mathbf{z}, \mathbf{x}] = 0, \quad [\mathbf{p}_y, \mathbf{p}_z] = 0$$

$$\textit{then}: [\mathbf{zp}_y, \mathbf{xp}_z] = \mathbf{x}[\mathbf{z}, \mathbf{p}_z]\mathbf{p}_y$$

$$\textit{hence}: [\mathbf{L}_x, \mathbf{L}_y] = \mathbf{y}[\mathbf{p}_z, \mathbf{z}]\mathbf{p}_x + \mathbf{x}[\mathbf{z}, \mathbf{p}_z]\mathbf{p}_y$$

From the commutators obtained from the matrix mechanics:

$$[\mathbf{P}, \mathbf{Q}] = -i\hbar\mathbf{I}$$

$$[\mathbf{Q}, \mathbf{P}] = i\hbar\mathbf{I}$$

we have:

$$[\mathbf{L}_x, \mathbf{L}_y] = -i\hbar\mathbf{yp}_x + i\hbar\mathbf{xp}_y = i\hbar(\mathbf{yp}_x + \mathbf{xp}_y)$$

$$\textit{where}: [\mathbf{p}_z, \mathbf{z}] = -i\hbar, \quad [\mathbf{z}, \mathbf{p}_z] = i\hbar$$

$$\textit{then}: [\mathbf{L}_x, \mathbf{L}_y] = i\hbar\mathbf{L}_z$$

The other commutators (see below) are obtained similarly:

$$[\mathbf{L}_y, \mathbf{L}_z] = i\hbar\mathbf{L}_x$$

$$[\mathbf{L}_z, \mathbf{L}_x] = i\hbar\mathbf{L}_y$$

Now, let us derive the commutative property between the square angular momentum, $\mathbf{L}^2$, and $\mathbf{L_x}$. The commutator is:

$[\mathbf{L}^2, \mathbf{L}_x] = 0$

We use the equation of the square angular momentum:

$$\left[\mathbf{L}^2,\mathbf{L}_x\right]=\left[\mathbf{L}_x^2+\mathbf{L}_y^2+\mathbf{L}_z^2,\mathbf{L}_x\right]$$

From the commutator identity (see Chapter three):

$$\left[\mathbf{A}+\mathbf{B},\mathbf{C}\right]=\left[\mathbf{A},\mathbf{C}\right]+\left[\mathbf{B},\mathbf{C}\right]$$

We obtain:

$$\left[\mathbf{L}^2,\mathbf{L}_x\right]=\left[\mathbf{L}_x^2,\mathbf{L}_x\right]+\left[\mathbf{L}_y^2,\mathbf{L}_x\right]+\left[\mathbf{L}_z^2,\mathbf{L}_x\right]$$

The first commutator on the right side of the above equation is zero:

$$\left[\mathbf{L}_x^2,\mathbf{L}_x\right]=\mathbf{L}_x^2\mathbf{L}_x-\mathbf{L}_x\mathbf{L}_x^2=\mathbf{L}_x\mathbf{L}_x\mathbf{L}_x-\mathbf{L}_x\mathbf{L}_x\mathbf{L}_x=0$$

Then, we have:

$$\left[\mathbf{L}^2,\mathbf{L}_x\right]=\left[\mathbf{L}_y^2,\mathbf{L}_x\right]+\left[\mathbf{L}_z^2,\mathbf{L}_x\right]$$

$$\left[\mathbf{L}^2,\mathbf{L}_x\right]=\mathbf{L}_y^2\mathbf{L}_x-\mathbf{L}_x\mathbf{L}_y^2+\mathbf{L}_z^2\mathbf{L}_x-\mathbf{L}_x\mathbf{L}_z^2$$

$$\left[\mathbf{L}^2,\mathbf{L}_x\right]=\mathbf{L}_y\mathbf{L}_y\mathbf{L}_x-\mathbf{L}_x\mathbf{L}_y\mathbf{L}_y+\mathbf{L}_z\mathbf{L}_z\mathbf{L}_x-\mathbf{L}_x\mathbf{L}_z\mathbf{L}_z$$

Let us find the commutators $[\mathbf{L}_y, \mathbf{L}_x]\,\mathbf{L}_y$ and $\mathbf{L}_y\,[\mathbf{L}_y, \mathbf{L}_x]$ and their sum:

$$\left[\mathbf{L}_y,\mathbf{L}_x\right]\mathbf{L}_y=\left(\mathbf{L}_y\mathbf{L}_x-\mathbf{L}_x\mathbf{L}_y\right)\mathbf{L}_y=\mathbf{L}_y\mathbf{L}_x\mathbf{L}_y-\mathbf{L}_x\mathbf{L}_y\mathbf{L}_y$$

$$\mathbf{L}_y\left[\mathbf{L}_y,\mathbf{L}_x\right]=\mathbf{L}_y\left(\mathbf{L}_y\mathbf{L}_x-\mathbf{L}_x\mathbf{L}_y\right)=\mathbf{L}_y\mathbf{L}_y\mathbf{L}_x-\mathbf{L}_y\mathbf{L}_x\mathbf{L}_y$$

$$\left[\mathbf{L}_y,\mathbf{L}_x\right]\mathbf{L}_y+\mathbf{L}_y\left[\mathbf{L}_y,\mathbf{L}_x\right]=\mathbf{L}_y\mathbf{L}_x\mathbf{L}_y-\mathbf{L}_x\mathbf{L}_y\mathbf{L}_y+$$
$$+\mathbf{L}_y\mathbf{L}_y\mathbf{L}_x-\mathbf{L}_y\mathbf{L}_x\mathbf{L}_y=\mathbf{L}_y\mathbf{L}_y\mathbf{L}_x-\mathbf{L}_x\mathbf{L}_y\mathbf{L}_y$$

In the same way we have for $[\mathbf{L}_z, \mathbf{L}_x]\,\mathbf{L}_z$ and $\mathbf{L}_z\,[\mathbf{L}_z, \mathbf{L}_x]$ and their sum:

$$\left[\mathbf{L}_z,\mathbf{L}_x\right]\mathbf{L}_z=\left(\mathbf{L}_z\mathbf{L}_x-\mathbf{L}_x\mathbf{L}_z\right)\mathbf{L}_z=\mathbf{L}_z\mathbf{L}_x\mathbf{L}_z-\mathbf{L}_x\mathbf{L}_z\mathbf{L}_z$$

$$\mathbf{L}_z\left[\mathbf{L}_z,\mathbf{L}_x\right]=\mathbf{L}_z\left(\mathbf{L}_z\mathbf{L}_x-\mathbf{L}_x\mathbf{L}_z\right)=\mathbf{L}_z\mathbf{L}_z\mathbf{L}_x-\mathbf{L}_z\mathbf{L}_x\mathbf{L}_z$$

$$\left[\mathbf{L}_z,\mathbf{L}_x\right]\mathbf{L}_z+\mathbf{L}_z\left[\mathbf{L}_z,\mathbf{L}_x\right]=\mathbf{L}_z\mathbf{L}_x\mathbf{L}_z-\mathbf{L}_x\mathbf{L}_z\mathbf{L}_z+$$
$$+\mathbf{L}_z\mathbf{L}_z\mathbf{L}_x-\mathbf{L}_z\mathbf{L}_x\mathbf{L}_z=\mathbf{L}_z\mathbf{L}_z\mathbf{L}_x-\mathbf{L}_x\mathbf{L}_z\mathbf{L}_z$$

Thus, the relation:

$$\left[\mathbf{L}^2,\mathbf{L}_x\right]=\mathbf{L}_y\mathbf{L}_y\mathbf{L}_x-\mathbf{L}_x\mathbf{L}_y\mathbf{L}_y+\mathbf{L}_z\mathbf{L}_z\mathbf{L}_x-\mathbf{L}_x\mathbf{L}_z\mathbf{L}_z$$

Becomes:

$$\left[\mathbf{L}^2,\mathbf{L}_x\right]=\left[\mathbf{L}_y,\mathbf{L}_x\right]\mathbf{L}_y+\mathbf{L}_y\left[\mathbf{L}_y,\mathbf{L}_x\right]+\left[\mathbf{L}_z,\mathbf{L}_x\right]\mathbf{L}_z+\mathbf{L}_z\left[\mathbf{L}_z,\mathbf{L}_x\right]$$

$$\left[\mathbf{L}_y,\mathbf{L}_x\right]=-i\hbar\mathbf{L}_z,\quad \left[\mathbf{L}_z,\mathbf{L}_x\right]=i\hbar\,\mathbf{L}_y$$

$$\left[\mathbf{L}^2,\mathbf{L}_x\right]=-i\hbar\mathbf{L}_z\mathbf{L}_y-i\hbar\mathbf{L}_y\mathbf{L}_z+i\hbar\mathbf{L}_y\mathbf{L}_z+i\hbar\mathbf{L}_z\mathbf{L}_y=0$$

Similar derivations can be used to demonstrate that $\mathbf{L}^2$ commutates with $\mathbf{L_y}$ and $\mathbf{L_z}$.

Further information about quantum angular momentum is given in chapter sixteen.

10. Exercises

(1) Demonstrate that a diagonalization process is equivalent to the eigenvalue problem, where A is the linear transformation operator, P is the eigenvector and D is the diagonal matrix.

$$\mathbf{P}^{-1}\mathbf{AP} = \mathbf{D} \Rightarrow \mathbf{AP} = \mathbf{PD}$$

$$\mathbf{D} = \begin{bmatrix} \lambda_1 & 0 \\ 0 & \lambda_2 \end{bmatrix}, \mathbf{P} = \begin{bmatrix} V_1 & W_1 \\ V_2 & W_2 \end{bmatrix} = \begin{bmatrix} \mathbf{V} & \mathbf{W} \end{bmatrix}$$

$$\mathbf{V} = \begin{bmatrix} V_1 \\ V_2 \end{bmatrix}, \mathbf{W} = \begin{bmatrix} W_1 \\ W_2 \end{bmatrix}$$

$$\mathbf{AP} = \mathbf{A}\begin{bmatrix} \mathbf{V} & \mathbf{W} \end{bmatrix} = \begin{bmatrix} \mathbf{AV} & \mathbf{AW} \end{bmatrix}$$

$$\mathbf{PD} = \begin{bmatrix} V_1 & W_1 \\ V_2 & W_2 \end{bmatrix}\begin{bmatrix} \lambda_1 & 0 \\ 0 & \lambda_2 \end{bmatrix} = \begin{bmatrix} \lambda_1 V_1 & \lambda_2 W_1 \\ \lambda_1 V_2 & \lambda_2 W_2 \end{bmatrix}$$

$$\mathbf{DP} = \begin{bmatrix} \lambda_1 & 0 \\ 0 & \lambda_2 \end{bmatrix}\begin{bmatrix} V_1 & W_1 \\ V_2 & W_2 \end{bmatrix} = \begin{bmatrix} \lambda_1 V_1 & \lambda_2 W_1 \\ \lambda_1 V_2 & \lambda_2 W_2 \end{bmatrix}$$

$$then: \mathbf{PD} = \mathbf{DP}$$

$$\mathbf{AP} = \mathbf{PD} \Rightarrow \mathbf{AP} = \mathbf{DP}$$

$$\mathbf{A}\begin{bmatrix} \mathbf{V} & \mathbf{W} \end{bmatrix} = \mathbf{D}\begin{bmatrix} \mathbf{V} & \mathbf{W} \end{bmatrix}$$

(2) The 3×3 matrix **T** is a linear transformation operator and **X** is the eigenvector. Find the corresponding eigenvalues.

$$\mathbf{T} = \begin{pmatrix} 1 & 0 & 0 \\ 0 & 1 & 2 \\ 0 & 2 & 1 \end{pmatrix}$$

$$\mathbf{TX} = \lambda\mathbf{X} \Rightarrow (\mathbf{T} - \lambda\mathbf{I})\mathbf{X} = 0$$

$$Then: \det(\mathbf{T} - \lambda\mathbf{I}) = 0$$

$$\begin{pmatrix} 1 & 0 & 0 \\ 0 & 1 & 2 \\ 0 & 2 & 1 \end{pmatrix} - \lambda\begin{pmatrix} 1 & 0 & 0 \\ 0 & 1 & 0 \\ 0 & 0 & 1 \end{pmatrix} = \begin{pmatrix} 1-\lambda & 0 & 0 \\ 0 & 1-\lambda & 2 \\ 0 & 2 & 1-\lambda \end{pmatrix}$$

$$\det\begin{pmatrix} 1-\lambda & 0 & 0 \\ 0 & 1-\lambda & 2 \\ 0 & 2 & 1-\lambda \end{pmatrix} = 0$$

$$S = \{1, -1, 3\}$$

(3) Obtain the commutation relations:

$$\left[\mathbf{L}_y, \mathbf{L}_z\right] = i\hbar\mathbf{L}_x$$

$$\left[\mathbf{L}_z, \mathbf{L}_x\right] = i\hbar\mathbf{L}_y$$

(4) By knowing that the square angular momentum is:

$$\mathbf{L}^2 = \mathbf{L}_x^2 + \mathbf{L}_y^2 + \mathbf{L}_z^2$$

Show that $\mathbf{L}^2$ commutates with **L**.

References cited

Aitchison, I.J.R. et al. 2004. Understanding Heisenberg's magical paper of July 1925: A new look at the calculation details. Am. J. Phys. 72: 1370–1379.

Birtwistle, G. 1928. The New Quantum Mechanics. Cambridge University Press, Cambridge.

Bohr, N. 1924. Zur polarisation des fluorescenzlightes. Naturwiss. 12: 1115–1117.

Born, M. 1924. Uber quantenmechanik. Z. Phys. 26. 379–395.

Born, M. and Heisenberg, W. 1923a. Über phasenbeziehungen bei den Bohrschen modellen von atomen and molekeln. Z. Phys. 14: 44–55.

Born, M. and Heisenberg, W. 1923b. Die elektronenbahnen im angeregten heliumatom. Z. Phys. 16: 229–243.

Born, M. and Heisenberg, W. 1924. Zur quantentheorie der molekeln. Ann. Phys. 74: 1–31.

Born, M. and Jordan, P. 1925. Zur quantenmechanik. Z. Phys. 34: 858–888.

Born, M., Heisenberg, W. and Jordan, P. 1925. Zur quantenmechanik II. Z. Phys. 35: 557–615.

Dirac, P.A.M. 1925. The fundamental equation of quantum mechanics. Proc. R. Soc. London A 109: 642–653.

Dirac, P.A.M. 1927. The quantum theory of dispersion. Proc. R. Soc. Lond. A. 114: 710–728.

Fedak, W.A. and Prentis, J.J. 2009. The 1925 Born and Jordan paper "On quantum mechanics". Am. J. Phys. 77: 128–139.

Heisenberg, W. 1922. Zurquantentheorie der linienstrukturund der anomalenZeemaneffekte. Z. Phys. 8: 273–297.

Heisenberg, W. 1924. Übereineabänderung der formalin regeln der quantentheorie beim problem der anomalen Zeeman-effekte. Z. Phys. 26: 291–307.

Heisenberg, W. 1925. Uber quantentheoretische undeutung kinematischer und mechanischer beziehungen. Z. Phys. 33: 879–893.

Kramers, H.A. 1924. The law of dispersion and Bohr's theory of spectra. Nature 113: 673–674.

Kramers, H.A. and Heisenberg, W. 1925. Uber die streuung von strahlen durch atome. Z. Phys. 31: 681–708.

Ladenburg, R. 1921. Die quantentheoretische deutung der zahl der dispersionselektronen. Z. Phys. 4: 451–468.

Landé, A. and Heisenberg, W. 1924. Termstruktur der multipletts höherer stufe. Z. Phys. 25: 279–286.

MacKinnon, E. 1977. Heisenberg, models and the rise of the matrix mechanics. Hist. Stud. Phys. Sci. 8: 137–188.

Pauli, W. 1925. Über den Einfluss der Geschwindigkeitsabhnangigkeit der Elektronenmasse auf den Zeemaneffekt. Z. Phys. 31: 373–385.

Pauli, W. 1926. Über das wasserstoffspektrum von standpunkt der neuen quantenmechanik. Z. Phys. 36: 336–363

Rechenberg, H. 2000. Max Born and molecular theory. Theor. Phys. Fin Siècle 539: 7–20.

Schrödinger, E. 1926. Uber das verhaltnis der Heisenberg-Born-Jordanschen quantenmechanik zu der meinen. Ann. Phys. 79: 734–756.

Sommerfeld, A. and Heisenberg, W. 1922. Eine bemerkung uber relativistische Röntgendubletts und linienscharfe. Z. Phys. 10: 391–398.

Wood, R.W. and Ellett, A. 1923. On the influence of magnetic fields on the polarization of resonance radiation. Proc. Royal Soc. 103: 396–403.

Wave Packet and de Broglie's Wave-particle Duality

9

1. Double-slit experiment

The debate about the nature of the light began in 1670 when Issac Newton stated that light was a collection of corpuscles while Christian Huygens stated it was a wave. In 1804, Thomas Young proved the wave-like nature of light by means of the double slit interference experiment whose pictorial representation can be seen in Fig. 9.1 (Young 1804). At the positions marked "Max" on the screen, the meeting waves are in-phase and the combined wave has enhanced amplitude. At the positions marked "Min", the combined wave amplitude is zero.

In 1927, Davisson and Germer were the first to prove the wave-like nature of electrons by an accidental discovery of the diffraction of the electrons (Davisson and Germer 1927). As stated by them: "These results are highly suggestive, of course, of the ideas underlying the theory of wave mechanics and we naturally inquire if the wave-length of the X-ray beam which we thus associate with a beam of electrons is in fact the h/mv of L. de Broglie" (Davisson and Germer 1927).

In 1961, in an experiment analogous to Young's experiment, Jönsson obtained the electron diffraction and interference phenomenon similarly to a wave, proving

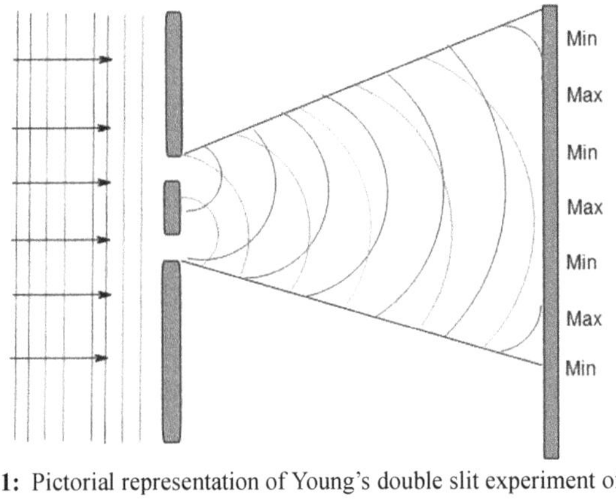

Fig. 9.1: Pictorial representation of Young's double slit experiment of light.

the wave-particle nature of the electron (Jönsson 1961). Since then, the particle interference has been demonstrated with neutrons, atoms and molecules, for instance, C_{60} molecules (Arndt 1999).

2. Special theory of relativity

The special theory of relativity has two postulates: (1) the laws of electrodynamics and optics are invariant in all inertial frames of reference, i.e., they are valid for all references in which the mechanics equations are valid; and (2) light always propagates in the empty space with a definite velocity c independently of the state of motion of the emitting body or observer (the constancy of the speed of light).The two postulates of special relativity predict the equivalence of mass (or rest mass) and energy (or rest energy):

$$E = mc^2$$

Where c is the speed of light.

In special theory of relativity, time and space cannot be treated separately as was previously thought to be the case. The space and time form the four-dimensional spacetime. In order to satisfy the second postulate of Einstein's special theory of relativity, Lorentz transformation is applied.

In special relativity, an event is defined in two frames: (1) the rest frame with spacetime coordinates (t,x,y,z); and (2) a reference frame moving at a velocity v with respect to the first frame with spacetime coordinates (t',x',y',z'). Then, the Lorentz transformation is a linear transformation of spacetime coordinates which relates the three spatial coordinates and time of the same event in two frames, S and S', where S' moves with velocity v with respect to S along x-axis.

$$x' = \gamma(x - vt), \quad \gamma = \frac{1}{\sqrt{1-\beta^2}}, \quad \beta = \frac{v}{c}$$

$$y' = y$$

$$z' = z$$

$$t' = \gamma\left(t - \frac{vx}{c^2}\right)$$

Where γ is the Lorentz factor.

The rest mass (invariant mass or intrinsic mass), m_0, is the total mass of a body that is independent of the overall motion of the system and it is a characteristic of the system's total energy and momentum that is the same in the rest frame (S). The rest mass is the Newtonian mass as measured by an observer moving along with the object. For the other reference frames where the system's momentum is nonzero, the total mass (i.e., the relativistic mass, m_{rel}) is greater than the rest mass:

$$m_{rel} = \frac{m_0}{\sqrt{1-\left(\frac{v}{c}\right)^2}}$$

In the center of momentum frame (where v = 0), the relativistic mass equals the rest mass. The relativistic mass depends on the observer's frame of reference. As a consequence, the momentum and the energy at the primed frame (S') are:

$$p' = \frac{m_0 v}{\sqrt{1-\left(\frac{v}{c}\right)^2}}, \qquad E' = \frac{m_0 c^2}{\sqrt{1-\left(\frac{v}{c}\right)^2}}$$

By combining the expression for the momentum at the primed frame, the rest energy and the energy at the primed frame, we have:

$$E^2 = p^2c^2 + m_0^2 c^4$$

3. Waves

A *periodic wave* is a wave with a repeating, continuous pattern. The sine and cosine waves are examples of periodic waves. For periodic waves, the frequency, ν, has an inverse relation with the wavelength, λ:

$$\nu = \frac{V}{\lambda}$$

Where V is the phase velocity of the wave. For the special case of electromagnetic waves moving in the vacuum, the speed of light, c, is its phase velocity.

$$\nu_{light} = \frac{c}{\lambda}$$

The *phase velocity* is the rate at which an individual wave propagates in some medium. The phase velocity can be understood as the ratio between the wavelength and the time period, T.

$$V = \frac{\lambda}{T}$$

The phase velocity can also be given in terms of the angular frequency, ω, and the wavenumber, k, as:

$$V = \frac{\omega}{k}, \quad \omega = 2\pi\nu$$

The *wavenumber* is the number of radians per unit distance:

$$k = \frac{2\pi}{\lambda}$$

Sine and cosine waves are example of periodic waves. The general formula of a sine wave is:

$$y(x) = A\sin\left(2\pi\nu x + \varphi\right), \quad 1\lambda = 2\pi$$

$$or: y(x) = A\sin\left(\frac{2\pi x}{\lambda} + \varphi\right)$$

Where 2π is a complete cycle that corresponds to the wavelength, A is the amplitude of the wave and φ is the phase constant that specifies (in radians) where in its cycle the oscillation is at x = 0.

The cosine wave is nearly identical to sine wave except that each point on the cosine wave occurs ¼ cycle earlier than that from sine wave.

The *number of waves*, n_x, and *phase of wave* at a given x,φ_x, are:

$$n_x = \frac{x}{\lambda}$$

$$\varphi_x = 2\pi n_x$$

Let a cosine wave be the one represented in Fig. 9.2(A) whose equation is given below:

$$y(x) = A\cos\left(\frac{2\pi x}{\lambda}\right)$$

$$A = 5,\quad \lambda = 5,\quad y(x) = 5\cos\left(\frac{2\pi x}{5}\right)$$

If this wave moves without distortion in the positive direction of the x-axis for a quantity x_0, it becomes the dotted-line wave in Fig. 9.2(B):

$$y(x) = A\cos\left(\frac{2\pi(x-x_0)}{\lambda}\right)$$

$$x_0 = 2, A = 5, \lambda = 5,\quad y(x) = 5\cos\left(\frac{2\pi(x-2)}{5}\right)$$

This second wave is 2 units of x ahead from the first wave and it can be understood as an example of travelling wave if:

$$x_0 = Vt = 2$$

The *travelling wave* moves at constant phase velocity in the x-axis where the following general property is respected:

$$y(x - Vt, t) = y(x, t)$$

That is, each (x – Vt,t) point is equal to (x,t) point (See Fig. 9.3).

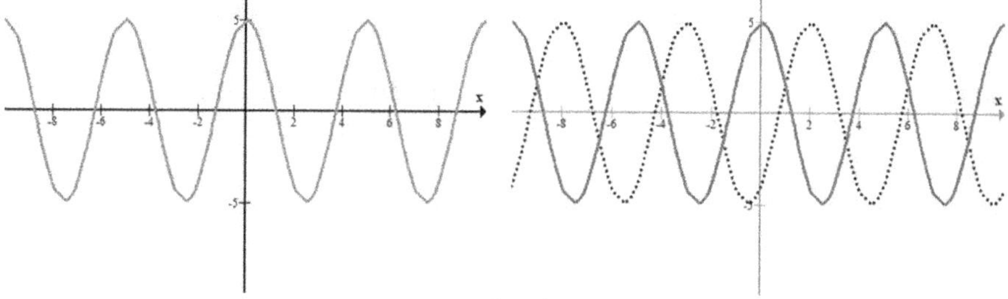

Fig. 9.2: Plot of (A) cosine wave $y(x) = 5\cos\left(\frac{2\pi x}{5}\right)$; (B) cosine wave $y(x) = 5\cos\left(\frac{2\pi(x-2)}{5}\right)$.

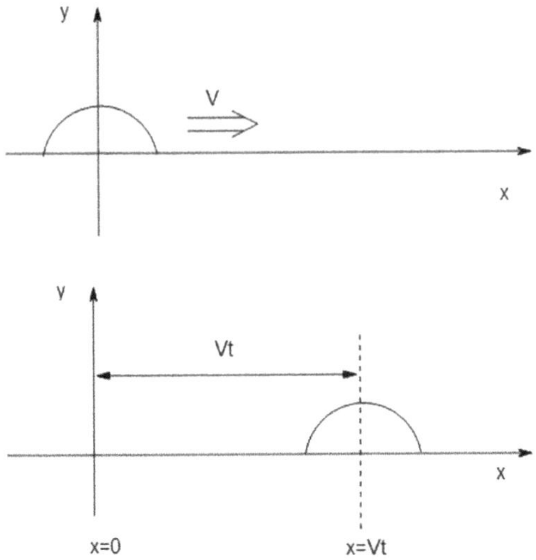

Fig. 9.3: Example of a travelling wave.

The travelling wave has the following general equation:

$$y(x,t) = \cos(kx - \omega t)$$

A travelling wave can be represented using Euler's formula:

$$y(x,t) = \cos(kx - \omega t) + i\sin(kx - \omega t) = e^{i(kx-\omega t)}$$

The Euler's formula was derived in chapter two as:

$$e^{ix} = \cos x + i\sin x$$

The *composite wave* is the sum (or superposition) of two or more periodic waves. In Fig. 9.4(A) there appears two different waves: $y(x) = 5\sin(x)$ and $y(x) = 4\cos(2x + (\pi/3))$. In Fig. 9.4(B), it is depicted the corresponding composite wave, $y(x) = 5\sin(x) + 4\cos(2x + (\pi/3))$, as a sum of the two waves of Fig. 9.4(A). The general equation for a composite wave is given below:

$$y(x) = \sum_{i=1}^{n} A_i e^{i(k_i x - \omega_i t)}$$

Where each ith wave is represented by Euler's formula.

The amplitude of a composite wave is not constant. The *envelope* of a wave can describe graphically the variation of the amplitude with respect to any variable (space, time or angle). The envelope of a wave is an imaginary smooth curve outlining its upper and lower extremes. The sine and cosine waves of Fig. 9.4(A) have straight, horizontal envelopes while their composite wave has upper and lower curved envelopes.

The *group velocity*, U, is defined as the derivation of the angular frequency with respect to k.

$$U = \frac{\partial \omega}{\partial k}$$

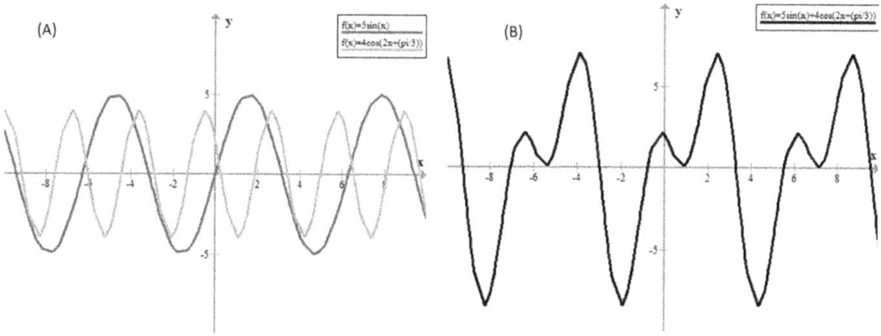

Fig. 9.4: (A) Plot of two waves: $y(x) = 5 \sin(x)$ and $y(x) = 4 \cos(2x + (\pi/3))$; (B) composite wave $y(x) = 5\sin(x) + 4\cos\left(2x + (\pi/3)\right)$.

The group velocity is the velocity of the overall envelope of the composite wave. The relation between phase velocity and group velocity is:

$$U = V + k\frac{dV}{dk}$$

The above relation is obtained from the differentiation of the equation of the phase velocity with respect to k.

$$V = \frac{\omega}{k}$$

$$then: \omega = Vk$$

$$\frac{d\omega}{dk} = V + k\frac{dV}{dk}$$

$$as: U = \frac{d\omega}{dk}$$

$$then: U = V + k\frac{dV}{dk}$$

If the angular velocity is directly proportional to the wavenumber, the group velocity is equal to the phase velocity. It is the case of a non-dispersive wave. If the angular velocity is a linear function of the wavenumber ($\omega = ak + b$), the envelope of the composite wave will travel at a single group velocity and the individual peaks and troughs will travel at the phase velocity. If the angular velocity is not a linear function of the wavenumber, the envelope of the composite wave will become distorted as it travels because the envelope will move at a single velocity, i.e., its wavenumber components (k_i) move at different velocities, distorting the envelope.

The *wave packet* is a composite wave which exhibits characteristic groups or packets. In Fig. 9.5(A) there is a composite wave (or wave packet) with three k values ($k_1 = 0.95$, $k_2 = 1.0$ and $k_3 = 1.05$), represented by $y(x) = 10 \sin(0.95x) + 10 \sin(x) + 10 \sin(1.05x)$, where one can see two different packets (a smaller packet between two large packets). In Fig. 9.5(B), there is a wave packet with five waves ($k_1 = 0.95$, $k_2 = 0.975$, $k_3 = 1.0$, $k_4 = 1.25$ and $k_5 = 1.05$), represented by $y(x) = 10 \sin(0.95x) +$

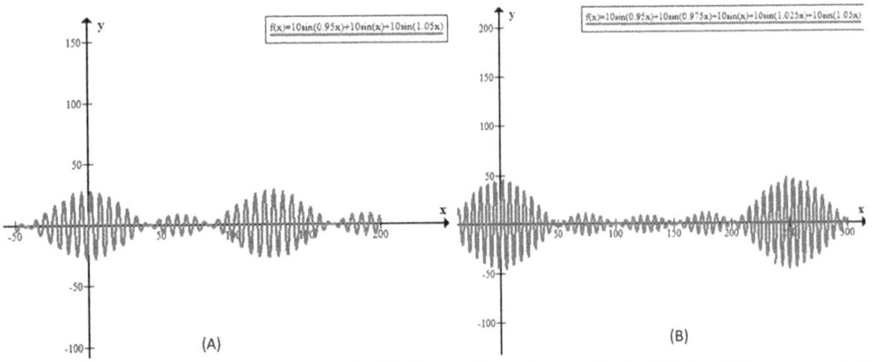

Fig. 9.5: (A) wave packet $y(x) = 10\ \sin(0.95x) + 10\ \sin(x) + 10\ \sin(1.05x)$; (B) wave packet $y(x) = 10\ \sin(0.95x) + 10\ \sin(0.975x) + 10\ \sin(x) + 10\ \sin(1.025x) + 10\ \sin(1.05x)$.

$10\ \sin(0.975x) + 10\ \sin(x) + 10\ \sin(1.025x) + 10\ \sin(1.05x)$, where there are three small packets between two large packets.

We can continue adding more and more waves and the distance between the large packets will further increase. However, the central packet (around x = 0) does not move. So, if the number of waves goes to infinity, only the central packet will remain. Then, by superposing an infinite number of constituent waves, we obtain a single wave packet. The *single wave packet* is represented by an integral instead of a sum.

$$y(x) = \frac{1}{\sqrt{2\pi}} \int_{-\infty}^{\infty} A_i e^{i(k_i x - \omega_i t)} dk$$

4. De Broglie's work

De Broglie proposed the wave-particle duality of the electron based on the quantum theory (Bohr's model and Einstein quantization hypothesis) and the special theory of relativity.

Before de Broglie, Einstein proposed the wave-particle duality of light (the light quanta). De Broglie assumed that in the wave-particle theory of light each photon is characterized by a distinctive frequency and that this theory needs the relativistic treatment since photons are travelling at the light speed. De Broglie began his 1924's paper citing Einstein and Compton: "The photoelectric effect, which is the chief mechanism of energy exchange between radiation and matter, seems with increasing probability to be always governed by Einstein's photoelectric law. Experiments on the photographic actions, the recent results of A.H. Compton on the change in wavelength of scattered X-rays, would be very difficult to explain without using the notion of the light quantum" (De Broglie 1924).

De Broglie proposed the matter waves where the mass m is corrected according to the equation from special relativity where he used V to represent the phase velocity [cms^{-1}] of the propagating waves. The phase velocity can be expressed in terms of the wavelength λ and the frequency ν.

$$V = \lambda \nu$$

De Broglie assumed that quanta are indivisible particles of light with rest mass different from zero and that all quanta are identical at rest. Then, he related both Einstein's equations of energy, E, and the rest mass:

$$E = h\nu$$

$$E = mc^2$$

$$m = \frac{m_0}{\sqrt{1-\beta^2}}$$

$$h\nu = \frac{m_0 c^2}{\sqrt{1-\beta^2}}, \qquad \beta = U/c$$

By squaring both sides of the last equation, he obtained the group velocity U as:

$$\frac{m_0^2 c^4}{1-\frac{U^2}{c^2}} = h^2\nu^2, \quad \frac{m_0^2 c^4}{\frac{c^2-U^2}{c^2}} = h^2\nu^2$$

$$\frac{m_0^2 c^6}{c^2-U^2} = h^2\nu^2$$

$$then: U = \sqrt{c^2 - \frac{m_0^2 c^6}{h^2\nu^2}}$$

According to the above equation, the greater the frequency of radiation, the greater its speed.
By establishing that V is close to c, de Broglie arrived to the relation:

$$\beta = \sqrt{1-\frac{m_0^2 c^4}{h^2\nu^2}} \simeq 1 - \frac{1}{2}\frac{m_0^2 c^4}{h^2\nu^2} \sim 1$$

Where he concluded that the rest mass of the photons has to be very small, at most of the order of 10^{-50} g. However, it is well known nowadays that photons have no mass.

Prior to de Broglie's work, the equation E = hν had only been applied to energy differences and not to rest mass energy. De Broglie then established the relation between the rest energy from Einstein's quantum theory and Einstein's special relativity in the rest frame S:

$$h\nu_0 = m_0 c^2$$

Then, the light atom (or photon) has an oscillatory motion whose frequency can be related to its rest mass:

$$\nu_0 = \frac{1}{h} m_0 c^2$$

De Broglie considered a moving particle of rest mass m_0, for example, an electron travelling a circular orbit with group velocity U:

$$U = \beta c, \beta < 1$$

having an internal energy m_0c^2 which is ascribed to a periodical phenomenon of frequency ν_0 within the frame S. In this statement, the particle is not associated with a single wave, but with a group of waves of similar frequencies. The derivation of the last equation is given ahead.

This internal frequency (or beat frequency)is transformed according to Lorentz transformation into:

$$\nu' = \nu_0\sqrt{1-\beta^2}$$

This is obtained from:

$$\nu = \frac{\nu_0}{\sqrt{1-\beta^2}} \to \nu^2 = \frac{\nu_0^2}{1-\beta^2} = \frac{\nu_0^2}{\left(1-\dfrac{c^2}{V^2}\right)}$$

$$\text{Since}: V = \lambda\nu$$

$$\frac{V^2}{\lambda^2}\left(1-\frac{c^2}{V^2}\right) = \nu_0^2$$

$$Then: V^2 = c^2 + \nu_0^2\lambda^2$$

Finally, the internal frequency, ν', is:

$$\nu' = \frac{\partial V}{\partial \lambda} = \frac{\nu_0^2\lambda}{\lambda\nu} = \frac{\nu_0^2}{\dfrac{\nu_0}{\sqrt{1-\beta^2}}} = \nu_0\sqrt{1-\beta^2}$$

For the fixed observer seeing the particle as a moving system (in the frame S') would attribute an energy and frequency as:

$$E = \frac{E_0}{\sqrt{1-\beta^2}}, \qquad \nu = \frac{1}{h}\frac{m_0c^2}{\sqrt{1-\beta^2}}$$

Then, we can say that the frequency of a wave associated with a particle is equivalent to the frequency associated with the energy of the particle for any inertial reference.

The comparison of the last two frequency equations yields:

$$\nu' = \nu\,(1-\beta^2)$$

Then, de Broglie assigned three different frequencies to one and same particle: the internal frequency in the rest system (ν_0) the internal frequency measured by an external observer (ν'); and the frequency this observer associates with the particle's total energy (ν).

The group of waves of similar frequencies which are associated with a particle has velocity cβ (the velocity of the particle), although each wave has velocity c/β, and the velocity of each wave does not vary according to the frequency transformation, but the beat frequency does. To demonstrate this, de Broglie used the group velocity (U):

$$\frac{1}{U} = \frac{d\left(\dfrac{\nu}{V}\right)}{d\nu}, \quad V = \frac{c}{\beta}, \quad \beta = \sqrt{1-\frac{m_0^2c^4}{h^2\nu^2}}$$

Where V is the phase velocity. Let us calculate ν/V:

$$\beta = \sqrt{1 - \frac{m_0^2 c^4}{h^2 \nu^2}} = \frac{\sqrt{h^2\nu^2 - m_0^2 c^4}}{h\nu}$$

$$\frac{\nu}{V} = \nu \frac{\sqrt{h^2\nu^2 - m_0^2 c^4}}{ch\nu} = \frac{\sqrt{h^2\nu^2 - m_0^2 c^4}}{ch}$$

Then, we have:

$$\frac{1}{U} = \frac{d\left(\frac{\sqrt{h^2\nu^2 - m_0^2 c^4}}{ch}\right)}{d\nu} = \frac{h\nu}{c\sqrt{h^2\nu^2 - m_0^2 c^4}} = \frac{1}{\beta c}$$

As a consequence, the group velocity of a wave is:

$$U = \beta c$$

See the relations of V and U with β:

$$V = \frac{c}{\beta}, \quad U = \beta c$$

Then we have the relation:

$$V = \frac{c}{\beta}, \quad \beta = \frac{U}{c}$$

$$V = \frac{c}{\frac{U}{c}} = \frac{c^2}{U}$$

As de Broglie said: "The velocity of the moving body is the energy velocity of a group of waves having frequencies $\nu = \frac{1}{h}\frac{m_0 c^2}{\sqrt{1-\beta^2}}$ and velocities c/β corresponding to very slightly different values of β" (de Broglie 1924).

In his 1924's paper, de Broglie stated: "We are inclined to admit that any moving body may be accompanied by a wave and that it is impossible to disjoin motion of body and propagation of wave" (de Broglie 1924). For instance, an electron moving in circular orbit emits a wave that travels the same circular orbit as the electron but with faster velocity.

Let us investigate how the particle's momentum relates to the associated wavelength using the equation of the energy at primed frame:

$$E' = \frac{m_0 c^2}{\sqrt{1 - \left(\frac{v}{c}\right)^2}}$$

$$E' = E, \quad v = U$$

$$E^2\left(1 - \frac{U^2}{c^2}\right) = m_0^2 c^4$$

We can rearrange it to:

$$E^2 - m_0^2c^4 = \frac{E^2U^2}{c^2}$$

Let us use the last equation of the previous section and E = hν in the equation above:

$$E^2 = p^2c^2 + m_0^2c^4$$
$$E = h\nu$$
$$p^2c^2 + m_0^2c^4 - m_0^2c^4 = \frac{U^2h^2\nu^2}{c^2}$$
$$p^2c^4 = U^2h^2\nu^2$$

Let us find another expression for the group velocity:

$$\text{since}: V = \frac{c^2}{U}$$
$$And: V = \lambda\nu$$
$$Then: \lambda\nu = \frac{c^2}{U}, \quad U = \frac{c^2}{\lambda\nu}$$

Let us replace the equation of group velocity in the expression:

$$p^2c^4 = U^2h^2\nu^2$$
$$U = \frac{c^2}{\lambda\nu}$$
$$Then: p^2c^4 = \left(\frac{c^2}{\lambda\nu}\right)^2 h^2\nu^2$$
$$p^2c^4 = \frac{c^4h^2\nu^2}{\lambda^2\nu^2}$$

By simplifying the last equation, we obtain the most famous equation of de Broglie:

$$p = \frac{h}{\lambda}$$

This is the equation that relates the momentum of any particle to the wavelength associated to it. One can see that the wave-particle duality is a consequence of the coexistence of phase velocity (V) and group wave velocity (U) in a moving particle. In addition, implicitly de Broglie's work leads to the existence of the superposition of wave functions.

De Broglie's work drew Einstein's attention for the association of a scalar wave (having phase velocity V and frequency ν) with any material particle of mass m and velocity v:

$$V = \frac{c^2}{\mathrm{v}}, \quad \mathrm{v} = U$$

$$\nu = \frac{\nu_0}{\sqrt{1 - \mathrm{v}^2/c^2}}$$

Einstein noticed that: "One now observes that it is possible to associate a scalar wave field with such a gas (…)". The idea of phase waves associated with gas molecules could provide a solution for the Gibbs paradox (Mehra 1987). The paradox allows the entropy of a system to decrease, violating the second law of thermodynamics.

Besides the connection between phase wave with material particles, de Broglie gave a new interpretation to the Bohr-Sommerfeld's quantum condition for the angular momentum (Chapter seven):

$$\int_0^{2\pi} p_\phi d\phi = nh$$

De Broglie found a new quantum condition:

$$\oint \frac{\nu}{V} dl = \frac{m}{h} \oint \mathrm{v} dl = \frac{2\pi m \mathrm{v} \mathrm{r}}{h} = n$$

Instead of a circular orbit of the electron in a hydrogen atom, there exist standing waves at discrete energies and frequencies. De Broglie assumed that if a wavelength is associated with the electron and an integral number of wavelengths must fit in the circumference of an orbit, the quantum condition is fulfilled.

$$2\pi r = n\lambda$$

$$\lambda = \frac{h}{mv}$$

$$2\pi r = n\left(\frac{h}{mv}\right)$$

From Bohr's relation of angular momentum, we have:

$$L = m\mathrm{v}r = nh$$

Then we have:

$$L = m\mathrm{v}r = \frac{nh}{2\pi}$$

Do-it-yourself-experience

Use the free software Graph (https://www.padowan.dk/) and plot a wave packet with eleven sine waves. Each sine wave (10 sin(kx)) has the same amplitude, A = 10, and different k = 0.95,0.96,...1,04,1.05.

References cited

Arndt, M. et al. 1999. Wave-particle duality of C_{60} molecules. Nature 401: 1252–1255.

Davisson, C. and Germer, L.H. 1927. The scattering of electrons by a single crystal of nickel. Nature 119: 558–560.

De Broglie, L. 1924. A tentative theory of light quanta. Phil. Mag. S. 6(46): 446–458.

Jönsson, C. 1961. Elektroneninterferenzen an mehreren künstlich hergestellten feinspalten. Z. Phys. 161: 454–474.

Mehra, J. 1987. Erwin Schrödinger and the rise of wave mechanics II. The creation of wave mechanics. Found. Phys. 17: 1141–1188.

Young, T. 1804. The bakerian lecture. Experiments and calculations relative to physical optics. Phil. Trans. Roy. Soc. London, 1–16.

Schrödinger's Wave Quantum Mechanics

10

1. Hamiltonian function

As already mentioned in Chapter seven, The Hamiltonian function, H, is the total energy of the point mass, m, that is, the sum of the kinetic energy, E_{kin}, and potential energy, E_{pot}, of this point mass.

$$H(q,p) = E_{kin} + E_{pot}$$

$$E_{kin} = \frac{m}{2}\left(\frac{dq}{dt}\right)^2 = \frac{p^2}{2m}$$

$$E_{pot} = V(q)$$

$$H(q,p) = \frac{p^2}{2m} + V(q)$$

The coordinate representation is a useful set of operators to represent a coordinate, r, and its corresponding momentum, p_r. In a single dimension, x, the operators in coordinate representation are:

$$\hat{p}_r = -i\hbar\frac{\partial}{\partial x_r}, \quad \hat{x}_r = x_r$$

A more general coordinate representation is:

$$\hat{p}_r = -i\hbar\frac{\partial}{\partial q_r}, \quad \hat{q}_r = q_r$$

In three dimensions, it becomes:

$$\hat{p}_r = -i\hbar\nabla_r, \quad \hat{q}_r = x_r, y_r, z_r$$

$$where: \nabla_r = \frac{\partial}{\partial x_r} + \frac{\partial}{\partial y_r} + \frac{\partial}{\partial z_r}$$

For a conservative, non-relativistic system having 3N degrees of freedom, where N is the number of particles of mass m, the Hamiltonian function is:

$$\widehat{H} = \sum_{n=1}^{3N} \frac{p_{in}^2}{2m} + V(x_1, y_1, z_1, \ldots, x_N, y_N, z_N)$$

$$where: i = \{x, y, z\}, p_{11} = p_{x1}, p_{21} = p_{y1}, p_{31} = p_{z1}, \ldots$$

The coordinate representation of this Hamiltonian is:

$$\widehat{p} = -i\hbar\nabla$$

$$\widehat{p}^2 = (-i\hbar)^2 \nabla^2 = -\hbar^2\nabla^2$$

$$r = x, y, z$$

$$\widehat{H} = \sum_{n=1}^{N} -\frac{\hbar^2\nabla_n^2}{2m} + V(r_1, r_2, \ldots, r_N)$$

$$where: \nabla^2 = \frac{\partial^2}{\partial x^2} + \frac{\partial^2}{\partial y^2} + \frac{\partial^2}{\partial z^2}$$

Where ∇^2 is the Laplacian operator.

2. Hamilton-Jacobi equation

The Hamilton-Jacobi equation is an alternative approach to other formulations of classical mechanics (Newton's law of motion, Lagrangian mechanics and Hamilton mechanics). The Hamilton-Jacobi equation is the closest approach of classical mechanics to Schrödinger's wave mechanics. It is useful in identifying conserved quantities and it is the only formulation in which the motion of a particle can be represented as a wave, uncovering the duality between trajectories and waves. The duality of the light (particle theory and wave theory) appeared in the eighteen century where light could be investigated by rays and geometric optics (Fermat's principle) or by wavefront (Huygens principle).

The wavefront is the spheric domain that the light covered at time t. If the propagation medium is homogeneous, the wave front is:

$$\|r - r_0\| = \frac{c}{n}(t - t_0)$$

$$or: S(r,t) = \|r - r_0\| - (c/n)t$$

Where light is emitted from a point r_0 at time t_0, c is the speed of light, n is the index of the propagating medium, S(r,t) is a collection of points that the light has reached at time t.

The action function, S(t), is a numerical description of how a physical system has changed over time. The action depends on the end points (t_0,q_0) and (t_1,q_1). In classical mechanics, the path followed by a physical system (system's trajectory describing evolution of the system over time) is that for which the action is minimized. The action can be represented as an integral over time.

$$S = \int_{t_1}^{t_2} L dt$$

$$L = T - V \quad or: L = pv - H$$

Where the integrand is the Lagrangian (i.e., the difference between kinetic and potential energies). The action is usually described by a functional with input function of the evolution of the generalized coordinate, q(t), between times t_1 and t_2.

$$S[q(t)] = \int_{t_1}^{t_2} L(q,v,t)dt$$

Where v is the velocity.

For the cases where the input function does not take into account the parametrization with time, we use the abbreviated action S_0 defined as integral of the linear momentum along a path in the coordinates:

$$S_0 = \int pdq$$

When the total energy E is conserved, the action function can be described by:

$$S(q,t) = S_0 - Et$$

The trajectory of a system must obey the Euler-Lagrange equation:

$$\frac{\partial L}{\partial q_i} = \frac{d}{dt}\frac{\partial L}{\partial v_i}$$

For example, a particle of mass m moving under the influence of a conservative force has the following Lagrangian:

$$L = \frac{1}{2}m\left(v_x^2 + v_y^2 + v_z^2\right) - V(x,y,z)$$

The equations of motion for the particle are found by applying the Euler-Lagrange equation for each coordinate (x, y and z) with the derivatives:

$$\frac{\partial L}{\partial x} = -\frac{\partial V}{\partial x}, \frac{\partial L}{\partial y} = -\frac{\partial V}{\partial y}, \frac{\partial L}{\partial z} = -\frac{\partial V}{\partial z}$$

$$\frac{\partial L}{\partial v_x} = mv_x, \frac{\partial L}{\partial v_y} = mv_y, \frac{\partial L}{\partial v_z} = mv_z$$

$$\frac{d}{dt}\left(\frac{\partial L}{\partial v_x}\right) = ma_x, \frac{d}{dt}\left(\frac{\partial L}{\partial v_y}\right) = ma_y, \frac{d}{dt}\left(\frac{\partial L}{\partial v_z}\right) = ma_z$$

$$\frac{\partial L}{\partial q_i} = \frac{d}{dt}\frac{\partial L}{\partial v_i}$$

$$-\frac{\partial V}{\partial x} = ma_x, -\frac{\partial V}{\partial y} = ma_y, -\frac{\partial V}{\partial z} = ma_z$$

Where one arrives at the Newton's second law of motion.

Since the trajectories must obey Euler-Lagrange equation, we can arrive at the relations below where the variation of the action can be computed as a function of the variation of its end points, that is, the derivatives of the action with respect to general coordinate and to time (Houchmandzadeh 2020).

$$\frac{\partial S}{\partial q} = \frac{\partial L}{\partial v}\bigg|_{t_1} = p(t_1)$$

$$-\frac{\partial S}{\partial t} = H = E$$

As a consequence of the first derivative, if we vary the end point q_1, the relative variation in the action is the linear momentum p at the end point.
Combining these relations, we have the Hamilton-Jacobi equation:

$$\frac{\partial S}{\partial t} + H\left(q, \frac{\partial S}{\partial q}, t\right) = 0$$

The fact that the derivative of S with respect to time is a constant, E, leads to the equation:

$$S(q,t) = W(q) - Et$$

Where W, the principal characteristic function, obeys the relation:

$$H\left(q, \frac{\partial W}{\partial q}\right) = E$$

This relation was used by Schrödinger in his papers on quantum mechanics which is a simplification of the Hamilton-Jacobi equation for conservative systems.

3. Schrödinger's wave mechanics

3.1 Introduction

In a comparison between the most important developers of the modern quantum mechanics, we see that Heisenberg had some important collaborators along his journey to the matrix mechanics, while Schrödinger had worked alone in his wave mechanics as well as most of his papers on other topics. On the other hand, Schrödinger was more realistic than Heisenberg. Schrödinger wanted a new quantum theory that could account for the reality of the atom which was not accomplished by matrix mechanics. Whereas Heisenberg had continuously worked on the quantum atomic theory since his doctorate in 1922, Schrödinger had only one work in 1923 on the quantum atomic theory (Schrödinger 1923) and only in 1926 he started a series of publications of his wave mechanics. Schrödinger worked with the quantum orbits of a single electron of the Bohr-Sommerfeld's quantum theory, but from 1922 to 1925 he had dedicated much more on quantum statistical mechanics.

Schrödinger's work on wave mechanics was strongly influenced by de Broglie. Schrödinger noticed: "The de Broglie interpretation of the quantum rules seems to me related in some ways to my note in Z. Phys 12,13 (1923), where a notable property of Weyl 'gauge factor' $\exp(-\int \phi_i dx_i)$ along the quasi-period [of the atomic system] is shown (…). After having recalled the laws of stability for the quantized trajectories (…) we have shown that one can interpret them as expressing the resonance of the phase wave on the length of the closed or quasi-closed trajectory. We believe that this is the first explanation which is physically plausible proposed for the stability conditions of Bohr and Sommerfeld" (Mehra 1987). Then, Schrödinger began to apply de Broglie waves to further atomic problems but he could not visualize the phase wave of an electron on Kepler orbits and he tried to establish a geometrical picture of the de Broglie waves for the cases of a hydrogen atom under the action of a static electric or magnetic field which were not considered by de Broglie (Mehra 1987).

3.2 Derivation of the wave equation and its use for hydrogen atom

In his first work of wave mechanics, Schrödinger stated that: "(...) the notion of whole numbers, merely as such, is not introduced. Rather when integralness does appear, it arises in the same natural way as it does in the case of the node-numbers of a vibrating string" (Schrödinger 1926a). As Mehra stated: "Schrödinger hoped to be able to remove the arbitrariness of the quantum conditions of the old quantum theory of atomic structure; he claimed that he could derive them from a deeper, more fundamental principle(...) connecting the Hamiltonian principle of classical dynamics with the idea of the matter wave equation" (Mehra 1987). Then, the whole number n appears naturally in the wave mechanics (and not by a postulate as in the old quantum mechanics) by the resolution of a differential equation from a variational problem which yields naturally to a function and the whole number as solutions. Schrödinger's first work deals with the radial solution of the hydrogen atom.

The starting point of Schrödinger's work is the Hamilton-Jacobi equation (see previous section):

$$H\left(q,\frac{\partial S}{\partial q}\right)=E$$

Where H is the Hamiltonian and the action S has an unknown function ψ which appears as a product of related functions of single coordinates:
A solution for the above equation is in the form of a sum of functions:

$$S=f_1(q_1)+f_2(q_2)+\ldots+f_s(q_s)$$

Schrödinger wrote S in the form

$$S=K\log\psi,\qquad K=\hbar$$

$$\psi=\psi_1(q_1)\psi_2(q_2)\ldots\psi_s(q_s)$$

Where K is a constant and ψ is a product of functions of single coordinates. Then, the Hamilton-Jacobi equation becomes:

$$S=K\log\psi$$

$$\frac{d}{dx}(\ln u)=\frac{1}{u}\frac{du}{dx}$$

$$\frac{\partial S}{\partial q}=\frac{\partial K\log\psi}{\partial q}=K\frac{\partial\log\psi}{\partial q}=\frac{K}{\psi}\frac{\partial\psi}{\partial q}$$

$$H\left(q,\frac{K}{\psi}\frac{\partial\psi}{\partial q}\right)=E$$

$$H(p,q)=T(p,q)+V(q)$$

$$T\left(\frac{K}{\psi}\frac{\partial\psi}{\partial q},q\right)+V-E=0$$

The above equation can be transformed to a quadratic form of ψ and its first derivatives are equated to zero when the relativistic variation of the mass is neglected. Then, the work is to seek a function ψ that is stationary (everywhere real, single-valued, finite and continuously differentiable up to the second order) for any arbitrary variation of the integral of the quadratic form. Then, the quantum condition of the old quantum mechanics is replaced by a variation problem.

Schrödinger used the Hamilton function for a Keplerian motion of the non-relativistic hydrogen atom where ψ can be chosen for all positive values of energy (corresponding to the energies of the hyperbolic orbits), E, and discrete set of negative values of E (corresponding to Balmer terms).

For an electron in the Cartesian coordinates, the Hamiltonian function is:

$$H = \frac{1}{2m}\left(p_x^2 + p_y^2 + p_z^2\right) + V$$

The Hamilton-Jacobi equation for a hydrogen atom becomes:

$$\frac{1}{2m}\left\{\left(\frac{\partial S}{\partial x}\right)^2 + \left(\frac{\partial S}{\partial y}\right)^2 + \left(\frac{\partial S}{\partial z}\right)^2\right\} + V - E = 0$$

$$S = K \log \psi$$

$$\frac{d}{dx}(\ln u) = \frac{1}{u}\frac{du}{dx}$$

$$\frac{\partial S}{\partial x} = \frac{\partial K \log \psi}{\partial x} = K\frac{\partial \log \psi}{\partial x} = \frac{K}{\psi}\frac{\partial \psi}{\partial x}$$

$$\frac{K^2}{2m\psi^2}\left\{\left(\frac{\partial \psi}{\partial x}\right)^2 + \left(\frac{\partial \psi}{\partial y}\right)^2 + \left(\frac{\partial \psi}{\partial z}\right)^2\right\} + V - E = 0$$

$$\times \frac{2m\psi^2}{K^2}$$

$$\left(\frac{\partial \psi}{\partial x}\right)^2 + \left(\frac{\partial \psi}{\partial y}\right)^2 + \left(\frac{\partial \psi}{\partial z}\right)^2 - \frac{2m}{K^2}\left(E + \frac{e^2}{r}\right)\psi^2 = 0$$

$$V = -\frac{e^2}{r}$$

Where e and m are the charge and mass of the electron, respectively, and r is the radius of the circular orbit.

The expression on the left-hand side (called F(x,y,z)) was subjected to a variation principle.

When the function F(x,y,z) is inserted in the variational principle, it becomes:

$$\delta J = \delta \iiint F(x, y, z)\,dxdydz = 0$$

$$where: F(x, y, z) = \left(\frac{\partial \psi}{\partial x}\right)^2 + \left(\frac{\partial \psi}{\partial y}\right)^2 + \left(\frac{\partial \psi}{\partial z}\right)^2 - \frac{2m}{K^2}\left(E + \frac{e^2}{r}\right)\psi^2$$

From the above equation, Schrödinger arrived at the wave equation for the hydrogen atom (Schrödinger 1926a):

$$\nabla^2\psi + \frac{2m}{\hbar^2}\left(E + \frac{e^2}{r}\right)\psi = 0$$

$$where: \nabla^2\psi = \left(\frac{\partial\psi}{\partial x}\right)^2 + \left(\frac{\partial\psi}{\partial y}\right)^2 + \left(\frac{\partial\psi}{\partial z}\right)^2$$

However, no particular reason has been given by Schrödinger for using the variational principle.

Important to mention that initially Schrödinger had not reduced the above equation to:

$$-\frac{\hbar^2}{2m}\nabla^2\psi - \frac{e^2}{r}\psi = E\psi$$

$$(T + V)\psi = E\psi$$

$$\widehat{\mathrm{H}}\psi = E\psi$$

The solution of the above wave equation can be in polar coordinates if ψ can be written as the product of the three functions (r, θ, ϕ). The function of angles turns out to be a surface harmonics and the radial function r is called χ (the radial function in polar coordinates)

$$\psi = S(\theta, \phi)\chi(r)$$

which leads to the equation:

$$\frac{d^2\chi}{dr^2} + \frac{2}{r}\frac{d\chi}{dr} + \left(\frac{2mE}{\hbar^2} + \frac{2me^2}{\hbar^2 r} - \frac{n(n+1)}{r^2}\right)\chi = 0$$

$$n = 0, 1, 2, 3, \ldots$$

The above equation is the radial Schrödinger equation of the hydrogen atom (see Chapter seventeen) where the radial function in polar coordinates becomes:

$$\chi = r^\alpha U$$

As a consequence, the radial wave equation becomes:

$$\frac{d^2U}{dr^2} + \frac{2(\alpha + 1)}{r}\frac{dU}{dr} + \frac{2m}{\hbar^2}\left(E + \frac{e^2}{r}\right)U = 0$$

No further details are given here, but can be found in Schrödinger's second paper on wave mechanics (Schrödinger 1926b) or can be found in some books (Birtwistle 1928). A different approach for the solution of the radial function of the hydrogen atom is given in Chapter seventeen. Although Schrödinger does not use the term wave function in his first paper of wave mechanics, he infers that some vibration process exists in the atom, which approaches more to the reality than the electronic orbits.

3.3 Hamilton's analogy between optics and mechanics

In his second paper on the wave mechanics (Schrödinger 1926b), Schrödinger established the foundations of his undulatory mechanics (which was popularly known as wave mechanics). In the first section, *he pursued the Hamilton's analogy between optics and mechanics*. Schrödinger stated: "Hamilton's variation principle can be shown to correspond to Fermat's principle for a wave propagation in configuration space (q-space) and the Hamilton-Jacobi equation expresses Huygens' principle for this wave propagation" (Schrödinger 1926b). Schrödinger used a different Hamilton-Jacobi equation

$$\frac{\partial W}{\partial t} + T\left(q_k, \frac{\partial W}{\partial q_k}\right) + V(q_k) = 0$$

Where he replaced the Hamilton function by kinetic (T) and potential (V) energies functions and used W (as he called it as the action function or the phase of the waves) instead of S from the relation:

$$W = S(q_k) - Et$$

To obtain:

$$2T\left(q_k, \frac{\partial W}{\partial q_k}\right) = 2(E - V)$$

After some derivations, Schrödinger established the relation between the step, ds, and an arbitrary surface W_0, given below,in order to find the new surface $W_0 + dW_0$.

$$ds = \frac{dW_0}{\sqrt{2(E-V)}} \Rightarrow \frac{Edt}{\sqrt{2(E-V)}}$$

And the surfaces move at a normal velocity (or wave velocity), u, as:

$$u = \frac{ds}{dt} = \frac{E}{\sqrt{2(E-V)}} \Rightarrow u = \frac{h\nu}{\sqrt{2(h\nu - V)}}$$

Then, the system of constant surfaces W is a system of wave surfaces of a progressive but stationary wave motion in q-space. Hence, the starting point for the undulatory representation of mechanics from a wave equation for q-space for all processes which only depend on time by a factor exp(2πiνt) is given by:

$$\nabla^2\psi + \frac{8\pi^2}{h^2}(E - V)\psi = 0$$

In his 'Handbuch der Experimentalphysik' Hans Thirring emphasized the great importance of Schrödinger and de Broglie's works in order to renew Hamilton's optical-mechanical analogy (Mehra 1987).

3.4 Application to one-particle harmonic oscillator

In his second paper of wave mechanics, Schrödinger applied his wave equation to the case of the Planck oscillator.

$$\frac{d^2\psi}{dq^2}+\frac{8\pi^2}{h^2}\left(E-2\pi^2 v_0^2 q^2\right)\psi=0$$

$$where: V=2\pi^2 v_0^2 q^2$$

This equation becomes:

$$\frac{d^2\psi}{dq^2}+\left(a-bq^2\right)\psi=0$$

$$where: a=\frac{8\pi^2 E}{h^2}, b=\frac{16\pi^4 v_0^2}{h^2}$$

After introducing x as independent variable, we have:

$$x=q\sqrt[4]{b}\therefore b=\left(x/q\right)^4, q=x/\sqrt[4]{b}$$

$$\frac{d^2\psi}{dx^2}+\left(\frac{a}{\sqrt{b}}-x^2\right)\psi=0$$

The above wave equation is the Hermite differential equation (Chapter four) whose proper values and functions (orthogonal functions of Hermite) are:

$$\frac{a}{\sqrt{b}}=1,3,5,...(2n+1)$$

$$e^{-\frac{x^2}{2}}H_n(x)$$

Where $H_n(x)$ are the Hermite polynomials given in Chapter four. From the proper values, we get the energy of the system as:

$$E_n=\frac{2n+1}{2}hv_0,\quad n=0,1,2,3,...$$

And the proper functions become:

$$\psi_n(q)=\exp\left(-\frac{2\pi^2 v_0 q^2}{h}\right)H_n\left(2\pi q\sqrt{\frac{v_0}{h}}\right)$$

In his second paper of wave mechanics (Schrödinger 1926b), Schrödinger also explored his wave equation for the cases of rigid rotator and non-rigid rotator of a diatomic molecule whose results are in accordance with those from matrix mechanics.

In another paper, Schrödinger complimented the study of harmonic oscillator by obtaining the wave function which aggregates all ψ_n into one wave group and obtaining the corresponding wave packet (Schrödinger 1926c).

$$\psi=\sum_{n=0}^{\infty}\left(\frac{A}{2}\right)^n\frac{\psi_n}{n!}$$

Schrödinger also used his wave mechanics for the study of the Stark effect of the Balmer lines where he found an excellent agreement between theory and experiment (Schrödinger 1926d).

Schrödinger extended his wave mechanics to non-relativistic time-dependent wave function (Schrödinger 1926e), which contains his famous one-particle equation:

$$i\hbar\frac{\partial\psi}{\partial t}=\left(-\frac{\hbar^2}{2m}\nabla^2+V\right)\psi$$

3.5 Wave equation applied to Compton effect

Schrödinger also studied the Compton effect (Schrödinger 1927a) and the exchange of the energy according to wave mechanics (Schrödinger 1927b) where he considered a conservative system with wave equation:

$$\nabla^2\psi-\frac{8\pi^2}{h^2}V\psi-\frac{4\pi i}{h}\psi=0$$

Which has normalized proper solutions:

$$\psi_k\exp\left(\frac{2\pi iE_kt}{h}\right)$$

Where ψ_k depends only on the coordinates of the system satisfying the equation:

$$\nabla^2\psi_k+\frac{8\pi^2}{h^2}(E_k-V)\psi_k=0$$

In which time does not appear. The general solution becomes:

$$\psi=\sum_k c_k\psi_k e^{\left(\frac{2\pi iE_kt}{h}\right)}$$

Where c_k's are arbitrary constants called amplitudes (the square of their absolute values is the squares of the amplitudes).

4. Dirac notation

The Dirac notation or bra-ket notation uses the angle brackets and a vertical bar to construct the bras and kets.

$$ket:|u\rangle$$

$$bra:\langle u|$$

The ket is the eigenvector or wave function. The bra is the starred wave function. In a matrix form, a ket is a row vector and a bra is a column vector:

$$\langle A|=\begin{pmatrix}A_1^* & A_2^* & \cdots & A_N^*\end{pmatrix}$$

$$|B\rangle=\begin{pmatrix}B_1\\B_2\\\vdots\\B_N\end{pmatrix}$$

The Dirac notation is particularly useful for the inner product operation of eigenvectors. It is represented as:

$$\langle A|B\rangle$$

Linear operators act on ket from the right side of the function. In the same state vector for bra and ket, the expression becomes:

$$\langle\psi|A|\psi\rangle$$

Where A is the linear operator and ψ is the eigenvector.
The last two equations can be interpreted from the integrals below:

$$\langle\psi|\psi\rangle = \int \psi^* \psi d\tau$$

$$\langle\psi|A|\psi\rangle = \int \psi^* A\psi d\tau$$

5. Expectation value

The expectation value is the probabilistic expected value of the measurement of a experiment. The expected value, E(X) or $\langle X\rangle$, of a random variable, X, is the weighted average of independent realizations of X. When the number of outcomes of X is finite, the expectation of X is:

$$E[X] = \langle X\rangle = \sum_{i=1}^{k} x_i p_i$$

$$where: p_1 + p_2 + \ldots + p_k = 1$$

and p_k is the probability of the occurrence of the value x_k.

If all outcomes x_i are equiprobable (i.e., $p_1 = p_2 = \ldots = p_k$), then the weighted average turns into the simple average.

If X is a random variable with a probability density function of $f(x)$, the expected value is given by:

$$E[X] = \langle X\rangle = \int_{\mathbb{R}} xf(x)dx$$

In quantum mechanics, the square of the wave function is the probability of finding the particle at a given position, r, defined as:

$$dP(r) = \psi^*(r)\psi(r)d^3r$$

Let us consider F(r) as an arbitrary function of the coordinates of the particle. Its expectation value is:

$$\langle F(r)\rangle = \int P(r)F(r)dr = \int \psi^*(r)F(r)\psi(r)dr$$

$$where: \int dP(r) = \int \psi^*(r)\psi(r)d^3r \Rightarrow$$

$$\Rightarrow P(r) = \psi^*(r)\psi(r)$$

Where the wave function used above is normalized, then:

$$\int_{-\infty}^{\infty} \psi_n^* \psi_m d\tau = \delta_{nm}$$

$$\delta_{nm} = \begin{cases} 1 \therefore n = m \Rightarrow normalization \\ 0 \therefore n \neq m \Rightarrow orthogonalization \end{cases}$$

Where δ_{nm} is the Kronecker delta.
The time-dependent expectation value < A > is given by:

$$\frac{d\langle A\rangle}{dt} = \frac{1}{i\hbar}\left\langle\left[\widehat{A}, \widehat{H}\right]\right\rangle + \left\langle\frac{\partial\widehat{A}}{\partial t}\right\rangle$$

Let us derive this equation by using the time-dependent Schrödinger equation and the derivation of the expectation value with time.

$$\langle A\rangle = \langle\psi(t)|\widehat{A}|\psi(t)\rangle$$

$$\frac{d\langle A\rangle}{dt} = \frac{d\langle\psi(t)|}{dt}\widehat{A}|\psi(t)\rangle + \langle\psi(t)|\frac{d\widehat{A}}{dt}|\psi(t)\rangle + \langle\psi(t)|\widehat{A}\frac{d|\psi(t)\rangle}{dt}$$

$$as: i\hbar\frac{d|\psi(t)\rangle}{dt} = \widehat{H}|\psi(t)\rangle, \quad \frac{d|\psi(t)\rangle}{dt} = \frac{1}{i\hbar}\widehat{H}|\psi(t)\rangle$$

$$and: \left(i\hbar\frac{d|\psi(t)\rangle}{dt}\right)^* = \left(\widehat{H}|\psi(t)\rangle\right)^*, -i\hbar\frac{d\langle\psi(t)|}{dt} = \langle\psi(t)|\widehat{H}$$

Then, we have:

$$\frac{d\langle A\rangle}{dt} = -\frac{1}{i\hbar}\langle\psi(t)|\widehat{H}\widehat{A}|\psi(t)\rangle + \frac{1}{i\hbar}\langle\psi(t)|\widehat{A}\widehat{H}|\psi(t)\rangle$$

$$+\langle\psi(t)|\frac{d\widehat{A}}{dt}|\psi(t)\rangle$$

$$\frac{d\langle A\rangle}{dt} = \frac{1}{i\hbar}\langle\psi(t)|\left[\widehat{A}\widehat{H} - \widehat{H}\widehat{A}\right]|\psi(t)\rangle + \langle\psi(t)|\frac{d\widehat{A}}{dt}|\psi(t)\rangle$$

$$\frac{d\langle A\rangle}{dt} = \frac{1}{i\hbar}\langle\psi(t)|\left[\widehat{A}, \widehat{H}\right]|\psi(t)\rangle + \left\langle\frac{\partial\widehat{A}}{\partial t}\right\rangle$$

6. Born's probability and superposition principle

Until 1926, Born has made great contributions from the old quantum mechanics to the matrix mechanics. But his major achievement was his statistical interpretation of the wave mechanics in his attempt to avoid the revitalization of the classical continuum theory in Schrödinger's wave theory. In his 1926's paper Born studied the quantum mechanics of the collision phenomena where the probability (of transition) for the scattering of the electron from z-direction into [θ, ϕ]-direction is given by the square of Φ_{nm} where n is the initial plane wave in the z-direction and m is the final plane wave in the [θ, ϕ]-direction (Born 1926a). In its sequel, Born considered a normalized stationary wave function ψ as a sum of discrete, non-degenerate proper functions ψ_n:

$$\psi(q) = \sum_n c_n \psi_n(q)$$

$$where: \int |\psi(q)|^2 dq = \sum_n |c_n|^2$$

And then $|c_n|^2$ is the probability for the system to be in the state n (Born 1926b).

Since a wave packet is a sum (or superposition) of large number of periodical waves (see chapter nine) described by the equation:

$$y(x) = \frac{1}{\sqrt{2\pi}} \int_{-\infty}^{\infty} A_i e^{i(k_i x - \omega_i t)} dk$$

Then, a wave function (a type of wave packet) is a superposition of its eigenvector components.

$$|\Psi\rangle = \sum_n C_n |\varphi\rangle_n$$

The superposition principle states that a combination of solutions for a linear equation (see Chapter three) is also a solution of it. For example, if state vectors f_1, f_2 and f_3 each solve the linear equation of an eigenvalue problem, then $\psi = c_1 f_1 + c_2 f_2 + c_3 f_3$ is also a solution of the same eigenvalue problem, where c is a coefficient. Consider an electron with two possible spin configurations (see Chapter twelve):

$$\varphi = c_1 |\uparrow\rangle + c_2 |\downarrow\rangle$$

The probability for a specified configuration is given by the square of the absolute value of the coefficient.

$$p_{up} = |c_1|^2$$

$$p_{down} = |c_2|^2$$

$$p_{total} = p_{up} + p_{down}$$

In a general form, any eigenvector can be described as:

$$|\Psi\rangle = \sum_n C_n |\varphi\rangle_n$$

The expectation value of a wave function described by the above equation is:

$$\widehat{A}\psi_n = \omega_n \psi_n$$

$$\langle A \rangle = \int \psi^* \widehat{A} \psi \, d\tau = \langle \psi | \widehat{A} | \psi \rangle$$

$$\langle A \rangle = \int \left(\sum_n C_n \psi_n \right)^* \widehat{A} \left(\sum_n C_n \psi_n \right) d\tau = \sum_n C_n^* C_n \int \psi_n^* \widehat{A} \psi_n d\tau$$

$$\langle A \rangle = \sum_n C_n^* C_n \omega_n \int \psi_n^* \psi_n d\tau = \sum_n C_n^* C_n \omega_n = \sum_n |C_n|^2 \omega_n$$

$$Then: \langle A \rangle = \sum_n |C_n|^2 \omega_n$$

where the wave function is normalized.

Consider again the general form of the superposition principle:

$$|\Psi\rangle = \sum_n C_n |u_n\rangle$$

$$\times \langle u_n|$$

$$\langle u_n|\Psi\rangle = \sum_n C_n \langle u_n|u_n\rangle$$

$$\langle u_n|u_n\rangle = 1$$

$$Then: C_n = \langle u_n|\Psi\rangle$$

7. Derivation of the quantum linear momentum operator

The expectation value of the position of a particle in x-axis for a time-dependent wave function is:

$$\langle x\rangle = \int_{-\infty}^{\infty} \psi^*(x,t)\hat{x}\psi(x,t)dx$$

Where the operator of x is x itself.

The expectation value of the linear momentum of a particle is:

$$\langle p\rangle = m\frac{\partial}{\partial t}\langle x\rangle$$

$$\langle p\rangle = m\frac{\partial}{\partial t}\int_{-\infty}^{\infty} \psi^*(x,t)\cdot x\cdot\psi(x,t)dx$$

$$\langle p\rangle = m\frac{\partial}{\partial t}\int_{-\infty}^{\infty} x\left[\psi^*(x,t)\psi(x,t)\right]dx$$

$$\langle p\rangle = m\int_{-\infty}^{\infty} x\left[\frac{\partial\psi^*(x,t)}{\partial t}\psi(x,t) + \psi^*(x,t)\frac{\partial\psi(x,t)}{\partial t}\right]dx$$

Let us use the time-dependent Schrödinger equation in order to obtain the partial derivative of the wave function with respect to time:

$$i\hbar\frac{\partial\psi(x,t)}{\partial t} = -\frac{\hbar^2}{2m}\frac{\partial^2\psi(x,t)}{\partial x^2} + V(x,t)\psi(x,t)$$

$$\frac{\partial\psi(x,t)}{\partial t} = -\frac{\hbar}{2mi}\frac{\partial^2\psi(x,t)}{\partial x^2} + \frac{V(x,t)}{i\hbar}\psi(x,t)$$

Now, let us obtain the partial derivative of the complex conjugate of the wave function with respect to time.

$$\frac{\partial\psi^*(x,t)}{\partial t} = \frac{\hbar}{2mi}\frac{\partial^2\psi^*(x,t)}{\partial x^2} - \frac{V(x,t)}{i\hbar}\psi^*(x,t)$$

Where the complex conjugate of i is –i and V(x, t) is real.

Let us replace both partial derivatives in the expectation value expression of the linear momentum:

$$\langle p\rangle = m\int_{-\infty}^{\infty} x\left[\begin{array}{l}\left\{\frac{\hbar}{2mi}\frac{\partial^2\psi^*(x,t)}{\partial x^2}-\frac{V(x,t)}{i\hbar}\psi^*(x,t)\right\}\psi(x,t)+\\ +\psi^*(x,t)\left\{-\frac{\hbar}{2mi}\frac{\partial^2\psi(x,t)}{\partial x^2}+\frac{V(x,t)}{i\hbar}\psi(x,t)\right\}\end{array}\right]dx$$

$$\langle p\rangle = m\int_{-\infty}^{\infty} x\left[\psi(x,t)\frac{\hbar}{2mi}\frac{\partial^2\psi^*(x,t)}{\partial x^2}-\psi^*(x,t)\frac{\hbar}{2mi}\frac{\partial^2\psi(x,t)}{\partial x^2}\right]dx$$

$$\langle p\rangle = \frac{\hbar}{2i}\int_{-\infty}^{\infty} x\left[\psi(x,t)\frac{\partial^2\psi^*(x,t)}{\partial x^2}-\psi^*(x,t)\frac{\partial^2\psi(x,t)}{\partial x^2}\right]dx$$

Let us separate both terms over integration above into terms A and B:

$$\langle p\rangle = \frac{\hbar}{2i}\left[\int_{-\infty}^{\infty} x\psi(x,t)\frac{\partial^2\psi^*(x,t)}{\partial x^2}dx-\int_{-\infty}^{\infty} x\psi^*(x,t)\frac{\partial^2\psi(x,t)}{\partial x^2}dx\right]$$

$$A=\int_{-\infty}^{\infty} x\psi(x,t)\frac{\partial^2\psi^*(x,t)}{\partial x^2}dx,\; B=\int_{-\infty}^{\infty} x\psi^*(x,t)\frac{\partial^2\psi(x,t)}{\partial x^2}dx$$

Let us now use the integration by parts of the general formula:

$$\int_a^b u(x)v'(x)dx=\left[u(x)v(x)\right]_a^b-\int_a^b u'(x)v(x)dx$$

In the term A of the previous above:

$$A=\int_{-\infty}^{\infty} x\psi(x,t)\frac{\partial^2\psi^*(x,t)}{\partial x^2}dx = x\psi(x,t)\frac{\partial\psi^*(x,t)}{\partial x}\Bigg|_{-\infty}^{\infty} -$$

$$-\int_{-\infty}^{\infty}\frac{\partial\left[x\psi(x,t)\right]}{\partial x}\frac{\partial\psi^*(x,t)}{\partial x}dx$$

$$x\psi(x,t)\frac{\partial\psi^*(x,t)}{\partial x}\Bigg|_{-\infty}^{\infty}=0$$

Then:

$$A=-\int_{-\infty}^{\infty}\frac{\partial\left[x\psi(x,t)\right]}{\partial x}\frac{\partial\psi^*(x,t)}{\partial x}dx$$

$$A=-\int_{-\infty}^{\infty}\left\{\psi(x,t)\frac{\partial\psi^*(x,t)}{\partial x}+x\frac{\partial\psi(x,t)}{\partial x}\frac{\partial\psi^*(x,t)}{\partial x}\right\}dx$$

Similarly, we use the same procedure for the term B, exchanging the complex conjugates:

$$B = \int_{-\infty}^{\infty} x\psi^*(x,t)\frac{\partial^2\psi(x,t)}{\partial x^2}dx =$$

$$B = -\int_{-\infty}^{\infty}\left\{\psi^*(x,t)\frac{\partial\psi(x,t)}{\partial x} + x\frac{\partial\psi^*(x,t)}{\partial x}\frac{\partial\psi(x,t)}{\partial x}\right\}dx$$

As a result, we have:

$$\langle p\rangle = \frac{\hbar}{2i}[A - B]$$

$$\langle p\rangle = \frac{\hbar}{2i}\int_{-\infty}^{\infty}\begin{bmatrix} -\psi(x,t)\frac{\partial\psi^*(x,t)}{\partial x} - x\frac{\partial\psi(x,t)}{\partial x}\frac{\partial\psi^*(x,t)}{\partial x} + \\ +\psi^*(x,t)\frac{\partial\psi(x,t)}{\partial x} + x\frac{\partial\psi^*(x,t)}{\partial x}\frac{\partial\psi(x,t)}{\partial x}\end{bmatrix}dx$$

$$\langle p\rangle = \frac{\hbar}{2i}\int_{-\infty}^{\infty}\left[\psi^*(x,t)\frac{\partial\psi(x,t)}{\partial x} - \psi(x,t)\frac{\partial\psi^*(x,t)}{\partial x}\right]dx$$

Once again, let us integrate by parts the second term of the above equation:

$$\int_{-\infty}^{\infty}\psi(x,t)\frac{\partial\psi^*(x,t)}{\partial x}dx = \psi(x,t)\psi^*(x,t)\Big|_{-\infty}^{\infty} - \int_{-\infty}^{\infty}\psi^*(x,t)\frac{\partial\psi(x,t)}{\partial x}dx$$

$$\psi(x,t)\psi^*(x,t)\Big|_{-\infty}^{\infty} = 0$$

$$Then: \int_{-\infty}^{\infty}\psi(x,t)\frac{\partial\psi^*(x,t)}{\partial x}dx = -\int_{-\infty}^{\infty}\frac{\partial\psi(x,t)}{\partial x}\psi^*(x,t)dx$$

Now, let us replace the result of the above equation in the expression of the expectation value of the linear momentum:

$$\langle p\rangle = \frac{\hbar}{2i}\int_{-\infty}^{\infty}\left[\psi^*(x,t)\frac{\partial\psi(x,t)}{\partial x} + \frac{\partial\psi(x,t)}{\partial x}\psi^*(x,t)\right]dx$$

$$\langle p\rangle = \frac{\hbar}{i}\int_{-\infty}^{\infty}\psi^*(x,t)\frac{\partial\psi(x,t)}{\partial x}dx = \int_{-\infty}^{\infty}\psi^*(x,t)\left(\frac{\hbar}{i}\frac{\partial}{\partial x}\right)\psi(x,t)dx$$

$$Then: \hat{p}_x = \frac{\hbar}{i}\frac{\partial}{\partial x}$$

$$or: \hat{p}_x = -i\hbar\frac{\partial}{\partial x}$$

In three-dimensional space, the linear momentum operator becomes:

$$\hat{p} = -i\hbar\nabla$$

$$\nabla = \frac{\partial}{\partial x} + \frac{\partial}{\partial y} + \frac{\partial}{\partial z}$$

Where ∇ is the gradient operator.

8. Normalization of a wave function

The normalization of a wave function, ψ(x), assures that the one particle represented by this wave function has to be found in the range of the system, for example, $-\infty$ to $+\infty$. The normalization is done from the product of the normalization constant, N, and the wave function so that:

$$\int_{-\infty}^{\infty} N\psi(x)N\psi(x)dx = 1$$

Which means that we have 100% probability to find the particle described by the wave function within the limits of the system.
Let the normalized wave function as:

$$\psi(x)_N = N\psi(x)$$

And let us apply the probability condition to find the equation for the normalization constant:

$$\int_{-\infty}^{+\infty} N\psi(x)N\psi(x)dx = 1$$

$$N^2\int_{-\infty}^{+\infty} [\psi(x)]^2\,dx = 1$$

$$N = \frac{1}{\sqrt{\int_{-\infty}^{+\infty} [\psi(x)]^2\,dx}}$$

For numerical integration we have that an integration is approximated to a sum, then we have

$$N = \frac{1}{\sqrt{sum|\psi(x)^2|}}$$

Since a state vector and a wave function can be quantum states, let \v> be a quantum state vector:

$$|v\rangle = 3|u_x\rangle + 4.2|u_y\rangle - 0.5|u_z\rangle, |u_x\rangle^2 = |u_y\rangle^2 = |u_z\rangle^2 = 1$$

Then the normalization constant for this quantum state vector is:

$$N = \frac{1}{\sqrt{|3|^2 + |4.2|^2 + |-0.5|^2}}$$

The Fortran source code that provides the normalization constant, the normalized state vector, the probability for each component and the corresponding sum (which must give one) for a general 3-dimensional state vector is given below.

```
!Program name: NORMALIZATION
real :: a, b, c, s, norm
integer::i
real, dimension (3)::psi, Npsi, P, Npsi2
func(a,b,c)=1./sqrt(sum(abs(psi)**2))
norm (a,b,c) = dot_product (psi, psi)
print *, "For the state vector aUx+bUy+cUz, give the values of a, b and c:"
read *, a, b, c
psi=[a,b,c]
Print *, psi
Npsi=func(a,b,c)*psi
P=abs(func(a,b,c)*psi)**2
Npsi2=psi/sqrt(norm(a,b,c))
Print *, "The normalization constant is:", func(a,b,c)
Print *, "The normalized state vector is:", Npsi, Npsi2
Print *, "The probability is:", P
s=P(1)+P(2)+P(3)
print *, s
stop
end
```

9. Orthogonality of a wave function

One of the properties of the eigenfunction is the orthogonality, where Kronecker delta, δ_{mn}, is 0 or 1.

$$\int_{-\infty}^{+\infty} \psi_m(x)\psi_n(x)dx = \delta_{mn} = \begin{cases} 1 & if \quad m = n \\ 0 & if \quad m \neq n \end{cases}$$

Let us take the particle in one-dimensional box (Chapter fifteen):

$$\psi_n(x) = \sqrt{\frac{2}{l}}\,\mathrm{in}\left(\frac{n\cdot\pi}{l}x\right), \qquad n = 0, \pm 1, \pm 2, \pm 3, \ldots$$

Let us use ψ_m and ψ_n where m ≠ n in the integral above:

$$I = \frac{2}{l}\int_0^l \sin\frac{m\pi x}{l}\sin\frac{n\pi x}{l}dx$$

Let us use the trigonometric relation:

$$\sin x \sin y = \frac{1}{2}\left[\cos(x-y) - \cos(x+y)\right]$$

In the previous integral:

$$\sin\frac{m\pi x}{l}\sin\frac{n\pi x}{l}=\frac{1}{2}\left[\cos\left(\frac{(m-n)\pi x}{l}\right)-\cos\left(\frac{(m+n)\pi x}{l}\right)\right]$$

$$I=\frac{1}{l}\int_0^l\left[\cos\left(\frac{(m-n)\pi x}{l}\right)-\cos\left(\frac{(m+n)\pi x}{l}\right)\right]dx$$

$$I=\frac{1}{l}\int_0^l\cos\left(\frac{(m-n)\pi x}{l}\right)dx-\frac{1}{l}\int_0^l\cos\left(\frac{(m+n)\pi x}{l}\right)dx$$

Each integral is zero because:

$$\frac{1}{l}\int_0^l\cos\left(\frac{(m+n)\pi x}{l}\right)dx=\left[\frac{1}{((m+n)\pi)}\sin\left(\frac{(m+n)\pi x}{l}\right)\right]_0^l$$

$$=\frac{1}{((m+n)\pi)}\left[\sin(m+n)\pi-\sin 0\right]=0$$

Then,

$$I=\frac{2}{l}\int_0^l\sin\frac{m\pi x}{l}\sin\frac{n\pi x}{l}dx=0,\qquad m\neq n$$

10. Hermitian operator

As we have seen in chapter three, an eigenvalue equation has an eigenvector **u** which changes only a scalar factor (an eigenvalue λ) after a linear transformation of an operator A acting on it.

$$Au=\lambda u$$

An operator can be represented by a matrix, A, acting on a vector space V in the vector **u** to yield another vector **v** which is a scalar, λ, of the vector **u**. Below, the linear transformation in a generic vector space with n elements.

$$\begin{bmatrix}a_{1,1} & a_{1,2} & \cdots & a_{1,n}\\ a_{2,1} & a_{2,2} & \cdots & a_{2,n}\\ \vdots & \vdots & \vdots & \vdots\\ a_{n,1} & a_{n,2} & \cdots & a_{n,n}\end{bmatrix}\begin{bmatrix}u_1\\ u_2\\ \vdots\\ u_n\end{bmatrix}=\begin{bmatrix}v_1\\ v_2\\ \vdots\\ v_n\end{bmatrix}=\lambda\begin{bmatrix}u_1\\ u_2\\ \vdots\\ u_n\end{bmatrix}$$

In Chapter three, we have seen that a Hermitian matrix A is a complex square matrix whose conjugate transpose, $(A^*)^T$, is equivalent to it. For example:

$$\begin{bmatrix}m & a-ib & c-id\\ a+ib & n & e-if\\ c+id & e+if & o\end{bmatrix}\therefore m,n,o\in\mathbb{R}$$

Physical pure states are represented as unit-norm vector or wave function in a special complex Hilbert space (a generalization of the Euclidian space with a finite or infinite number of dimensions).

The operators that yield real eigenvalues (observables that can also be obtained from a physical experiment) are called Hermitian. All operators of an eigenvalue equation are Hermitian. Hermitian operators have real eigenvalues and orthonormal eigenfunctions. A Hermitian operator, A, is a linear operator that satisfies:

$$\int \varphi^* \hat{A} \psi d\tau = \int \psi \left(\hat{A}\varphi\right)^* d\tau$$

$$or: \int \varphi^* \hat{A} \psi d\tau = \left(\int \psi^* \hat{A} \varphi d\tau\right)^*$$

$$where: \left(\psi^*\right)^* = \psi$$

Let us assume the following eigenvalue equation of a normalized eigenfunction ψ:

$$\widehat{A}\psi = a\psi$$

The corresponding expectation eigenvalue is:

$$\langle A \rangle = \int \psi^* \widehat{A} \psi d\tau = \int \psi^* a\psi d\tau = a \int \psi^* \psi d\tau = a$$

$$where: \int \psi^* \psi d\tau = 1$$

Since physical observables are real numbers, then:

$$a = a^*$$

Then, we have the conjugate of the expectation value as:

$$\langle A \rangle^* = \left(\int \psi^* \widehat{A} \psi d\tau\right)^* = \int \psi \widehat{A}^* \psi^* d\tau$$

$$as: \widehat{A}^* \psi^* = a^* \psi^*$$

$$then: \langle A \rangle^* = \int \psi a^* \psi^* d\tau = a^* \int \psi^* \psi d\tau = a^* = a$$

As a consequence, any Hermitian operator follows the condition:

$$\int \psi^* \widehat{A} \psi d\tau = \int \psi \widehat{A}^* \psi^* d\tau = \int \psi \widehat{A} \psi^* d\tau$$

$$where: \widehat{A}^* = \widehat{A}$$

Properties of Hermitian operators:

(1) The eigenvalues of a Hermitian operator are real (proof: example 2);

(2) The eigenfunctions of a Hermitian operator whose eigenvalues are distinguished (non-degenerate) are orthogonal (proof: example 3);

(3) When k eigenfunctions correspond to the same eigenvalue (k-degenerate eigenfunctions) and they are a complete set of another eigenfunction, then this eigenfunction has the same eigenvalue (proof: example 4).

11. Examples

(1) Show that the quantum linear momentum operator is Hermitian.

Solution: We have to demonstrate the following condition:

$$\int \psi^* p_x \psi \, dx = \int \psi \, p_x^* \psi^* dx$$

Let us use the expression for the quantum linear momentum operator in the left hand of the above equation:

$$\int \psi^* p_x \psi \, dx = -i\hbar \int \psi^* \frac{d}{dx} \psi \, dx$$

By using the general formula for the integration by parts:

$$\int u dv = uv - \int v du$$

$$or: \int u \frac{dv}{dx} dx = uv - \int v \frac{du}{dx} dx$$

we have:

$$\int \psi^* p_x \psi \, dx = -i\hbar \int_{-\infty}^{\infty} \psi^* \frac{d\psi}{dx} dx = -i\hbar \, \psi^* \psi \Big|_{-\infty}^{\infty} - i\hbar \int_{-\infty}^{\infty} \psi \frac{d\psi^*}{dx} dx$$

$$-i\hbar \, \psi^* \psi \Big|_{-\infty}^{\infty} = 0$$

$$\int \psi^* p_x \psi \, dx = -i\hbar \int_{-\infty}^{\infty} \psi \frac{d\psi^*}{dx} dx = \int_{-\infty}^{\infty} \psi \left(-i\hbar \frac{d\psi^*}{dx} \right) dx = \int \psi \, p_x^* \psi^* dx$$

(2) Show that the eigenvalues of a Hermitian operator are real.

Solution: Let us consider the following eigenvalue equations:

$$\hat{A} \psi_n = a_n \psi_n$$

$$\hat{A} \psi_m = a_m \psi_m$$

And let us use the Hermitian condition:

$$\left\langle \psi_n \middle| \hat{A} \middle| \psi_m \right\rangle = \left\langle \psi_m \middle| \hat{A} \middle| \psi_n \right\rangle^*$$

Let us replace the previous equations in the equation above:

$$\left\langle \psi_n \middle| a_m \middle| \psi_m \right\rangle = \left\langle \psi_m \middle| a_n \middle| \psi_n \right\rangle^*$$

$$a_m \left\langle \psi_n \middle| \psi_m \right\rangle = a_n^* \left\langle \psi_m \middle| \psi_n \right\rangle^*$$

$$\left\langle \psi_n \middle| \psi_m \right\rangle = \left\langle \psi_m \middle| \psi_n \right\rangle^* = \delta_{ij}$$

$$if: i = j$$

$$then: \delta_{ij} = 1$$

$$and: a_m = a_n^*$$

(3) Show that the eigenfunctions of a Hermitian operator whose eigenvalues are distinguished are orthogonal.

Solution: Consider the eigenvectors ψ and φ of the Hermitian operator Â:

$$\hat{A}\psi = a\psi \equiv \hat{A}|\psi\rangle = a|\psi\rangle$$

$$\hat{A}\varphi = b\varphi \equiv \hat{A}|\varphi\rangle = b|\varphi\rangle$$

And their corresponding Hermitian conjugates:

$$\hat{A}^*\psi^* = a^*\psi^*$$

$$\hat{A}^*\varphi^* = b^*\varphi^*$$

$$\hat{A}^* = \hat{A}, \quad a = a^*, b = b^*$$

$$then:$$

$$\hat{A}\psi^* = a\psi^* \equiv \langle\psi|\hat{A} = a\langle\psi|$$

$$\hat{A}\varphi^* = b\varphi^* \equiv \langle\varphi|\hat{A} = b\langle\varphi|$$

Multiplying $\hat{A}\psi^* = a\psi^*$ by φ and $\hat{A}\varphi^* = b\varphi^*$ by ψ, we have:

$$\langle\psi|\hat{A}|\varphi\rangle = a\langle\psi|\varphi\rangle$$

$$\langle\varphi|\hat{A}|\psi\rangle = b\langle\varphi|\psi\rangle$$

By subtracting both equations, we have:

$$As: \langle\psi|\hat{A}|\varphi\rangle = \langle\varphi|\hat{A}|\psi\rangle$$

$$And: \langle\psi|\varphi\rangle = \langle\varphi|\psi\rangle$$

$$then: (a-b)\langle\psi|\varphi\rangle = 0$$

If (a–b) ≠ 0, then $\langle\psi|\varphi\rangle = 0$, i.e., the eigenfunctions ψ and φ are orthogonal.

(4) Show that the eigenfunctions φ_i that form a complete set for the expansion of ψ and when they have the same eigenvalue, the eigenfunction ψ has the same eigenvalue.

Solution: Let us suppose that the complete set of ψ is:

$$\psi = c_1\varphi_1 + c_2\varphi_2$$

And that:

$$\hat{A}\varphi_1 = a\varphi_1$$

$$\hat{A}\varphi_2 = a\varphi_2$$

Then, we have to show that:

$$\hat{A}\psi = a\psi$$

Let us replace the expansion in the left side of the equation above:

$$\hat{A}\psi = \hat{A}\left(c_1\varphi_1 + c_2\varphi_2\right) = \hat{A}c_1\varphi_1 + \hat{A}c_2\varphi_2 = c_1\hat{A}\varphi_1 + c_2\hat{A}\varphi_2$$

Now, let us replace the eigenvalue equations of φ_i in the equation above:

$$c_1\hat{A}\varphi_1 + c_2\hat{A}\varphi_2 = c_1 a\varphi_1 + c_2 a\varphi_2 = a\left(c_1\varphi_1 + c_2\varphi_2\right) = a\psi$$

By comparing the last two equations, we have:

$$\hat{A}\psi = a\psi$$

(5) Show that the probability density, p, of an eigenfunction is time-invariant.

Solution: Let us derivate the norm of the state vector with time.

$$norm: \langle\psi(t)|\psi(t)\rangle$$

$$\frac{d\langle\psi(t)|\psi(t)\rangle}{dt} = \frac{d\langle\psi(t)|}{dt}|\psi(t)\rangle + \langle\psi(t)|\frac{d|\psi(t)\rangle}{dt}$$

Let us use the time-dependent Schrödinger equation in the above equation:

$$ih\frac{d|\psi(t)\rangle}{dt} = \hat{H}|\psi(t)\rangle, \quad \frac{d|\psi(t)\rangle}{dt} = \frac{1}{i\hbar}\hat{H}|\psi(t)\rangle$$

$$and: \left(ih\frac{d|\psi(t)\rangle}{dt}\right)^* = \left(\hat{H}|\psi(t)\rangle\right)^*, -ih\frac{d\langle\psi(t)|}{dt} = \langle\psi(t)|\hat{H}$$

$$then: \frac{d\langle\psi(t)|\psi(t)\rangle}{dt} = -\frac{1}{i\hbar}\langle\psi(t)|\hat{H}|\psi(t)\rangle + \langle\psi(t)|\hat{H}|\psi(t)\rangle = 0$$

$$\frac{d\langle\psi(t)|\psi(t)\rangle}{dt} = 0$$

If the derivative of the norm of the eigenfunction with time is zero, then derivative of the square of the norm (the probability density, p) with time is also zero.

$$p = \left|\langle\psi(t)|\psi(t)\rangle\right|^2$$

$$\frac{d\left|\langle\psi(t)|\psi(t)\rangle\right|^2}{dt} = 0$$

(6) A system described by the Hamiltonian (Johnson and Pedersen 1986):

$$H = \left(-\frac{d^2}{dx^2} + x^2\right)$$

(a) Show that $Ax.\exp(-x^2/2)$ is an eigenfunction of the Hamiltonian function given above.

Solution:

$$-\frac{d^2}{dx^2}\left(Axe^{-x^2/2}\right)=-A\frac{d^2}{dx^2}\left(xe^{-x^2/2}\right)=$$

$$\frac{d}{dx}\left(xe^{-x^2/2}\right)=e^{-x^2/2}\frac{dx}{dx}+x\frac{d}{dx}\left(e^{-x^2/2}\right)$$

$$\frac{d}{dx}e^{u'}=\frac{de^{u'}}{du'}\frac{du'}{dx}=e^{u'}\frac{du'}{dx}$$

$$\frac{d}{dx}\left(e^{-x^2/2}\right)=\frac{d\left(e^{-x^2/2}\right)}{d\left(-x^2/2\right)}\frac{d\left(-x^2/2\right)}{dx}=-e^{-x^2/2}\frac{2}{2}x$$

$$\frac{d}{dx}\left(xe^{-x^2/2}\right)=\left(1-x^2\right)e^{-x^2/2}$$

$$-A\frac{d}{dx}\left[\left(1-x^2\right)e^{-x^2/2}\right]=A\left(3-x^2\right)xe^{-x^2/2}=3Axe^{-x^2/2}-Ax^3e^{-x^2/2}$$

$$-\frac{d^2}{dx^2}\left(Axe^{-x^2/2}\right)=A\left(3-x^2\right)xe^{-x^2/2}$$

$$\left(-\frac{d^2}{dx^2}+x^2\right)Axe^{-x^2/2}=3Axe^{-x^2/2}-Ax^3e^{-x^2/2}+Ax^3e^{-x^2/2}$$

$$Hu=3u$$

(b) Determine the coefficient A

Solution:

$$\int u^* u dx=1$$

$$\int_{-\infty}^{+\infty}A^2x^2e^{-x^2}dx=1;$$

$$A^2\frac{\sqrt{\pi}}{2}=1\therefore A=\left(4/\pi\right)^{1/4}$$

(7) Transform the time-dependent Schrödinger equation into the time-independent Schrödinger equation.

Solution: Let us assume a time-dependent wave function

$$\psi\left(q,t\right)=\phi(t)\xi(q)$$

And let us obtain the time-independent Schrödinger equation from the time-dependent Schrödinger equation. The time-dependent Schrödinger equation is:

$$\widehat{H}\psi\left(q,t\right)=i\hbar\frac{\partial\psi\left(q,t\right)}{\partial t}$$

Let us explicit ψ(q,t) in the time-dependent Schrödinger equation:

$$\widehat{H}\phi(t)\xi(q)=i\hbar\frac{\partial\phi(t)\xi(q)}{\partial t}$$

Since the Hamiltonian function operates only on spatial coordinates, the above equation can be rearranged into:

$$\phi(t)\widehat{H}\xi(q) = i\hbar\xi(q)\frac{\partial\phi(t)}{\partial t}$$

And now, let us divide the above equation by the explicit form of ψ(q,t):

$$\frac{\phi(t)\widehat{H}\xi(q)}{\phi(t)\xi(q)} = \frac{i\hbar\xi(q)}{\phi(t)\xi(q)}\frac{\partial\phi(t)}{\partial t}$$

$$\frac{\widehat{H}\xi(q)}{\xi(q)} = \frac{i\hbar}{\phi(t)}\frac{\partial\phi(t)}{\partial t}$$

The two sides of the above equation depend on separate independent variables. For this equality to hold, each side of the equation must be equal to a constant (this constant is E, the total energy of the system). Then, we have:

$$\frac{\widehat{H}\xi(q)}{\xi(q)} = E, \quad \widehat{H}\xi(q) = E\,\xi(q)$$

and

$$\frac{i\hbar}{\phi(t)}\frac{\partial\phi(t)}{\partial t} = E, \quad \frac{\partial\phi(t)}{\partial t} = -\frac{iE}{\hbar}\phi(t)$$

Then, we arrived at the famous time-independent Schrödinger equation:

$$\widehat{H}\Psi = E\Psi$$

Where Ψ is the time-independent wave function.

12. Exercises

(1) For the wave function Ax.exp($-x^2/2$) give the expectation value of x.

(2) For the Hamiltonian operator in one dimension

$$\widehat{H} = \frac{\hat{p}^2}{2m} + V(x)$$

Derive the commutators:

$$\left[\widehat{H}, x\right] = -i\hbar\frac{d}{dx}$$

$$\left[\widehat{H}, \hat{p}\right] = i\hbar\frac{dV(x)}{dx}$$

Where:

(i) V(x) can be represented in a power series which gives [V(x), x] = 0

(ii) The commutator $[p^2, p] = p[p, p] = 0$

(iii) The commutator $[p^2, x] = [p, x]p + p[p, x]$

Hint: Let each commutator operate on an arbitrary function u. For example, to evaluate the commutator [x. d/dx], we use:

$$\left[x, d/dx\right]u = \left[x\frac{d}{dx} - \frac{d}{dx}x\right]u = x\frac{du}{dx} - \frac{du}{dx}x - \frac{dx}{dx}u = -u$$

$$Then: \left[x, d/dx\right] = -1$$

(2) In time t = 0 the state of a harmonic oscillator is given by the normalized wave function (Johnson and Pedersen 1986):

$$\psi = \left(1/5\right)^{1/2} u_1(x) + \left(1/2\right)^{1/2} u_2(x) + \left(3/10\right)^{1/2} u_3(x)$$

Where $u_n(x)$ is a time-independent component of the wave function and $(1/5)^{1/2}$, $(1/2)^{1/2}$, $(3/10)^{1/2}$ are coefficients, $c_n(t_0)$ at time t =0. Give the corresponding time dependent wave function, where now $u_n(x,t)$ becomes a time-dependent component of the wave function

Hint: Use the follow relations:

$$\widehat{H}\left|\varphi_{n,x}\right\rangle = E_n \left|\varphi_{n,x}\right\rangle$$

$$\left|\psi(t)\right\rangle = \sum_{n,x} c_{n,x}(t)\left|\varphi_{n,x}\right\rangle$$

$$c_{n,x}(t) = \left\langle \varphi_{n,x} \middle| \psi(t)\right\rangle$$

Answer:

$$c_1(t) = \left(1/5\right)^{1/2} e^{-iE_1(t-t_0)/\hbar}$$

$$c_2(t) = \left(1/2\right)^{1/2} e^{-iE_2(t-t_0)/\hbar}$$

$$c_3(t) = \left(3/10\right)^{1/2} e^{-iE_3(t-t_0)/\hbar}$$

$$E_n = \left(n + 1/2\right)h\nu, E_1 = \frac{3}{2}h\nu, E_2 = \frac{5}{2}h\nu, E_3 = \frac{7}{2}h\nu$$

$$h = 2\pi\hbar, E_1 = 3\pi\hbar\nu, E_2 = 5\pi\hbar\nu, E_3 = 7\pi\hbar\nu$$

$$\psi = \left(1/5\right)^{1/2} e^{-3i\pi\nu(t-t_0)} + \left(1/2\right)^{1/2} e^{-5i\pi\nu(t-t_0)} + \left(3/10\right)^{1/2} e^{-7i\pi\nu(t-t_0)}$$

(3) By knowing that the operator p_x is $-i\hbar D_x$, where D_x is d/dx, derive the commutator relation: $[p_x x - x p_x]f(x) = -i\hbar f(x)$.

(4) Find the expectation value of a particle in a one-dimensional box described by:

$$\psi_n(x) = \sqrt{\frac{2}{a}}\cos\left(k_n x\right), \quad k_n = \frac{n\pi}{a}, \quad -a/2 \le x \le a/2$$

Answer: $\langle x \rangle = 0$

(5) By knowing the equation for the linear momentum operator, derive the equation for the kinetic energy operator:

Data:

$$\hat{p} = -i\hbar\nabla$$

$$\hat{T} = -\frac{\hbar^2}{2m}\nabla^2$$

References cited

Birtwistle, G. 1928. The new quantum mechanics. Cambridge University Press, Cambridge.

Born, M. 1926a. Zurquantenmechanik der stobvorgange. Z. Phys. 37: 863–867.

Born, M. 1926b. Zurquantenmechanik der stobvorgange (part II). Z. Phys. 38: 803–827.

Houchmandzadeh, B. 2020. The Hamilton-Jacobi equation: An alternative approach. Am. J. Phys. 88: 353–359.

Johnson, C.S. and Pedersen, L.G. 1986. Problems and Solutions in Quantum Chemistry and Physics. Dover Publications, New York.

Mehra, J. 1987. Erwin Schrödinger and the rise of wave mechanics II. The creation of wave mechanics. Found. Phys. 17: 1141–1188.

Schrödinger, E. 1923. Uber einebemerkenswerteeigenschaft der quantenbahneneineseinzelnenelektrons. Z. Phys. 12: 13–23.

Schrödinger, E. 1926a. Quantisierungalseigenwertproblem (part I). Ann. Phys. 79: 361–376.

Schrödinger, E. 1926b. Quantisierungalseigenwertproblem (part II). Ann. Phys. 79: 489–527.

Schrödinger, E. 1926c. Der stetigeubergang von der mikro- zurmakromechanik. Naturwissenschaften 14: 664–666.

Schrödinger, E. 1926d. Quantisierungalseigenwertproblem (part III). Ann. Phys. 80: 437–490.

Schrödinger, E. 1926e. Quantisierungalseigenwertproblem (part IV). Ann. Phys. 81: 109–139.

Schrödinger, E. 1927a. Uber den Comptoneffekt. Ann. Phys. 82: 257–264.

Schrödinger, E. 1927b. Energieaustauschnach der wellenmechanik. Ann. Phys. 83: 956–968.

Applications of Matrix and Wave Quantum Mechanics

11

1. Equivalence between matrix mechanics and wave mechanics

In order to test the equivalence between matrix mechanics and wave mechanics, let us use the example of the one-dimensional free motion of particle (Chapter fourteen) in the interval [0,1].

$$\psi(x) = \sqrt{2}\sin(\nu\pi x)$$

From mathematics, every continuous function $f(x)$ that is also continuously differentiable can be represented by a series with these eigenfunctions:

$$f(x) = \sum_{\nu=1}^{\infty} a_\nu \sqrt{2}\sin(\nu\pi x)$$

$$a_\nu = \int_{-\infty}^{\infty} \psi_\nu^*(x) f(x) dx$$

The coefficient a_ν is obtained after the multiplication of the previous equation with the starred wave function on the left of both sides of the equation followed by the integration. Alternatively, the function $f(x)$ can be written as:

$$f(x) = x \cdot \psi_\mu(x) = \sum_{\nu=1}^{\infty} a_\nu \psi_\nu(x)$$

Then, the coefficient a_ν is:

$$a_\nu = \int_{-\infty}^{\infty} \psi_\nu^*(x) \cdot x \cdot \psi_\mu(x)\, dx$$

Which gives:

$$f(x) = \sum_{\nu=1}^{\infty} \psi_\nu(x) \int_{-\infty}^{\infty} \psi_\nu^*(y) \cdot y \cdot \psi_\mu(y)\, dy$$

where y is a general representation of the operator. The components of the matrices **X**(or **Q**) and **P**, $(X)_{\nu\mu}$ and $(P)_{\nu\mu}$, respectively, can be calculated from the eigenfunctions $\varphi(x)$, where they will represent the coefficient a_ν.

$$(X)_{\nu\mu} = \int_{-\infty}^{\infty} \psi_\nu^*(x) \cdot x \cdot \psi_\mu(x) dx$$

$$(P)_{\nu\mu} = \int_{-\infty}^{\infty} \psi_\nu^*(x) \cdot \frac{\hbar}{i}\frac{d}{dx} \cdot \psi_\mu(x) dx$$

We can see that the function $f(x)$ can be written in the form:

$$f(x) = \sum_{\nu=1}^{\infty} \psi_\nu(x)(X)_{\nu\mu}$$

Then, we have to show that the matrices **X** and **P**, whose elements are defined above satisfy the commutation relation:

$$\mathbf{PX} - \mathbf{XP} = \frac{\hbar}{i}\mathbf{I}$$

Let us obtain the elements of the matrix product **PX**:

$$(PX)_{\nu\mu} = \sum_\kappa (P)_{\nu\kappa}(X)_{\kappa\mu} = \frac{\hbar}{i}\int_{-\infty}^{\infty} \psi_\nu^*(x) \cdot \frac{d}{dx}\sum_\kappa \psi_\mu(x)(X)_{\kappa\mu}\, dx$$

By replacing the equation of the function $f(x)$ into the equation above, we have:

$$(PX)_{\nu\mu} = \frac{\hbar}{i}\int_{-\infty}^{\infty} \psi_\nu^*(x) \cdot \frac{d}{dx}\sum_\kappa f(x) dx$$

$$(PX)_{\nu\mu} = \frac{\hbar}{i}\int_{-\infty}^{\infty} \psi_\nu^*(x) \cdot \frac{d}{dx}\sum_\kappa x \cdot \psi_\mu(x) dx$$

For the elements of the matrix product **XP**, we have:

$$(XP)_{\nu\mu} = \sum_\kappa (X)_{\nu\kappa}(P)_{\kappa\mu} = \frac{\hbar}{i}\int_{-\infty}^{\infty} \psi_\nu^*(x) \cdot x \sum_\kappa \psi_\mu(x)(P)_{\kappa\mu}\, dx$$

A similar function to function $f(x)$, g(x), is defined as:

$$g(x) = \frac{\hbar}{i}\frac{d}{dx}\psi(x) = \frac{\hbar}{i}\sum_\nu b_\nu \psi_\nu(x)$$

$$b_\nu = \int_{-\infty}^{\infty} \psi_\nu^*(x) \cdot \frac{d}{dx} \cdot \psi(x) dx$$

Then, we have:

$$g(x) = \frac{\hbar}{i}\sum_\nu \psi_\nu(x)\int_{-\infty}^{\infty} \psi_\nu^*(y) \cdot \frac{d}{dy} \cdot \psi(y) dy$$

$$g(x) = \frac{\hbar}{i}\sum_\nu \psi_\nu(x)(P)_{\nu\mu}$$

The equation of the elements of the matrix product **XP** can be written as:

$$(XP)_{\nu\mu} = \frac{\hbar}{i}\int_{-\infty}^{\infty}\psi_\nu^*(x)\cdot x\frac{d}{dx}\cdot\psi_\mu(x)\,dx$$

Now let us combine the equations of the elements of the matrix products **XP** and **PX**:

$$(PX - XP)_{\nu\mu} = \frac{\hbar}{i}\int_{-\infty}^{\infty}\psi_\nu^*(x)\left[\frac{d}{dx}x - x\frac{d}{dx}\right]\psi_\mu(x)\,dx$$

Now, let us obtain the commutation relation [d/dx,x]:

$$\left[\frac{d}{dx},x\right] = \left(\frac{d}{dx}x - x\frac{d}{dx}\right)$$

We know that:

$$p = -i\hbar\frac{d}{dx}$$
$$[p,x] = -i\hbar$$

Then, we have:

$$[p,x] = -i\hbar\frac{d}{dx}x + i\hbar x\frac{d}{dx} = -i\hbar$$
$$\div -i\hbar$$
$$\frac{d}{dx}x - x\frac{d}{dx} = 1$$
$$Then:$$
$$\left[\frac{d}{dx},x\right] = \left(\frac{d}{dx}x - x\frac{d}{dx}\right) = 1$$

As a consequence, we obtain:

$$(PX - XP)_{\nu\mu} = \frac{\hbar}{i}\int_{-\infty}^{\infty}\psi_\nu^*(x)\psi_\mu(x)\,dx = \frac{\hbar}{i}\delta_{\nu\mu}$$

$$when: \nu = \mu : (\mathbf{PX} - \mathbf{XP}) = \frac{\hbar}{i}$$

Schrödinger himself showed the equivalence between matrix mechanics and wave mechanics by replacing the quantum operator of the linear momentum in the commutator (q_r,p_r) and applying the wave function on the same commutator (Schrödinger 1926):

$$(q_r p_r - p_r q_r)\psi = -\frac{i\hbar}{2\pi}\left\{q_r\frac{\partial}{\partial q_r} - \frac{\partial}{\partial q_r}q_r\right\}\psi$$

$$where: p_r = -\frac{i\hbar}{2\pi}\frac{\partial}{\partial q_r}$$

$$(q_r p_r - p_r q_r)\psi = -\frac{i\hbar}{2\pi}\left\{q_r \frac{\partial\psi}{\partial q_r} - \frac{\partial}{\partial q_r} q_r \psi\right\}$$

$$(q_r p_r - p_r q_r)\psi = -\frac{i\hbar}{2\pi}\left\{q_r \frac{\partial\psi}{\partial q_r} - \psi - q_r \frac{\partial\psi}{\partial q_r}\right\}$$

$$(q_r p_r - p_r q_r)\psi = \frac{i\hbar}{2\pi}\psi$$

2. Quantum virial theorem and QTAIM

As stated in chapter eight, the Heisenberg equation of motion is:

$$\frac{d\mathbf{A}}{dt} = \frac{i}{\hbar}[\mathbf{H}, \mathbf{A}] = \frac{i}{\hbar}(\mathbf{HA} - \mathbf{AH})$$

Where **A** is the observable and **H** is the Hamiltonian. In Heisenberg equation of motion (HEM), the state vectors are time-independent while the operators (and observables) are time-dependent.

Heisenberg equation of motion is equivalent to Schrödinger equation of motion (SEM) by only changing the time-dependency between operators (variable in HEM and constant in SEM) and vector states (constant in HEM and variable in SEM). Then, it is a matter of convenience to use either of them.

The observables for SEM and HEM are:

$$\text{SEM}: \langle A\rangle = \langle \psi(t)|\hat{\mathrm{A}}|\psi(t)\rangle$$

$$\text{HEM}: \langle A\rangle = \langle \psi(0)|U^{t}(t)\hat{\mathrm{A}}U(t)|\psi(0)\rangle$$

By applying the Heisenberg equation of motion for a time-independent arbitrary linear operator, **F**, one finds:

$$\frac{d}{dt}\mathbf{F} = \frac{i}{\hbar}[\mathbf{H}, \mathbf{F}]$$

The corresponding eigenvalues of the operator **F** are time-invariant on the stationary state. This leads to the hypervirial theorem. This result is achieved by considering the Hermitian property of Hamiltonian operator ($\mathbf{H}\psi^*\psi = \psi^*\mathbf{H}\psi$), the restriction on **F** that preserves its Hermiticity and the use of Schrödinger equation for stationary state ($\mathbf{H}\psi = \mathrm{E}\psi$ or $\mathbf{H}\psi^* = \mathrm{E}\psi^*$).

$$\langle[\mathbf{H}, \mathbf{F}]\rangle = \langle\psi|[\mathbf{H}, \mathbf{F}]|\psi\rangle$$

$$\langle\psi|[\mathbf{H}, \mathbf{F}]|\psi\rangle = \langle\psi|\mathbf{HF} - \mathbf{FH}|\psi\rangle =$$

$$= \langle\psi|\mathbf{HF}|\psi\rangle - \langle\psi|\mathbf{FH}|\psi\rangle = \langle\mathbf{H}\psi^*\mathbf{F}\psi\rangle - \langle\psi^*\mathbf{FH}\psi\rangle =$$

$$= \langle E\psi^*\mathbf{F}\psi\rangle - \langle\psi^*\mathbf{F}E\psi\rangle = E\left[\langle\psi^*\mathbf{F}\psi\rangle - \langle\psi^*\mathbf{F}\psi\rangle\right] = 0$$

Then, the corresponding expectation value of dF/dt is:

$$\frac{d}{dt}\langle \mathbf{F} \rangle = \frac{i}{\hbar}\langle [\mathbf{H}, \mathbf{F}] \rangle = 0$$

The equation above is the hypervirial theorem (Hirschfelder 1960). The operator **F** is called hypervirial operator. Chen defined it as: "it is a time-independent linear operator with an arbitrary functional structure expressed in terms of dynamical variables of the system under consideration (...) In the energy representation, the diagonal matrix elements of the hypervirial operator are constant in time; this is known as the hypervirial theorem" (Chen 1964).

The quantum virial theorem can be obtained from the hypervirial theorem for an arbitrary hypervirial operator (linear time-independent operator), F, depending only on the coordinates and momenta of the system where its expectation value, <F>, is time-invariant.

By considering **F** as a product of coordinates and linear momentum, the operator **F** is:

$$\hat{r} = x + y + z, \qquad \hat{p} = -i\hbar\left(\frac{\partial}{\partial x} + \frac{\partial}{\partial y} + \frac{\partial}{\partial z}\right)$$

$$F = \hat{r} \cdot \hat{p} = -i\hbar\left(x\frac{\partial}{\partial x} + y\frac{\partial}{\partial y} + z\frac{\partial}{\partial z}\right) = -i\hbar r \cdot \nabla$$

And H is given by:

$$H = -\frac{\hbar^2}{2m}\nabla^2 + V(r)$$

Then, the commutator [**H**, **F**] is:

$$[\mathbf{H}, \mathbf{F}] = -i\hbar\left(-\frac{\hbar^2}{m}\nabla^2 - r \cdot \nabla(V)\right) = -i\hbar\left(2T - r \cdot \nabla(V)\right)$$

$$or: [\mathbf{H}, \mathbf{F}] = \frac{\hbar}{i}\left(-\frac{\hbar^2}{m}\nabla^2 - r \cdot \nabla(V)\right) = \frac{\hbar}{i}\left(2T - r \cdot \nabla(V)\right)$$

Remember that:

$$[\mathbf{P}, \mathbf{Q}] = \frac{\hbar}{i} I = -i\hbar I$$

$$where: \frac{1}{i} = \frac{1}{i} \cdot \frac{i}{i} = \frac{i}{i^2} = \frac{i}{-1} = -i$$

Since F is a hypervirial operator, its expectation value is time-invariant. Then, the last expression below is the quantum mechanical virial theorem.

$$\frac{d}{dt}\langle \mathbf{F} \rangle = \frac{i}{\hbar}\langle [\mathbf{H}, \mathbf{F}] \rangle = 0$$

$$\frac{\hbar}{i}\left(2\langle T \rangle - \langle r \cdot \nabla(V) \rangle\right) = 0$$

$$2\langle T \rangle = \langle r \cdot \nabla(V) \rangle$$

The classical virial theorem states that:

$$\overline{T} = \frac{1}{2}\sum_i m_i v_i^2 = -\frac{1}{2}\sum_i x_i \frac{dV_i(x)}{dx}$$

The quantum mechanical counterpart for a single particle in one dimension is:

$$\frac{1}{2m}\langle \psi | \hat{p}_x^2 | \psi \rangle = \frac{1}{2}\langle \psi | x \frac{dV(x)}{dx} | \psi \rangle$$

If $V = kr^n$ where k is a constant, then we have:

$$2\langle T \rangle = \left\langle r \cdot \nabla\left(kr^n\right)\right\rangle = \left\langle r \cdot nkr^{n-1}\right\rangle = n\langle V \rangle$$

Where <T> and <V> are the expectation values of kinetic and potential energy operators.

For a system of interacting electrons, the potential energy term contains the term $1/r_{ij}$ where n = –1. Then, the virial theorem becomes:

$$2\langle T \rangle = -\langle V \rangle$$

In Quantum Theory of Atoms in Molecules (QTAIM), Bader used the above expression to a subsystem (or open system), i.e., for an atomic basin, Ω, in a molecule in a stationary state, where the Hermiticity of H is lost and by assuming a one-electron Hamiltonian he arrived at the expression below (Bader 1994):

$$\langle \psi | [\mathbf{H}, \mathbf{F}] | \psi \rangle_\Omega = \left(-\hbar^2 / 2m\right) \oint dS(r;\Omega) \mathbf{j}_F(r) \cdot \mathbf{n}(r)$$

$$\mathbf{j}_F(r) = \psi^* \nabla\left(\mathbf{F}\psi\right) - \nabla\psi^*\left(\mathbf{F}\psi\right)$$

Where $\mathbf{j}_F(r)$ is the vector current, giving the velocity of density of F at position r, and $\mathbf{j}_F(r) \bullet \mathbf{n}(r)$ is called *flux*.

*While Heisenberg equation of motion vanishes for a total system, for an open one-electron system is equal to the surface integral of the normal component of the current property of **F**. The same expression is found for an open N-electron system by multiplying to N.* This is the great achievement of Bader's theory, where it was possible to obtain any atomic property (atomic energy, atomic dipole moment, etc.) of an atom in a molecule. As Bader stated: "this approach enables one to define all properties, including those that depend upon inter-particle coordinates such as energy, in terms of real-space density distribution" (Bader 2007).

3. Quantum resonance for helium atom and chemical bond

Resonance is an important concept (or model) in chemistry. To our knowledge, the concept of resonance applied to atoms (and molecules as a consequence) was firstly developed by Heisenberg in order to rationalize the helium atom.

Heisenberg used the model of two oscillators to describe helium atom (Heisenberg 1926a). As already mentioned in Chapter five that Planck's model, Slater's virtual oscillator model and matrix mechanics are based on Lorentz model of the electronic motion represented by a harmonic oscillation of the electron under the influence of the nucleus in the center of the atom. He was aware that his model must be in accordance with the Bose-Einstein statistics (Bose 1924, Einstein 1924) and Pauli's exclusion principle (Pauli 1925).

The Hamiltonian for this system of two harmonic oscillators representing two identical electrons 1 and 2 subject to identical forces is:

$$H = \frac{1}{2m} p_1^2 + \frac{m}{2}\omega^2 q_1^2 + \frac{1}{2m} p_2^2 + \frac{m}{2}\omega^2 q_2^2 + m\lambda q_1 q_2$$

$$where:$$

$$E_P = \frac{1}{2}kq^2, \quad \omega = \sqrt{\frac{k}{m}}, \quad k = m\omega^2$$

$$Then: E_P = \frac{1}{2}m\omega^2 q^2$$

Where m and ω are mass and frequencies of the coupled oscillators and λ is the interaction constant.

By using known transformations, the Hamiltonian becomes a sum of two oscillators corresponding to a "principle resonance frequency", as stated by Heisenberg.

$$H = \frac{1}{2m} p'^2_1 + \frac{m}{2}\omega'^2_1 q'^2_1 + \frac{1}{2m} p'^2_2 + \frac{m}{2}\omega'^2_2 q'^2_2$$

$$where: q'_1 = \frac{1}{\sqrt{2}}(q_1 + q_2), q'_2 = \frac{1}{\sqrt{2}}(q_1 - q_2)$$

$$\omega'^2_1 = \omega^2 + \lambda, \omega'^2_2 = \omega^2 - \lambda$$

And the energies for the combined system according to quantum mechanics is:

$$H_{n'_1 n'_2} = \frac{\omega'_1 h}{2\pi}\left(n'_1 + \frac{1}{2}\right) + \frac{\omega'_2 h}{2\pi}\left(n'_2 + \frac{1}{2}\right)$$

Where n' is the quantum number. Then, the electrons execute the same motion but in different phase.

Heisenberg stated that the energies of both electrons are degenerate when their states n and m are different due to the resonance phenomenon, but when their states are the same (i.e., n = m) their energies are different due to the lack of resonance. When there is resonance, the degenerate energy is lifted by additional energy W^1 of the perturbated systems which is related to the energy of the unperturbed system H^1 as follows:

$$\mathbf{W}^{-1} = \mathbf{S}^{-1}\overline{\mathbf{H}^1}\mathbf{S}$$

Where all the matrices of the above equation are diagonal and S is the transformation matrix according to Heisenberg and his coworkers' previous paper (Born et al. 1925). For nondegenerate states, the values of the matrix elements are all equal to 1, which requires the solution of two linear equations:

$$\begin{bmatrix} W^1 - H^1(nm, nm) & H^1(nm, mn) \\ -H^1(nm, mn) & W^1 - H^1(nm, nm) \end{bmatrix} \begin{bmatrix} S_{nm} \\ S_{mn} \end{bmatrix} = \begin{bmatrix} 0 \\ 0 \end{bmatrix}$$

The solutions of the above equation are:

$$W_{nm}^{1} = H^{1}(nm,nm) + H^{1}(nm,nm)$$

$$S_{nm,nm} = 1/\sqrt{2}, S_{mn,nm} = 1/\sqrt{2}$$

$$W_{mn}^{1} = H^{1}(nm,nm) - H^{1}(nm,mn)$$

$$S_{nm,mn} = 1/\sqrt{2}, S_{mn,mn} = -1/\sqrt{2}$$

Afterwards, Heisenberg incorporated the resonance phenomenon in Schrödinger quantum mechanics using normalized wave functions φ_n^a and φ_n^b in his Matrix Mechanics. Then, the equivalent solutions were:

$$W_{nm}^{1} = 1/\sqrt{2}\left(\varphi_n^a \varphi_m^b + \varphi_m^a \varphi_n^b\right)$$

$$W_{mn}^{1} = 1/\sqrt{2}\left(\varphi_n^a \varphi_m^b - \varphi_m^a \varphi_n^b\right)$$

The first eigenfunction, W^1_{nm}, is symmetric and the second one, W^1_{mn}, is antisymmetric. Heisenberg continued this work in a second paper where he found numerical results which, for the first time ever, gave comparatively good agreement with experimental results of the helium spectrum (Heisenberg 1926b).

The last equations above were the basis for the formulation of the chemical bond by Heitler and London. They used the same Heisenberg's reasoning changing two interacting oscillators into two electrons in a bonding region between two hydrogen nuclei, yielding the two guess wave functions for H_2: $\psi = \phi_A(1)\phi_B(2)$ and $\psi = \phi_A(2)\,\phi_B(1)$, where the numbers are the electrons and the letters are the hydrogen nuclei (Heitler and London 1927). When considering the indistinguishability of identical particles and doing normalization and orthogonalization processes of these wave functions, two final wave functions, α and β, were obtained:

$$\alpha = \frac{1}{\sqrt{2+2S}}\left[\phi_A(1)\phi_B(2) + \phi_A(2)\phi_B(1)\right]$$

$$\beta = \frac{1}{\sqrt{2-2S}}\left[\phi_A(1)\phi_B(2) - \phi_A(2)\phi_B(1)\right]$$

Where S is the overlap integral, α is the H_2 symmetric repulsive wave function and β the H_2 antisymmetric attractive wave function. These equations are equivalent to the solution for the interaction between n and m oscillators (W_{nm}^1 and W_{mn}^1) in terms of wave mechanics. *Then, resonance is the basis of the formation of both H_2 wave functions.*

4. Uncertainty principle

In his paper about the uncertainty principle, Heisenberg showed the difference between the kinematics of a body in a continuous theory (i.e., the classical physics) and a particle in a discontinuous theory (i.e., the quantum mechanics). He demonstrated that from a body moving along the x-axis where the position changes with the time, x(t), one has the information of the instant velocity from the tangent at each point of x(t) curve and, on the other hand, in a similar discontinuous picture where there is a series of points at finite separation it is meaningless to speak about the velocity at one position (Heisenberg 1927). He added that: "At the instant at which the position of the electron is known, its momentum therefore can be known up to magnitudes which correspond to that discontinuous change. Thus, the more precisely the position is determined, the less precisely the momentum is known, and conversely. In this circumstance we see a direct physical interpretation of the equation $pq - qp = -i\hbar$." (Heisenberg 1927). He stated q_1 as the precision of the value q (or the mean error of q) where the position is governed by the wavelength of the light in the γ-ray microscope carrying the determination of position. Additionally, he established that p_1 is the precision with which the value of p is determinable (i.e., the discontinuous change of p in the Compton effect). Under the elementary laws of the Compton effect, q_1 and p_1 stand in the relation:

$$p_1 q_1 \sim \hbar$$

$$\varepsilon(p)\varepsilon(q) \sim \hbar$$

Where ε(q) and ε(p) are the mean error of the position measurement and momentum measurement, respectively.

Then, Heisenberg derived the relation for the standard deviations of the position and momentum for a class of Gaussian wave functions.

$$\sigma(p)\sigma(p) = \frac{\hbar}{2}$$

Later, Kennan proved the inequality:

$$\sigma(p)\sigma(p) \geq \frac{\hbar}{2}$$

for arbitrary wave function (Kennard 1927). The prerequisite for this is that Schrödinger's wave packet is reinterpreted as a probability packet.

Let us use two general linear operators **A** and **B**, whose average eigenvalues are:

$$\langle \mathbf{A} \rangle = \xi^\dagger \mathbf{A} \xi$$

$$\langle \mathbf{B} \rangle = \xi^\dagger \mathbf{B} \xi$$

The square deviation from the average value of **A** and **B** is;

$$(\Delta \mathbf{A})^2 = \xi^\dagger (\mathbf{A} - \langle \mathbf{A} \rangle \mathbf{I})^2 \xi$$

$$(\Delta \mathbf{B})^2 = \xi^\dagger (\mathbf{B} - \langle \mathbf{B} \rangle \mathbf{I})^2 \xi$$

Let us define a complex matrix M as:

$$\mathbf{M} = (\mathbf{A} - \langle \mathbf{A} \rangle \mathbf{I}) + i\alpha (\mathbf{B} - \langle \mathbf{B} \rangle \mathbf{I})$$

Where α > 0 and real. Then:

$$(\mathbf{M}\xi)^{\dagger}(\mathbf{M}\xi) \geq 0$$

$$\xi^{\dagger}\mathbf{M}^{\dagger}\mathbf{M}\xi = \xi^{\dagger}\begin{bmatrix}(\mathbf{A}-\langle\mathbf{A}\rangle\mathbf{I})^2+\alpha^2(\mathbf{B}-\langle\mathbf{B}\rangle\mathbf{I})^2\\ +i\alpha(\mathbf{AB}-\mathbf{BA})\end{bmatrix}\xi$$

$$\xi^{\dagger}\mathbf{M}^{\dagger}\mathbf{M}\xi = (\Delta\mathbf{A})^2+\alpha^2(\Delta\mathbf{B})^2+i\alpha\xi^{\dagger}(\mathbf{AB}-\mathbf{BA})\xi \geq 0$$

Then:

$$(\Delta\mathbf{A})^2+\alpha^2(\Delta\mathbf{B})^2+i\alpha\xi^{\dagger}(\mathbf{AB}-\mathbf{BA})\xi \geq 0$$

$$\times\alpha^{-1}$$

$$\alpha^{-1}(\Delta\mathbf{A})^2+\alpha(\Delta\mathbf{B})^2 \geq -i\xi^{\dagger}[\mathbf{A},\mathbf{B}]\xi$$

$$\alpha^{-1}(\Delta\mathbf{A})^2+\alpha(\Delta\mathbf{B})^2 \geq -i\langle[\mathbf{A},\mathbf{B}]\rangle$$

Varying α for fixed $\Delta\mathbf{A}$ and $\Delta\mathbf{B}$, we find that the left side of the above inequality has its minimum value when α satisfies the equation:

$$\textbf{minimum of}: \alpha^{-1}(\Delta\mathbf{A})^2+\alpha(\Delta\mathbf{B})^2$$

$$\frac{d\left\{\alpha^{-1}(\Delta\mathbf{A})^2+\alpha(\Delta\mathbf{B})^2\right\}}{d\alpha} = -\alpha^{-2}(\Delta\mathbf{A})^2+(\Delta\mathbf{B})^2 = 0$$

$$Then: \alpha = \frac{\Delta\mathbf{A}}{\Delta\mathbf{B}}$$

As a consequence, we have the inequality equation:

$$\left(\frac{\Delta\mathbf{A}}{\Delta\mathbf{B}}\right)^{-1}(\Delta\mathbf{A})^2+\frac{\Delta\mathbf{A}}{\Delta\mathbf{B}}(\Delta\mathbf{B})^2 \geq -i\langle[\mathbf{A},\mathbf{B}]\rangle$$

$$2\Delta\mathbf{A}\Delta\mathbf{B} \geq -i\langle[\mathbf{A},\mathbf{B}]\rangle$$

$$\Delta\mathbf{A}\Delta\mathbf{B} \geq \frac{-i\langle[\mathbf{A},\mathbf{B}]\rangle}{2}$$

For the particular case where $\mathbf{A} = \mathbf{Q}$ and $\mathbf{B} = \mathbf{P}$, we have:

$$[\mathbf{Q},\mathbf{P}] = i\hbar\mathbf{I}$$

$$\langle[\mathbf{Q},\mathbf{P}]\rangle = \langle i\hbar\mathbf{I}\rangle = i\hbar$$

$$\Delta\mathbf{Q}\Delta\mathbf{P} \geq \frac{\hbar}{2}$$

Then, we derived the Heisenberg's uncertainty principle.

Now, let us show that when if there is a simultaneous eigenfunction of **P** and **Q**, their commutation relation is zero.

$$\left[\hat{Q},\hat{P}\right]=0$$

Suppose that $\{u_i\}$ constitutes a complete set of eigenfunctions for the two linear operators of the momentum and position, i.e., $\{u_i\}$ is a simultaneous eigenfunctions of the linear momentum and position operators. Since the set $\{u_i\}$ is a simultaneous set of eigenfunctions of linear momentum and position operators, we have:

$$\hat{Q}u_i = q_i u_i, \qquad \hat{P}u_i = p_i u_i, \qquad i = 1,2,\ldots,n$$

Where r_i and p_i are the eigenvalues of the position and linear momentum operators, respectively. Then, we have that:

$$\left[\hat{Q},\hat{P}\right]=\left(\hat{Q}\hat{P}-\hat{P}\hat{Q}\right)f=0$$

$$where: f=\sum_i c_i u_i$$

Then, by replacing the eigenequations in the above equation, we have:

$$\left(\hat{Q}\hat{P}-\hat{P}\hat{Q}\right)\sum_i c_i u_i=\sum_i c_i\left(\hat{Q}\hat{P}-\hat{P}\hat{Q}\right)u_i=$$

$$=\sum_i c_i\left(\hat{Q}p_i-\hat{P}q_i\right)u_i=\sum_i c_i\left(q_i p_i-p_i q_i\right)u_i=0$$

Since:

$$f=\sum_i c_i u_i\neq 0$$

Then:

$$\sum_i\left(q_i p_i-p_i q_i\right)=0$$

$$Or:\left(\hat{Q}\hat{P}-\hat{P}\hat{Q}\right)=\left[\hat{Q},\hat{P}\right]=0$$

Then, it is impossible to have a simultaneous set of eigenfunctions of **P** and **Q** because it violates the most important postulate of the quantum mechanics.

5. Dirac's quantum mechanics

The English physicist Dirac based on both matrix mechanics and wave mechanics to develop his own quantum mechanics: a mix of matrix mechanics, wave mechanics and the algebra that he implemented. For example, he distinguished q-numbers from c-numbers. The former are quantum numbers/variables and the latter are numbers/variables of the classical mathematics. He stated that is not possible to affirm that one q-number is greater or less than another and the q-numbers satisfy all the ordinary laws of algebra except for the commutative law of multiplication (as already observed in matrix mechanics). For example, he used matrix mechanics q-numbers of coordinates represented by a set of harmonic components of the type

x(nm).exp{iω(nm)t, where x(nm) and ω(nm) are q-numbers and n, m and t are c-numbers (Dirac 1926a).

Dirac defined a dynamical system as a multiply periodic with a set of arbitrary variables J and w when it follows the properties (where the Hamiltonian H is a function of J's only):

$$[J_r, J_s] = 0, \quad [w_r, w_s] = 0$$
$$[w_r, J_s] = 0(r \neq s), or: 1(r = s)$$
$$\sum_\alpha C_\alpha \exp i(\alpha_1 w_1 + \alpha_2 w_2 + \ldots + \alpha_n w_n) = \sum_\alpha C_\alpha \exp i(\alpha w)$$
$$w_k = v_k t$$
$$\sum_\alpha C_\alpha \exp i(\alpha w) = \sum_\alpha C_\alpha \exp 2\pi i(\alpha v)t$$
$$\dot{J}_r = [J_r, H] = 0$$
$$\dot{w}_r = [w_r, H] = \partial H / \partial J_r$$
$$where: [J_r, J_s] = J_r J_s - J_s J_r$$

Where v_k are the linear frequencies and the product $v_k t$ are the angle variables. The quantities $\dot{w}_r$ correspond to orbital frequencies from Bohr's theory.

In quantum theory, x and y are functions of the coordinates q_1, q_2, ...q_s and momenta p_1, p_2, ...,o_s of a multiply periodic system of s degrees of freedom. Suppose x can be expanded as:

$$x = \sum_\alpha x(\alpha, J) \exp 2\pi i(\alpha v)t$$

Where α_1, α_2,..., α_s are integers and J's and vt's (the product of frequency and time) are the action and angle variables, respectively. On the quantum theory, x is an aggregate of terms

$$x = x(n, n-\alpha) \exp 2\pi i v(n, n-\alpha)t$$

Where x(n, n – α) is the amplitude for the corresponding frequency ν(n,n – α) due to the transition from n to n – α state.

In classical theory, x is a function of p's and q's, so that:

$$\dot{x} = \sum_k \left(\frac{\partial x}{\partial q_k} \dot{q}_k + \frac{\partial x}{\partial p_k} \dot{p}_k \right)$$

But if H is the Hamiltonian function, then we have:

$$\frac{\partial H}{\partial p_k} = \dot{q}_k, \quad \frac{\partial H}{\partial q_k} = -\dot{p}_k$$

$$then: \dot{x} = \sum_k \left(\frac{\partial x}{\partial q_k} \frac{\partial H}{\partial p_k} - \frac{\partial x}{\partial p_k} \frac{\partial H}{\partial q_k} \right)$$

$$\dot{x} = [x, H]$$

Dirac used Schrödinger's equation below in his works:

$$\{H(q_r, p_r) - W\}\psi = 0$$

$$p_r = i\hbar\frac{\partial}{\partial q_r}, \quad -W = -i\hbar\frac{\partial}{\partial t}$$

Where for the values of the parameter W exists a wave function satisfying the above equation.

Dirac stated that the dynamical system is specified by a Hamiltonian equation between the variables (Dirac 1926b):

$$H(q_r, p_r, t) - W = 0$$

Or more generally by:

$$F(q_r, p_r, t, W) = 0$$

Then, one has:

$$F\psi = 0$$

If 'a' is a constant of integration of the system, we have:

$$[a, F] = 0$$

$$Fa\psi = aF\psi = 0$$

6. Importance of the commutation to the quantum mechanics

The operators act on a quantum mechanical state (eigenvector or eigenfunction) and transform it into a new state under a linear transformation.

When a quantum mechanical measurement is performed, an eigenvalue of an operator is measured and after the measurement the initial eigenstate is returned.

If two or more operators commute,they have the same eigenfunction and they can have the same set of eigenfunctions. Then, the corresponding physical quantities can be evaluated or measured simultaneously without uncertainty. For example, if **A** and **B** commute and ψ is their common eigenfunction, we have:

$$\mathbf{A}\psi = a\psi$$

$$\mathbf{B}\psi = b\psi$$

As a consequence of their commutative property, **A** acting on ψ first is the same as **B** acting on ψ first.

$$\mathbf{BA}\psi = \mathbf{AB}\psi$$

Then, we have:

$$\mathbf{BA}\psi = \mathbf{AB}\psi = \mathbf{A}b\psi = b\mathbf{A}\psi$$

$$\mathbf{B}(\mathbf{A}\psi) = b(\mathbf{A}\psi)$$

The above equation indicates that **A**ψ is an eigenfunction of **B** with eigenvalue b. Then, when **A** operates on ψ it cannot change ψ, but it produces a constant times ψ. As a consequence, when two operators act on their common eigenfunction the result is the product of their corresponding eigenvalues.

$$\mathbf{B}(\mathbf{A}\psi) = \mathbf{B}a\psi = a\mathbf{B}\psi = ab\psi$$

On the other hand, when two operators do not commute, their corresponding eigenvalues cannot be obtained simultaneously due to the Heisenberg's uncertainty principle.

For a complete set of commuting operators it is possible to find a unique unitary transformation (see Chapter three) that diagonalizes simultaneously all the operators in order to obtain their corresponding eigenvalues. For example, $\mathbf{L}^2$, $\mathbf{L_z}$ and $\mathbf{H}$ form a complete set of commuting operators for the hydrogen atom and the set of their corresponding eigenvalues (see Chapters sixteen and seventeen) completely specifies a unique eigenstate of the hydrogen atom.

7. Copenhagen interpretation

Copenhagen interpretation was neither a formal conference (such as Solvay conference) nor any original published paper. It is a designation for a set of ideas probably discussed between Bohr and Heisenberg from 1925 to 1927. Heisenberg coined the term "Copenhagen interpretation" in 1955 and firstly published it in 1958 (Heisenberg 1958). However, Bohr was the first to defend the ideas of the "Copenhagen interpretation" in Solvay conference in 1927 after the attacks of Einstein against these ideas. By the end of 1927, the ideas of the Copenhagen interpretation were accepted by Pauli, Dirac, Born, Jordan and Ehrenfest ending the philosophical dispute of the new results of the quantum mechanics at that time, despite the persistent criticisms of Einstein, Schrodinger and Planck in 1930s (Camilleri 2009). In addition, there were some disagreements among the founders of quantum mechanics, mainly between Bohr and Heisenberg, in 1920–1930s which lead to the conclusion that Copenhagen interpretation was not a coherent philosophical framework. In 1940–50s, the Russian physicists criticized the Copenhagen interpretation and this probably triggered Heisenberg's agenda for defending the orthodox interpretations of quantum mechanics from 1955 on.

Some basic principles of the Copenhagen interpretation are:

1. The quantum mechanics is intrinsically indeterministic (owing to Heisenberg's uncertainty principle);
2. In the appropriate limit, the quantum mechanics approaches classical physics and reproduces classical predictions (owing to Bohr's correspondence principle);
3. The square of the amplitude of the wave function gives the probability for the outcomes of measurements of the system (Born 1926), owing to Born's probability statement (Born rule),;
4. The objects have certain pairs of complementary properties which cannot be observed or measured simultaneously (Bohr 1928), owing to Bohr's complementary theory.
5. During the observation, the system must interact with the measurement device and the corresponding wave function (a superposition of several eigenstates) collapses, reducing to an eigenstate of the observable that is registered.

8. Examples

(1) By knowing that the time-dependent wave function is:

$$|\psi(t)\rangle = U(t)|\psi(0)\rangle$$

$$where: U(t) = e^{-iHt/\hbar}$$

Give the time-dependent expectation value (Chapter ten).

Solution: For obtaining the observable of a given time-dependent operator, A(t), one has to use the Heisenberg equation of motion.

$$\langle A\rangle = \langle\psi(0)|U^{\dagger}(t)\widehat{A}U(t)|\psi(0)\rangle$$

Where U(t) becomes the time-dependent operator.

$$\langle A\rangle_t = \langle\psi(0)|e^{iHt/\hbar} Ae^{-iHt/\hbar}|\psi(0)\rangle$$

The equation of motion for A(t) is:

$$\frac{d}{dt}\langle A(t)\rangle = \langle\psi(0)\left|\begin{array}{l}\frac{i}{\hbar}He^{iHt/\hbar}Ae^{-iHt/\hbar} + e^{iHt/\hbar}\left(\frac{\partial A}{\partial t}\right)e^{-iHt/\hbar} + \\ +\frac{i}{\hbar}e^{iHt/\hbar}A(-H)e^{-iHt/\hbar}\end{array}\right|\psi(0)\rangle$$

Then, time-dependent expectation value becomes:

$$\frac{d}{dt}\langle A(t)\rangle = \langle\psi(0)\left|\frac{i}{\hbar}[H, A(t)] + e^{iHt/\hbar}\left(\frac{\partial A}{\partial t}\right)e^{-iHt/\hbar}\right|\psi(0)\rangle$$

$$\frac{d\langle A\rangle}{dt} = \frac{1}{i\hbar}\langle\psi(0)|\left[\widehat{A}, \widehat{H}\right]|\psi(0)\rangle + \left\langle e^{iHt/\hbar}\frac{\partial\widehat{A}}{\partial t}e^{-iHt/\hbar}\right\rangle$$

This equation is similar to the time-dependent expectation value obtained in Chapter ten:

$$\frac{d\langle A\rangle}{dt} = \frac{1}{i\hbar}\langle\psi(t)|\left[\widehat{A}, \widehat{H}\right]|\psi(t)\rangle + \left\langle\frac{\partial\widehat{A}}{\partial t}\right\rangle$$

(2) Derive the quantum mechanical virial theorem (Johnson and Pedersen 1986):

$$\frac{1}{2m}\langle\psi|\widehat{p}_x^{\,2}|\psi\rangle = \frac{1}{2}\langle\psi|x\frac{dV(x)}{dx}|\psi\rangle$$

By knowing the commutators:

$$\left[\widehat{p}_x, \widehat{\mathrm{H}}\right] = -i\hbar\frac{dV(x)}{dx}, \quad \left[x, \widehat{\mathrm{H}}\right] = \frac{\hbar^2}{m}\frac{d}{dx}$$

Solution (for simplicity: p = px):

$$\frac{dV(x)}{dx} = -\frac{1}{i\hbar}\left[\hat{p}, \hat{\mathrm{H}}\right]$$

$$\frac{1}{2}\langle\psi|x\frac{dV(x)}{dx}|\psi\rangle = \frac{i}{2\hbar}\langle\psi|x\left[\hat{p}, \hat{\mathrm{H}}\right]|\psi\rangle =$$

$$= \frac{i}{2\hbar}\langle\psi|x\left(\hat{p}\hat{\mathrm{H}} - \hat{\mathrm{H}}\hat{p}\right)|\psi\rangle = \frac{i}{2\hbar}\langle\psi|x\hat{p}\hat{\mathrm{H}} - \hat{\mathrm{H}}x\hat{p} + \hat{p}x\hat{\mathrm{H}} - x\hat{\mathrm{H}}\hat{p}|\psi\rangle =$$

$$= \frac{i}{2\hbar}\langle\psi|x\hat{p}\hat{\mathrm{H}} - \hat{\mathrm{H}}x\hat{p} + \left[x, \hat{\mathrm{H}}\right]\hat{p}|\psi\rangle = \frac{i}{2\hbar}\langle\psi|x\hat{p}\hat{\mathrm{H}} - \hat{\mathrm{H}}x\hat{p} + \frac{\hbar^2}{m}\frac{d}{dx}\hat{p}|\psi\rangle =$$

$$as: \hat{p} = -i\hbar\frac{d}{dx}, \quad \hbar\frac{d}{dx} = i\,\hat{p}$$

$$then: \frac{1}{2}\langle\psi|x\frac{dV(x)}{dx}|\psi\rangle = \frac{i}{2\hbar}\langle\psi|x\hat{p}\hat{\mathrm{H}} - \hat{\mathrm{H}}x\hat{p} - \frac{\hbar}{m}i\hat{p}^2|\psi\rangle =$$

$$as: \langle\psi|\hat{\mathrm{H}}x\hat{p}|\psi\rangle = \sum\langle\psi|\hat{\mathrm{H}}|\psi'\rangle\langle\psi'|x\hat{p}|\psi\rangle = E\langle\psi|x\hat{p}|\psi\rangle$$

$$then: \frac{1}{2}\langle\psi|x\frac{dV(x)}{dx}|\psi\rangle = \frac{i}{2\hbar}\left\{(E-E)\langle\psi|x\hat{p}|\psi\rangle - \frac{\hbar i}{m}\langle\psi|\hat{p}^2|\psi\rangle\right\}$$

$$\frac{1}{2}\langle\psi|x\frac{dV(x)}{dx}|\psi\rangle = \frac{1}{2m}\langle\psi|\hat{p}^2|\psi\rangle$$

9. Exercises

(1) For one particle in a symmetric box whose potential energy, V, is:

$$V = \begin{cases} 0, & -a/2 < x < a/2 \\ \infty, & \text{otherwise} \end{cases}$$

Show that the expectation value of $\delta \times \delta p$ is

$$\langle \delta x \delta p \rangle = i\hbar/2$$

Where:

$$\delta_x = x - \langle x \rangle$$

$$\delta_p = p - \langle p \rangle$$

And the ground state wave function is:

$$\psi(x) = \sqrt{2/a}\cos\left(\pi x/a\right)$$

By knowing that for a symmetrical box, < x > = < p > = 0 and:

$$\langle \delta x \delta p \rangle = \int \psi^*(x)\hat{\delta}x\hat{\delta}p\psi(x)dx$$

$$\hat{\delta}p = -i\hbar\frac{d}{x}, \quad \hat{\delta}x = x$$

(2) By knowing that the uncertainties in x and p are defined as:

$$\Delta x = \left[\int \psi^*(x)\left(x - \langle x \rangle\right)^2 \psi(x)dx \right]^{1/2}$$

$$\Delta p = \left[\int \psi^*(x)\left(\hat{p} - \langle \hat{p} \rangle\right)^2 \psi(x)dx \right]^{1/2}$$

Determine the uncertainty product, ΔxΔp, of the harmonic oscillator with the wave function:

$$\psi(x) = (\alpha/\pi)^{1/4} e^{(-\alpha x^2/2)}$$

Tips: (i) set < x > = < p > = 0 since the integrands in both cases are odd functions of x and thus integrate to zero; (b) calculate firstly (Δx)2 and (Δp)2.

Answer: $\Delta x \Delta p = \hbar/2$

(3) By using the identity of the commuting kinetic energy operator and linear momentum operator:

$$\mathbf{T}(\mathbf{P}\psi) = \mathbf{P}(\mathbf{T}\psi)$$

Show that they commute, i.e.,

$$[\mathbf{T}, \mathbf{P}] = 0$$

Tip: Use the expressions of the corresponding operators in the first equation to arrive the second equation.

References cited

Bader, R.F.W. 1994. Atoms in Molecules. A quantum theory. Oxford university press, New York.

Bader, R.F.W. 2007. Everyman's derivation of the theory of atoms in molecules. J. Phys. Chem. 111: 7966–7972.

Bohr, N. 1928. The quantum postulate and the recent development of atomic theory. Nature 121: 580–590.

Born, M. 1926. Quantenmechanik der Stoßvorgange. Z. Phys. 38: 803–827.

Born, M., Heisenberg, W. and Jordan, P. 1925. Zurquantenmechanik II. Z. Phys. 35: 557–615.

Bose, N.S. 1924. Plancksgesetz und lichtquantenhypothese. Z. Phys. 26: 178–181.

Camilleri, K. 2009. Constructing the myth of the Copenhagen interpretation. Perspect. Sci. 17: 26–57.

Chen, J.C.Y. 1964. Off-diagonalhypervirial theorem and its applications. J. Chem. Phys. 40: 615–621.

Dirac, P.M.A. 1926a. Quantum mechanics and a preliminary investigation of the hydrogen atom. Proc. R. Soc. Lond. A 110: 561–579.

Dirac, P.M.A. 1926b. On the theory of quantum mechanics. Proc. R. Soc. Lond. A. 112: 661–667.

Einstein, A. 1924. Quantentheorie des einatomigenidealen gases. Sitzungsber. Preuss. Akad. Wiss. 261.

Heisenberg, W. 1926a. Mehrkörperproblem und resonanz in der quantenmechanik. Z. Phys. 38: 411–425.

Heisenberg, W. 1926b. Über die spectra von atomsystemenmitzweielektronen. Z. Phys. 39: 499–518.

Heisenberg, W. 1927. Über den anschaulicheninhalt der quantentheoretischenkinematik und mechanic. Z. Phys. 43: 172–198.

Heisenberg, W. 1958. The Copenhagen interpretation of quantum theory. Chapter 3 of Physics and Philosophy: The revolution in modern science. Harper and Row. New York.

Heitler, W. and London, F. 1927. Wechselwirkungneutraleratome und homoopolarebindungnach der quantenmechanik. Z. Phys. 44: 455–472.

Hirschfelder, J.O. 1960. Classical and quantum mechanical hypervirial theorem. J. Chem. Phys. 33: 1462–1466.

Johnson, C.S. and Pedersen, L.G. 1986. Problems and solutions in quantum chemistry and physics. Dover Publications, New York.

Kennard, E.H. 1927. Zurquantenmechanikeinfacherbewegungstypen. Z. Phys. 44: 326–352.

Pauli, W.Z. 1925. Über den zusammenhang des abschlusses der elektronengruppenim atom mit der komplexstruktur der spektren. Phys. 31: 765–783.

Schrödinger, E. 1926. Uber das verhaltnis der Heisenberg-Born-Jordanschenquantenmechanikzu der meinen. Ann. Phys. 79: 734–756.

Landé, Pauli, Dirac and Spin

12

1. The magnetic moment, Larmor's theorem, magnetic quantum number and Landé's core model

The magnetic moment, m,is the magnetic strength and orientation of a magnet (a material that produces a magnetic field). The magnetic moment usually refers to a system's magnetic dipole moment, μ (the component of the magnetic moment equivalent to a magnetic dipole). The magnetic dipole moment is formed by the passage of electrons through a conducting loop of wire and it is the product of the electric current I and the area that the loop encloses, A.

$$\mu = IA$$

As for the case of the hydrogen atom, the electron creates its own current as it goes around the atom where its nucleus is at the center of the loop. Then,the electron magnetic dipole moment is (in SI unit):

$$\mu_e = -\frac{q_e \upsilon r}{2}$$

$$where: L = m_e \upsilon r \therefore \upsilon = \frac{L}{m_e r}$$

$$\mu_e = -\frac{q_e \left(\frac{L}{mr}\right) r}{2} = -\frac{q_e}{2m_e} L$$

Where q_e is the charge of the electron and m_e is the electron mass.

The square of the angular momentum, L^2, is given below where we can obtain the equation for the angular momentum L (see Chapter sixteen):

$$L^2 = \hbar^2 l(l+1) \Rightarrow L = \hbar\sqrt{l(l+1)}$$

Where l is the quantum number of the angular momentum L. Then, the electron magnetic dipole moment is:

$$\mu_e = -\frac{q_e}{2m_e} \hbar\sqrt{l(l+1)}$$

The magnetic moment of an electron is directly proportional to the angular momentum L or the resultant of the angular momentum, J. The proportionality constant is the gyromagnetic ratio, γ.

$$\mu = \gamma J$$

$$\gamma = \frac{\mu}{J} = \frac{q_e}{2m_0 c}$$

Where m_0 is the electron's rest mass. The above equation of the gyromagnetic ratio is associated with atoms showing normal Zeeman effect (Chapter five).
Hence, the magnetic dipole moment of the electron becomes (in CGS unit):

$$\mu = -\frac{q_e}{2m_0 c} J$$

When an external magnetic field is applied to the hydrogen atom, it exerts a torque on the electron's angular momentum so that it is aligned to the magnetic field axis (in Oz axis, for example). Then, we have L_z instead of L, where its value is:

$$L_Z = m\hbar$$

Then, the electron magnetic dipole moment, μ_B, becomes (in two different unit systems):

$$SI: \mu_B = -\frac{q_e}{2m_0} L_Z = -\frac{q_e \hbar}{2m_0} m$$

$$CGS: \mu_B = -\frac{q_e}{2m_0 c} L_Z = -\frac{q_e \hbar}{2m_0 c} m$$

Where the constant $q_e\hbar/2m_0$ is the Bohr magneton discussed in Chapter seven (Heilbron and Kuhn 1969).

The torque, τ, resultant of the action of an external magnetic field over a magnet (for example, a hydrogen atom) is similar to a precession of a classical gyroscope:

$$\tau = \mu \times B = \gamma J \times B$$

The angular momentum precesses about the external field axis with an angular frequency (known as Larmor frequency). The Larmor frequency, ν_L, is:

$$\nu_L = \frac{q_e B}{4\pi m_0 c}$$

$$where: \omega = 2\pi \nu_L$$

Where ω, is the angular frequency about the Oz axis.

According to Larmor's frequency equation, the presence of an external magnetic field does not change the angle of the resultant angular momentum J around the field B, but it alters the orientation of the orbit with respect to the field direction. In addition, the vector J does not project itself on the magnetic field in a continuous way, but only in discrete values of the angle denoted by the discrete values of the magnetic quantum number m (m = ±1, ±2, ±3,.., ±k, for k = 0,1,2,3,4,..n) where n is the principal quantum number and k is the azimuthal quantum number (see Chapter seven). For example, for n = 2 and k = 2, m has four values (m = ±1, ±2) corresponding to the four projections of the orbital angular momentum J on the field direction. Then, m is the magnetic quantum number that is observed when the atom is under magnetic field.

Sommerfeld created a third quantum number j (called inner quantum number) in order to account for anomalous Zeeman effect (see Chapter seven) and Landé empirically ascribed half-integral values to the magnetic quantum number (or the inner quantum number j) so that he could obtain the doublets and the quartets (Landé 1921). In this model, so-called atomic core model, there is a vector sum of K (angular momentum vector of valence electrons, later modified to the electron's angular momentum) and R (angular momentum vector of core electrons and nucleus, later modified to electron spin) to yield the resultant J (total angular momentum of the whole atom, later modified to the electron's total angular momentum).

$$J = K + R$$

The limits of the quantum number of the total angular momentum, j, are limited by:

$$|k - r| \leq j \leq k + r$$
$$k = 0, 1, 2, 3, \ldots$$
$$r = 1/2, 3/2, \ldots$$

Where k (k = 0, 1, 2,...) and r are the corresponding quantum numbers of K and R (Landé 1921). In the singlets, r = 0. In the triplets, r = 1. In the doublets, r = 1/2 and in the quartets, r = 3/2.
In modern notation, r and k are replaced by s and l, respectively. Then, we have:

$$k \Rightarrow l, K \Rightarrow L$$
$$r \Rightarrow s, R \Rightarrow S$$
$$J = L + S$$
$$|l - s| \leq j \leq l + s$$
$$l = 0, 1, 2, 3, \ldots$$
$$s = 1/2, 3/2, \ldots$$

See in Table 12.1 the Landé scheme of quantum numbers (j, l and s) for singlets and doublets (Landé 1923a).

2. Zeeman effect and Landé g-factor

The experiment of the Zeeman effect and Lorentz model to account for the normal Zeeman effect was discussed in Chapter five. When a magnetic field is applied to an atom, each spectral line from its emission spectroscopy is split into three lines (the central one is at the place of the original line and the outer ones are equidistant from it). The effect of the external magnetic field to the atom is to cause the Larmor

Table 12.1: Landé scheme of quantum numbers (l, j and s) for singlets and doublets.

		Singlets (s = 0)				
		Quantum number j				
		0	1	2	3	...
Quantum Number l	0	s				
	1		p			
	2			d		
	3				f	
	...					...
		Doublets (s = 1/2)				
		Quantum number j				
		1/2	3/2	5/2	7/2	...
Quantum Number l	0	s				
	1	p_1	p_2			
	2		d_1	d_2		
	3			f_1	f_2	
	...					....

precession of the orbit of its electron around the axis of the field. The frequency of this precession, v_L, is:

$$v_L = \frac{q_e B}{4\pi m_0 c}$$

The possible transitions for the Zeeman effect occur for m changing ±1 or 0. If $\Delta m = 0$ there is a linear polarization parallel to the magnetic field.

The change of energy due to the action of the magnetic field (ΔE) it is an energy difference for $\Delta m = \pm 1$. From quantum mechanics, the energy is:

$$E = hv$$

$$Then: \Delta E = h\Delta v$$

The frequency difference, Δv, is:

$$\Delta v = m v_L$$

Then, the energy difference from the action of the magnetic field from a normal Zeeman effect is:

$$\Delta E = h m v_L$$

Since the resultant of the angular momentum about the magnetic field axis, J_B (or J_Z), is given by:

$$J_B = \frac{mh}{2\pi}$$

$$J_B = J\cos(J, B)$$

Where (J,B) is the angle between the resultant angular momentum and the magnetic field. Then, ΔE becomes:

$$\Delta E = E_{n,m+1} - E_{n,m} = 2\pi v_L J \cos(J,B)$$

$$\Delta E = E_{n,m+1} - E_{n,m} = 2\pi v_L J_B$$

Since the magnetic dipole moment resolved along the magnetic field, μ_B, is:

$$\mu_B = \mu \cos(B,\mu)$$

$$\mu_B = m\left(\frac{q_e h}{4\pi m_0 c}\right)$$

Then, ΔE can also be written as:

$$\Delta E = hmv_L$$

$$v_L = \frac{q_e B}{4\pi m_0 c}$$

$$\Delta E = hm\frac{q_e B}{4\pi m_0 c}$$

$$\Delta E = B\left(\frac{hmq_e}{4\pi m_0 c}\right)$$

$$\Delta E = B\mu_B$$

$$\Delta E = B\mu \cos(B,\mu)$$

Where m is the quantum magnetic quantum number, μ is the atom's magnetic moment in the direction of the resultant angular momentum J and (B,μ) is the angle between B and μ. Both equations of ΔE are associated with the normal Zeeman effect.

However, for the change of energy for the anomalous Zeeman effect, the Larmor frequency, v_L, is replaced by gv_L. Where g (Landé g-factor) is the separation factor (Landé 1923a,b). The change of energy due to the action of the magnetic field in the anomalous Zeeman effect is:

$$\Delta E = h\Delta v$$

$$\Delta v = mgv_L$$

$$\Delta E = hmgv_L$$

$$\Delta E = 2\pi g v_L J \cos(JB) = 2\pi g v_L J_B$$

The equation of the gyromagnetic ratio associated with atoms showing anomalous Zeeman effect is:

$$\gamma = g\frac{q_e}{2m_0 c}$$

Then, the atomic magnetic moment around the magnetic field B under anomalous Zeeman effect is:

$$\mu_B = \gamma J_B = g\frac{q_e}{2m_0 c}\frac{mh}{2\pi} = mg\left(\frac{q_e h}{4\pi m_0 c}\right)$$

The values of the magnetic quantum number are m = ± j, ± (j–1), ± (j–2). … The g-factor was empirically introduced by Landé to determine the Zeeman terms to any multiplet term (see equation below). In addition, the product mg gave the right splitting factors for the Zeeman terms (normal and anomalous multiplets). For example, the d_2 doublet term for which l = 2, s = 1/2 and j = 5/2, the Landé g-factor is 6/5.

$$g = 1 + \frac{j(j+1) + s(s+1) - l(l+1)}{2j(j+1)}$$

See Table 12.2 for the Landé g-factors according to Zeeman terms.

Table 12.2: Landé g-factor values for singlet and doublet Zeeman terms.

	Singlets				Doublets				
Term	s	p	d	S	p_1	p_2	d_1	d_2	…
g	1	1	1	2	2/3	4/3	4/5	6/5	…

See Table 12.3 for the relation between mg and Stern-Gerlach deviation (see next section).

Table 12.3: The mg term (product of the magnetic quantum number and Landé g-factor) and Stern-Gerlach deviation for some normal states.

Normal state	mg	Stern-Gerlach deviation
Singlet s term	0	|
Doublet s term	–1 1	| |
Doublet p_1 term	–1/3 1/3	| |
Doublet p_2 term	–6/3 –2/3 2/3 6/3	| | | |
Triplet s term	–2 0 2	| | |
Quartet s term	–3 –1 1 3	| | | |

According to the core model, the product mg should correspond to the sum of the projection of the vector R (or L) on the external field B, |R| cos (R,B), and the projection of the vector K (or S) on the external field B, |K| cos (K,B).

$$mg = \frac{2\pi}{h}\left[|K|\cos(K,B) + |R|\cos(R,B)\right]$$

However, the observed result was an expected value 2 of the g factor which yielded:

$$mg = \frac{2\pi}{h}\left[|K|\cos(K,B) + 2|R|\cos(R,B)\right]$$

This result suggested that the vector R precessed twice faster around B than K did. Due to this anomaly, either the Larmor's theorem needed a modification or a further rotation of the core electrons (associated with R) should be included. Moreover, the observed change in energy is:

$$\Delta E = hmg\nu_L$$

$$\nu_L = \frac{q_e B}{4\pi m_0 c}$$

$$\Delta E = \frac{q_e B}{2m_0 c}\left[|K|\cos(K,B) + 2|R|\cos(R,B)\right]$$

Which means that the electron has additional energy due to the field B as $(q_e/2m_0c)L$ and $2(q_e/2m_0c)S$.

3. Heisenberg's core models

After Landé's work, Heisenberg developed his own core model (Heisenberg 1922) where he established that the valence electron should share half of its angular momentum with the core (supposedly not having any net angular momentum). He assigned half-integral values for the azimuthal quantum number while Bohr and Landé assigned integral values to the secondary quantum number. The core had ½ angular momentum (somehow equivalent to the electron spin of the alkali doublets). In his model, the electron's angular momentum vector should move around the core's angular momentum vector in a way that an internal magnetic field (H_{int}) was created. The interaction energy of the precession of the core was:

$$\Delta E = \frac{1}{2} h\nu_L \cos\beta$$

Where ν_L is the Larmor frequency and β is the angle between electron's angular momentum vector and the core's angular momentum vector.

This model had a relative success in describing the Paschen-Back effect and the anomalous Zeeman effect, but it failed for intermediate fields. Then, Heisenberg, along with Landé, developed his second core model starting from Landé's analysis of the neon spectrum (Landé and Heisenberg 1924). They developed the branching process which is a building-up process of the angular momentum, but this process violated the Bohr's building-up principle.

4. Pauli's electron's two foldness

In some papers, Wolfgang Pauli is referred as Wolfgang Pauli Jr. since Pauli had his father's name. Pauli and Heisenberg had the same doctoral advisor: Sommerfeld. Likewise Heisenberg, Pauli also worked with Born after his doctorate (Born and Pauli 1922). Pauli and Heisenberg worked together in the study of the quantum dynamics of wave field (Heisenberg and Pauli 1929, 1930). Pauli also gave a very important contribution to the matrix mechanics when he studied the spectrum of the hydrogen atom (Pauli 1926). Pauli also invented the 2 × 2 Pauli matrices as a basis of spin operators (Pauli 1927), but his most important contribution to the quantum mechanics was the exclusion principle (Pauli 1925a,b).

In 1923, Pauli followed a different path with respect to that Landé did in order to rationalize the anomalous Zeeman effect (Pauli 1923a,b). In this different path, Pauli did not use any model (such as the core model) of the empirical data and recovered the term values for the anomalous Zeeman effect from known values of the Paschen-Back's study (Paschen and Back 1912). Pauli calculated the mg splitting factors for several multiplets in strong fields where the magnetic quantum number is:

$$m = m_k + m_r$$
$$m_k = 0, \pm 1, \ldots, \pm (k-1)$$
$$m_r = \pm 1/2, \ldots, \pm (r - 1/2), \quad or: m_k = 0, \pm 1, \ldots, \pm (r - 1/2)$$

Where m_k is the quantum number of the orbital magnetic moment and m_r is the quantum number of the core magnetic moment.
Pauli calculated the Zeeman energy as:

$$\Delta E = (m_k + 2m_r) h\omega_L = (2m - m_k) h\nu_L$$

Where once again the puzzling value 2 appeared in Zeeman energy which seems to violate Larmor's theorem (in which an electron will not have any change in motion other than the Larmor precession when placed in a magnetic field).

Pauli inferred that the deviation of the Larmor's theorem would occur in heavy atoms where the relativistic change of mass has to be taken into account (Pauli 1925a).

$$m = \frac{m_0}{\sqrt{1 - \upsilon^2/c^2}}$$

The gyromagnetic ratio deviates from its normal values by a correction factor γ:

$$\gamma = \frac{\overline{m_0}}{m} = \sqrt{1 - \frac{\upsilon^2}{c^2}} = 1 + \frac{E}{m_0 c^2} =$$

$$\gamma = \left\{ 1 + \frac{\alpha^2 Z^2}{\left[n - k + \sqrt{k^2 - \alpha^2 Z^2} \right]^2} \right\}^{-1/2}$$

$$where: \alpha = \frac{2\pi q_e^2}{hc}$$

For n = k = 1, the correction factor γ reduces to (Pauli 1925a):

$$\gamma = \sqrt{1 - \alpha^2 Z^2}$$

For the hydrogen atom (Z = 1), where the $\alpha^2 Z^2$ is negligible and the correction factor does not differ from 1 and hence there is no influence on the Zeeman splitting. But for high Z, the correction factor differs significantly from 1 and it leads to the violation of the Larmor frequency. By neglecting higher powers of $\alpha^2 Z^2$ (for any values of n and k), we have γ as (Pauli 1925a):

$$\gamma = 1 - \frac{1}{2}\frac{\alpha^2 Z^2}{n^2}$$

Pauli's g-factor for the anomalous Zeeman effect becomes:

$$g = 1 + (2\gamma - 1)\delta$$

$$or: g = 1 + \gamma\delta$$

$$\delta = \frac{j(j+1) + s(s+1) - l(l+1)}{2j(j+1)}$$

However, Landé announced that thallium (Z = 90) has negligible relativistic correction factor which spoke against the assumption of the atomic core angular momentum. Other facts against the core angular momentum were listed by Pauli in a letter to Landé. Pauli stated: "In the alkalis, the valence electron alone makes the complex structure as well as the anomalous Zeeman effect. The contribution of the atomic core is out of question (also in other elements). In a puzzling, non-mechanical way, the valence electron (in the alkalis) manages to run about in two states with the same k but with different angular momenta" (Pauli 1979). **Then, Pauli was sure that the atomic core did not influence the Zeeman effect.** The way in which the valence electron could take two different states in alkalis with the same angular momentum k was referred as the electron's two foldness (which later was known as the two-valuedness of the spin angular momentum). Pauli did not speak in terms of spin in 1924, but he was the first to abandon the concept of the core model for the explanation of the anomalous Zeeman effect. **He was the first to consider that the valence electron alone could be responsible for the anomalous Zeeman effect.** Pauli denoted k_1 as the azimuthal quantum number and k_2 as the magnitude of the relativistic correction, γ. He introduced the magnetic number, m, for the atom under a strong magnetic field. Since the core angular momentum was no more regarded, the doubling of the number of states was ascribed to the 'two foldness' of electron.

5. Pauli's exclusion principle

In 1925, based on Stoner's work (Stoner 1924) and the 'two foldness' of electrons, Pauli established his own building-up principle for the atom's electronic configuration (Pauli 1925b). Most importantly, in this work Pauli also established the exclusion principle where no electron can have equal values for all quantum numbers. In Pauli's words: "There can never be two or more equivalent electrons of all quantum numbers n, k_1, k_2, m_1 are the same. If an electron is present in the atom for which these quantum numbers (in an external field) have definite values, this state is 'occupied'" (Pauli 1925b). In the next year, Heisenberg successfully used the exclusion principle for the helium atom (Heisenberg 1926).

Pauli admitted that the exclusion rule lacked a firm theoretical foundation, i.e., it could not be proved even in 1948, when Pauli received the Nobel Prize. However, it was a natural observation of the anomalous Zeeman effect and Stoner's work of the building-up the electronic configuration of atoms (Stoner 1924). Fifteen years later, Pauli extended this principle to all fermions (Pauli 1940) and it took the same time to consolidate the exclusion principle. The history of the creation of the quantum electrodynamics, quantum field theory and quantum statistics is intertwined with the history of the exclusion principle.

On the Landé's core model, the core electrons of the alkalis would have angular momentum r = s = ½ which would explain the doublet on the Zeeman effect. If one suggests that instead of atomic core angular momentum, there is a spin which corresponds to one quantum rotation and there are two of these states differing in the inclination of the spin axis to the orbital plane, this explains the Zeeman effect for the alkalis (with a single valence electron). Then, we have:

$$|l-1/2| \le j \le l+1/2$$

$$l = 0,1,2,3,...$$

$$j = l \pm \frac{1}{2}$$

For a single electron, j has two values, except for l = 0. For l = 1, j = 1/2 and 3/2 give the separation of doublet terms p_1 and p_2. For l = 2, j = 3/2 and 5/2 give the separation of doublets of the terms d_1, d_2, and so on.

6. The discovery of the spin

In 1922, Stern and Gerlach devised an experiment in which a gaseous silver beam (with a single valence electron) reached a non-uniform magnetic field which caused deflection of the original beam into two new ones (Fig. 12.1). As the title of their paper stated, they proved the directional quantization in a magnetic field (Gerlach and Stern 1922). As a result of their experiment, the existence of the two-valuedness electron has been proven. A very similar result was obtained by Phipps and Taylor using a hydrogen beam (with detailed information on how to obtain the atomic hydrogen) which reinforced Stern-Gerlach result (Phipps and Taylor 1927). However, Stern-Gerlach's result was misunderstood in 1922. It was credited as confirming the space-quantization and Bohr-Sommerfeld model (the distance between the two lines corresponded to a value of one Bohr magneton) and not the electron spin. Then, the experiment was not interpreted correctly at that time.

In 1925, Uhlenbeck and Goudsmit pointed out that the application of the spinning of the electron could interpret the anomalous Zeeman effect (Uhlenbeck and Goudsmit 1925,1926). The gyromagnetic ratio for the electron spin was twice the corresponding ratio for the orbital motion (that from the orbital angular momentum), that is:

$$\gamma_s = 2\frac{q_e}{2m_0c}$$

In their hypotheses, the spin corresponded to a one-quantum rotation that led to two quantum states and the energy difference of these states is proportional to the fourth power of the nuclear charge. The doublets are a consequence of the coupling between the orbital angular momentum and the intrinsic angular momentum (spin).

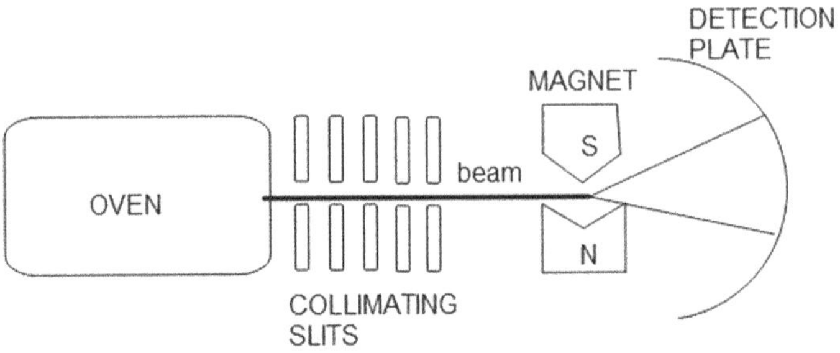

Fig. 12.1: Schematic representation of Stern-Gerlach experiment.

In 1926, Heisenberg and Jordan analyzed the doublet fine structure and the anomalous Zeeman effect using matrix mechanics (Heisenberg and Jordan 1926). The Hamiltonian function is:

$$H = H_0 + H_1 + H_2 + H_3$$

$$H_0 = \frac{1}{2m}\left(p_x^2 + p_y^2\right) - \frac{q_e^2 Z}{r}$$

$$where: r^2 = x^2 + y^2$$

$$H_1 = \frac{q_e}{2m_0 c} H\left(\mathbf{k} + 2\mathbf{s}\right)$$

$$H_2 = \frac{q_e^2 Z}{2m_0^2 c^2}\frac{1}{r^3}\mathbf{ks}$$

$$H_3 = -\frac{1}{2m_0 c^2}\left[E_0^2 + 2q_e^2 Z E_0 \frac{1}{r} + q_e^4 Z^2 \frac{1}{r^2}\right]$$

The vector **s** is associated with the spin angular momentum and the vector **k** is associated with the orbital angular momentum. The Hamiltonian H_0 is relative to the non-perturbated system. The Hamiltonian H_1 is the Larmor-violating magnetic interaction energy with the anomalous factor 2, while the Hamiltonian H_2 is associated with the Larmor precession and the Hamiltonian H_3 is the relativistic mass variability additional energy. Then, they artificially included the spin operator and the anomalous factor 2 in the Hamiltonian in order to obtain the expected result for the anomalous Zeeman effect.

At first, Pauli disliked the idea of the electron spin because it brought back the classical model to the quantum mechanics. But in the next year, he recognized that the work of Uhlenbeck and Goudsmit helped to connect the exclusion principle to the idea of spin.

7. Pauli's work on spin

In 1927, Pauli repeated Fermi's calculations using quantized ideal gas (Fermi 1926) and investigated its magnetic behavior where the results were in accordance with experimental values for alkali metals (Pauli 1927). Pauli introduced the s_z spin variable along with q_k spatial coordinates in the wave function $\psi(q_k,s_z)$ leading to two components $\psi_\alpha(q_k)$ and $\psi_\beta(q_k)$ corresponding to the wave functions with +1/2 and –1/2 spin values, respectively. The probability density of the quantum states in the interval $(q_k,q_k + dq_k)$ with spin either up or down is:

$$P(\uparrow)=\left|\psi_\alpha(q_k)\right|^2 dq_\tau$$
$$P(\downarrow)=\left|\psi_\beta(q_k)\right|^2 dq_\tau$$

The electron spin state, ψ, can be written as a linear combination of two states ψ_α and ψ_β.

$$\psi = c_\alpha\psi_\alpha + c_\beta\psi_\beta \rightarrow \psi = \begin{pmatrix} c_\alpha \\ c_\beta \end{pmatrix}$$

The normalization condition for the wave function is:

$$\int \psi^*\psi\, dq_\tau = \int\left\{\left|\psi_\alpha\right|^2 + \left|\psi_\beta\right|^2\right\}dq_\tau = 1$$
$$where: \psi_\alpha = \psi_\alpha(q_k), \psi_\beta = \psi_\beta(q_k)$$
$$or: \psi^*\psi = \begin{pmatrix} c_\alpha^* & c_\beta^* \end{pmatrix}\begin{pmatrix} c_\alpha \\ c_\beta \end{pmatrix} = \left|c_\alpha\right|^2 + \left|c_\beta\right|^2 = 1$$

And the orthogonality relation is:

$$\int\left(\psi_{\alpha,n}\psi_{\alpha,m}^* + \psi_{\beta,n}\psi_{\beta,m}^*\right)dq_\tau = 0, \quad n \neq m$$

The spin operator **S** is a vector matrix that has to respect the following relation:

$$\mathbf{S}^2 = \hbar^2 s(s+1), s = ½$$

The relations involving the components of the spin operator **S**, known as spin projection operators $\mathbf{S}_x$, $\mathbf{S}_y$ and $\mathbf{S}_z$, are:

$$\mathbf{S}_x\mathbf{S}_y - \mathbf{S}_y\mathbf{S}_x = 2i\mathbf{S}_z, \ldots,$$
$$where: \mathbf{S}_x\mathbf{S}_y = -\mathbf{S}_y\mathbf{S}_x = i\mathbf{S}_z$$
$$\mathbf{S}_x^2 + \mathbf{S}_y^2 + \mathbf{S}_z^2 = 3$$

They are measured in the unit h/4π. The linear transformations of the components of the operator **S** in the wave function are:

$$\mathbf{S}_x(\psi_\alpha) = \psi_\beta, \quad \mathbf{S}_x(\psi_\beta) = \psi_\alpha$$
$$\mathbf{S}_y(\psi_\alpha) = -i\psi_\beta, \quad \mathbf{S}_y(\psi_\beta) = i\psi_\alpha$$
$$\mathbf{S}_z(\psi_\alpha) = \psi_\alpha, \quad \mathbf{S}_z(\psi_\beta) = -\psi_\beta$$

Which can be represented in matrix form (Pauli matrix) as:

$$\mathbf{S}_x(\psi) = \begin{pmatrix} 0 & 1 \\ 1 & 0 \end{pmatrix}\psi$$

$$\mathbf{S}_y(\psi) = \begin{pmatrix} 0 & -i \\ i & 0 \end{pmatrix}\psi$$

$$\mathbf{S}_z(\psi) = \begin{pmatrix} 1 & 0 \\ 0 & -1 \end{pmatrix}\psi$$

Another way to represent the Pauli matrices are given below where the unit of each component is included.

$$\mathbf{S}_x = \frac{\hbar}{2}\boldsymbol{\sigma}_x = \frac{\hbar}{2}\begin{pmatrix} 0 & 1 \\ 1 & 0 \end{pmatrix}$$

$$\mathbf{S}_y = \frac{\hbar}{2}\boldsymbol{\sigma}_y = \frac{\hbar}{2}\begin{pmatrix} 0 & -i \\ i & 0 \end{pmatrix}$$

$$\mathbf{S}_z = \frac{\hbar}{2}\boldsymbol{\sigma}_z = \frac{\hbar}{2}\begin{pmatrix} 1 & 0 \\ 0 & -1 \end{pmatrix}$$

And they satisfy the relations:

$$\boldsymbol{\sigma}_\mu^2 = \begin{pmatrix} 1 & 0 \\ 0 & 1 \end{pmatrix} = I, \quad (\mu = x, y, z)$$

$$\boldsymbol{\sigma}_\mu\boldsymbol{\sigma}_\nu + \boldsymbol{\sigma}_\nu\boldsymbol{\sigma}_\mu = 0, \quad (\mu \neq \nu; \mu, \nu = x, y, z)$$

Their determinants and traces are:

$$\det \boldsymbol{\sigma}_\mu = -1$$

$$\mathbf{tr}\boldsymbol{\sigma}_\mu = 0$$

The commutators of the Pauli matrices are:

$$\left[\boldsymbol{\sigma}_x, \boldsymbol{\sigma}_y\right] = 2i\boldsymbol{\sigma}_z$$

$$\left[\boldsymbol{\sigma}_y, \boldsymbol{\sigma}_z\right] = 2i\boldsymbol{\sigma}_x$$

$$\left[\boldsymbol{\sigma}_z, \boldsymbol{\sigma}_x\right] = 2i\boldsymbol{\sigma}_y$$

$$\left[\boldsymbol{\sigma}_x, \boldsymbol{\sigma}_x\right] = 0$$

The commutative property between $\mathbf{S_x}$ and $\mathbf{S_y}$

$$\left[\mathbf{S}_x,\mathbf{S}_y\right]=\mathbf{S}_x\mathbf{S}_y-\mathbf{S}_y\mathbf{S}_x=\frac{1}{4}\hbar^2\begin{bmatrix}0&1\\1&0\end{bmatrix}\begin{bmatrix}0&-i\\i&0\end{bmatrix}-\frac{1}{4}\hbar^2\begin{bmatrix}0&-i\\i&0\end{bmatrix}\begin{bmatrix}0&1\\1&0\end{bmatrix}$$

$$\left[\mathbf{S}_x,\mathbf{S}_y\right]=\frac{1}{4}\hbar^2\begin{bmatrix}i&0\\0&-i\end{bmatrix}-\frac{1}{4}\hbar^2\begin{bmatrix}-i&0\\0&i\end{bmatrix}=\frac{1}{4}\hbar^2\left\{\begin{bmatrix}i&0\\0&-i\end{bmatrix}-\begin{bmatrix}-i&0\\0&i\end{bmatrix}\right\}$$

$$\left[\mathbf{S}_x,\mathbf{S}_y\right]=\frac{1}{4}\hbar^2\begin{bmatrix}i-(-i)&0\\0&-i-i\end{bmatrix}=\frac{1}{4}\hbar^2\begin{bmatrix}2i&0\\0&-2i\end{bmatrix}=\frac{1}{2}\hbar^2\begin{bmatrix}i&0\\0&-i\end{bmatrix}$$

$$\left[\mathbf{S}_x,\mathbf{S}_y\right]=\frac{i}{2}\hbar^2\begin{bmatrix}1&0\\0&-1\end{bmatrix}=i\hbar\mathbf{S}_z$$

The derivation of the Pauli matrix in z-axis, $\boldsymbol{\sigma_z}$, is simpler than in x-axis and y-axis, $\boldsymbol{\sigma_x}$ and $\boldsymbol{\sigma_y}$, where it is necessary to use raising and lowering operators.

$$\widehat{S}_z\alpha=\frac{\hbar}{2}\alpha\Rightarrow\alpha=\begin{pmatrix}1\\0\end{pmatrix}$$

$$\begin{pmatrix}a&b\\c&d\end{pmatrix}\begin{pmatrix}1\\0\end{pmatrix}=\frac{\hbar}{2}\begin{pmatrix}1\\0\end{pmatrix}$$

$$\begin{pmatrix}a\\c\end{pmatrix}=\begin{pmatrix}\frac{\hbar}{2}\\0\end{pmatrix}\Rightarrow a=\frac{\hbar}{2},c=0$$

$$\widehat{S}_z\beta=-\frac{\hbar}{2}\beta\Rightarrow\beta=\begin{pmatrix}0\\1\end{pmatrix}$$

$$\begin{pmatrix}a&b\\c&d\end{pmatrix}\begin{pmatrix}0\\1\end{pmatrix}=-\frac{\hbar}{2}\begin{pmatrix}0\\1\end{pmatrix}$$

$$\begin{pmatrix}b\\d\end{pmatrix}=\begin{pmatrix}0\\-\frac{\hbar}{2}\end{pmatrix}\Rightarrow b=0,c=-\frac{\hbar}{2}$$

$$\mathbf{S}_Z=\frac{\hbar}{2}\begin{pmatrix}1&0\\0&-1\end{pmatrix}$$

The Schrödinger equations for the two spin states are:

$$H\left(\frac{h}{2\pi i}\frac{\partial}{\partial q_k},q_k,s_x,s_y,s_z\right)\psi_{E,\alpha}=E\psi_\alpha$$

$$H\left(\frac{h}{2\pi i}\frac{\partial}{\partial q_k},q_k,s_x,s_y,s_z\right)\psi_{E,\beta}=E\psi_\beta$$

Pauli used the same Hamiltonian function from Heisenberg and Jordan's paper (Heisenberg and Jordan 1926).

$$H = H_0 + H_1 + H_2 + H_3$$

$$H_0 = -\frac{1}{2m}\frac{h^2}{4\pi^2}\nabla^2 - \frac{q_e^2 Z}{r}$$

$$H_1 = \frac{q_e}{2m_0 c} H(\mathbf{k} + 2\mathbf{s})$$

$$H_2 = \frac{q_e^2 Z}{2m_0^2 c^2}\frac{1}{r^3}\mathbf{ks}$$

$$H_3 = -\frac{1}{2m_0 c^2}\left[E_0^2 + 2q_e^2 Z E_0 \frac{1}{r} + q_e^4 Z^2 \frac{1}{r^2}\right]$$

$$where: k_x = y\frac{\partial}{\partial z} - z\frac{\partial}{\partial y}, k_y = z\frac{\partial}{\partial x} - x\frac{\partial}{\partial z}$$

$$k_z = x\frac{\partial}{\partial y} - y\frac{\partial}{\partial x}$$

Although Pauli's introduction of the spin half-integral in the wave function and the relativistic correction of the Hamiltonian were done artificially, this work was very important to Dirac's work. He used the Pauli matrices to form the relativistic wave equation where the spin appeared naturally (see next section).

8. Dirac's work on spin and Dirac equation

The Dirac equation is a relativistic wave equation which describes the electron-spin. Whereas Heisenberg's matrix mechanics and Schrödinger's wave mechanics can be considered as free field theory, Dirac included both the electromagnetic field and electric charge matter as quantum mechanical variables. The wave functions of Dirac equation are vector of four complex numbers, bispinors. The Dirac equation can be written as:

$$\left(\beta mc^2 + c\sum_{n=1}^{3}\alpha_n p_n\right)\psi(q,t) = i\hbar\frac{\partial\psi(q,t)}{\partial t}$$

Where β, α_1, α_2, and α_3 are 4×4 Hermitian matrices of the wave function and they are the spin-up electron, spin-down electron, spin-up positron and spin-down positron.

Dirac stated in 1928: "In the present paper it is shown that this is the case, the incompleteness of the previous theories (Pauli's work in 1927 and Heisenberg and Jordan's work in 1926 of the previous sections) lying in their disagreement with relativity, or alternatively, with the general transformation theory of quantum mechanics. It appears that the simplest Hamiltonian for a point-charge electron (without spin) satisfying the requirements of both relativity and the general transformation theory leads to an explanation of all duplexity without further assumption. All the same there is a great deal of truth in the spinning electron model" (Dirac 1928a).

The relativity Hamiltonian for a point electron moving in an arbitrary electromagnetic field with scalar potential A_0 and vector potential A is (see Chapter eleven for the Dirac's usage of F instead of H for the Hamiltonian):

$$F = \left(\frac{W}{c} + \frac{q_e}{c}A_0\right)^2 + \left(p + \frac{q_e}{c}A\right)^2 + m_0^2c^2$$

$$W = i\hbar\frac{\partial}{\partial t}, \quad p_r = -i\hbar\frac{\partial}{\partial x_r}, \quad r = 1,2,3$$

Where p is the momentum vector operator and W is the energy operator from Schrödinger's time-dependent wave equation. The corresponding wave equation is.

$$F\psi = \left[\left(i\hbar\frac{\partial}{c\partial t} + \frac{q_e}{c}A_0\right)^2 + \left(-i\hbar\frac{\partial}{\partial x_r} + \frac{q_e}{c}A\right)^2 + m_0^2c^2\right]\psi = 0$$

If the wave functions ψ_n and ψ_m are the solutions, then the charge density and current density associated with the transition m → n during emission and absorption radiation are (known as Gordon-Klein theorem):

$$\rho_{mn} = -\frac{e}{2mc^2}\left\{i\hbar\left(\psi_m\frac{\partial\psi_n}{\partial t} - \overline{\psi}_n\frac{\partial\psi_m}{\partial t}\right) + 2eA_0\psi_m\overline{\psi}_n\right\}$$

$$I_{nm} = -\frac{e}{2m}\left\{-i\hbar\left(\psi_m\nabla\overline{\psi}_n - \overline{\psi}_n\nabla\psi_m\right) + 2\frac{e}{c}A_m\psi_m\overline{\psi}_n\right\}$$

The corresponding Klein-Gordon equation is also a relativistic wave equation, but it is second-order in space and time (which brings some difficulties to its solution) and is invariant to Lorentz transformation and it describes the spinless particle field (Gordon 1926, Klein 1926).

$$\left(-\frac{1}{c^2}\frac{\partial^2}{\partial t^2} + \nabla^2\right)\phi = \frac{m^2c^2}{\hbar^2}\phi$$

Then, Dirac tried an equation that was first-order in both space and time. Then, he expanded the square root of energy in an infinite series of space (x,y,z) and time (t).

$$E^2 = c^2p^2 + m^2c^4 = c^2\left(p^2 + m^2c^2\right)$$

$$E = c\sqrt{p^2 + m^2c^2}$$

$$\nabla^2 - \frac{1}{c^2}\frac{\partial^2}{\partial t^2} = \left(\mathbf{A}\partial_x + \mathbf{B}\partial_y + \mathbf{C}\partial_z + \frac{i}{c}\mathbf{D}\partial_t\right)\left(\mathbf{A}\partial_x + \mathbf{B}\partial_y + \mathbf{C}\partial_z + \frac{i}{c}\mathbf{D}\partial_t\right)$$

In order to get all cross-terms such as $\partial \times \partial y$ to vanish is necessary to assume:

$$\mathbf{AB} + \mathbf{BA} = 0, \mathbf{AC} + \mathbf{CA} = 0, \mathbf{AD} + \mathbf{DA} = 0, \ldots$$

$$\mathbf{A}^2 = \mathbf{B}^2 = \ldots = 1$$

These conditions are met if **A**, **B**, **C** and **D** are matrices and, as a consequence, the wave function has four components. The corresponding wave equation becomes:

$$\left[\left(\mathbf{A}\partial_x+\mathbf{B}\partial_y+\mathbf{C}\partial_z+\frac{i}{c}\mathbf{D}\partial_t\right)\left(\mathbf{A}\partial_x+\mathbf{B}\partial_y+\mathbf{C}\partial_z+\frac{i}{c}\mathbf{D}\partial_t\right)\right]\psi^2=\frac{m^2c^2}{\hbar^2}\psi^2$$

$$\left(\mathbf{A}\partial_x+\mathbf{B}\partial_y+\mathbf{C}\partial_z+\frac{i}{c}\mathbf{D}\partial_t\right)\psi=\frac{mc}{\hbar}\psi$$

$$\mathbf{A}=i\beta\alpha_1,\mathbf{B}=i\beta\alpha_2,\mathbf{C}=i\beta\alpha_3,\mathbf{D}=\beta$$

Subsequently, Dirac questioned: "what is the probability of any dynamical variable at any specified time having a value lying between any specified limits when the system is represented by a given function ψ_m?" and he completed: "We should expect the interpretation of the relativity theory to be just as general as that of the non-relativity theory". After he stated: "The wave equation of the relativity theory must also be linear in W if the general interpretation is to be possible". Then, Dirac wanted to remove the difficulty that Gordon-Klein theorem can only answer about the position of the electron (by the use of ρ_{mn}) and not its momentum or angular momentum or any other dynamical variable.

Then, Dirac stated in his paper that: "Our problem is to obtain a wave equation of the form (shown above) which shall be invariant under a Lorentz transformation and shall be equivalent to the time-dependent Schrödinger equation in the limit of large quantum numbers" where he paved his own way to solve this problem and found the spin naturally. Then he started with a simpler Hamiltonian without a field whose wave equation is:

$$\left(-\mathbf{p}_0^2+\mathbf{p}^2+m_0^2c^2\right)\psi=0$$

$$where:\mathbf{p}_0=\frac{W}{c}=i\hbar\frac{\partial}{c\partial t}$$

Then, due to "the symmetry between p_0 and $p_r(r=1,2,3)$ required by the relativity shows that since the Hamiltonian we want is linear in p_0, it must also be linear in p_r $(r=1,2,3)$", Dirac affirmed. The wave equation becomes:

$$\left(p_0+\alpha_1p_1+\alpha_2p_2+\alpha_3p_3+\beta\right)\psi=0$$

Which is similar to:

$$\left(\mathbf{A}\partial_x+\mathbf{B}\partial_y+\mathbf{C}\partial_z+\frac{i}{c}\mathbf{D}\partial_t\right)\psi=\frac{mc}{\hbar}\psi$$

$$\mathbf{A}=i\beta\alpha_1,\mathbf{B}=i\beta\alpha_2,\mathbf{C}=i\beta\alpha_3,\mathbf{D}=\beta$$

Dirac used the dynamical variables $\alpha_r(r=1,2,3)$ and β which are independent of p_0 and $p_r(r=1,2,3)$, that is, they commute with x,y,z,t. The $\alpha_r(r=1,2,3)$ and β are also independent of x,y,z,t, and commute with p_0 and p_r. Then, it is necessary to have other dynamical variables in the wave function besides the coordinates and momentum of the electron in order that $\alpha_r(r = 1,2,3)$ and β may be function of them. The above equation becomes:

$$0 = (-p_0 + \alpha_1 p_1 + \alpha_2 p_2 + \alpha_3 p_3 + \beta)(p_0 + \alpha_1 p_1 + \alpha_2 p_2 + \alpha_3 p_3 + \beta)\psi =$$

$$\Rightarrow \left[-p_0^2 + \sum \alpha_1^2 p_1^2 + \sum (\alpha_1\alpha_2 + \alpha_2\alpha_1) p_1 p_2 + \beta^2 + \sum (\alpha_1\beta + \beta\alpha_1) p_1\right]\psi = 0$$

$$\alpha_r^2 = 1, \qquad \alpha_r\alpha_s + \alpha_s\alpha_r = 0 (r \neq s), \qquad r,s = 1,2,3$$

$$\beta^2 = m^2c^2, \qquad \alpha_r\beta + \beta\alpha_r = 0$$

The conditions above are established so that the cross-terms are vanished (as above-mentioned).

which are used in the wave function along with the ordinary variables x_r and t. The variables α_r's have some conditions that are similar to those of Pauli's matrices which are then used in Dirac's work. However, since there are three variables α_r (r = 1,2,3), it is needed to extend the Pauli's matrices in a diagonal manner in two more rows and columns in order to introduce three more matrices. They are:

$$\boldsymbol{\sigma}_1 = \begin{pmatrix} 0 & 1 & 0 & 0 \\ 1 & 0 & 0 & 0 \\ 0 & 0 & 0 & 1 \\ 0 & 0 & 1 & 0 \end{pmatrix}, \boldsymbol{\sigma}_2 = \begin{pmatrix} 0 & -i & 0 & 0 \\ i & 0 & 0 & 0 \\ 0 & 0 & 0 & -i \\ 0 & 0 & i & 0 \end{pmatrix}, \boldsymbol{\sigma}_3 = \begin{pmatrix} 1 & 0 & 0 & 0 \\ 0 & -1 & 0 & 0 \\ 0 & 0 & 1 & 0 \\ 0 & 0 & 0 & -1 \end{pmatrix}$$

$$\rho_1 = \begin{pmatrix} 0 & 0 & 1 & 0 \\ 0 & 0 & 0 & 1 \\ 1 & 0 & 0 & 0 \\ 0 & 1 & 0 & 0 \end{pmatrix}, \rho_2 = \begin{pmatrix} 0 & 0 & -i & 0 \\ 0 & 0 & 0 & -i \\ i & 0 & 0 & 0 \\ 0 & i & 0 & 0 \end{pmatrix}, \rho_3 = \begin{pmatrix} 1 & 0 & 0 & 0 \\ 0 & 1 & 0 & 0 \\ 0 & 0 & -1 & 0 \\ 0 & 0 & 0 & -1 \end{pmatrix}$$

The wave equation takes the form:

$$\left[\mathbf{p}_0 + \rho_1(\sigma, p) + \rho_3 m_0 c\right]\psi = 0$$

Where **p** and **ρ** are 4 × 4 matrices.

In the next section of his paper, Dirac proved the invariance under Lorentz transformation (see chapter nine) for the Hamiltonian without the field **A**:
Hereafter, he used the Hamiltonian for an arbitrary electro-magnetic field **A**:

$$\left[\mathbf{p}_0 + \frac{q_e}{c}\mathbf{A}_0 + \rho_1\left(\boldsymbol{\sigma}, \mathbf{p} + \frac{q_e}{c}\mathbf{A}\right) + \rho_3 m_0 c\right]\psi = 0$$

The invariance of the Lorentz transformation also occurred in the Hamiltonian above. When compared to the first relativity wave equation presented in his work, a derived version of the Hamiltonian above had two the additional terms in F:

$$\frac{q_e h}{c}(\sigma, H) + \frac{iq_e h}{c}\rho_1(\sigma, E)$$

that when divided by a factor $2m_0$, can be regarded as the additional potential energy of the electron due to its new degree of freedom (i.e., the spin). The electron will therefore have a magnetic moment, M_B and an electric moment, M_E:

$$M_B = \frac{q_e h}{2m_0 c}\sigma$$

$$M_E = \frac{iq_e h}{2m_0 c}\rho_1\sigma$$

And this magnetic moment is assumed as the spinning of the electron and the electric moment is purely imaginary and does not appear in the model.

Recently, four-spinor Dirac equation has been used for the helium atom (Nascimento and Fonseca 2016). The authors used approximate solutions for this equation.

References cited

Bohr, N. 1923. The structure of the atom. Nature 112: 29–44.

Born, M. and Pauli, Jr., W. 1922. Über die quantelunggestörtermechanischersysteme. 10: 137–158.

Dirac, P.A.M. 1928. The quantum theory of the electron. Proc. Roy. Soc. Lond. A. 117: 610–624.

Darwin, C. 1928. The wave equations of the electron. Proc. Roy. Soc. Lond. A. 118: 654–680.

Gerlach, W. and Stern, O. 1922. Der experiment ellenachweis der richtungs quantelungim magnetfeld. Z. Phys. 9: 349–352.

Gordon, W. 1926. Der Comptoneffektnach der Schrödingerschen theory. Z. Phys. 40: 177–133.

Fermi, E. 1926. Sullaquantizzazionedelgasperfetto monoatômico. RendicontidellaRealeAccademia dei Lincei 3: 145–149.

Heilbron, J.L. and Kuhn, T.S. 1969. The genesis of the Bohr atom. Hist. Stud. Phys. Sci. 1: vi–290.

Heisenberg, W. 1922. Zurquantentheorie der linienstruktur und der anomalenZeemaneffekte. Z. Phys. 8: 273–297.

Heisenberg, W. 1926. Mehrkörperproblem und resonanz in der quantenmechanik. Z. Phys. 38: 411–425.

Heisenberg, W. and Jordan, P. 1926. Anwendung der quantenmechanik auf das problem der anomalen Zeemaneffekte. Z. Phys. 37: 263–277.

Heisenberg, W. and Pauli, W. 1929. Zurquantendynamik der wellenfelder. Z. Phys. 56: 1–61.

Heisenberg, W. and Pauli, W. 1930. Zurquantendynamik der wellenfelder II. Z. Phys. 59: 168–190.

Klein, O. 1926. Quantentheorie und fünfdimensionalerelativitatstheorie. Z. Phys. 37: 895–906.

Landé, A. 1921. Uber den anomalenZeemaneffekt. Z. Phys. 5: 231–241.

Landé, A. 1923a. Termstruktur und Zeemaneffekt der multipletts. Z. Phys. 15: 189–205.

Landé, A. 1923b. Termstruktur und Zeemaneffekt der multipletts. ZweiteMitteilung. Z. Phys. 19: 112–123.

Landé, A. and Heisenberg, W. 1924. Termstruktur der multiplettshöhererstufe. Z. Phys. 25: 279–286.

Nascimento, D.L. and Fonseca, A.L.A. 2016. A 2D formulation of the helium atom using four-spinor Dirac-like equation and the discussion of an approximate ground state solution. Phys. Sci. Int. J. 9: 1–12.

Paschen, F. and Back, E. 1912. Normale und anomaleZeemaneffekte. Ann. Phys. 39: 897–932.

Pauli, W. 1923a. Uber die gesetzmäβigkeiten des anomalenZeemaneffektes. Z. Phys. 16: 155–164.

Pauli, W. 1923b. ZurFrage der Zuordnung der Komplexstrukturterme in starken und in schwachenausseren. Feldern', Z. Phys. 20: 371–387.

Pauli, W. 1925a. Über den einflub der geschwindigkeitsabhängigkeit der elektronenmasse auf den Zeemaneffekt. Z. Phys. 31: 373–385.

Pauli, W. 1925b. Über den Zusammenhang des Abschlusses der Elektronengruppenim Atom mit der Komplexstruktur der Spektren. Z. Phys. 31: 765–783.

Pauli, W. 1926. Über das wasserstoffspektrum von standpunkt der neuenquantenmechanik. Z. Phys. 36: 336–363.

Pauli, W. 1927. Zurquantenmechanik des magnetischenelektronszeitschriftfürphysik. Z. Phys. 43: 601–623.

Pauli, W. 1940. The connection between spin and statistics. Phys. Rev. 58: 716–722.

Pauli, W. 1979. Wissenschaftlicherbriefwechselmit Bohr, Einstein, Heisenberg u.a. Band 1: 1919–1929. Edited by A. Hermann et al., Springer Verlag, Berlin.

Phipps, T.E. and Taylor, J.B. 1927. The magnetic moment of the hydrogen atom. Phys. Rev. 29: 309–320.

Stoner, E.C. 1924. The distribution of electrons among atomic levels. Phil. Mag. 48: 719–736.

Uhlenbeck,G.E,Goudsmit,S.1925.Ersetzungderhypothesevomunmechanischerzwangdurcheineforderung bezuglich des innerenverhaltensjedeseinzelnenelektrons. Naturwissenschaften13: 953–954.

Uhlenbeck, G.E. and Goudsmit, S. 1926. Spining electrons and the structure of spectra. Nature 117: 264–265.

Boltzmann and Fermi-Dirac Statistics

13

1. Boltzmann distribution (for distinguished particles)

Boltzmann distribution played an important role to obtain the Planck's law of the spectral radiance of the thermal black-body experiment and its derivation from Einstein's equipartition (see Chapter six).

A macrostate is a thermally isolated system with a thermal equilibrium at temperature T and it is an assembly of N identical distinguishable (localized) particles contained in a fixed volume V and having a fixed internal energy U. Important to add that N is also the total number of weakly interacting particles.

One axiom of Boltzmann statistics is that identical particles that cannot be infinitely close to each other can be distinguished by their initial conditions and by the continuity of their motions. This axiom leads to the term of a gas with energy U consisting of "N identical distinguishable molecules". Let ε_1 be the energy of n_1 particles, ..., ε_i be the energy of n_i particles, so that the total sum of $n_i\varepsilon_i$ (total energy, U) and total sum of n_i (N) are constant.

$$\sum_i n_i \varepsilon_i = U \therefore \sum_i n_i = N$$

For a given (N, V, U, T, p) macrostate there are huge amount of microstates (a cell with a number of particles with the same energy). The microstate specifies the state of each particle of the system while the macrostate specifies the state of the whole system (N, V, U, T, p).

Each microstate has a distribution of the particles $\{n_i\}$, where n_i is the number of particles in the ith state having energy ε_i. The distribution $\{n_i\}$ indicates the number of particles in each level of energy 0 (n_0), ε (n_1), 2ε (n_2), and so on. But, there is a distribution, $\{n_i^*\}$, that is overwhelmingly more probable than any other distribution. The statistical weight of the distribution $\{n_i\}$, W, is given below:

$$W = \frac{N!}{\prod_i n_i!}, \qquad N = \sum_i n_i$$

Let us take the logarithm of W:

$$\ln W = \ln N! - \sum_i \ln n_i!$$

By using Stirling's approximation in order to eliminate the factorials, we have:

$$\ln W = (N\ln N - N) - \sum_i (n_i \ln n_i - n_i)$$

For a particular fluctuation of configurations, there is one with largest W. The configuration with largest W is dominant and usually the properties of the system are determined by the dominant configuration. At maximum value of W, named W*, any infinitesimal addition vanishes, that is, $d \ln W^* = 0$.

$$for: d\ln W^* = 0$$

$$d\ln W^* = 0 - \sum_i dn_i (\ln n_i + n_i/n_i - 1) = -\sum_i \ln n_i^* dn_i = 0$$

$$where: dN = 0$$

Where n_i* is the distribution for W*. The above relation has to follow two restrictions:

$$d(U) = \sum_i \varepsilon_i dn_i = 0$$

$$d(N) = \sum_i dn_i = 0$$

Then, we use Lagrange method of undetermined multipliers to deal with a restricted maximum. Since the addition of any arbitrary multiples (say α and $\beta\varepsilon_i$) do not change the result zero, then we have:

$$\sum_i (-\ln n_i^* + \alpha + \beta\varepsilon_i) dn_i = 0$$

Where α and β are constants. For the above sum gives result zero, each individual term of the sum must have a result zero. Then, we have:

$$-\ln n_i^* + \alpha + \beta\varepsilon_i = 0$$

The above equation can be written as:

$$n_i^* = \exp(\alpha + \beta\varepsilon_i)$$

The above relation is the Boltzmann distribution.
Now, let us determine α using the above expression and the expression for N:

$$N = \sum_i n_i = \sum_i \exp(\alpha + \beta\varepsilon_i) = \exp\alpha \sum_i \exp(\beta\varepsilon_i)$$

$$A = \exp\alpha$$

$$N = A\sum_i \exp(\beta\varepsilon_i)$$

Where A is the normalization constant for the distribution so that the distribution describes the correct number of N particles. From the above equation, we have:

$$A = \frac{N}{\sum_i \exp(\beta\varepsilon_i)} = \frac{N}{Z}$$

$$where: Z = \sum_i \exp(\beta\varepsilon_i)$$

Then, the Boltzmann distribution becomes:

$$n_i^* = A\exp(\beta\varepsilon_i) = \frac{N\exp(\beta\varepsilon_i)}{Z}$$

Now, let us determine the value of β:

$$d(\ln W^*) = -\sum_i \ln n_i^* dn_i = -\sum_i (\alpha + \beta\varepsilon_i) dn_i$$

$$where: \ln n_i^* = \alpha + \beta\varepsilon_i$$

$$d(\ln W^*) = -\beta\sum_i \varepsilon_i dn_i, N = fixed$$

$$d(\ln W^*) = -\beta dU = -\beta(TdS)$$

$$S = k_B \ln W \rightarrow dS = k_B d(\ln W)$$

$$d(\ln W^*) = -\beta T k_B d(\ln W)$$

$$then: \beta = -\frac{1}{Tk_B}$$

The Boltzmann distribution becomes:

$$n_i^* = \frac{N\exp(-\varepsilon_i/k_B T)}{\sum_i \exp(-\varepsilon_i/k_B T)}$$

2. Fermi-Dirac statistics (for undistinguishable particles)

Unlike Boltzmann's approach, Fermi ascribed quantum numbers to molecules of an ideal gas and also used Pauli's exclusion principle which led to a different statistical distribution of the particles.

Fermi calculated the specific heat and the energy distribution of an ideal gas whose motion was quantized as a collection of harmonic oscillators where the monoatomic elements (where he used the term 'molecule') were not distinguishable from each other (Fermi 1926). He performed the quantization of the motion by applying Sommerfeld's rules to the monoatomic gases (molecules). An attractive force toward a fixed point O acts on the molecules so that each atom becomes a single spatial harmonic oscillator with frequency ν. The orbit of each molecule is characterized by three quantum numbers s_1, s_2 and s_3 and Fermi used Pauli's exclusion principle to state that no pair of molecules could have the same quantum numbers s_1, s_2 and s_3. In this model, the energy of each set of molecules according to the number of modes Qs (number of molecules with the same energy) are:

$$E = h\nu s$$

$$s = s_1 + s_2 + s_3$$

$$Q_s = \tfrac{1}{2}(s+1)(s+2)$$

$$if: s = 1, \quad E = h\nu \quad, Q_s = 3$$

$$if: s = 2, \quad E = 2h\nu \quad, Q_s = 6$$

Fermi considered the limiting case of having N molecules at zero temperature. If there was no restriction on the number of molecules that can have the same energy, all molecules would be in the state of zero energy. Instead, since its not possible to have more than one molecule with all three quantum numbers equal to zero ($s = 0$ and $E = 0$), only one molecule will have $E = 0$. If $N = 10$, one molecule will have $E = 0$, three molecules will have $E = h\nu$ (each molecule with, at least, two different values of s_1, s_2 and s_3), and six molecules will have six places of energy $2\ h\nu$. While Fermi used monoatomic gases, Dirac worked on a single many-electron atom.

In 1926, Dirac gave an excellent explanation with respect to the treatment of many-particle system that it is worthy to reproduce it here paying attention that n and m refer to the principal quantum number related to the 'orbit', the term most used at that time and he referred to the Heisenberg quantum mechanics as a theoretical reference. Dirac stated (Dirac 1926):

"Consider now a system that contains two or more similar particles, say, for definiteness, an atom with two electrons. Denote by (nm) that state of the atom in which one electron is in an orbit labelled m, and the other in the orbit n. The question arises whether the two states (mn) and (nm), which are physically indistinguishable as they differ only by the interchange of the two electrons, are to be counted as two different states or as only one state, i.e., do they give rise to two rows and columns in the matrices (from matrix mechanics) or to only one? If the first alternative is right, then the theory would enable one to calculate the intensities due to the two transitions $(mn) \rightarrow (m'n')$ and $(nm) \rightarrow (n'm')$ separately (...). The two transitions are, however, physically indistinguishable, and only the sum of the intensities for the two together could be determined experimentally. Hence, in order to keep the essential characteristic of the theory (Heisenberg's matrix mechanics) that it shall enable one to calculate only observable quantities, one must adopt the second alternative that (mn) and (nm) count as only one state.

This alternative, though, also leads to difficulties. The symmetry between the two electrons require that the amplitude associated with the transition $(mn) \rightarrow (m'n')$ of x_1, a coordinate of one of the electrons, shall equal the amplitude with the transition $(nm) \rightarrow (n'm')$ of x_2, the corresponding coordinate of the other electron (...). If we now count (mn) and (nm) as both defining the same row and column of the matrices, and similarly for (m′n′) and (n′m′), equation: $x_1(mn; m'n') = x_2\ (nm; n'm')$ shows that each element of the matrix x_1 equal the corresponding element of the matrix x_2 so that should have the matrix equation: $x_1 = x_2$. This relation is obviously impossible, as, amongst other things, it is inconsistent with the quantum conditions. We must infer that unsymmetrical functions of the coordinates (and momenta) of the two electrons cannot be represented by matrices. Symmetrical functions, such as the total polarization of the atom, can be considered to be represented by matrices without inconsistency, and these matrices are by themselves sufficient to determine all the physical properties of the system.

(...)If we neglect the interaction between the two electrons, then we can obtain the eigenfunctions for the whole atom simply by multiplying the eigenfunctions for one electron when it exists alone in the atom by the eigenfunctions for the other electron alone, and taking the same time variable for each. Thus if $\psi_n(x,y,z,t)$ is the eigenfunction for a single electron in the orbit n, then the eigenfunction for the whole atom in the state (mn) is: $\psi_m(x_1,y_1,z_1,t)\psi_n(x_2,y_2,z_2,t) = \psi_m(1)\psi_n(2)$, say, where x_1, y_1, z_1 and x_2, y_2, z_2 are the coordinates of the two electrons, $\psi(r)$ means $\psi(x_r,y_r,z_r,t)$. The eigenfunction $\psi_m(2)\psi_n(1)$, however, corresponds to the same state of the atom if we count the (mn) and (nm) states as identical. (…) If we are to have only one row and column in the matrices corresponding to both (mn) and (nm), we must find a set of eigenfunctions ψ_{mn} of the form: $\psi_{mn} = a_{mn}\psi_m(1)\psi_n(2) + b_{mn}\psi_m(2)\psi_n(1)$, where a_{mn}'s and b_{mn}'s are constants, which set must contain only one ψ_{mn} corresponding to both (mn) and (nm), and must be sufficient. (…) we may take $a_{mn} = -b_{mn}$ which makes ψ_{mn}= antisymmetrical (…) and only antisymmetrical eigenfunctions will be required for its expansion. Thus the symmetrical eigenfunctions alone or the antisymmetrical eigenfunctions alone give a complete solution to the problem. The theory at present is uncapable of deciding which solution is the correct one. (…) For r non-interacting electrons with coordinates $x_1,y_1,z_1,\ldots,x_r,y_r,z_r$, the symmetrical eigenfunctions are: $\Sigma\alpha_1,\ldots,\alpha_r\psi_{n1}(\alpha_1),\ldots,\psi_{nr}(\alpha_r)$, where $\alpha_1,\ldots,\alpha_r$ are any permutation of the integers 1, 2, …, r, while the unsymmetrical ones may be written in the determinantal form:

$$\begin{vmatrix} \psi_{n_1}(1) & \psi_{n_1}(2) & \cdots & \psi_{n_1}(r) \\ \psi_{n_2}(1) & \psi_{n_2}(2) & \cdots & \psi_{n_2}(r) \\ \vdots & \vdots & \cdots & \\ \psi_{n_r}(1) & \psi_{n_r}(2) & \cdots & \psi_{n_r}(r) \end{vmatrix}$$

(…) An antisymmetrical eigenfunction vanishes identically when two of the electrons are in the same orbit. This means that in the solution of the problem with antisymmetrical eigenfunctions there can be no stationary state with two or more electrons in the same orbit, which is just Pauli's exclusion principle. The solution with symmetrical eigenfunctions, on the other hand, allows any number of electrons to be in the same orbit, so that this solution cannot be the correct one for the problem of electrons in an atom" (Dirac 1926).

Dirac, along with Heisenberg, was the first to mention the determinantal form of the wave function for many-electron atoms, although it is popularly known as Slater's determinant who used it three years later (Slater 1929). Dirac also applied this result to ideal monoatomic gas molecules which are represented as a product of single molecules. If the assembly of molecules was symmetrized, it follows Bose-Einstein statistics. Otherwise, the assembly of molecules was antisymmetrized and it follows Fermi-Dirac statistics.

References cited

Boltzmann, L. 1877. Über die Beziehungzwischen dem zweitenHauptsatz der mechanischenWärmetheorie und der Wahrscheinlichkeitsrechnungrespektive den Sätzenüber das Wärmegleichgewicht. Sitzungsberichte der Kaiserlichen Akademie der Wissenschaften in Wien, Mathematisch-NaturwissenschaftlicheClasse LXXVI Abt. II, 76: 373–435.

Dirac, P.M.A. 1926. On the theory of quantum mechanics. Proc. R. Soc. Lond. A. 112: 661–667.

Fermi, E. 1926. Sullaquantizzazionedelgasperfetto monoatômico. RendicontidellaReale Accademia deiLincei 3: 145–149.

Slater, J. 1929. The theory of complex spectra. Phys. Rev. 34: 1293–1322.

Part Three

Schrödinger's Solutions to One and Two-electron Problems

One-particle Quantum Harmonic Oscillator

14

1. Classical harmonic oscillator

The classical harmonic oscillator is a physical system where a body of mass m is fixed to a spring whose other extreme is fixed to a wall or two bodies of masses m_1 and m_2 each one fixed at each extreme of the spring and having a one-dimensional displacement (Fig. 14.1).

In the simple classical harmonic oscillator, the restoring force depends on the spring constant, k, and the displacement x from the equilibrium point (where this force is zero).

$$\vec{F} = -k\vec{x}$$

The potential energy formula, V(x), of the simple classical harmonic oscillator is:

$$\frac{dV(x)}{dx} = -F(x)$$

$$V(x) = \int_{x=0}^{x=-x} (-kx)\,dx = \frac{k}{2}x^2$$

At the equilibrium position, x_0, $V(x_0)$ is zero and the kinetic energy is maximum. At the extremes of the oscillation, the potential energy is maximum while kinetic energy is zero. The total energy (potential energy plus kinetic energy) along the whole range of oscillation is constant, but each component describes a parabola along the x axis.

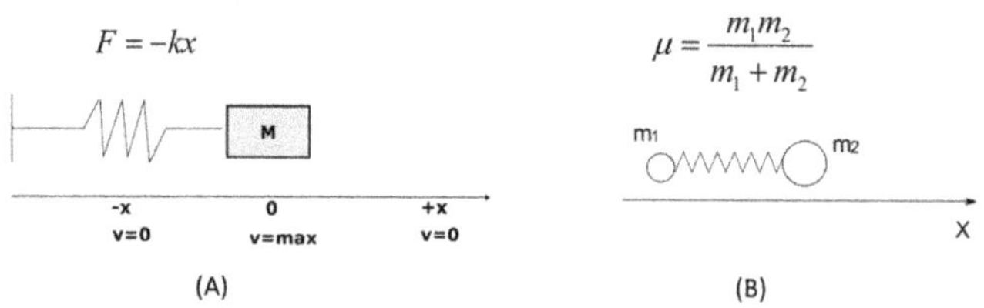

Fig. 14.1: Schematic representations of (a) one-body harmonic oscillator and two-body harmonic oscillator.

Since the force is also the product of the mass of the body and the acceleration, then we have:

$$\vec{F} = m\vec{a} = m\frac{d^2x}{dt^2}$$

$$\vec{F} = -k\vec{x}$$

$$m\frac{d^2x}{dt^2} = -kx$$

$$\frac{d^2x}{dt^2} = -\frac{k}{m}x$$

$$\frac{d^2x}{dt^2} + \frac{k}{m}x = 0$$

The last equation above is a second-order differential equation of the general type below (See Section 3 of the Chapter four):

$$\frac{d^2y}{dx^2} + a\frac{dy}{dx} + by = 0$$

Whose general solution is:

$$y = e^{\lambda x}$$

$$\lambda = \frac{1}{2}\left(-a \pm \sqrt{a^2 - 4b}\right)$$

Where the discriminant for the general solution (a^2–4b) is lower than zero :

$$a = 0, \quad b > 0, \quad a^2 - 4b < 0$$

Then, for negative discriminant, we have the following general solutions:

$$for: a^2 - 4b < 0$$

$$y = e^x\left(c_1e^{ix} + c_2e^{-ix}\right)$$

$$y = e^x\left(d_1\cos x + d_2\sin x\right)$$

In the specific case where the ratio k/m can be replaced by ω^2, we have the following general solution. Remember that in the case of the harmonic oscillator, y is replaced by x and x is replaced by t (see Section 3 of Chapter four):

$$\omega^2 = \frac{k}{m}$$

$$y \to x, \quad x \to t$$

$$x(t) = c_1e^{i\omega t} + c_2e^{-i\omega t}$$

$$x(t) = d_1\cos\omega t + d_2\sin\omega t$$

From the trigonometric relations below:

$$\sin(A+B) = \sin A\cos B + \cos A\sin B$$

$$a\sin\theta + b\sin\theta = R(\sin\theta + \alpha)$$

$$R = \sqrt{a^2+b^2}, \qquad \alpha = \arctan\frac{b}{a}$$

We have the last general solution can be rewritten as:

$$x(t) = R\sin(\omega\cdot t + \alpha)$$

$$\omega = 2\pi\nu$$

Where R is the amplitude (or the maximum displacement from the equilibrium position), ω is the angular frequency, ν is the vibrational frequency and α is the phase angle.

Let us establish an important relation involving the vibrational frequency from the important relation $\omega^2 = k/m$ to solve the corresponding differential equation and the relation between linear and angular frequencies.

$$As:$$

$$\omega^2 = \frac{k}{m}$$

$$\omega = 2\pi\nu$$

$$Then:$$

$$\omega^2 = 4\pi^2\nu^2$$

$$and:$$

$$\frac{k}{m} = 4\pi^2\nu^2$$

$$k = 4\pi^2\nu^2 m$$

$$V = \frac{1}{2}kx^2 = 2\pi^2\nu^2 mx^2 = \frac{1}{2}\omega^2 mx^2$$

The total energy is the sum of the kinetic energy and potential energy of the classical harmonic oscillator. This is the classical Hamiltonian, H.

$$H = T + U = \frac{p^2}{2m} + \frac{kx^2}{2} = \frac{p^2}{2m} + \frac{m\omega^2x^2}{2}$$

Where m is the mass of the body attached to the massless spring and p is the linear momentum.

2. One-particle quantum harmonic oscillator

In Chapter nine, we have shown the solution for the quantum harmonic oscillator using matrix mechanics. The one-particle harmonic oscillator is the Lorentz model for the hydrogen-like atom discussed in the chapters of the old quantum mechanics and matrix mechanics. Here, we show the solution using wave quantum mechanics.

In the one-particle quantum harmonic oscillator, we change the one body attached to spring into one particle under the influence of the parabolic potential energy in its displacement along the axis x.
The quantum Hamiltonian of harmonic oscillator is (see Chapter ten):

$$H=-\frac{\hbar^2}{2m}\frac{d^2}{dx^2}+\frac{1}{2}kx^2, \qquad p=-\hbar i\frac{d}{dx}$$

Then, the Schrödinger equation of the one-particle quantum harmonic oscillator is:

$$\left(-\frac{\hbar^2}{2m}\frac{d^2}{dx^2}+\frac{1}{2}kx^2\right)\psi(x)=E\psi(x)$$

or:

$$\left(-\frac{\hbar^2}{2m}\frac{d^2}{dx^2}+2\pi^2 m\nu^2 x^2\right)\psi(x)=E\psi(x)$$

The above equation has two dimensional physical quantities: energy E and length x. Let us undimensionalize the above equation. Table 14.1 shows the physical quantities and corresponding units (in SI) related to the one particle quantum harmonic oscillator.
The dimensionless energy, ε, is given by the expression:

$$\varepsilon=\frac{E}{\hbar 2\pi\nu}=\frac{E}{\hbar\omega}$$

$$\varepsilon=\frac{\left[kg\cdot m^2\cdot s^{-2}\right]}{\left[kg\cdot m^2\cdot(2\pi)^{-1}\cdot s^{-1}\right]\cdot\left[rad\cdot s^{-1}\right]}=\left[\frac{2\pi}{rad}\right]$$

The dimensionless length, y, is given by the expression:

$$y=\sqrt{\frac{2\pi\nu m}{\hbar}}x=\sqrt{\frac{\omega m}{\hbar}}x$$

$$y=\sqrt{\frac{\left[rad\cdot s^{-1}\right]\left[kg\right]}{\left[m^2\cdot kg\cdot(2\pi)^{-1}\, s^{-1}\right]}}\cdot[m]$$

$$y=\sqrt{\left[\frac{2\pi\cdot rad}{m^2}\right]}\cdot[m]=\left[\sqrt{2\pi\cdot rad}\right]$$

Then, we have to rearrange the Schrödinger equation in order to replace E into ε and x into y.
By multiplying the above equation by $-2\,m/\hbar^2$, we have:

$$\left(\frac{d^2}{dx^2}-\frac{4\pi^2 m^2\nu^2 x^2}{\hbar^2}\right)\psi(x)=-\frac{2m}{\hbar^2}E\psi(x)$$

Table 14.1: Physical quantities and corresponding units in SI.

Physical quantity	unit
Energy (E)	$kg.m^2.s^{-2}$
Angular frequency (w)	$rad.s^{-1}$
Length (x)	m
Mass (m)	kg
Planck's constant (h)	$kg.m^2.s^{-1}$

Let us rearrange the equation above:

$$\left(\frac{d^2}{dx^2}+\frac{2m}{\hbar^2}E-\frac{4\pi^2m^2v^2x^2}{\hbar^2}\right)\psi(x)=0$$

$$\left(\frac{d^2}{dx^2}+\left[\frac{2m}{\hbar^2}E-\left(\frac{2\pi mv}{\hbar}\right)^2x^2\right]\right)\psi(x)=0$$

Let us use now two dummy variables: α and β.

$$\beta=\frac{2m}{\hbar^2}E,\qquad \alpha=\frac{2\pi mv}{\hbar}$$

$$\left(\frac{d^2}{dx^2}+\left[\beta-\alpha^2x^2\right]\right)\psi(x)=0$$

Let us divide the above equation by α to obtain:

$$\left[\frac{1}{\alpha}\frac{d^2}{dx^2}+\left(\frac{\beta}{\alpha}-\alpha x^2\right)\right]\Psi(x)=0$$

The ratio β/2α is equivalent to the dimensionless energy:

$$\varepsilon=\tfrac{1}{2}\frac{\beta}{\alpha}=\frac{\dfrac{2mE}{\hbar^2}}{\dfrac{4\pi vm}{\hbar}}$$

$$\varepsilon=\frac{E}{2\pi v\hbar}$$

Then, we have:

$$\left[\frac{1}{\alpha}\frac{d^2}{dx^2}+\left(2\varepsilon-\alpha x^2\right)\right]\Psi(x)=0$$

Let us change the variable x into $y = \alpha^{1/2}x$ to obtain:

$$y=\sqrt{\alpha}\cdot x,\qquad y^2=\alpha\cdot x^2$$

then:

$$\left[\frac{1}{\alpha}\frac{d^2}{dx^2}+\left(2\varepsilon-y^2\right)\right]\Psi(x)=0$$

Let us complete the substitution of x into y in the whole equation:

$$\frac{d\psi(x)}{dx} = \frac{d\psi(y)}{dy}\frac{dy}{dx}$$

$$y = \sqrt{\alpha} \cdot x$$

$$\frac{d\psi(x)}{dx} = \sqrt{\alpha}\frac{d\psi(y)}{dy}$$

$$\left[\frac{d\psi(x)}{dx}\right]^2 = \frac{d^2\psi(x)}{dx^2} = \alpha\frac{d^2\psi(y)}{dy^2}$$

Then, we have:

$$\psi''(y) + \left(2\varepsilon - y^2\right)\psi(y) = 0$$

Let us now analyze the extreme positions of the above differential equation. For small values of energy E, ε is small and the term y^2 dominates. Then, we can approximate the equation above to:

$$\psi''(y) - y^2\psi(y) = 0$$

The same approximation above is obtained for the amplitude of the vibration, y, going to the infinite.

The last equation is a second order differential equation of the type:

$$\frac{d^2y}{dx^2} + a\frac{dy}{dx} + by = 0$$

Where:

$$a = 0, \quad b = -y^2$$

$$then: a^2 - 4b > 0$$

The general solution for the discriminant a^2–4b > 0 is (see Chapter four):

$$y = c_1e^{\lambda_1 x} + c_2e^{\lambda_2 x}$$

Then, for:

$$\psi''(y) - y^2\psi(y) = 0$$

We have the general solution (see Chapter four):

$$\psi(y) = e^{ty}$$

Where t is similar to λ in Chapter four. Then, we have:

$$\frac{d^2\left(e^{ty}\right)}{dy^2} = t^2e^{ty}$$

Replace the relations above into the differential equation:

$$\psi''(y) - y^2\psi(y) = 0$$

$$t^2 e^{ty} - y^2 e^{ty} = 0$$

$$t^2 - y^2 = 0$$

$$(t-y)(t+y) = 0$$

$$t = y, \qquad t = -y$$

$$then:$$

$$\psi(y) = Ae^{y^2} + Be^{-y^2}$$

$$A = c_1, \qquad B = c_2$$

The coefficient A must be zero because ψ(y) must remain finite for all values of y in order to allow the normalization. Then, the general solution is:

$$\psi(y) = Be^{-y^2}$$

The coefficient B must guarantee that the wave function is zero when y goes to infinity. Then, B must be a function (g(y)) instead of a constant. However, we can anticipate that we won't find g(y) with this general solution. Then, we have to change slightly the general solution to:

$$\psi(y) = g(y)e^{-\frac{y^2}{2}}$$

In order to find g(y), let us differentiate the equation above:

$$\psi'(y) = g'(y)e^{-\frac{y^2}{2}} - yg(y)e^{-\frac{y^2}{2}}$$

Let us differentiate the above equation:

$$\psi''(y) = g''(y)e^{-\frac{y^2}{2}} - yg'(y)e^{-\frac{y^2}{2}} - g(y)e^{-\frac{y^2}{2}}$$

$$-yg'(y)e^{-\frac{y^2}{2}} + y^2 g(y)e^{-\frac{y^2}{2}}$$

$$\psi''(y) = g''(y)e^{-\frac{y^2}{2}} - 2yg'(y)e^{-\frac{y^2}{2}} - g(y)e^{-\frac{y^2}{2}} + y^2 g(y)e^{-\frac{y^2}{2}}$$

$$\psi''(y) = \left[g''(y) - 2yg'(y) - g(y) + y^2 g(y)\right]e^{-\frac{y^2}{2}}$$

$$\psi''(y) = \left[g''(y) - 2yg'(y) + (y^2 - 1)g(y)\right]e^{-\frac{y^2}{2}}$$

Let us substitute the above ψ(y) and ψ''(y) equations into:

$$\psi''(y) + \left(2\varepsilon - y^2\right)\psi(y) = 0$$

Then, we get:

$$\left[g''(y)-2yg'(y)+(y^2-1)g(y)\right]e^{-\frac{y^2}{2}}+$$
$$+\left(2\varepsilon-y^2\right)\left[g(y)e^{-\frac{y^2}{2}}\right]=0$$

$$\left[g''(y)-2yg'(y)+(2\varepsilon-1)g(y)\right]e^{-\frac{y^2}{2}}=0$$

The general Hermite equation (see Chapter four) is:

$$y''-2xy'+2my=0$$

We can see that:

$$g''(y)-2yg'(y)+(2\varepsilon-1)g(y)=0$$

is a Hermite equation, H(y), where m = 2ε–1. Then, the solution is:

$$\psi(y)=H(y)e^{-\frac{y^2}{2}}$$
$$H(y)=g''(y)-2yg'(y)+(2\varepsilon-1)g(y)=0$$

The normalized wave function becomes:

$$\psi(y)=N_nH(y)e^{-\frac{y^2}{2}}$$
$$N_n=\frac{1}{2^n n!}\sqrt{\frac{\alpha}{\pi}}$$
$$y^2=\alpha x^2$$

The solutions for the Hermite differential equations are:

$$n=0,\quad H(y)=1$$
$$n=1,\quad H(y)=2y$$
$$n=2,\quad H(y)=4y^2-2$$
$$n=3,\quad H(y)=8y^3-12y$$
$$n=4,\quad H(y)=16y^4-48y^2+12$$

Let us now find the expression for the dimensionless energy ε by using power series method (Chapter two). Then, we go back to the Hermitian differential equation below to find its solution for the power series method.

$$\psi(y)=H(y)e^{-\frac{y^2}{2}}$$
$$H(y)=g''(y)-2yg'(y)+(2\varepsilon-1)g(y)=0$$

Then, H(y) can be represented as:

$$H(y)=\sum_{n=0}^{\infty}a_ny^n,\qquad n=0,1,2,3,...$$

Where n is an integer number. Let us find its first derivatives:

$$H'(y) = \sum_{n=1}^{\infty} n a_n y^{n-1}$$

$$H''(y) = \sum_{n=2}^{\infty} n(n-1) a_n y^{n-2}$$

Let us do the power equalization for the expressions above. For the summation of the first derivative, H'(y), if n starts at 0 or at 1 does not change the result since for n = 0 the second the corresponding term is zero. Then:

$$\sum_{n=1}^{\infty} n a_n y^{n-1} = \sum_{n=0}^{\infty} n a_n y^n$$

For the second derivative, H''(y), let us change n to start at zero.

$$for: n(n-1)$$
$$n = 2 \rightarrow 2(2-1) = 2$$
$$n = 3 \rightarrow 3(3-1) = 6$$
$$for: (n+2)(n+1)$$
$$n = 0 \rightarrow (0+2)(0+1) = 2$$
$$n = 1 \rightarrow (1+2)(1+1) = 6$$
$$then: n(n-1)\big|_{n=2} = (n+2)(n+1)\big|_{n=0}$$

$$finally: \sum_{n=2}^{\infty} n(n-1) = \sum_{n=0}^{\infty} (n+2)(n+1)$$

Hence:

$$H''(y) = \sum_{n=0}^{\infty} (n+2)(n+1) a_{n+2} y^n$$

Let us substitute the derivatives above into the Hermite equation below:

$$H(y) = g''(y) - 2yg'(y) + (2\varepsilon - 1) g(y) = 0$$

$$\sum_{n=0}^{\infty} (n+2)(n+1) a_{n+2} y^n - 2\sum_{n=0}^{\infty} n a_n y^n + (2\varepsilon - 1) \sum_{n=0}^{\infty} a_n y^n = 0$$

$$\sum_{n=0}^{\infty} \left[(n+2)(n+1) a_{n+2} - 2n a_n + (2\varepsilon - 1) a_n \right] y^n = 0$$

Since y can be any value, then:

$$(n+2)(n+1) a_{n+2} - 2n a_n + (2\varepsilon - 1) a_n = 0$$

As a consequence, the recursion relation is:

$$a_{n+2} = \frac{(2n+1-2\varepsilon)}{(n+1)(n+2)} a_n$$

Then, we have the following H(y) equation:

$$H(y) = a_0 + a_1 y + a_2 y^2 + a_3 y^3 + a_4 y^4 + a_5 y^5 + \dots$$

If a_0 and a_1 are known, the whole equation H(y) is solved by the use of the above recursion relation of the power series method.

From this point on, we need to concern whether the previous powers terminates/ converges or not. In last case, its sum goes to the infinity. Then, let us assume that this series does not terminate and check whether it will provide a reasonable wave function or not. Assuming the power series going to the infinity, i.e., for extremely large values of n, the recursion relation becomes:

$$hypothesis: sum \to \infty$$

$$i.e.: n \to \infty$$

$$then: a_{n+2} \approx \frac{2n}{n^2} a_n = \frac{2}{n} a_n$$

The ratio between two successive even (or odd) terms in this series (for large n) is similar to that from the power series of the function $\exp(y^2)$.

$$\frac{a_{n+2} y^{n+2}}{a_n y^n} = \frac{(2/n) a_n}{a_n} y^2 = \frac{2}{n} y^2$$

$$\exp(y^2) = \sum_{n=0}^{\infty} \frac{y^{2n}}{n!} = \sum_{n:even}^{\infty} \frac{y^n}{(n/2)!}$$

$$\frac{y^{n+2} / ((n+2)/2)!}{y^n / (n/2)!} = \frac{2}{n+2} y^2$$

Then, we can assume that in the hypothesis that the power series of the H(y) does not terminate (goes to the infinite), it approaches the function $\exp(y^2)$.

$$hypothesis: sum \to \infty$$

$$i.e.: n \to \infty$$

$$H(y) = \exp(y^2)$$

In this case, the overall wave function will be:

$$hypothesis: sum \to \infty$$

$$i.e.: n \to \infty$$

$$\psi(y) = e^{y^2} \cdot e^{-\frac{y^2}{2}} = e^{\frac{y^2}{2}}$$

This is a serious problem because the wave function will diverge and it cannot be normalized. Than, the hypothesis of the power series for H(y) does not terminate cannot be considered as a reasonable solution.

Therefore, we have to assure that the power series for H(y) must terminate. Then, we can arrange that the term a_{n+2}, after a finite number of n, will be zero, the next terms will also be zero, which means that the series will terminate. Then, for the series of H(y) terminate we need:

$$a_{n+2} = \frac{(2n+1-2\varepsilon)}{(n+1)(n+2)} a_n = 0$$

$$then: 2n+1-2\varepsilon = 0$$

$$\varepsilon = \frac{2n+1}{2}$$

Since the relation between ε and E is:

$$\varepsilon = \frac{E}{\hbar\omega}, \qquad E = \varepsilon\hbar\omega$$

We have the solution of the energy of the one-particle quantum harmonic oscillator:

$$\varepsilon = \frac{2n+1}{2}$$

$$E = \left(\frac{2n+1}{2}\right)\omega\hbar = \left(n+\frac{1}{2}\right)\omega\hbar$$

$$n = 0,1,2,3,\ldots$$

Figure 14.2 depicts the wave function of the one-particle quantum harmonic oscillator for n = 0,1 and 2. One can see that n also corresponds to the number of nodes of the wave function.

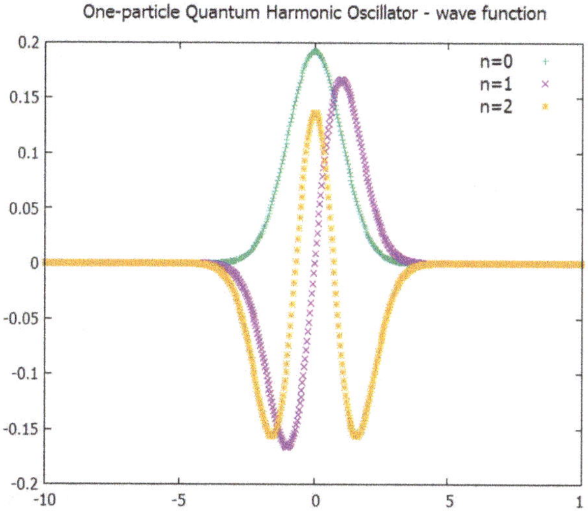

Fig. 14.2: One-particle quantum harmonic oscillator for n = 0,1 and 2.

3. Numerical analysis and the Fortran code for the one-particle quantum harmonic oscillator

In the problem of the quantum harmonic oscillator of one particle of mass m, the single particle is described by a wave function or eigenfunction in x axis, ψ(x). It is moving in a one-dimensional potential along the x direction, so-called V(x), as a quadratic function in the range of two limits (or the walls of the system): x_{min} and x_{max}. The corresponding time-independent Schrödinger equation and boundary conditions are:

$$\frac{d^2\psi(x)}{dx^2} + K^2(x)\psi(x) = 0, \qquad \psi(x_{min}) = \psi(x_{max}) = 0$$

$$K^2(x) = \frac{2m}{h^2}\left[E - V(x)\right]$$

$$-\frac{\hbar}{2m}\frac{d^2\psi(x)}{dx^2} + \frac{1}{2}m\omega^2 x^2\psi(x) = E\psi(x)$$

Where E is the energy eigenvalues, h is the Planck's constant, $\hbar = h/2\pi$, ω the angular frequency and ν is the frequency.
Let us firstly undimensionalize the above equation.

The dimensional solutions, along with the dimensional potential energy, are given below:

$$E_n = \left(n + \frac{1}{2}\right)\hbar\omega = \left(n + \frac{1}{2}\right)h\nu$$

$$V = \frac{1}{2}kx^2, \qquad k = m\omega^2$$

Where k is the force constant and ν is the frequency.

Do not confuse the h in the equation above, the Planck's constant, with the h in the Numerov's algorithm. Since the energy has $\hbar\omega$ dimension, then $\varepsilon = E/\hbar\omega$ is the dimensionless energy and since $m\omega/\hbar$ has the dimension of $[\text{length}]^{-2}$, then $u = (m\omega/\hbar)^{1/2}$ is the dimensionless length. Then, the dimensionless quantum harmonic oscillator is:

$$-\frac{d^2\psi}{du^2} + u^2\psi(x) = \varepsilon\psi(x)$$

$$u = \sqrt{\frac{m\omega}{\hbar}}x, \qquad u^2 = \frac{m\omega}{\hbar}x^2, \qquad \frac{\hbar}{m\omega}\frac{d^2}{dx^2} = \frac{d^2}{du^2}$$

$$\varepsilon = \frac{2E}{\hbar\omega}$$

The quantum harmonic oscillator has two natural scales, i.e., two dimensionless physical quantities: length (position) and energy. In the dimensionless quantum harmonic oscillator, the Hamiltonian and the energy solutions are given below. These are the equations for the potential energy and energy used in the code.

$$V = 0.5x^2, \quad E_n = n + \frac{1}{2}$$

Having the dimensionless quantum harmonic oscillator, we need to search for numerical non-zero solutions (non-zero E_n) for this problem. In the code, the energy E_n is adjusted according to the midpoint value of the potential energy range (Vupper and Vlower) where the value of the desired E_n is located. The wave function is integrated point by point according to the grid (number of points for the integration along the axis x). After the integration finishes in the whole range of x, if the number of nodes, n, (number of changes of sign of the function) is larger than the predetermined number of nodes (given by the user), E is adjusted to a new value of a lower half interval of energy. If the number of changes of the sign of the function is smaller than the predetermined n, then E is adjusted to the upper half interval of energy. This process cycles until it reaches the threshold where the convergence is reached.

The code takes into account the symmetry of the potential energy for constructing the wave function.

As a quantum system, the wave function has classical allowed regions and classical forbidden regions (where there is a tunneling effect). The classical allowed regions are within the limits of the wall, i.e., the lower (x_{cl1}) and upper (x_{cl2}) classical limits are:

$$\left.\begin{array}{l} x_{cl1} > x_{\min}, \quad x_{cl2} < x_{\max} \\ x_{cl1} \leq x_{classical} \leq x_{cl2} \end{array}\right\} \text{classical region}$$

There are two classical forbidden regions which are two extreme regions in the system. Each of either forbidden region is located between the classical limit (x_{cl}) and the limit of the wall (x_{min} or x_{mas}).

$$\left.\begin{array}{l} x_{\min} \leq x_{forbidden1} < x_{cl1} \\ x_{cl2} < x_{forbbiden2} \leq x_{\max} \end{array}\right\} \textbf{forbidden region}$$

As already mentioned in a previous section, when k > 0, the function has classical oscillations and when $K^2 < 0$ it decays or grows exponentially. Then, when E > V(x), the eigenfunction $\psi(x)$ has oscillatory behavior in the classically allowed regions. When E < V(x), the wave function behaves exponentially in the classical forbidden regions of the system. Since $V(x) \leq 0$ (and it is multiplied by –1), if the eigenvalue E < 0 then $K^2 < 0$ and the wave function has a exponential behavior towards the walls of the system. If E > 0, then $K^2 > 0$ and the wave function has a classical oscillatory behavior.

Important to add that at both $x = x_{cl1}$ and $x = x_{cl2}$, the wave function has a maximum where each derivative at each point is zero.

$$\left.\frac{d\psi(x)}{dx}\right|_{x_{cl1}} = \left.\frac{d\psi(x)}{dx}\right|_{x_{cl2}} = 0$$

There are some numerical methods to solve this equation. One of them is the shooting method where we search for trial eigenvalues. Let us begin with one negative trial eigenvalue (of a exponential behavior) in which the integration in the forbidden region (let be from x_{cl1} to x_{min}) gives:

$$\int_{x_{cl1}}^{x_{\min}} (QHDE)\, dx = \psi_{<}$$

Where QHDE is the quantum harmonic differential equation.

If we increase the limit of this integration before x_{cl1} towards the upper limit x_{max}, it would yield an inaccurate result (an exponentially growing solution). Then, at each eigenvalue, it is recommended to generate a second wave function:

$$\int_{x_{max}}^{x_{cl1}} (QHDE)\,dx = \psi_>$$

Both wave functions, $\psi_<$ and $\psi_>$, satisfy a homogeneous equation (homogeneous because S(x) = 0). Henceforth, one of the classical limits, x_{cl1}, will be the matching point, x_m so that

$$x_m = x_{cl1}$$
$$\psi_<(x_m) = \psi_>(x_m)$$

Then, a normalization is done to that $\psi_<(x_m) = \psi_>(x_m)$.

By using the finite difference approximation method (Chapter two) from the derivatives below:

$$\left.\frac{d\psi_<}{dx}\right|_{x_m} = \left.\frac{d\psi_>}{dx}\right|_{x_m} = 0$$

$$\left.\frac{d\psi_<}{dx}\right|_{x_m} - \left.\frac{d\psi_>}{dx}\right|_{x_m} = 0$$

We have:

$$\frac{\psi_<(x_m - h) - \psi_<(x_m)}{h} - \frac{\psi_>(x_m - h) - \psi_>(x_m)}{h} = 0$$

$$\psi_<(x_m - h) - \psi_>(x_m - h) = 0$$

Then, one has to find the root of the function *f*. The secant or bisection methods can be used. In the code below, it is used the bisection method whose algorithm is very similar to that from BISECTION3 code (see Chapter two). To find *f* for a given energy, the Numerov method (Chapter two) is used to integrate leftward and rightward and the solutions are matched at leftmost turning point, $x_m = x_{cl1}$. The search stops when *f* goes near zero.

Since h is a real number and each ith wave function is located at ith integral step, it is not possible to use in the code the following data: $\psi_<(x_m - h) \therefore \psi_>(x_m - h)$

In the second iteration process, the two parts of the wave function will have a discontinuity at the matching point. For good solution, this discontinuity must be zero.

In the code below, it calculated the discontinuity between $\psi_<(x_m)$ and $\psi_>(x_m)$, given by the variable DJUMP. When DJUMP*Y(x_m) is zero, the iteration process stops.

$$djump = \frac{\left[y(x_{cl1}+1) + y(x_{cl1}-1) - \left(14 - 12 f_n(x_{cl1})\right)\right] y(x_{cl1})}{h}$$

The boundary conditions are $x_{min} = 0$ (y = 0) and $x_{max} = 10$, where y is the wave function. The grid (the number of points in the x axis) is set to 300. Since h = x_{max}/grid, then h = 10/300. Each point in x direction is given by: x(i) = i*h, where i = 0 to 300. Each corresponding potential (of a parabolic potential well).

The objective of this code is to obtain the y(i), that is, the wave function at the point i corresponding to x(i) using the Numerov equation for the wave function (Chapter two).

The code makes the function f_n of the Numerov's algorithm explicit. The first equation below is the Numerov equation for the wave function. The second equation is the term f_n of the first equation. Be aware to not confuse f and f_n. The explicit function f_n is also highlighted in the code's commentaries.

$$\psi_{n+1} = \frac{(12 - 10 f_n)\psi_n - f_{n-1}\psi_{n-1}}{f_{n+1}}$$

$$f_n = 1 + K_n^2 \frac{h^2}{12}, \qquad K^2(x) = \frac{2m}{\hbar^2}(E - V(x))$$

The sign of f_n(i) and Y(i) plays a major role in the adjustment of the wave function since the number of crossings (i.e., how many times the sign of the wave function changes) has to be equal to the number of nodes. When this is not the case, the energy E is adjusted in order to yield a different value of f_n(i) and Y(i). This iteration process continues until the equality ncross = n is found. This is the first iteration process in the recursive cycle of the code. When ncross = n, the recursive cycle goes to the second iteration process to find DJUMP*Y(x_m) = 0 (see above). When the second equality is met, the recursive cycle stops.

Important to add that the energy E is dependent on the number of nodes, n, while the potential energy V is dependent only on the x variable. We show this on the results of the code printing the values of V and E. We see that there is an initial value of E, which is 25 and it decays to the final value according to the iteration process (using ITER variable). The dimensionless analytical solutions of the energy levels, E_n, for the quantum harmonic oscillator are 0.5 (for n = 0), 1.5 (for n = 1), 2.5 (for n = 2), 3.5 (for n = 3), 4.5 (for n = 4), 5.5 (for n = 5), and so on.

At two distinguished steps of the code, there is a section to determine whether the number of nodes, n, is even or odd. This is because an even number of the node corresponds to an even wave function (sum of phases is even) while an odd number of the node corresponds to an odd wave function (sum of phases is odd). There is a routine for determining whether a number is even or odd in Fortran and it is shown in Chapter One (name of the program: EVEN-ODD). This is the same routine used here.

The recursive process of Numerov's algorithm goes from 'x_{min}' to the match point, x_m, and backwards from 'x_{max}' to this match point. In another words, two integrations are performed: the forward recursion (from x_{min} to x_m) and the backward recursion (from x_{max} to x_m).

The forward integration is performed until the grid point as close as possible to the match point, yielding the wave function, $\psi_<$. Then, the backward integration goes to the grid point as close as possible to the match point yielding the wave function, $\psi_>$. In general, the two parts of the wave function $\psi_<$ and $\psi_>$ have different values of the match point. Then, the wave function $\psi_<$ is rescaled in a iteration process so that both functions give the same value at match point. All the wave function points, $Y(x_i)$, from the icl point (the classical turning point) to the last point (x_{grid}) are rescaled.

$$y(icl) = y(icl)\,\frac{y(icl_{old})}{y(icl_{new})}$$

Henceforth, the wave function is normalized (See the NORMALIZATION code).

When the code is run, it requires the number of nodes, n (n = 0, 1, 2,3…) and the trial energy (which should be given zero value). The output file contain two columns x, y(x) – actually the wave function and y(x)^2. The plot can is x vs y(x). One can use the multiplatform, free GNUPLOT (Willians et al. 2018) to plot the results from the code.

The results of this code (the wave function for n = 0, 1 and 2) are depicted in Fig. 15.1.

The algorithm of the code can be summed up as: *integration process to find values of wave function y at each point of the grid from two different regions giving the inward and outward integration until the point that divides both regions called matching point (icl). The guess to start integration is the value of y at both extremes (shooting method) is given according to the particularities of the wave function. The energy E is adjusted in the recursive process which concomitantly produces the icl.* This is the same algorithm used in the code of the Chapter seventeen.

```
!-------------------------------------------------------------
!Program name: ONE_PART_QHO
!Original code: Giannozzi,P., Ercolessi, F. and Gironcoli, S.
!Modified by Caio Lima Firme
!One particle quantum harmonic oscillator
!PARABOLIC POTENTIAL WELL
!Objective: to plot the wave function for each node
!-------------------------------------------------------------
!The dimensionless physical quantities:
!Potential energy: V(x)=0.5X^2
! Energy: E=1/2+n
!-------------------------------------------------------------
! Use of Numerov method to find the wave function
! Function fn from Numerov equation is
! Fn = 1 + K2*h^2/12 , where K2=2m/((H/2pi)^2)*(E-V)
! H=Planck constant
!-------------------------------------------------------------
  integer :: grid, i, icl
  integer :: n, hn, ncross, j, iter
double precision :: xmax, h, fh12, norm, djump, fac
double precision :: Vupper, Vlower, e, k
  double precision :: x(0:300), y(0:300), V(0:300), fn(0:300)
  character (len=20) :: fileout
  character (len=1) :: A
!-------------------------------------------------------
```

```
! Boundary conditions:
  y(0) = 0.0d0
  x(0) = 0.0d0
  xmax=10
  grid=300
! The number 300 refers to the grid everywhere in the code
  h = xmax/grid
! Make explicit one term of the fn function from the Numerov method: fh12
fh12=h*h/12.0d0
!-----------------------------------------------------
!Input data file
print *, 'output file name is (include .dat at the end): '
  read *, fileout
!-----------------------------------------------------------
! Record in the computer memory each value of x(i) and V(i)
  do i = 0, grid
     x(i) = i * h
     V(i) = 0.5d0 * x(i)*x(i)
     write (*,*) i, 'X(i)=', x(i),'V(i)=', V(i)
  end do
!-------------------------------------------------------------
! Beginning of searching process
10 search: do
     !Input the number of nodes, n
     20 print *, 'give the number of nodes, n (n=0,1,2,3,): '
        read *, n
        if (n < 0) then
        go to 20
        end if
!-------------------------------------------------------------
     ! set initial lower and upper bounds of the potential V(x)
        ! Set trial energy E
           Vupper=maxval (V(0:300))
           Vlower=minval (V(0:300))
           e = 0.5d0 * (Vlower + Vupper)
           iter = 100
     Write (*,*) "Vupper=", Vupper, "e(initial)=", e
!---------------------------------------------------------------
! BEGINNING OF THE RECURSIVE PROCESS (STEP=j)
  iterate: do j = 1, iter
! set up the fn-function used by the Numerov method
! fn=1+[2(V-E)]h^2/12 without m (mass) and H/2pi (Planck constant)
```

```
! determine the position of its last crossing, i.e. change of sign
              ! fn < 0 means classically allowed region
              ! fn > 0 means classically forbidden region
      fn(0)=fh12*(2.0d0*(V(0)-e))
      icl=-1
      do i=1,grid
                fn(i)=fh12*2.0d0*(V(i)-e)
      if ( fn(i) == 0.0d0) fn(i)=1.d-15
! store the index 'icl' where the last change of sign has been found
                if ( fn(i) /= sign(fn(i),fn(i-1)) ) icl=i
              end do
      ! fn as required by the Numerov method
              fn = 1.0d0 - fn
 ! determination of the wave-function in the first two points
!-----------------------------------------------------------------
! Routine to determine whether n (number of nodes) is even or odd
      ! Check the program EVEN-ODD
         hn = n/2
      ! if n is even, the wave function is even
         ! if n is odd, the wave function is odd
         if (2*hn == n) then
      ! even number of n: wavefunction is even
              y(0) = 1.0d0
      ! assume f(-1) = f(1)
              y(1) = 0.5d0*(12.0d0-10.0d0*fn(0))*y(0)/fn(1)
         else
      ! odd number of n: wavefunction is odd
         y(0) = 0.0d0
         y(1) = h
      end if
!-----------------------------------------------
      ! outward integration and count number of crossings
      ncross=0
      do i =1,icl-1
y(i+1)=((12.0d0-10.0d0*fn(i))*y(i)-fn(i-1)*y(i-1))/fn(i+1)
      if ( y(i) /= sign(y(i),y(i+1)) ) ncross=ncross+1
      end do
      fac = y(icl)
!-----------------------------------------------
! Routine to determine whether n (number of nodes) is even or odd
      ! Check the program EVEN-ODD
         if (2*hn == n) then
      ! even number of n: no node in x=0
              ncross = 2*ncross
```

```
        else
     ! odd number of n: node in x=0
              ncross = 2*ncross+1
        end if
!------------------------------------------
        ! check number of crossings
        ! FIRST ITERATION PROCESS
!THE NUMBER OF CROSSINGS HAS TO BE EQUAL TO THE NUMBER
OF NODES
!Bisection method (see BISECTION3 code in chapter two)
        if ( iter > 1 ) then
              if (ncross /= n) then
     ! Incorrect number of crossings: adjust energy
if ( j == 1) & print '("step Energy n Discontinuity Vupper Vlower icl")'
     write (*,5) j, e, ncross, Vupper, Vlower, icl
     5 format (i5,f25.15,i5,f35.15,f35.15,i40)
              if (ncross > n) then
     ! Too many crossings: current energy is too high,
        ! lower the upper bound
                   Vupper = e
              else
     ! Too few crossings: current energy is too low,
     ! raise the lower bound
                   Vlower = e
              end if
     ! New trial value:
              e = 0.5d0 * (Vupper+Vlower)
! go to beginning of do loop, don't perform inward integration
              cycle
          end if
       end if
     !The equality ncross = n was reached
     !END OF THE FIRST ITERATION PROCESS
!------------------------------------------------------------------
! Correct number of crossings: proceed to inward integration
! Determination of the wave-function in the last two points
     ! assuming y(grid+1) = 0
     y(grid) = h
  y(grid-1) = (12.0d0-10.0d0*fn(grid))*y(grid)/fn(grid-1)
     do i = grid-1,icl+1,-1
  y(i-1)=((12.0d0-10.0d0*fn(i))*y(i)-fn(i+1)*y(i+1))/fn(i-1)
     end do
! ----------------------------------------------------------------
```

```
! rescale function to match at the classical turning point (icl)
      fac = fac/y(icl)
      Write (*,*) "fac=", fac
      y(icl:) = y(icl:)*fac
!---------------------------------------------------------------------
      !Normalization of the wave function
      !DOT_PRODUCT : scalar product of vectors
      !See the code: NORMALIZATION
      norm = dot_product (y, y)
      y = y / sqrt(norm)
!------------------------------------------------------------------
 ! SECOND ITERATION PROCESS
 ! Bisection method (See BISECTION3 code in chapter one)
      if ( iter > 1 ) then
     ! calculate the discontinuity in the first derivative
        ! y'(i;RIGHT) - y'(i;LEFT)
djump = ( y(icl+1) + y(icl-1) - (14.0d0-12.0d0*fn(icl))*y(icl) ) / h
      write (*,30) j, e, n, djump, Vupper, Vlower,icl
      30 format (i5,f25.15,i5,f14.8,f22.15,f34.15,i40)
              if (djump*y(icl) > 0.0d0) then
     ! Energy is too high --> choose lower energy range
              Vupper = e
          else
     ! Energy is too low --> choose upper energy range
              Vlower = e
          end if
          e = 0.5d0 * (Vupper+Vlower)
! -------------------- convergence test ----------------------------
          if ( Vupper-Vlower < 1.d-10) exit iterate
        end if
     ! END OF THE SECOND ITERATION PROCESS
      end do iterate
     ! END OF THE RECURSIVE PROCESS
!-------------Convergence achieved----------------------
print *, 'Do you want to do another calc (Y/N)?'
      read *, A
      Select case (A)
         Case ('Y','y')
             go to 10
         case default
             continue
         end select
!--------------------------------------------------------------
```

```
!Printing results x(i) and y(i) in the 'fileout'
! x<0 region:
open(7,file=fileout,status='replace')
do i=grid,1,-1
      write (7,'(f7.3,3e16.8,f12.6)') -x(i), ((-1)**n)*y(i)
  end do
  ! x>0 region:
  do i=0,grid
write (7,'(f7.3,3e16.8,f12.6)') x(i), y(i)
enddo
  write (7,'(/)')
  close(7)
  stop
!------------------------------------------------------------------------
end do search
 stop
end
```

Exercises

(1) Give the approximated Hamiltonian and the wave function of the two-particle harmonic oscillator of reduced mass, μ, within the parabolic potential energy, that is, the system vibrating in the bound region. The initial value of x (at the equilibrium) is the equilibrium interatomic distance of the molecule and it vibrates symmetrically to left and right.

(2) Change the code above in order to describe the vibration of the diatomic molecule beyond the limits of the parabolic potential energy, i.e., assuming the Morse potential (Morse 1929, Taseli 1998).

$$E_n = h\nu\left(n+\tfrac{1}{2}\right) - \frac{(h\nu)^2}{4D_e}\left(n+\tfrac{1}{2}\right)^2$$

$$V(x) = D_e\left[e^{-2ax} - 2e^{-ax} + 1\right], \qquad x = r - r_e$$

$$a = \sqrt{k_e/2D_e}$$

Where D_e is the dissociation energy, r_e is the equilibrium interatomic distance and r is the distance between two atoms. The parameter a controls the width of the potential well. This potential breaks the inversion symmetry of the potential associated to one-particle quantum harmonic oscillator.

(3) By knowing that probability density is:

$$\rho(x) = \sum_{n=1}^{grid} [\psi(x)]^2$$

Include in the code of the harmonic oscillator the "DO loop" to calculate and plot the probability density of the wave function.

References cited

Giannozzi, P., Ercolessi, F. and Gironcoli, S. 2013. Lecture notes: Numerical methods in quantum mechanics, Udine and Trieste, Italy (http://www.fisica.uniud.it/~giannozz/Didattica/MQ/mq.html).

Morse, P.M. 1929. Diatomic molecules according to the wave mechanics. II. Vibrational levels. Phys. Rev. 34: 57–64.

Taseli, H. 1998. Exact solutions for vibrational levels of the Morse potential. J. Phys. A: Math. Gen. 31: 779–788.

Williams, T., Kelley, C., Merritt, E.A., Bersch, C., Bröker, H.-B., Campbell, J., Cunningham, R., Denholm, D., Elber, G., Fearick, R., Grammes, C., Hart, L., Hecking, L., Juhász, P., Koenig, T., Kotz, D., Kubaitis, E., Lang, R., Lecomte, T., Lehmann, A., Lodewyck, J, Mai, A., Märkisch, B., Mikulik, P., Sebald, D., Steger, C., Takeno, S., Tkacik, T., der Woude, J.V., Zandt, J.R.V., Woo, Alex and Zellner, J. 2018. Gnuplot 5.2—an interactive plotting program.

Particle in a Box

15

1. Particle in unidimensional box

The particle in a one-dimensional box is the system where one particle is allowed to move freely along the x direction from x > 0 to x < l (l is abbreviation for length) where the potential energy, V(x), is zero in order to ensure that no external force acts on this particle in the box. In addition, the potential energy is infinite in the walls (x = 0 and x = l) and beyond the limits of the walls (x < 0 and x > l) to ensure that the particle cannot leave the box.
The Schrödinger equation for this system is:

$$\left[-\frac{\hbar^2}{2m}\frac{d^2}{dx^2}+V(x)\right]\psi(x)=E\psi(x)$$

The boundary conditions are:

$$x\begin{cases}=0, & \text{lower limit}\\ =l, & \text{upper limit}\end{cases}$$

$$\psi(x)=0\begin{cases}x=0\\ x=l\end{cases}$$

$$V(x)\begin{cases}=\infty\Rightarrow x\leq 0 \therefore x\geq l\\ =0\Rightarrow \quad 0<x<l\end{cases}$$

Then, the Schrödinger equation becomes:

$$-\frac{\hbar^2}{2m}\frac{d^2}{dx^2}\psi(x)=E\psi(x)$$

$$\frac{d^2}{dx^2}\psi(x)=\frac{2mE}{\hbar^2}\psi(x)$$

The particle in the box has similar equation for the one-particle quantum harmonic oscillator except for the potential energy (see Chapter fourteen).

When setting:

$$\omega^2 = \frac{2mE}{\hbar^2}$$

We have the equation:

$$\frac{d^2\psi(x)}{dx^2} + \omega^2\psi(x) = 0$$

or

$$y'' + \omega^2 = 0$$

Then, the general solution for this elliptic second-order differential equation (as discussed in Chapter four) is:

$$\psi(x) = d_1 \cos \omega x + d_2 \sin \omega x$$

By applying the periodic boundary conditions at x = 0, we have:

$$x = 0, \quad \psi(0) = 0$$

$$\psi(0) = d_1 \cos 0 + d_2 \sin 0 = 0 \therefore d_1 = 0$$

$$x = l, \quad \psi(l) = 0$$

$$\psi(i) = d_2 \sin(\omega \cdot l) = 0$$

The function sine(x) is only zero at multiples of π, i.e., nπ (n = 0, ±1, ±2, ±3,…). Then:

$$n \cdot \pi = \omega \cdot l, \qquad \omega = \frac{n \cdot \pi}{l}, n = 0, \pm 1, \pm 2, \pm 3, ...$$

and

$$\psi_n(x) = d_2 \sin\left(\frac{n \cdot \pi}{l} x\right), n = 0, \pm 1, \pm 2, \pm 3, ...$$

The coefficient d_2 is the normalization constant. We need to find this coefficient to obtain the normalized function.

The total probability to find the wave function between 0 and l in x coordinate is a unit (i.e., it has 100% of likelihood). Then:

$$\int_0^l \psi^2(x)dx = 1, \quad \int_0^l \left[d_2 \sin^2\left(\frac{n\pi x}{l}\right)\right] dx = 1,$$

$$d_2^2 \int_0^l \left[\sin^2\left(\frac{n\pi x}{l}\right)\right] dx = 1$$

Remember the trigonometric identity:

$$\sin^2 x = \tfrac{1}{2}(1 - \cos 2x)$$

By replacing the above equation in the previous integral, we have:

$$\frac{1}{2}d_2^2\int_0^l\left[1-\cos\left(\frac{2n\pi x}{l}\right)\right]dx=1$$

$$u=2x,\qquad du=2dx$$

$$\frac{1}{4}d_2^2\int_0^l\left[1-\cos\left(\frac{n\pi u}{l}\right)\right]du=1$$

$$\frac{1}{4}d_2^2\left[u-\sin\left(\frac{n\pi u}{l}\right)\right]\Bigg|_{x=0}^{x=l}=1$$

$$\frac{1}{4}d_2^2\left[2x-\sin\left(\frac{2n\pi x}{l}\right)\right]\Bigg|_0^l=1$$

$$\frac{1}{2}d_2^2\left[x-\sin\left(\frac{n\pi x}{l}\right)\right]\Bigg|_0^l=1$$

$$\frac{1}{2}d_2^2\left[(l-\sin n\pi)-(0-\sin 0)\right]=1$$

$$\frac{l}{2}d_2^2=1\therefore d_2=\sqrt{\frac{2}{l}}$$

Then, the normalized wave function is:

$$\psi_n(x)=\sqrt{\frac{2}{l}}\sin\left(\frac{n\cdot\pi}{l}x\right),\qquad n=0,\pm1,\pm2,\pm3,\ldots$$

This wave function is similar to that from one-particle harmonic oscillator wave function (Chapter fourteen).
Remember that:

$$\omega^2=\frac{2mE}{\hbar^2},\omega=\frac{n\pi}{l},\qquad n=0,\pm1,\pm2,\ldots$$

Then, we have:

$$E=\frac{n^2\pi^2\hbar^2}{2ml^2},\qquad \hbar=\frac{h}{2\pi}$$

$$E=\frac{n^2\pi^2h^2}{2m4\pi^2l^2}=\frac{n^2h^2}{8ml^2},\qquad n=0,\pm1,\pm2,\ldots$$

2. Particle in bidimensional box

The Fig. 15.1 shows the representation of a particle of mass m in a bidimensional box in the similar boundary conditions of the unidimensional box.

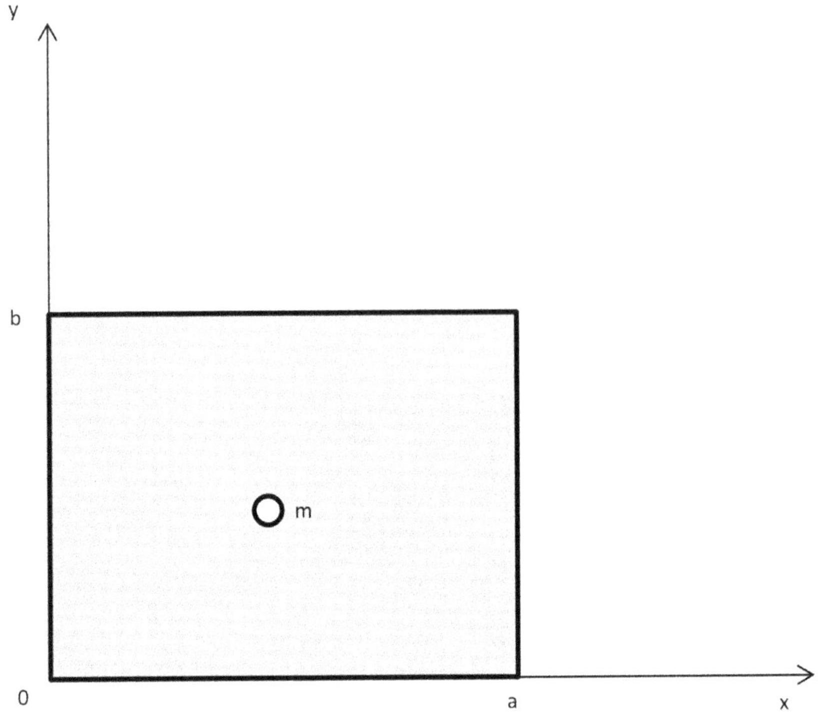

Fig. 15.1: Representation of the particle in bidimensional box.

The classical kinetic and potential energies, K and V, respectively, of this system are:

$$K = \frac{1}{2}mv_x^2 + \frac{1}{2}mv_y^2$$

$$p = mv, Then: K = \frac{1}{2m}p_x^2 + \frac{1}{2m}p_y^2$$

$$K = \frac{1}{2m}\left(p_x^2 + p_y^2\right)$$

$$V(x,y) = 0\begin{cases} 0 < x < a \\ 0 < y < b \end{cases}$$

Then, the quantum Hamiltonian becomes:

$$H = -\frac{\hbar^2}{2m}\left[\frac{\partial^2}{\partial x^2} + \frac{\partial^2}{\partial y^2}\right]$$

Since the Hamiltonian is the sum of two terms with independent, separate variables (x, y), we can use the method of the separation of the variables (see Chapter four) in order to provide a straightforward solution. Then, the total wave function is a product of two wave functions:

$$\psi(x,y) = X(x) \cdot Y(y)$$

And the Hamiltonian can be split into two Hamiltonians:

$$H_X = -\frac{\hbar^2}{2m}\frac{d^2}{dx^2}, \qquad H_Y = -\frac{\hbar^2}{2m}\frac{d^2}{dy^2}$$

Likewise, there are two separate Schrödinger equations:

$$-\frac{\hbar^2}{2m}\frac{d^2X(x)}{dx^2} = E_xX(x)$$

$$-\frac{\hbar^2}{2m}\frac{d^2Y(y)}{dy^2} = E_yY(y)$$

The total energy is the sum of the two components:

$E = E_x + E_y$

Where:

$$E_x = E_{n_x} = \frac{n_x^2h^2}{8ma^2}$$

$$E_y = E_{n_y} = \frac{n_y^2h^2}{8mb^2}$$

The two components of the wave function are:

$$X(x) = \sqrt{\frac{2}{a}}\sin\frac{n_x\pi x}{a}$$

$$Y(y) = \sqrt{\frac{2}{b}}\sin\frac{n_y\pi y}{b}$$

For a symmetric box (a = b), we have:

$$E_{n_x} + E_{n_y} = \frac{h^2}{8ma^2}\left(n_x^2 + n_y^2\right)$$

Let us evaluate the energy of the symmetric box for some values of n_x and n_y:

$$n_x = 1 \therefore n_y = 1, \qquad E = \frac{h^2}{4ma^2}$$

$$n_x = 1 \therefore n_y = 2, \qquad E = \frac{5h^2}{8ma^2}$$

$$n_x = 2 \therefore n_y = 1, \qquad E = \frac{5h^2}{8ma^2}$$

We can see that the wave function $X_1(x)\ Y_2(y)$ and $X_2(x)\ Y_1(y)$ have the same energy, that is, these wave functions are degenerate. This is the first case of degeneracy so far in this book. *Degeneracy means two or more eigenfunctions or eigenvectors with the same eigenvalue.*

By supposing that the symmetric box has length a = 10. Let us find the wave function for some values of n_x and n_y:

$$\psi(x,y) = \frac{2}{a}\sin\frac{n_x \pi x}{a}\sin\frac{n_y \pi y}{a}, \qquad a = 10$$

$$n_x = 1 \therefore n_y = 1, \qquad \psi(x,y) = \frac{1}{5}\sin\frac{\pi x}{10}\sin\frac{\pi y}{10}$$

$$n_x = 1 \therefore n_y = 2, \qquad \psi(x,y) = \frac{1}{5}\sin\frac{\pi x}{10}\sin\frac{\pi y}{5}$$

$$n_x = 2 \therefore n_y = 1, \qquad \psi(x,y) = \frac{1}{5}\sin\frac{\pi x}{5}\sin\frac{\pi y}{10}$$

$$n_x = 2 \therefore n_y = 2, \qquad \psi(x,y) = \frac{1}{5}\sin\frac{\pi x}{5}\sin\frac{\pi y}{5}$$

The Fig. 15.2 shows the plots of the above wave functions.

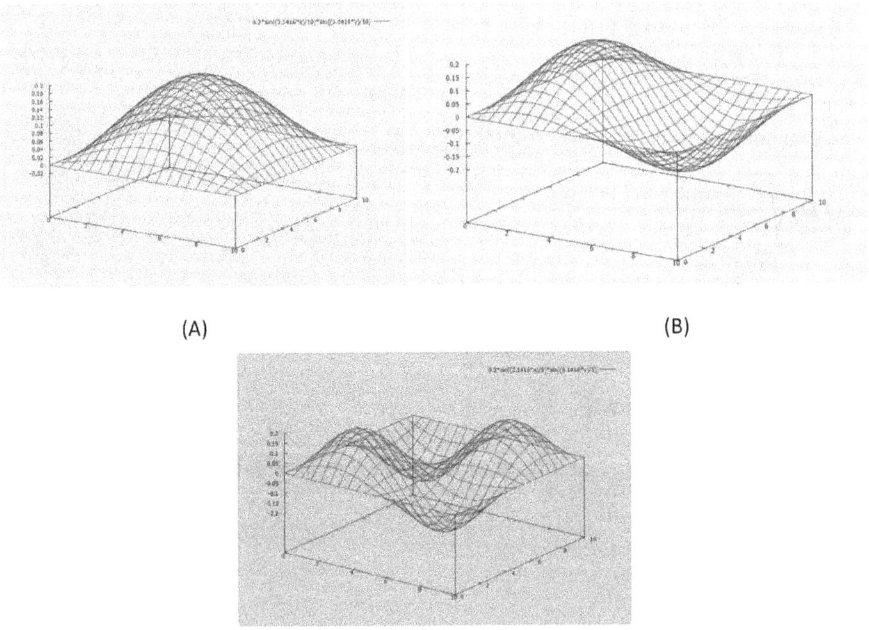

(A) (B) (C)

Fig. 15.2: Plots of the wave functions of the particle in bidimensional, symmetrical box: (A) ψ_{11}; (B) ψ_{12}; and (C) ψ_{22}.

Exercises

(1) Plot the wave functions ψ_{11}; ψ_{12}; and ψ_{22}. for a rectangular box (a = 5 and b = 10).

(2) Plot the probability density of the wave functions ψ_{11}; ψ_{12}; and ψ_{22}. for a symmetrical box.

Tip: Use the free program GNUPLOT (Williams 2018).

Reference cited

Williams, T., Kelley, C., Merritt, E.A., Bersch, C., Bröker, H.-B., Campbell, J. Cunningham, R., Denholm, D., Elber, G., Fearick, R., Grammes, C., Hart, L., Hecking, L., Juhász, P., Koenig, T., Kotz, D., Kubaitis, E., Lang, R., Lecomte, T., Lehmann, A., Lodewyck, J, Mai, A., Märkisch, B., Mikulik, P., Sebald, D., Steger, C., Takeno, S., Tkacik, T., der Woude, J.V., Zandt, J.R.V., Woo, Alex and Zellner, J. 2018. Gnuplot 5.2—an interactive plotting program.

Particle in a Circular Motion and Angular Momentum 16

1. Classic circular motion

The circular motion is a movement of a body in a circumference or rotation in a circular path. Although we cannot infer that the electrons describe circular motion around nuclei, the mechanics of circular motion is necessary for the rotation of a molecule (rigid rotor) and to obtain the solution for the hydrogen atom.

Let us consider a body moving at constant velocity around a circle of radius r in the plane (x,y). Let us convert this problem from Cartesian coordinate system into polar coordinate system. See Fig. 16.1.

$$x = r \cdot \cos\varphi, \qquad y = r \cdot \sin\varphi$$

The mean angular velocity, $\bar{\omega}$, is given by the formula:

$$\bar{\omega} = \frac{\Delta\varphi}{\Delta t}$$

$$if : t_0 = 0, \qquad \varphi = \omega \cdot t$$

And the angular velocity, ω, is then defined by:

$$\omega = \frac{d\varphi}{dt}$$

On the other hand, the linear velocity, v, has two components: v_x and v_y

$$\vec{v} = \vec{v}_x + \vec{v}_y$$

$$\vec{v}_x = \frac{dx}{dt} = \frac{d(r \cdot \cos\varphi)}{dt} = \frac{d[r \cdot \cos(\omega \cdot t)]}{dt} = -r\omega \cdot \sin(\omega \cdot t)$$

$$\vec{v}_y = \frac{dy}{dt} = \frac{d(r \cdot \sin\varphi)}{dt} = \frac{d[r \cdot \sin(\omega \cdot t)]}{dt} = r\omega \cdot \cos(\omega \cdot t)$$

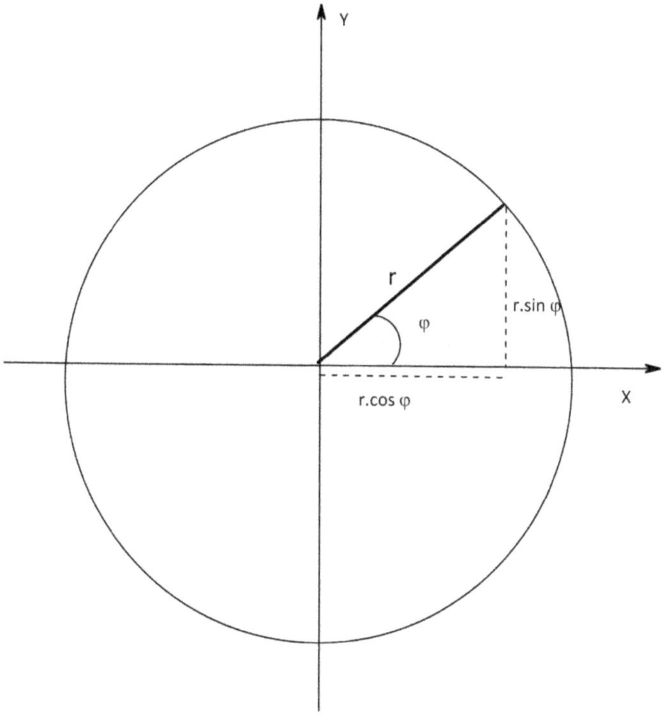

Fig. 16.1: Schematic representation of components of a circular motion in polar coordinate system.

The modulus of the linear velocity is the dot (scalar) product of radius r and angular velocity.

$$\left|\vec{v}\right|^2 = \left|\vec{v_x}\right|^2 + \left|\vec{v_y}\right|^2$$

$$\left|\vec{v}\right|^2 = \left[-r\omega \sin(\omega \cdot t)\right]^2 + \left[r\omega \cos(\omega \cdot t)\right]^2$$

$$v = r\omega\sqrt{\sin^2(\omega \cdot t) + \cos^2(\omega \cdot t)}$$

$$\sin^2(\omega \cdot t) + \cos^2(\omega \cdot t) = 1$$

$$v = r \cdot \omega$$

2. Bidimensional Laplace operator in polar coordinate system

The bidimensional Laplace operator in polar coordinate system is important to solve Schrödinger equation in next section. Then, we change the function in Cartesian coordinate into polar coordinate.

$$f(x, y) \rightarrow g(r, \phi)$$

$$x = r \cdot \cos\phi, \qquad y = r \cdot \sin\phi$$

$$r^2 = x^2 + y^2, \qquad \phi = \arctan\left(y/x\right)$$

Then, we have to transform the bidimensional Laplace operator:

$$\frac{\partial^2 f(x,y)}{\partial x^2}+\frac{\partial^2 f(x,y)}{\partial y^2}$$

into r and φ independent variables.
Let us write two partial derivatives of r with respect to x and y:

$$\frac{\partial r}{\partial x}=\frac{\partial\left(x^2+y^2\right)^{-1/2}}{\partial x}=\frac{x}{\left(x^2+y^2\right)^{1/2}}$$

$$\frac{\partial r}{\partial x}=\frac{r\cdot\cos\varphi}{\left[\left(r^2\cdot\cos^2\varphi\right)+\left(r^2\cdot\sin^2\varphi\right)\right]^{1/2}}=\cos\varphi$$

$$\frac{\partial r}{\partial y}=\frac{\partial\left(x^2+y^2\right)^{-1/2}}{\partial y}=\frac{y}{\left(x^2+y^2\right)^{1/2}}$$

$$\frac{\partial r}{\partial x}=\frac{r\cdot\sin\varphi}{\left[\left(r^2\cdot\cos^2\varphi\right)+\left(r^2\cdot\sin^2\varphi\right)\right]^{1/2}}=\sin\varphi$$

By knowing the derivative of arctan x as:

$$\frac{d\left(\arctan x\right)}{dx}=\frac{1}{1+x^2}$$

Then, the two partial derivatives of φ with respect to x and y are:

$$\frac{\partial\varphi}{\partial x}=\frac{-\left(y/x^2\right)}{1+\left(y/x\right)^2}=\frac{-\dfrac{r\cdot\sin\varphi}{r^2\cdot\cos^2\varphi}}{1+\dfrac{r^2\cdot\sin^2\varphi}{r^2\cdot\cos^2\varphi}}=\frac{-\dfrac{r\cdot\sin\varphi}{r^2\cdot\cos^2\varphi}}{\dfrac{r^2\cdot\cos^2\varphi+r^2\cdot\sin^2\varphi}{r^2\cdot\cos^2\varphi}}$$

$$\frac{\partial\varphi}{\partial x}=-\frac{r\cdot\sin\varphi}{r^2}=-\frac{\sin\varphi}{r}$$

$$\frac{\partial\varphi}{\partial y}=\frac{1/x}{1+\left(y/x\right)^2}=\frac{\dfrac{1}{r\cdot\cos\varphi}}{\dfrac{r^2\cdot\cos^2\varphi+r^2\cdot\sin^2\varphi}{r^2\cdot\cos^2\varphi}}=\frac{r^2\cdot\cos^2\varphi}{\left(r\cdot\cos\varphi\right)r^2}$$

$$\frac{\partial\varphi}{\partial y}=\frac{\cos\varphi}{r}$$

Now, let us use the chain rule to the first partial derivative of the function g with respect to variable x and y:

$$\frac{\partial g(r,\varphi)}{\partial x}=\frac{\partial\varphi}{\partial x}\frac{\partial g(r,\varphi)}{\partial\varphi}+\frac{\partial r}{\partial x}\frac{\partial g(r,\varphi)}{\partial r}=\left(\frac{\partial\varphi}{\partial x}\frac{\partial}{\partial\varphi}+\frac{\partial r}{\partial x}\frac{\partial}{\partial r}\right)g(r,\varphi)$$

$$\frac{\partial g(r,\varphi)}{\partial x}=\left(-\frac{\sin\varphi}{r}\frac{\partial}{\partial\varphi}+\cos\varphi\frac{\partial}{\partial r}\right)g(r,\varphi)$$

$$\frac{\partial g(r,\varphi)}{\partial y}=\frac{\partial \varphi}{\partial y}\frac{\partial g(r,\varphi)}{\partial \varphi}+\frac{\partial r}{\partial y}\frac{\partial g(r,\varphi)}{\partial r}=\left(\frac{\partial \varphi}{\partial y}\frac{\partial}{\partial \varphi}+\frac{\partial r}{\partial y}\frac{\partial}{\partial r}\right)g(r,\varphi)$$

$$\frac{\partial g(r,\varphi)}{\partial y}=\left(\frac{\cos\varphi}{r}\frac{\partial}{\partial \varphi}+\sin\varphi\frac{\partial}{\partial r}\right)g(r,\varphi)$$

Then, we can see that:

$$\frac{\partial}{\partial x}=-\frac{\sin\varphi}{r}\frac{\partial}{\partial \varphi}+\cos\varphi\frac{\partial}{\partial r}$$

$$\frac{\partial}{\partial y}=\frac{\cos\varphi}{r}\frac{\partial}{\partial \varphi}+\sin\varphi\frac{\partial}{\partial r}$$

By knowing that:

$$\frac{\partial^2 f(x,y)}{\partial x^2}=\frac{\partial}{\partial x}\left(\frac{\partial f(x,y)}{\partial x}\right)=\frac{\partial}{\partial x}\left(\frac{\partial}{\partial x}\right)f(x,y)$$

Let us do the proper substitution on the previous equation:

$$\frac{\partial^2 f(x,y)}{\partial x^2}=\frac{\partial}{\partial x}\left(\frac{\partial}{\partial x}\right)f(x,y)$$

$$\frac{\partial^2 f(x,y)}{\partial x^2}=\left(\cos\varphi\frac{\partial}{\partial r}-\frac{\sin\varphi}{r}\frac{\partial}{\partial \varphi}\right)\left(\cos\varphi\frac{\partial}{\partial r}-\frac{\sin\varphi}{r}\frac{\partial}{\partial \varphi}\right)g(r,\varphi)$$

$$\frac{\partial^2 f(x,y)}{\partial x^2}=\left(\cos\varphi\frac{\partial}{\partial r}-\frac{\sin\varphi}{r}\frac{\partial}{\partial \varphi}\right)\left(\cos\varphi\frac{\partial g(r,\varphi)}{\partial r}-\frac{\sin\varphi}{r}\frac{\partial g(r,\varphi)}{\partial \varphi}\right)$$

$$\frac{\partial^2 f(x,y)}{\partial x^2}=\cos\varphi\frac{\partial}{\partial r}\cos\varphi\frac{\partial g(r,\varphi)}{\partial r}-\cos\varphi\frac{\partial}{\partial r}\frac{\sin\varphi}{r}\frac{\partial g(r,\varphi)}{\partial \varphi}$$

$$+\frac{\sin\varphi}{r}\frac{\partial}{\partial \varphi}\frac{\sin\varphi}{r}\frac{\partial g(r,\varphi)}{\partial \varphi}-\frac{\sin\varphi}{r}\frac{\partial}{\partial \varphi}\cos\varphi\frac{\partial g(r,\varphi)}{\partial r}$$

Let us analyze each term of the above equation:

$$\cos\varphi\frac{\partial}{\partial r}\cos\varphi\frac{\partial g(r,\varphi)}{\partial r}=\cos^2\varphi\frac{\partial^2 g(r,\varphi)}{\partial r^2}$$

$$-\cos\varphi\frac{\partial}{\partial r}\frac{\sin\varphi}{r}\frac{\partial g(r,\varphi)}{\partial \varphi}=\frac{\cos\varphi\sin\varphi}{r^2}\frac{\partial g(r,\varphi)}{\partial \varphi}-\frac{\cos\varphi\sin\varphi}{r}\frac{\partial^2 g(r,\varphi)}{\partial r\partial \varphi}$$

$$\frac{\sin\varphi}{r}\frac{\partial}{\partial \varphi}\frac{\sin\varphi}{r}\frac{\partial g(r,\varphi)}{\partial \varphi}=\frac{\cos\varphi\sin\varphi}{r^2}\frac{\partial g(r,\varphi)}{\partial \varphi}+\frac{\sin^2\varphi}{r^2}\frac{\partial^2 g(r,\varphi)}{\partial \varphi^2}$$

$$-\frac{\sin\varphi}{r}\frac{\partial}{\partial \varphi}\cos\varphi\frac{\partial g(r,\varphi)}{\partial r}=\frac{\sin^2\varphi}{r}\frac{\partial}{\partial \varphi}\frac{\partial g(r,\varphi)}{\partial r}-\frac{\sin\varphi\cos\varphi}{r}\frac{\partial^2 g(r,\varphi)}{\partial \varphi\partial r}$$

Then, we have:

$$\frac{\partial^2 f(x,y)}{\partial x^2} = \cos^2\varphi\frac{\partial^2 g(r,\varphi)}{\partial r^2} + \frac{2\cos\varphi\sin\varphi}{r^2}\frac{\partial g(r,\varphi)}{\partial\varphi}$$

$$-\frac{2\cos\varphi\sin\varphi}{r}\frac{\partial^2 g(r,\varphi)}{\partial\varphi\partial r} + \frac{\sin^2\varphi}{r}\frac{\partial g(r,\varphi)}{\partial r} + \frac{\sin^2\varphi}{r^2}\frac{\partial^2 g(r,\varphi)}{\partial\varphi^2}$$

A similar expression we have for:

$$\frac{\partial^2 f(x,y)}{\partial y^2} = \frac{\partial}{\partial y}\left(\frac{\partial}{\partial y}\right) f(x,y)$$

$$\frac{\partial^2 f(x,y)}{\partial y^2} = \left(\sin\varphi\frac{\partial}{\partial r} + \frac{\cos\varphi}{r}\frac{\partial}{\partial\varphi}\right)\left(\sin\varphi\frac{\partial}{\partial r} + \frac{\cos\varphi}{r}\frac{\partial}{\partial\varphi}\right) g(r,\varphi)$$

$$\frac{\partial^2 f(x,y)}{\partial y^2} = \sin^2\varphi\frac{\partial^2 g(r,\varphi)}{\partial r^2} - \frac{2\cos\varphi\sin\varphi}{r^2}\frac{\partial g(r,\varphi)}{\partial\varphi}$$

$$+\frac{2\cos\varphi\sin\varphi}{r}\frac{\partial^2 g(r,\varphi)}{\partial\varphi\partial r} + \frac{\cos^2\varphi}{r}\frac{\partial g(r,\varphi)}{\partial r} + \frac{\cos^2\varphi}{r^2}\frac{\partial^2 g(r,\varphi)}{\partial\varphi^2}$$

When adding both expressions, we have the bidimensional Laplace operator in polar coordinate system:

$$\frac{\partial^2 f(x,y)}{\partial x^2} + \frac{\partial^2 f(x,y)}{\partial y^2} = \frac{\partial^2 g(r,\varphi)}{\partial r^2} + \frac{1}{r}\frac{\partial g(r,\varphi)}{\partial r} + \frac{1}{r^2}\frac{\partial^2 g(r,\varphi)}{\partial\varphi^2}$$

3. Particle in a ring and quantum angular momentum

In this system, a particle is restricted to move in a (x, y)-planar ring with zero potential energy (no external force acting on it) with constant radius. In order to solve this problem, we use the Schrödinger equation in polar coordinate system. The only component of the Hamiltonian is the quantum kinetic energy operator, T. Below, it is shown the quantum 2-dimensional T in Cartesian coordinate and in polar coordinate.

$$T = -\frac{\hbar^2}{2m}\left(\frac{\partial^2}{\partial x^2} + \frac{\partial^2}{\partial y^2}\right), \text{Cartesian coordinate}$$

$$T = -\frac{\hbar^2}{2m}\left(\frac{\partial^2}{\partial r^2} + \frac{1}{r}\frac{\partial}{\partial r} + \frac{1}{r^2}\frac{\partial^2}{\partial\phi^2}\right), \textit{polar}\ \text{coordinate}$$

Since the particle is restricted to a planar circular motion with constant radius, r, the derivatives of the wave function with respect to r are zero.

$$r = const$$

$$\frac{\partial\psi}{\partial r} = \nexists, \quad \frac{\partial^2\psi}{\partial r^2} = \nexists$$

$$then:$$

$$T = -\frac{\hbar^2}{2mr^2}\frac{\partial^2}{\partial\phi^2}$$

The Schrödinger equation becomes:

$$-\frac{\hbar^2}{2mr^2}\frac{\partial^2\psi}{\partial\varphi^2} = E\psi$$

$$I = mr^2$$

$$-\frac{\hbar^2}{2I}\frac{\partial^2\psi}{\partial\varphi^2} = E\psi$$

Let us rearrange the above equation:

$$\frac{\partial^2\psi}{\partial\varphi^2} = -\frac{2EI}{\hbar^2}\psi$$

$$\frac{\partial^2\psi}{\partial\varphi^2} + \frac{2EI}{\hbar^2}\psi = 0$$

$$\omega^2 = \frac{2EI}{\hbar^2}$$

$$\frac{\partial^2\psi}{\partial\varphi^2} + \omega^2\psi = 0$$

$$y'' + \omega^2 = 0$$

This equation is similar to that from the particle in one-dimensional box (Chapter fifteen) and the one-particle harmonic oscillator (Chapter fourteen), whose general solution for this elliptic second-order differential equation (as discussed in chapter four) is:

$$y = c_1 e^{i\omega x} + c_2 e^{-i\omega x}$$

or

$$y = d_1 \cos \omega x + d_2 \sin \omega x$$

Do not confuse ω^2 in the previous equations with square angular velocity. It is just a conventional symbol used in chapter four. Let us change now ω^2 into m^2 to avoid further confusions.

$$\frac{\partial^2\psi}{\partial\varphi^2} = -\frac{2EI}{\hbar^2}\psi$$

$$m^2 = \frac{2EI}{\hbar^2}$$

$$\frac{\partial^2\psi}{\partial\varphi^2} = -m^2\psi$$

Actually, x becomes φ in the particle in a ring. Then, the general solution becomes:

$$\psi(\varphi) = c_1 e^{im\varphi} + c_2 e^{-im\varphi}$$

or

$$\psi(\varphi) = d_1 \cos m\varphi + d_2 \sin m\varphi$$

The only difference with respect to the particle in a one-dimensional box lies in the independent angular variable φ that repeats itself at each 2π radian. Then, the boundary condition is:

$$\psi(\varphi + 2\pi) = \psi(\varphi)$$

By applying the periodic boundary conditions we have:

$$\psi(\varphi) = N\cos m\varphi$$

$$\psi(\varphi) = N\sin m\varphi$$

$$\psi(\varphi) = Ne^{im\varphi}$$

This requires that:

$$e^{im\varphi} = e^{i\omega(\varphi+2\pi)} \Rightarrow e^{2\pi\cdot im} = 1$$

Where N is the normalization constant. This is only true when:

$$m = 0, \pm 1, \pm 2, \pm 3, \ldots$$

The proof for the previous statement comes from Euler's identity.

$$e^{xi} = \cos(x) + i\sin(x)$$

$$m = 1 \therefore x = 2\pi$$

$$e^{2\pi i} = \cos(2\pi) + i\sin(2\pi) = 1 + 0$$

$$e^{2\pi i} = 1$$

$$m = 2 \therefore x = 4\pi$$

$$e^{4\pi i} = \cos(2(2\pi)) + i\sin(2(2\pi)) = 1 + 0$$

$$e^{4\pi i} = 1$$

$$m = 3 \therefore x = 6\pi$$

$$e^{6\pi i} = \cos(3(2\pi)) + i\sin(3(2\pi)) = 1 + 0$$

$$e^{6\pi i} = 1$$

Let us substitute this wave function in the Schrödinger equation:

$$-\frac{\hbar^2}{2I}\frac{d^2}{d\varphi^2}\left(Ne^{im\varphi}\right) = E\psi$$

$$-\frac{\hbar^2}{2I}\frac{d}{d\varphi}\left(imNe^{im\varphi}\right) = E\psi$$

$$-\frac{\hbar^2}{2I}\left(i^2m^2Ne^{im\varphi}\right) = E\psi$$

$$\frac{m^2\hbar^2}{2I}Ne^{im\varphi} = E\psi$$

Then, the energy of this system is:

$$E = \frac{m^2\hbar^2}{2I}, \qquad m = 0, \pm 1, \pm 2, \ldots$$

By knowing the relation between the (orbital) angular momentum and the kinetic energy (see Chapter eight) and the quantum kinetic energy:

$$T = \frac{L^2}{2I}, \qquad \textit{classic}$$

$$T = \frac{m^2\hbar^2}{2I}, \qquad \textit{quantum}$$

We have the equation for the (orbital) angular momentum:

$$\mathbf{L}^2 = m^2\hbar^2$$

$$\mathbf{L} = m\hbar, \qquad m = 0, \pm 1, \pm 2, \ldots$$

The Schrödinger equation of the particle in the ring is part of the solution to the hydrogen atom. The wave function of the hydrogen atom can be separated into three wave functions:

$$\psi(r,\theta,\varphi) = R(r)\Theta(\theta)\Phi(\varphi)$$

Where one of these wave functions, the $\Phi(\varphi)$ wave function, is associated with the projection of the angular momentum in the z axis, L_z, and it has the following differential equation (as a part of the three decomposed differential equations of the hydrogen atom):

$$\frac{d^2\Phi}{d\varphi^2} = -m^2\Phi$$

$$\textit{where}:$$

$$\Phi = \sin(m\varphi) \quad \textit{or} \quad \cos(m\varphi) \quad \textit{or} \quad \exp(\pm im\varphi)$$

This is a similar differential equation and wave function for the particle in the ring:

$$\frac{\partial^2\psi}{\partial\phi} + m^2\psi = 0, \qquad \frac{\partial^2\psi}{\partial\phi} = -m^2\psi$$

$$\textit{where}:$$

$$\psi \equiv \Phi$$

The wave function of the hydrogen atom can also be written as:

$$\psi(r,\theta,\phi) = R_{n,l}(r)Y_l^{ml}$$

$$\textit{and}: Y_l^{ml} = Y(\theta,\phi) = \Theta(\theta)\Phi(\phi)$$

where $Y(\theta,\varphi)$ or $Y(\theta,\phi)$ is the spherical harmonics function, l is the azimuthal quantum number (or orbital quantum number) and m_l is the magnetic quantum number.

In the hydrogen atom, the $\Phi(\varphi)$ wave function is associated with the Schrödinger's wave equation of the particle in the ring. Then, the quantum number m (from the particle in the ring) becomes m_l, **L** becomes $\mathbf{L_z}$ and their Schrödinger solution becomes:

$$\mathbf{L}_z Y_l^{m_l} = m\hbar Y_l^{m_l}$$

$$\textit{where}: -l \le m_l \le l$$

As a consequence, the z-component of the angular momentum is quantized, i.e., it only assumes a discrete set of values.

4. Angular momentum in spherical polar coordinate system

The Fig. 16.2 shows the position of an arbitrary point with the following spherical polar coordinates: r, θ, φ, where φ is the azimuthal angle (in horizontal plane) and θ is the zenith angle or polar angle (in vertical plane). The relation between spherical polar coordinates and Cartesian coordinates is given below:

$$x = r\sin\theta\cos\varphi$$

$$y = r\sin\theta\sin\varphi$$

$$z = r\cos\theta$$

$$r = \sqrt{x^2 + y^2 + z^2}$$

$$\tan\theta = \frac{\sqrt{x^2 + y^2}}{z}$$

$$\tan\varphi = \frac{y}{x}$$

The (orbital) angular momentum operator in spherical polar coordinates can be obtained in seven steps.

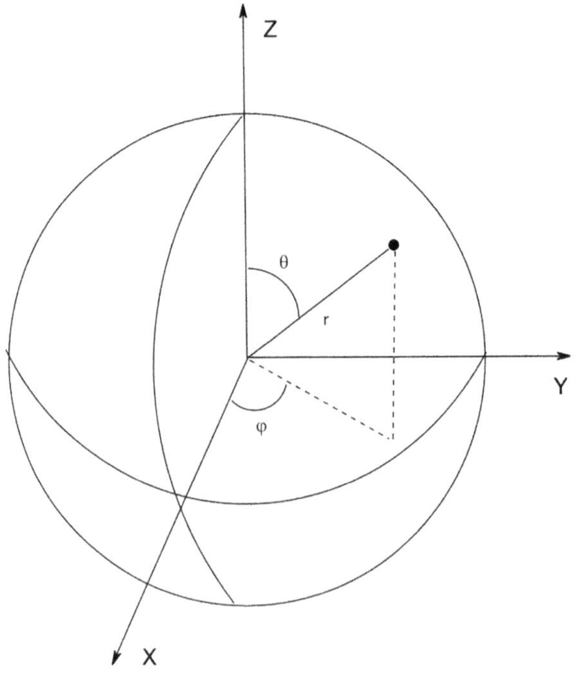

Fig. 16.2: Schematic representation of spherical polar coordinates.

(1) First step: let us obtain dx, dy and dz in terms of spherical polar coordinates:

$$x = r\sin\theta\cos\varphi, \qquad dx = \frac{\partial x}{\partial r}dr + \frac{\partial x}{\partial \theta}d\theta + \frac{\partial x}{\partial \varphi}d\varphi$$

$$dx = \sin\theta\cos\varphi dr + r\cos\theta\cos\varphi d\theta - r\sin\theta\sin\varphi d\varphi$$

$$y = r\sin\theta\sin\varphi, \qquad dy = \frac{\partial y}{\partial r}dr + \frac{\partial y}{\partial \theta}d\theta + \frac{\partial y}{\partial \varphi}d\varphi$$

$$dy = \sin\theta\sin\varphi dr + r\cos\theta\sin\varphi d\theta + r\sin\theta\cos\varphi d\varphi$$

$$z = r\cos\theta, \qquad dz = \cos\theta dr - r\sin\theta d\theta$$

(2) Second step: we obtain the derivatives ∂x/∂r, ∂x/∂φ, ∂x/∂θ, ∂y/∂r, ∂y/∂φ, ∂y/∂θ, ∂z/∂r, ∂z/∂φ and ∂z/∂θ:

$$dx = \sin\theta\cos\varphi dr + r\cos\theta\cos\varphi d\theta - r\sin\theta\sin\varphi d\varphi$$

$$\frac{\partial x}{\partial r} = \sin\theta\cos\varphi\frac{dr}{\partial r} + r\cos\theta\cos\varphi\frac{d\theta}{\partial r} - r\sin\theta\sin\varphi\frac{d\varphi}{\partial r}$$

$$\frac{dr}{\partial r} = 1, \qquad \frac{d\theta}{\partial r} = 0, \qquad \frac{d\varphi}{\partial r} = 0$$

$$\frac{\partial x}{\partial r} = \sin\theta\cos\varphi$$

$$dx = \sin\theta\cos\varphi dr + r\cos\theta\cos\varphi d\theta - r\sin\theta\sin\varphi d\varphi$$

$$\frac{\partial x}{\partial \theta} = r\cos\theta\cos\varphi, \qquad \frac{\partial x}{\partial \varphi} = -r\sin\theta\sin\varphi$$

$$dy = \sin\theta\sin\varphi dr + r\cos\theta\sin\varphi d\theta + r\sin\theta\cos\varphi d\varphi$$

$$\frac{\partial y}{\partial r} = \sin\theta\sin\varphi, \frac{\partial y}{\partial \theta} = r\cos\theta\sin\varphi, \frac{\partial y}{\partial \varphi} = r\sin\theta\cos\varphi$$

$$dz = \cos\theta dr - r\sin\theta d\theta$$

$$\frac{\partial z}{\partial r} = \cos\theta, \qquad \frac{\partial z}{\partial \theta} = -r\sin\theta$$

(3) Third step: we obtain the inverse of the above derivatives (i.e., ∂r/∂x, ∂φ/∂x, ∂θ/∂x, ∂r/∂y, ∂φ/∂y, ∂θ/∂y, ∂r/∂z and ∂θ/∂z). Let us begin by obtaining ∂r/∂x:

$$\frac{\partial r}{\partial x} = \frac{\partial}{\partial x}\left(x^2 + y^2 + z^2\right)^{1/2}, \qquad u = x^2 + y^2 + z^2$$

$$\frac{\partial r}{\partial x} = \frac{\partial u^{1/2}}{\partial u}\frac{\partial u}{\partial x} = \frac{x}{\left(x^2 + y^2 + z^2\right)^{1/2}}$$

$$\frac{\partial r}{\partial x} = \frac{r\sin\theta\cos\varphi}{\left(r^2\sin^2\theta\cos^2\varphi + r^2\sin^2\theta\sin^2\varphi + r^2\cos^2\theta\right)^{1/2}}$$

$$\frac{\partial r}{\partial x}=\frac{r\sin\theta\cos\varphi}{\left(r^2\sin^2\theta\left[\cos^2\varphi+\sin^2\varphi\right]+r^2\cos^2\theta\right)^{1/2}}$$

$$\frac{\partial r}{\partial x}=\frac{r\sin\theta\cos\varphi}{\left(r^2\left[\sin^2\theta+\cos^2\theta\right]\right)^{1/2}}=\sin\theta\cos\varphi$$

The same procedure is used for the partial derivatives of r with respect to y and z ($\partial r/\partial y$ and $\partial r/\partial z$). Then, we have:

$$\frac{\partial r}{\partial y}=\sin\theta\sin\varphi,\qquad \frac{\partial r}{\partial z}=\cos\varphi$$

Let us now obtain $\partial\varphi/\partial x$:

$$\mathrm{tg}\varphi=\frac{y}{x},\qquad \frac{\partial}{\partial x}(\mathrm{tg}\varphi)=\frac{\partial}{\partial x}(y/x)=\frac{-y}{x^2}$$

$$\frac{\partial}{\partial x}(\mathrm{tg}\varphi)=\frac{-r\sin\theta\sin\varphi}{r^2\sin^2\theta\cos^2\varphi}=\frac{-\sin\varphi}{r\sin\theta\cos^2\varphi}$$

By knowing that:

$$\frac{\partial}{\partial x}(\mathrm{tg}\varphi)=\frac{1}{\cos^2\varphi}\frac{\partial\varphi}{\partial x}$$

Then, we have:

$$\frac{1}{\cos^2\varphi}\frac{\partial\varphi}{\partial x}=\frac{-\sin\varphi}{r\sin\theta\cos^2\varphi}$$

$$\frac{\partial\varphi}{\partial x}=-\frac{\sin\varphi}{r\sin\theta}$$

By using the same procedure, we obtain $\partial\varphi/\partial y$:

$$\frac{\partial\varphi}{\partial y}=\frac{\cos\varphi}{r\sin\theta}$$

Let us now obtain $\partial\theta/\partial x$.

$$tg\theta=\frac{\sqrt{x^2+y^2}}{z}$$

$$\frac{\partial}{\partial x}\left(\mathrm{tg}^2\theta\right)=\frac{\partial}{\partial x}\left(\frac{x^2+y^2}{z^2}\right)=\frac{\partial}{\partial x}\left(\frac{x^2}{z^2}+\frac{y^2}{z^2}\right)$$

$$\frac{\partial}{\partial x}\left(\mathrm{tg}^2\theta\right)=\frac{2x}{z^2}$$

$$\frac{\partial}{\partial x}\left(\mathrm{tg}^2\theta\right)=\frac{2r\sin\theta\cos\varphi}{r^2\cos^2\theta}=\frac{2\sin\theta\cos\varphi}{r\cos^2\theta}$$

By knowing that:

$$\frac{\partial}{\partial x}\left(\mathrm{tg}^2\theta\right)=2\frac{\partial\theta}{\partial x}\frac{\sin\theta}{\cos^3\theta}$$

We have:

$$2\frac{\partial\theta}{\partial x}\frac{\sin\theta}{\cos^3\theta}=\frac{2\sin\theta\cos\varphi}{r\cos^2\theta}$$

$$\frac{\partial\theta}{\partial x}=\frac{\cos\varphi\cos\theta}{r}$$

We use the same procedure to obtain $\partial\theta/\partial y$ and $\partial\theta/\partial z$.

$$\frac{\partial\theta}{\partial y}=\frac{\cos\theta\cos\varphi}{r}$$

$$\frac{\partial\theta}{\partial y}=-\frac{\sin\theta}{r}$$

(4) Fourth step: we obtain $\partial/\partial x$, $\partial/\partial y$ and $\partial/\partial z$:

$$\frac{\partial}{\partial x}=\frac{\partial}{\partial r}\frac{\partial r}{\partial x}+\frac{\partial}{\partial\theta}\frac{\partial\theta}{\partial x}+\frac{\partial}{\partial\varphi}\frac{\partial\varphi}{\partial x}$$

$$\frac{\partial r}{\partial x}=\sin\theta\cos\varphi\therefore\frac{\partial\theta}{\partial x}=\frac{1}{r}\cos\theta\cos\varphi,\frac{\partial\varphi}{\partial x}=-\frac{\sin\varphi}{r\sin\theta}$$

$$\frac{\partial}{\partial x}=\sin\theta\cos\varphi\frac{\partial}{\partial r}+\frac{1}{r}\cos\theta\cos\varphi\frac{\partial}{\partial\theta}-\frac{\sin\varphi}{r\sin\theta}\frac{\partial}{\partial\varphi}$$

$$\frac{\partial}{\partial y}=\frac{\partial}{\partial r}\frac{\partial r}{\partial y}+\frac{\partial}{\partial\theta}\frac{\partial\theta}{\partial y}+\frac{\partial}{\partial\varphi}\frac{\partial\varphi}{\partial y}$$

$$\frac{\partial r}{\partial y}=\sin\theta\sin\varphi\therefore\frac{\partial\theta}{\partial y}=\frac{1}{r}\cos\theta\sin\varphi,\frac{\partial\varphi}{\partial y}=\frac{\cos\varphi}{r\sin\theta}$$

$$\frac{\partial}{\partial y}=\sin\theta\sin\varphi\frac{\partial}{\partial r}+\frac{1}{r}\cos\theta\sin\varphi\frac{\partial}{\partial\theta}+\frac{\cos\varphi}{r\sin\theta}\frac{\partial}{\partial\varphi}$$

$$\frac{\partial}{\partial z}=\frac{\partial}{\partial r}\frac{\partial r}{\partial z}+\frac{\partial}{\partial\theta}\frac{\partial\theta}{\partial z}+\frac{\partial}{\partial\varphi}\frac{\partial\varphi}{\partial z}=\cos\theta\frac{\partial}{\partial r}-\frac{\sin\theta}{r}\frac{\partial}{\partial\theta}$$

Then, we have:

$$\frac{\partial}{\partial x}=\sin\theta\cos\varphi\frac{\partial}{\partial r}+\frac{1}{r}\cos\theta\cos\varphi\frac{\partial}{\partial\theta}-\frac{\sin\varphi}{r\sin\theta}\frac{\partial}{\partial\varphi}$$

$$\frac{\partial}{\partial y}=\sin\theta\sin\varphi\frac{\partial}{\partial r}+\frac{1}{r}\cos\theta\sin\varphi\frac{\partial}{\partial\theta}+\frac{\cos\varphi}{r\sin\theta}\frac{\partial}{\partial\varphi}$$

$$\frac{\partial}{\partial z}=\cos\theta\frac{\partial}{\partial r}-\frac{\sin\theta}{r}\frac{\partial}{\partial\theta}$$

(5) Fifth step: we obtain L_x, L_y and L_z. Let us begin with L_x:

$$L_x = \frac{\hbar}{i}\left(y\frac{\partial}{\partial z} - z\frac{\partial}{\partial y}\right)$$

$$L_x = \frac{\hbar}{i}\left[\begin{array}{l} y\left(\cos\theta\frac{\partial}{\partial r} - \frac{\sin\theta}{r}\frac{\partial}{\partial\theta}\right) - \\ -z\left(\sin\theta\sin\varphi\frac{\partial}{\partial r} + \frac{1}{r}\cos\theta\sin\varphi\frac{\partial}{\partial\theta} + \frac{\cos\varphi}{r\sin\theta}\frac{\partial}{\partial\varphi}\right)\end{array}\right]$$

$$L_x = \frac{\hbar}{i}\left[\begin{array}{l} \left(y\cos\theta - z\sin\theta\sin\varphi\right)\frac{\partial}{\partial r} + \\ +\left(-\frac{y\sin\theta}{r} - \frac{z}{r}\cos\theta\sin\varphi\right)\frac{\partial}{\partial\theta} - \frac{z\cos\varphi}{r\sin\theta}\frac{\partial}{\partial\varphi}\end{array}\right]$$

$$z = r\cos\theta, \qquad y = r\sin\theta\sin\varphi$$

$$L_x = \frac{\hbar}{i}\left[\begin{array}{l} \left(r\sin\theta\sin\varphi\cos\theta - r\cos\theta\sin\theta\sin\varphi\right)\frac{\partial}{\partial r} \\ +\left(-\frac{r\sin\theta\sin\varphi\sin\theta}{r} - \frac{r\cos\theta}{r}\cos\theta\sin\varphi\right)\frac{\partial}{\partial\theta} - \\ -\frac{r\cos\theta\cos\varphi}{r\sin\theta}\frac{\partial}{\partial\varphi}\end{array}\right]$$

$$L_x = \frac{\hbar}{i}\left[\left(-\sin\varphi\sin^2\theta - \cos^2\theta\sin\varphi\right)\frac{\partial}{\partial\theta} - \cot\theta\cos\varphi\frac{\partial}{\partial\varphi}\right]$$

$$L_x = \frac{\hbar}{i}\left[\left(-\sin\varphi\left\{\sin^2\theta + \cos^2\theta\right\}\right)\frac{\partial}{\partial\theta} - \cot\theta\cos\varphi\frac{\partial}{\partial\varphi}\right]$$

$$L_x = \frac{\hbar}{i}\left[-\sin\varphi\frac{\partial}{\partial\theta} - \cot\theta\cos\varphi\frac{\partial}{\partial\varphi}\right]$$

$$L_x = i\hbar\left[\sin\varphi\frac{\partial}{\partial\theta} + \cot\theta\cos\varphi\frac{\partial}{\partial\varphi}\right]$$

The same procedure is used to obtain L_y and L_z:

$$L_y = -\frac{\hbar}{i}\left(\cos\varphi\frac{\partial}{\partial\theta} - \cot\theta\sin\varphi\frac{\partial}{\partial\varphi}\right)$$

$$L_y = i\hbar\left(-\cos\varphi\frac{\partial}{\partial\theta} + \cot\theta\sin\varphi\frac{\partial}{\partial\varphi}\right)$$

$$L_z = \frac{\hbar}{i}\frac{\partial}{\partial\varphi}$$

(6) Sixth step: we square L_x, L_y and L_z.

$$L_x^2 = \left(\frac{\hbar}{i}\right)^2 \left(-\sin\varphi\frac{\partial}{\partial\theta} - \cot\theta\cos\varphi\frac{\partial}{\partial\varphi}\right)^2 =$$

$$L_x^2 = -\hbar^2 \left[\begin{array}{l} \sin^2\varphi\dfrac{\partial^2}{\partial\theta^2} + \sin\varphi\dfrac{\partial}{\partial\theta}\left(\cot\theta\cos\varphi\dfrac{\partial}{\partial\varphi}\right) + \\ +\cot\theta\cos\varphi\dfrac{\partial}{\partial\varphi}\left(\sin\varphi\dfrac{\partial}{\partial\theta}\right) + \cot^2\theta\cos^2\varphi\dfrac{\partial^2}{\partial\varphi^2} \end{array}\right]$$

$$L_x^2 = -\hbar^2 \left[\begin{array}{l} \sin^2\varphi\dfrac{\partial^2}{\partial\theta^2} + \sin\varphi\left(-\csc^2\theta\cos\varphi\dfrac{\partial}{\partial\varphi} + \cot\theta\cos\varphi\dfrac{\partial}{\partial\varphi}\right) \\ +\cot\theta\cos\varphi\left(\cos\varphi\dfrac{\partial}{\partial\theta} + \sin\varphi\dfrac{\partial}{\partial\varphi}\dfrac{\partial}{\partial\theta}\right) + \cot^2\theta\cos^2\varphi\dfrac{\partial^2}{\partial\varphi^2} \end{array}\right]$$

$$L_x^2 = -\hbar^2 \left[\begin{array}{l} \sin^2\varphi\dfrac{\partial^2}{\partial\theta^2} - \sin\varphi\cos\varphi\csc^2\theta\dfrac{\partial}{\partial\varphi} + \sin\varphi\cot\theta\cos\varphi\dfrac{\partial}{\partial\varphi} + \\ +\cot\theta\cos^2\varphi\dfrac{\partial}{\partial\theta} + \cot\theta\cos\varphi\sin\varphi\dfrac{\partial}{\partial\varphi}\dfrac{\partial}{\partial\theta} + \cot^2\theta\cos^2\varphi\dfrac{\partial^2}{\partial\varphi^2} \end{array}\right]$$

The same procedure is used to obtain the square of L_y and L_z. This can be done as exercise.

$$L_y^2 = -\hbar^2\left(-\cos\varphi\frac{\partial}{\partial\theta} + \cot\theta\sin\varphi\frac{\partial}{\partial\varphi}\right)^2$$

$$L_y^2 = -\hbar^2 \left[\begin{array}{l} \cos^2\varphi\dfrac{\partial^2}{\partial\theta^2} + \cos\varphi\sin\varphi\csc^2\theta\dfrac{\partial}{\partial\varphi} - \cos\varphi\cot\theta\sin\varphi\dfrac{\partial}{\partial\theta}\dfrac{\partial}{\partial\varphi} + \\ +\cot\theta\sin^2\varphi\dfrac{\partial}{\partial\theta} - \cot\theta\sin\varphi\cos\varphi\dfrac{\partial}{\partial\varphi}\dfrac{\partial}{\partial\theta} + \cot^2\theta\cos^2\varphi\dfrac{\partial^2}{\partial\varphi^2} \end{array}\right]$$

$$L_z^2 = -\hbar^2\frac{\partial^2}{\partial\varphi^2}$$

(7) Seventh step: we obtain L^2 by adding the square of L_x, L_y and L_z. We begin by emphasizing the similar terms for the square of L_x and L_y

$$L_x^2 = -\hbar^2 \left[\begin{array}{l} \sin^2\varphi\dfrac{\partial^2}{\partial\theta^2} - \cos\varphi\sin\varphi\csc^2\theta\dfrac{\partial}{\partial\varphi} + \cos\varphi\cot\theta\sin\varphi\dfrac{\partial}{\partial\theta}\dfrac{\partial}{\partial\varphi} + \\ \cot\theta\sin^2\varphi\dfrac{\partial}{\partial\theta} + \cot\theta\sin\varphi\cos\varphi\dfrac{\partial}{\partial\varphi}\dfrac{\partial}{\partial\theta} + \cot^2\theta\cos^2\varphi\dfrac{\partial^2}{\partial\varphi^2} \end{array}\right]$$

$$L_y^2 = -\hbar^2 \left[\begin{array}{l} \cos^2\varphi \dfrac{\partial^2}{\partial\theta^2} + \cos\varphi\sin\varphi\csc^2\theta \dfrac{\partial}{\partial\varphi} - \cos\varphi\cot\theta\sin\varphi \dfrac{\partial}{\partial\theta}\dfrac{\partial}{\partial\varphi} + \\ + \cot\theta\sin^2\varphi \dfrac{\partial}{\partial\theta} - \cot\theta\sin\varphi\cos\varphi \dfrac{\partial}{\partial\varphi}\dfrac{\partial}{\partial\theta} + \cot^2\theta\cos^2\varphi \dfrac{\partial^2}{\partial\varphi^2} \end{array} \right]$$

$$L_z^2 = -\hbar^2 \frac{\partial^2}{\partial\varphi^2}$$

By eliminating the similar terms with opposite signs and summing up all the remaining terms, we have:

$$L^2 = L_x^2 + L_y^2 + L_z^2$$

$$L^2 = -\hbar^2 \left[\frac{\partial^2}{\partial\theta^2} + \cot\theta \frac{\partial}{\partial\theta} + \left(\cot^2\theta + 1\right) \frac{\partial^2}{\partial\varphi^2} \right]$$

$$L^2 = -\hbar^2 \left[\frac{\partial^2}{\partial\theta^2} + \cot\theta \frac{\partial}{\partial\theta} + \left(\frac{\cos^2\theta + \sin^2\theta}{\sin^2\theta} \right) \frac{\partial^2}{\partial\varphi^2} \right]$$

$$L^2 = -\hbar^2 \left[\frac{\partial^2}{\partial\theta^2} + \cot\theta \frac{\partial}{\partial\theta} + \left(\frac{1}{\sin^2\theta} \right) \frac{\partial^2}{\partial\varphi^2} \right]$$

$$L^2 = -\hbar^2 \left[\frac{1}{\sin\theta}\frac{\partial}{\partial\theta} \sin\theta \frac{\partial}{\partial\theta} + \left(\frac{1}{\sin^2\theta} \right) \frac{\partial^2}{\partial\varphi^2} \right]$$

4. Tridimensional Laplace operator in spherical polar coordinate system

The tridimensional Laplace operator in Cartesian coordinate system is:

$$\nabla^2 = \frac{\partial^2}{\partial x^2} + \frac{\partial^2}{\partial y^2} + \frac{\partial^2}{\partial z^2}$$

The tridimensional Laplace operator in spherical polar coordinate system is:

$$\nabla^2 = \frac{1}{r^2}\frac{\partial}{\partial r}\left(r^2 \frac{\partial}{\partial r} \right) + \frac{1}{r^2}\left[\frac{1}{\sin\theta}\frac{\partial}{\partial\theta}\left(\sin\theta \frac{\partial}{\partial\theta} \right) + \frac{1}{\sin^2\theta}\left(\frac{\partial^2}{\partial\varphi^2} \right) \right]$$

$$\nabla^2 = \frac{\partial^2}{\partial r^2} + \frac{2}{r}\frac{\partial}{\partial r} + \frac{1}{r^2}\Lambda^2$$

$$\Lambda^2 = \frac{1}{\sin\theta}\frac{\partial}{\partial\theta}\left(\sin\theta \frac{\partial}{\partial\theta} \right) + \frac{1}{\sin^2\theta}\left(\frac{\partial^2}{\partial\varphi^2} \right)$$

Where Λ^2 is the Legendrian operator. The above equation is equivalent to:

$$\nabla^2 = \frac{1}{r^2}\frac{\partial}{\partial r}\left(r^2 \frac{\partial}{\partial r} \right) - \frac{1}{\hbar^2 r^2} L^2$$

In the fourth step of the last section, we have obtained the following relations:

$$\frac{\partial}{\partial x} = \sin\theta\cos\varphi\frac{\partial}{\partial r} + \frac{1}{r}\cos\theta\cos\varphi\frac{\partial}{\partial\theta} - \frac{\sin\varphi}{r\sin\theta}\frac{\partial}{\partial\varphi}$$

$$\frac{\partial}{\partial y} = \sin\theta\sin\varphi\frac{\partial}{\partial r} + \frac{1}{r}\cos\theta\sin\varphi\frac{\partial}{\partial\theta} + \frac{\cos\varphi}{r\sin\theta}\frac{\partial}{\partial\varphi}$$

$$\frac{\partial}{\partial z} = \cos\theta\frac{\partial}{\partial r} - \frac{\sin\theta}{r}\frac{\partial}{\partial\theta}$$

In order to obtain the tridimensional Laplacian in spherical polar coordinates, the next steps are: (1) to square $\partial/\partial x$, $\partial/\partial y$ and $\partial/\partial z$ and; (2) to sum these square terms. This can be done as exercise.

Another method to obtain tridimensional Laplacian in spherical polar coordinates is based on tensor analysis (which is beyond the scope of this book). See, for example, the book of Boas in chapter 10 (Boas 2006).

The Laplacian can also be expressed as:

$$\nabla^2 = \frac{\partial^2}{\partial r^2} + \frac{2}{r}\frac{\partial}{\partial r} + \frac{1}{r^2}\left(\frac{\partial^2}{\partial\theta^2} + \frac{1}{\tan\theta}\frac{\partial}{\partial\theta} + \frac{1}{\sin^2\theta}\frac{\partial^2}{\partial\varphi^2}\right)$$

Proof:

$$\nabla^2 = \frac{1}{r^2}\frac{\partial}{\partial r}\left(r^2\frac{\partial}{\partial r}\right) + \frac{1}{r^2\sin\theta}\frac{\partial}{\partial\theta}\left(\sin\theta\frac{\partial}{\partial\theta}\right) + \frac{1}{r^2\sin^2\theta}\frac{\partial^2}{\partial\varphi^2}$$

$$\nabla^2 = \frac{1}{r^2}\left(2r\frac{\partial}{\partial r} + r^2\frac{\partial}{\partial r^2}\right) + \frac{1}{r^2\sin\theta}\left(\cos\theta\frac{\partial}{\partial\theta} + \sin\theta\frac{\partial^2}{\partial\theta^2}\right) +$$

$$+\frac{1}{r^2\sin^2\theta}\frac{\partial^2}{\partial\varphi^2}$$

$$\nabla^2 = \frac{2}{r}\frac{\partial}{\partial r} + \frac{\partial^2}{\partial r^2} + \frac{1}{r^2\tan\theta}\frac{\partial}{\partial\theta} + \frac{1}{r^2}\frac{\partial^2}{\partial\theta^2} + \frac{1}{r^2\sin^2\theta}\frac{\partial^2}{\partial\varphi^2}$$

then:

$$\nabla^2 = \frac{\partial^2}{\partial r^2} + \frac{2}{r}\frac{\partial}{\partial r} + \frac{1}{r^2}\left(\frac{\partial^2}{\partial\theta^2} + \frac{1}{\tan\theta}\frac{\partial}{\partial\theta} + \frac{1}{\sin^2\theta}\frac{\partial^2}{\partial\varphi^2}\right)$$

5. Particle in a sphere and quantum angular momentum

For a particle moving in the surface of a sphere of radius *a* (or simply a particle in a sphere), the radial function is constant and the boundary conditions are:

$$V(a) = 0$$

$$R(a) = a$$

Then, there is no radial solution and the particle moving in the surface of a sphere is restricted only to the angular function. Then, the Schrödinger equation of one particle in the surface of a sphere can be written as:

$$-\frac{\hbar^2}{2\mu}\nabla^2 Y(\theta,\varphi) = EY(\theta,\varphi)$$

$$\nabla^2 = \frac{1}{r^2}\frac{\partial}{\partial r}\left(r^2\frac{\partial}{\partial r}\right) + \frac{1}{r^2}\left[\frac{1}{\sin\theta}\frac{\partial}{\partial\theta}\left(\sin\theta\frac{\partial}{\partial\theta}\right) + \frac{1}{\sin^2\theta}\left(\frac{\partial^2}{\partial\varphi^2}\right)\right]$$

$$\times r^2:$$

$$-\frac{\hbar^2}{2\mu}r^2\nabla^2 Y(\theta,\varphi) = r^2 EY(\theta,\varphi)$$

$$then:$$

$$r^2\nabla^2 Y(\theta,\varphi) = -\frac{2\mu r^2 E}{\hbar^2}Y(\theta,\varphi)$$

$$\left[r^2\nabla^2 + \frac{2\mu r^2 E}{\hbar^2}\right]Y(\theta,\varphi) = 0$$

Since r = constant, the first term in the Laplacian does not contribute (the derivative of a function over a constant does not exist) and it can be removed from the Schrödinger equation:

$$\nabla^2 = \frac{1}{r^2}\frac{\partial}{\partial r}\left(r^2\frac{\partial}{\partial r}\right) + \frac{1}{r^2}\left[\frac{1}{\sin\theta}\frac{\partial}{\partial\theta}\left(\sin\theta\frac{\partial}{\partial\theta}\right) + \frac{1}{\sin^2\theta}\left(\frac{\partial^2}{\partial\varphi^2}\right)\right]$$

$$\frac{1}{r^2}\frac{\partial}{\partial r}\left(r^2\frac{\partial}{\partial r}\right): removed$$

$$\left[r^2\nabla^2 + \frac{2\mu r^2 E}{\hbar^2}\right]Y(\theta,\varphi) = 0$$

$$\left[\frac{1}{\sin\theta}\frac{\partial}{\partial\theta}\left(\sin\theta\frac{\partial}{\partial\theta}\right) + \frac{1}{\sin^2\theta}\left(\frac{\partial^2}{\partial\varphi^2}\right) + \lambda\right]Y(\theta,\varphi) = 0$$

$$\lambda = \frac{2\mu r^2 E}{\hbar^2}$$

The function Y(θ,φ) is known as spherical harmonics and it is a product of the angular or polar equation (from the polar angle θ) and the azimuthal equation (from the azimuthal angle φ). The proof for the equation below is given in chapter seventeen (next chapter).

$$Y(\theta,\varphi) = \Theta(\theta)\Phi(\varphi)$$

Let us replace the above equation in the previous Schrödinger equation:

$$\frac{\Phi(\varphi)}{\sin\theta}\frac{\partial}{\partial\theta}\left(\sin\theta\frac{\partial\Theta(\theta)}{\partial\theta}\right)+\frac{\Theta(\theta)}{\sin^2\theta}\left(\frac{\partial^2\Phi(\varphi)}{\partial\varphi^2}\right)+\lambda\Theta(\theta)\Phi(\varphi)=0$$

$$\times\frac{\sin^2\theta}{\Theta(\theta)\Phi(\varphi)}$$

$$\frac{\sin\theta}{\Theta(\theta)}\frac{\partial}{\partial\theta}\left(\sin\theta\frac{\partial\Theta(\theta)}{\partial\theta}\right)+\frac{1}{\Phi(\varphi)}\left(\frac{\partial^2\Phi(\varphi)}{\partial\varphi^2}\right)+\lambda\sin^2\theta=0$$

The first and third terms depend only on θ whereas the second term depends only on φ. Then:

$$\frac{\sin\theta}{\Theta(\theta)}\frac{\partial}{\partial\theta}\left(\sin\theta\frac{\partial\Theta(\theta)}{\partial\theta}\right)+\lambda\sin^2\theta=-\frac{1}{\Phi(\varphi)}\left(\frac{\partial^2\Phi(\varphi)}{\partial\varphi^2}\right)$$

This equality can only be satisfied if both sides are equal to the same constant. Since the function Φ(φ) is already known as the function of the particle in a ring, then we have:

$$-\frac{1}{\Phi(\varphi)}\left(\frac{\partial^2\Phi(\varphi)}{\partial\varphi^2}\right)=m^2$$

As a consequence:

$$\frac{\sin\theta}{\Theta(\theta)}\frac{\partial}{\partial\theta}\left(\sin\theta\frac{\partial\Theta(\theta)}{\partial\theta}\right)+\lambda\sin^2\theta=m^2$$

The above equation is known as associated Legendre equation (Chapter four). A derivation of similar equation is given in detail in next chapter (Chapter seventeen). Then, the parameters λ and m are restricted to the values (see Chapter four):

$$\lambda=l(l+1),\quad l=0,1,2,\ldots$$

$$m=0,\pm1,\pm2,\ldots,\pm l$$

As already mentioned in the section of the particle in the ring, l is the azimuthal quantum number (or the orbital quantum number) and m (or m_l) is the magnetic quantum number.

Remember that λ was firstly attributed to:

$$\lambda=\frac{2\mu r^2E}{\hbar^2}$$

then:

$$\frac{2\mu r^2E}{\hbar^2}=l(l+1)$$

$$E_l=\frac{\hbar^2}{2\mu r^2}l(l+1)$$

The levels of energy of the particle in a sphere are independent of the second quantum number m. Then, the levels of energy are (2l + 1)-fold degenerate.
Since:

$$T = \frac{L^2}{2\mu r^2}$$

and:

$$E = T = \frac{\hbar^2}{2\mu r^2} l(l+1)$$

Then:

$$\mathbf{L}^2 = l(l+1)\hbar^2$$

The amplitude of the (orbital) angular momentum is obtained from the squared angular momentum and it is shown below:

$$\mathbf{L} = \hbar\sqrt{l(l+1)}$$

Since l (the azimuthal quantum number) is an integer, the quantum angular momentum of the spherical harmonics function (particle in a sphere) is quantized, i.e., it can only assume a discrete set of values.

Since the squared angular momentum and the z-component of the angular momentum commute (see Chapter eight):

$$[\mathbf{L}^2, \mathbf{L}_z] = 0$$

the functions of spherical harmonics, $Y(\theta,\varphi)$, are simultaneous eigenfunctions of $\mathbf{L}^2$ and $\mathbf{L}_z$:

$$\mathbf{L}^2 Y_{lm}(\theta,\phi) = l(l+1)\hbar^2 Y_{lm}(\theta,\phi)$$
$$\mathbf{L}_z Y_{lm}(\theta,\phi) = m\hbar Y_{lm}(\theta,\phi)$$

The former equation comes from the Schrödinger equation of the particle in a sphere and the latter equation comes from the Schrödinger equation of the particle in a ring (Section 3).

Figure 16.3 shows the discrete values of the orbital angular momentum for two azimuthal quantum numbers (l = 1 and 2).

6. Commutative property of Hamiltonian and squared angular momentum operators

Let us show that the Hamiltonian and the squared angular momentum have a commutative property, that is:

$$[H, L] = 0$$

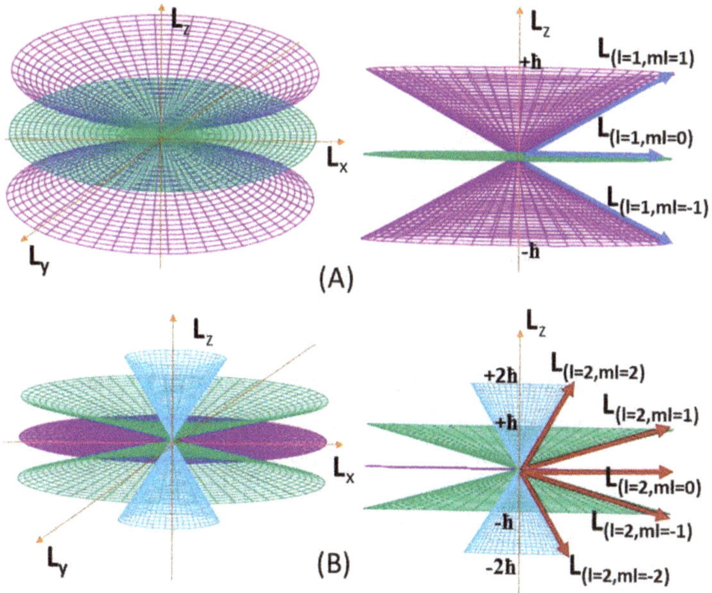

Fig. 16.3: Vector model representation of the orbital angular momentum L for (A) $l = 1$, $m_l = -1,0,1$ and (B) $l = 2$, $m_l = -2,-1,0,1,2$.

Proof:

$$\left[H, L^2\right] = HL^2 - L^2H =$$

$$= \left[-\frac{\hbar^2}{2m}\left(\frac{\partial^2}{\partial r^2} + \frac{2}{r}\frac{\partial}{\partial r} - \frac{L^2}{r^2\hbar^2}\right) + V(r)\right]L^2 -$$

$$-L^2\left[-\frac{\hbar^2}{2m}\left(\frac{\partial^2}{\partial r^2} + \frac{2}{r}\frac{\partial}{\partial r} - \frac{L^2}{r^2\hbar^2}\right) + V(r)\right]$$

$$= -\frac{\hbar^2}{2m}\frac{\partial^2}{\partial r^2}L^2 - \frac{\hbar^2}{2m}\frac{2}{r}\frac{\partial}{\partial r}L^2 + \frac{\hbar^2}{2m}\frac{L^4}{r^2\hbar^2} + \frac{\hbar^2}{2m}V(r)L^2 +$$

$$+\frac{\hbar^2}{2m}L^2\frac{\partial^2}{\partial r^2} + \frac{\hbar^2}{2m}L^2\frac{2}{r}\frac{\partial}{\partial r} - \frac{\hbar^2}{2m}\frac{L^4}{r^2\hbar^2} - \frac{\hbar^2}{2m}L^2V(r)$$

$$= -\frac{\hbar^2}{2m}\frac{\partial^2}{\partial r^2}L^2 - \frac{\hbar^2}{2m}\frac{2}{r}\frac{\partial}{\partial r}L^2 + \frac{\hbar^2}{2m}V(r)L^2 + \frac{\hbar^2}{2m}L^2\frac{\partial^2}{\partial r^2} +$$

$$+\frac{\hbar^2}{2m}L^2\frac{2}{r}\frac{\partial}{\partial r} - \frac{\hbar^2}{2m}L^2V(r)$$

The total angular momentum operator L^2 has only angular dependence. Then, the order of the operators products above is interchangeable. Hence, we have:

$$= -\frac{\hbar^2}{2m}\frac{\partial^2}{\partial r^2}L^2 - \frac{\hbar^2}{2m}\frac{2}{r}\frac{\partial}{\partial r}L^2 + \frac{\hbar^2}{2m}V(r)L^2 + \frac{\hbar^2}{2m}\frac{\partial^2}{\partial r^2}L^2 +$$

$$+\frac{\hbar^2}{2m}\frac{2}{r}\frac{\partial}{\partial r}L^2 - \frac{\hbar^2}{2m}V(r)L^2$$

Which leads to a zero result.

Exercises

(1) Give the normalization constant of the wave function of the particle in the ring.

(2) Improve the code of the exercise from chapter one (See Exercise section of Chapter one) by including: (a) the normalized function; (b) the probability density of this wave function.

(3) Obtain the tridimensional Laplace operator in spherical polar coordinates from the relations $\partial/\partial x$, $\partial/\partial y$ and $\partial/\partial z$ obtained in the fourth step of the Section 4. See the tips in the Section 5.

(4) Show that:

$[H, L_z] = 0$

Reference cited

Boas, M.L. 2006. Mathematical methods in the physical sciences. John Wiley & Sons, Inc. Third edition, Hoboken.

Hydrogen-like Atom and Atomic Orbitals

17

1. Introduction

The hydrogen atom has one proton of charge +Q = 1.60×10^{-19}C and one electron of charge –Q = 1.60×10^{-19}C and the charge of the electron per unit of volume is:

$$\rho(r) = -\left(Q/\pi a_0^3\right)e^{-2r/a_0}$$

Where $a_0 = 5.29 \times 10^{-11}$m is the Bohr radius.

The proton may be regarded at the center of the atom while the electron moves hugely fast 'around' the proton (1/137th of the speed of light). The proton can be regarded as a stationary particle since it is nearly 1836 times heavier than the electron and the proton moves very, very slowly in response to the force exerted by the electron. The proton generates a spherical electric field and the potential energy of the electron does not vary in time (only in space):

$$U(r) = -K\frac{e^2}{r}, \qquad K = 1/4\pi \in_0$$

As a consequence, we can use time independent Schrödinger equation for the hydrogen atom.

In the ground state of the hydrogen atom, the motion of the electron is smeared out in a spherical distribution. Since U(r) is spherically symmetrical, we can solve Schrödinger equation in terms of spherical coordinates (r,θ,φ) instead of rectangular coordinates (x,y,z).

We will use the reduced mass, μ, of these two particles to solve the problem of the hydrogen atom. Notice that we use the same symbol (μ) to represent the mass of a single particle in the last chapters. The reduced mas of the hydrogen atom is nearly the mass of the electron. Then, we can consider the hydrogen atom as a particle in a box with spherical, soft walls.

$$\mu = \frac{m_p m_e}{m_p + m_e} = \frac{1836 m_e^2}{1837 m_e} \approx m_e$$

Due to the spherical symmetry of the Coulomb potential, there are two commuting operators for the angular momentum: L^2 and L_z. Since the angular momentum operators do not include the radial variable, another operator needs to be included. The third operator is the Hamiltonian, H, because it includes the angle and radial variables and it commutes with both L^2 and L_z (see previous chapter). The complete set of commuting operators (observables) for the hydrogen atom is: H, L^2 and L_z (excluding the spin which is artificially incorporated in wave quantum mechanics). The Schrödinger equation in spherical coordinates is:

$$H\psi = E\psi, \quad \left(-\frac{\hbar^2}{2m}\nabla^2 + V\right)\psi = E\psi, \quad V = -K\frac{e^2}{r}$$

$$\nabla^2 = \left[\frac{1}{r^2}\frac{\partial}{\partial r}\left(r^2\frac{\partial}{\partial r}\right) + \frac{1}{r^2\sin\theta}\frac{\partial}{\partial\theta}\left(\sin\theta\frac{\partial}{\partial\theta}\right) + \frac{1}{r^2\sin^2\theta}\frac{\partial^2}{\partial\phi^2}\right]$$

Where K is the electrostatic constant or Coulomb constant.

The hydrogen-like atoms are any atomic nucleus bound to one electron like He^+, Li^{+2}, Be^{+3}, so on, having a similar solution as hydrogen atom using Schrödinger equation.

2. Particle in a spherically symmetric potential

The particle of mass μ confined to a spherically symmetric potential with radius *a* has the following boundary conditions:

$$V(r) = \begin{cases} 0 & if: r \le a \\ \infty & f: r > a \end{cases}$$

The time-independent Schrödinger equation in spherical polar coordinate is:

$$\left[-\frac{\hbar^2}{2\mu}\nabla^2 + V(r)\right]\psi = E\psi$$

$$\nabla^2 = \frac{1}{r^2}\frac{\partial}{\partial r}\left(r^2\frac{\partial}{\partial r}\right) - \frac{1}{\hbar^2 r^2}L^2$$

$$\frac{1}{2\mu r^2}\left[-\hbar^2\frac{\partial}{\partial r}\left(r^2\frac{\partial}{\partial r}\right) + L^2\right]\psi + V(r)\psi = E\psi$$

$$\frac{-\hbar^2}{2\mu}\left[\begin{array}{l}\frac{1}{r^2}\frac{\partial}{\partial r}\left(r^2\frac{\partial}{\partial r}\right) + \\ \frac{1}{r^2\sin\theta}\frac{\partial}{\partial\theta}\left(\sin\theta\frac{\partial}{\partial\theta}\right) + \frac{1}{r^2\sin^2\theta}\frac{\partial^2}{\partial\varphi^2}\end{array}\right]\psi + V(r)\psi = E\psi$$

$$\times\left(-2\mu/\hbar^2\right)$$

$$\frac{1}{r^2}\frac{\partial}{\partial r}\left(r^2\frac{\partial\psi}{\partial r}\right) + \frac{1}{r^2\sin\theta}\frac{\partial}{\partial\theta}\left(\sin\theta\frac{\partial\psi}{\partial\theta}\right) + \frac{1}{r^2\sin^2\theta}\frac{\partial^2\psi}{\partial\varphi^2} -$$

$$-\frac{2\mu}{\hbar^2}\left[V(r) - E\right]\psi = 0$$

We assume a solution for the above equation based on independent, separate variables of the form:

$$\psi\left(r,\theta,\varphi\right)=R(r)Y(\theta,\varphi)$$

Let us substitute the above equation on the previous Schrödinger equation:

$$\frac{1}{r^2}\frac{\partial}{\partial r}\left(r^2\frac{\partial\left[R(r)Y(\theta,\varphi)\right]}{\partial r}\right)+\frac{1}{r^2\sin\theta}\frac{\partial}{\partial\theta}\left(\sin\theta\frac{\partial\left[R(r)Y(\theta,\varphi)\right]}{\partial\theta}\right)+$$

$$\frac{1}{r^2\sin^2\theta}\frac{\partial^2\left[R(r)Y(\theta,\varphi)\right]}{\partial\varphi^2}-\frac{2\mu}{\hbar^2}\left[V(r)-E\right]R(r)Y(\theta,\varphi)=0$$

The above equation becomes:

$$\frac{Y(\theta,\varphi)}{r^2}\frac{\partial}{\partial r}\left(r^2\frac{\partial\left[R(r)\right]}{\partial r}\right)+\frac{R(r)}{r^2\sin\theta}\frac{\partial}{\partial\theta}\left(\sin\theta\frac{\partial\left[Y(\theta,\varphi)\right]}{\partial\theta}\right)+$$

$$\frac{R(r)}{r^2\sin^2\theta}\frac{\partial^2\left[Y(\theta,\varphi)\right]}{\partial\varphi^2}-\frac{2\mu}{\hbar^2}\left[V(r)-E\right]R(r)Y(\theta,\varphi)=0$$

Dividing the above equation by R(r)Y(θ,φ), multiplying by r^2, we have:

$$\frac{1}{R(r)}\frac{\partial}{\partial r}\left(r^2\frac{\partial\left[R(r)\right]}{\partial r}\right)+\frac{1}{Y(\theta,\varphi)\sin\theta}\frac{\partial}{\partial\theta}\left(\sin\theta\frac{\partial\left[Y(\theta,\varphi)\right]}{\partial\theta}\right)+$$

$$\frac{1}{Y(\theta,\varphi)\sin^2\theta}\frac{\partial^2\left[Y(\theta,\varphi)\right]}{\partial\varphi^2}-\frac{2\mu r^2}{\hbar^2}\left[V(r)-E\right]=0$$

Rearranging the terms, we have:

$$\left\{\frac{1}{R(r)}\frac{\partial}{\partial r}\left(r^2\frac{\partial\left[R(r)\right]}{\partial r}\right)-\frac{2\mu r^2}{\hbar^2}\left[V(r)-E\right]\right\}+$$

$$+\left\{\begin{matrix}\dfrac{1}{Y(\theta,\varphi)\sin\theta}\dfrac{\partial}{\partial\theta}\left(\sin\theta\dfrac{\partial\left[Y(\theta,\varphi)\right]}{\partial\theta}\right)\\+\dfrac{1}{Y(\theta,\varphi)\sin^2\theta}\dfrac{\partial^2\left[Y(\theta,\varphi)\right]}{\partial\varphi^2}\end{matrix}\right\}=0$$

The first two terms of the above equation depend only on the variable r, while the last two terms depend only on the angle variables. Then, we have a sum of two separate, independent equations: radial equation and angular equation. Except for the trivial solution, the only way the sum of the first two terms and the sum of the last two terms be zero is when they assume the same constant with opposite signs. Let us choose this separation constant as l(l+1), which comes from Legendre equation (chapter four). Then, we have the following radial equation:

$$\frac{1}{R(r)}\frac{\partial}{\partial r}\left(r^2\frac{\partial\left[R(r)\right]}{\partial r}\right)-\frac{2\mu r^2}{\hbar^2}\left[V(r)-E\right]=l(l+1)$$

And the following angular equation:

$$\frac{1}{Y(\theta,\varphi)\sin\theta}\frac{\partial}{\partial\theta}\left(\sin\theta\frac{\partial[Y(\theta,\varphi)]}{\partial\theta}\right)+$$
$$+\frac{1}{Y(\theta,\varphi)\sin^2\theta}\frac{\partial^2[Y(\theta,\varphi)]}{\partial\varphi^2}=-l(l+1)$$

3. Solution to the angular equation

As it was derived in the Section 2, the angular equation is:

$$\frac{1}{Y(\theta,\varphi)\sin\theta}\frac{\partial}{\partial\theta}\left(\sin\theta\frac{\partial[Y(\theta,\varphi)]}{\partial\theta}\right)+\frac{1}{Y(\theta,\varphi)\sin^2\theta}\frac{\partial^2[Y(\theta,\varphi)]}{\partial\varphi^2}=-l(l+1)$$

Let us substitute the trial function $\Theta(\theta)\Phi(\varphi)$ in the above equation:

$$\frac{1}{\Theta(\theta)\Phi(\varphi)\sin\theta}\frac{\partial}{\partial\theta}\left(\sin\theta\frac{\partial[\Theta(\theta)\Phi(\varphi)]}{\partial\theta}\right)+$$
$$+\frac{1}{\Theta(\theta)\Phi(\varphi)\sin^2\theta}\frac{\partial^2[\Theta(\theta)\Phi(\varphi)]}{\partial\varphi^2}=-l(l+1)$$

Hence, we get:

$$\frac{1}{\Theta(\theta)\sin\theta}\frac{\partial}{\partial\theta}\left(\sin\theta\frac{\partial[\Theta(\theta)]}{\partial\theta}\right)+\frac{1}{\Phi(\varphi)\sin^2\theta}\frac{\partial^2[\Phi(\varphi)]}{\partial\varphi^2}=-l(l+1)$$

Next, we multiply the above equation in both sides by $\sin^2\theta$ and rearranging:

$$\frac{1}{\Phi(\varphi)}\frac{\partial^2[\Phi(\varphi)]}{\partial\varphi^2}+\frac{\sin\theta}{\Theta(\theta)}\frac{\partial}{\partial\theta}\left(\sin\theta\frac{\partial[\Theta(\theta)]}{\partial\theta}\right)+l(l+1)\sin^2\theta=0$$

Then, we observe that the Hamiltonian is the sum of two independent separate variables: θ and φ. The first term depends only on φ and the other two terms depend only on θ. Hence, it is possible to separate the above equation in two independent equations: the azimuthal angle equation (depending only on φ) and the polar angle equation (depending only on θ). The product of these two angle equations, $Y(\theta,\varphi)$, is called spherical harmonics.

$$Y(\theta,\varphi)=\Theta(\theta)\cdot\Phi(\varphi)$$

3.1 The azimuthal angle equation

The first part of the above equation (the azimuthal angle equation) is associated with a number, let us say $-m^2$, which has the opposite sign of the number associated with other two terms (of the polar angle equation) so that the whole equation is zero. Then,

$$\frac{1}{\Phi(\varphi)}\frac{d^2\Phi(\varphi)}{d\varphi^2} = -m^2$$

$$\frac{\sin\theta}{\Theta(\theta)}\frac{d}{d\theta}\left(\sin\theta\frac{d\Theta(\theta)}{d\theta}\right) + l(l+1)\sin^2\theta = +m^2$$

The azimuthal angle equation (first equation) is similar to that of the particle in ring:

$$\frac{\partial^2\psi}{\partial\varphi^2} = -m^2\psi$$

Then, the solution of the azimuthal angle function, $\Phi(\varphi)$, is similar to that from the particle in a ring.

$$\frac{d^2\Phi(\varphi)}{d\varphi^2} = -m^2\Phi(\varphi)$$

$$\Phi(\varphi) \approx \exp(im\varphi)$$

The normalized wave function is:

$$\Phi(\varphi) = \sqrt{\frac{1}{2\pi}}\exp(im\varphi)$$

From the particle in a ring problem, the energy of this system is:

$$E = \frac{m^2\hbar^2}{2I}, \qquad m = 0, \pm 1, \pm 2, \ldots$$

And by knowing the relation between the angular momentum and kinetic energy (Section two) and the quantum kinetic energy:

$$T = \frac{L_z^2}{2I}, \qquad \textit{classic}$$

$$T = \frac{m^2\hbar^2}{2I}, \qquad \textit{quantum}$$

We have the equation for the quantum angular momentum in the z axis:

$$L_z^2 = m^2\hbar^2$$

$$L_z = \pm m\hbar$$

3.2 The polar angle equation

Let us now consider the polar angle equation (the second equation with respect to θ). After multiplication by $\Theta(\theta)$, it becomes:

$$\frac{\sin\theta}{\Theta(\theta)}\frac{d}{d\theta}\left(\sin\theta\frac{d}{d\theta}\right) + l(l+1)\sin^2\theta = +m^2$$

$$\frac{\sin\theta}{\Theta(\theta)}\frac{d}{d\theta}\left(\sin\theta\frac{d\Theta(\theta)}{d\theta}\right) + l(l+1)\sin^2\theta - m^2 = 0$$

$$\times\Theta(\theta)$$

$$\sin\theta\frac{d}{d\theta}\left(\sin\theta\frac{d\Theta(\theta)}{d\theta}\right)+\left(l(l+1)\sin\theta-\frac{m^2}{\sin\theta}\right)\Theta(\theta)=0$$

Evaluating the first term of the above equation, we have:

$$\sin\theta\frac{d}{d\theta}\left(\sin\theta\frac{d\Theta(\theta)}{d\theta}\right)=\sin\theta\left(\cos\theta\frac{d\Theta(\theta)}{d\theta}+\sin\theta\frac{d^2\Theta(\theta)}{d\theta^2}\right)$$

$$\sin\theta\frac{d}{d\theta}\left(\sin\theta\frac{d\Theta(\theta)}{d\theta}\right)=\sin\theta\cos\theta\frac{d\Theta(\theta)}{d\theta}+\sin^2\theta\frac{d^2\Theta(\theta)}{d\theta^2}$$

Then, the polar angle equation becomes:

$$\sin\theta\cos\theta\frac{d\Theta(\theta)}{d\theta}+\sin^2\theta\frac{d^2\Theta(\theta)}{d\theta^2}+\left[l(l+1)\sin^2\theta-m^2\right]\Theta(\theta)=0$$

Let us now change the variable θ into x and assume x=cos θ. Firstly, we have to change the derivatives of the above equation:

$$x=\cos\theta$$

$$\frac{d\Theta(\theta)}{d\theta}=\frac{d\Theta(x)}{dx}\frac{dx}{d\theta}=-\sin\theta\frac{d\Theta(x)}{dx}$$

$$\frac{d^2\Theta(\theta)}{d\theta^2}=\frac{d}{d\theta}\left(-\sin\theta\frac{d\Theta(x)}{dx}\right)=-\cos\theta\frac{d\Theta(x)}{dx}-\sin\theta\frac{d}{d\theta}\frac{d\Theta(x)}{dx}$$

$$\frac{d^2\Theta(\theta)}{d\theta^2}=-\cos\theta\frac{d\Theta(x)}{dx}-\sin\theta\frac{d}{dx}\frac{dx}{d\theta}\frac{d\Theta(x)}{dx}$$

$$\frac{d^2\Theta(\theta)}{d\theta^2}=-\cos\theta\frac{d\Theta(x)}{dx}-\sin\theta\frac{d}{dx}(-\sin\theta)\frac{d\Theta(x)}{dx}$$

$$\frac{d^2\Theta(\theta)}{d\theta^2}=-\cos\theta\frac{d\Theta(x)}{dx}-\sin^2\theta\frac{d^2\Theta(x)}{dx^2}$$

Secondly, we replace the above derivatives in the previous polar angle equation:

$$\sin\theta\cos\theta\left[-\sin\theta\frac{d\Theta(x)}{dx}\right]+\sin^2\theta\left[-\cos\theta\frac{d\Theta(x)}{dx}+\sin^2\theta\frac{d^2\Theta(x)}{dx^2}\right]+$$
$$+\left[l(l+1)\sin^2\theta-m^2\right]\Theta(x)=0$$

Thirdly, we divide the above equation by $\sin^2\theta$:

$$-\cos\theta\frac{d\Theta(x)}{dx}-\cos\theta\frac{d\Theta(x)}{dx}+\sin^2\theta\frac{d^2\Theta(x)}{dx^2}+\left[l(l+1)-\frac{m^2}{\sin^2\theta}\right]\Theta(x)=0$$

$$-2\cos\theta\frac{d\Theta(x)}{dx}+\sin^2\theta\frac{d^2\Theta(x)}{dx^2}+\left[l(l+1)-\frac{m^2}{\sin^2\theta}\right]\Theta(x)=0$$

Since we assume x = cosθ, we have:

$$x = \cos\theta, \quad \sin^2\theta = 1 - x^2$$

$$-2x\frac{d\Theta(x)}{dx} + \left(1-x^2\right)\frac{d^2\Theta(x)}{dx^2} + \left[l(l+1) - \frac{m^2}{1-x^2}\right]\Theta(x) = 0$$

$$\frac{d}{dx}\left[\left(1-x^2\right)\frac{d}{dx}\Theta(x)\right] + \left[l(l+1) - \frac{m^2}{1-x^2}\right]\Theta(x) = 0$$

which is similar to the associated Legendre equation:

$$\frac{d}{dx}\left[\left(1-x^2\right)\frac{d}{dx}P_l^m(x)\right] + \left[l(l+1) - \frac{m^2}{1-x^2}\right]P_l^m(x) = 0$$

The solutions for the associated Legendre equation are known as the associated Legendre polynomials (see Chapter four).

$$p_l^m(x) = \frac{(-1)^m}{2^l l!}(1-x^2)^{m/2}\frac{d^{l+m}}{dx^{l+m}}(x^2-1)^l$$

$$p_l^{-m}(x) = (-1)^m\frac{(l-m)!}{(l+m)!}p_l^m(x)$$

Then, the solution for the polar angle function, Θ(θ), is the set of the associated Legendre polynomials.

$$\Theta(\theta) = P_l^m(x)$$

The orthogonality relation for the associated Legendre functions is shown in chapter four to yield:

$$\int_{-1}^{1} P_l^m(x)P_k^m(x)dx = \frac{2}{2l+1}\frac{(l+m)!}{(l-m)!}\delta_{lk}$$

Then, the normalized polar angle function, Θ(θ), is:

$$N = (-1)^{m+|m|}\left[\frac{2l+1}{2}\frac{(l-m)!}{(l+m)!}\right]^{1/2}$$

3.3 The spherical harmonics

As previously mentioned, the spherical harmonics is given by product of the azimuthal angle function and the polar angle function.

$$Y(\theta,\varphi) = \Theta(\theta)\cdot\Phi(\varphi)$$

Then, the normalized spherical harmonics is:

$$Y(\theta,\varphi) = N'\exp(im\,\varphi)\cdot N''P_l^m(\cos\theta)$$

Where

$$N' = \frac{1}{\sqrt{2\pi}}, \quad N'' = (-1)^{m+|m|}\left[\frac{2l+1}{2}\frac{(l-m)!}{(l+m)!}\right]^{1/2}$$

Then, the normalization constant of the spherical harmonics is the product N'N", giving the following expression for the normalized total angular function, Y(θ,φ):

$$Y(\theta,\varphi) = (-1)^{m+|m|}\left[\frac{2l+1}{4\pi}\frac{(l-m)!}{(l+m)!}\right]^{1/2} P_l^m(\cos\theta)e^{im\varphi}$$

The first spherical harmonics, Y(θ,φ), are shown below:

$$Y_0^0 = \sqrt{\frac{1}{4\pi}}$$

$$Y_1^0 = \sqrt{\frac{3}{4\pi}}\cos\theta$$

$$Y_1^{\pm1} = \pm\sqrt{\frac{3}{8\pi}}\sin\theta\exp(\pm i\varphi)$$

$$Y_2^0 = \sqrt{\frac{5}{16\pi}}\left(3\cos^2\theta - 1\right)$$

$$Y_2^{\pm1} = \pm\sqrt{\frac{15}{8\pi}}\sin\theta\cos\theta\exp(\pm i\varphi)$$

$$Y_2^{\pm2} = \pm\sqrt{\frac{15}{32\pi}}\sin^2\theta\exp(\pm 2i\varphi)$$

The Fig. 17.1 shows the plot of some spherical harmonics. See in the section do-it-yourself activity how to plot these graphs.

4. Solution to the radial equation of the hydrogen-like atom

As it was derived in Section 2, the radial equation is:

$$\frac{1}{R(r)}\frac{\partial}{\partial r}\left(r^2\frac{\partial[R(r)]}{\partial r}\right) - \frac{2\mu r^2}{\hbar^2}[V(r) - E] = l(l+1)$$

Let us make explicit the expression of the Coulomb potential in the radial equation as shown below. See that Z represents the atomic number (number of protons in the nucleus). For example, the atomic number of some hydrogen-like atoms are: for Z = 1 (for H), Z = 2(for He^{+2}), Z = 3 (for Li^{+3}), so on.

$$\frac{1}{R(r)}\frac{d}{dr}\left(r^2\frac{d}{dr}\right)R(r) - \frac{2\mu r^2}{\hbar^2}\left[-\frac{KZe^2}{r} - E\right] - l(l+1) = 0$$

$$V(r) = -K\frac{Ze^2}{r}, K = \frac{1}{4\pi\in_0} = 9\times10^9\,Nm^2C^{-2}$$

$$V(r) = -2.3\times10^{-28}\frac{Z}{r}$$

where e is the electron charge (e = 1.6021 × 10^{-19} C) and $\in_0$ is the vacuum permittivity ($\in_0$ = 8.854 × 10^{-12} C/Vm).

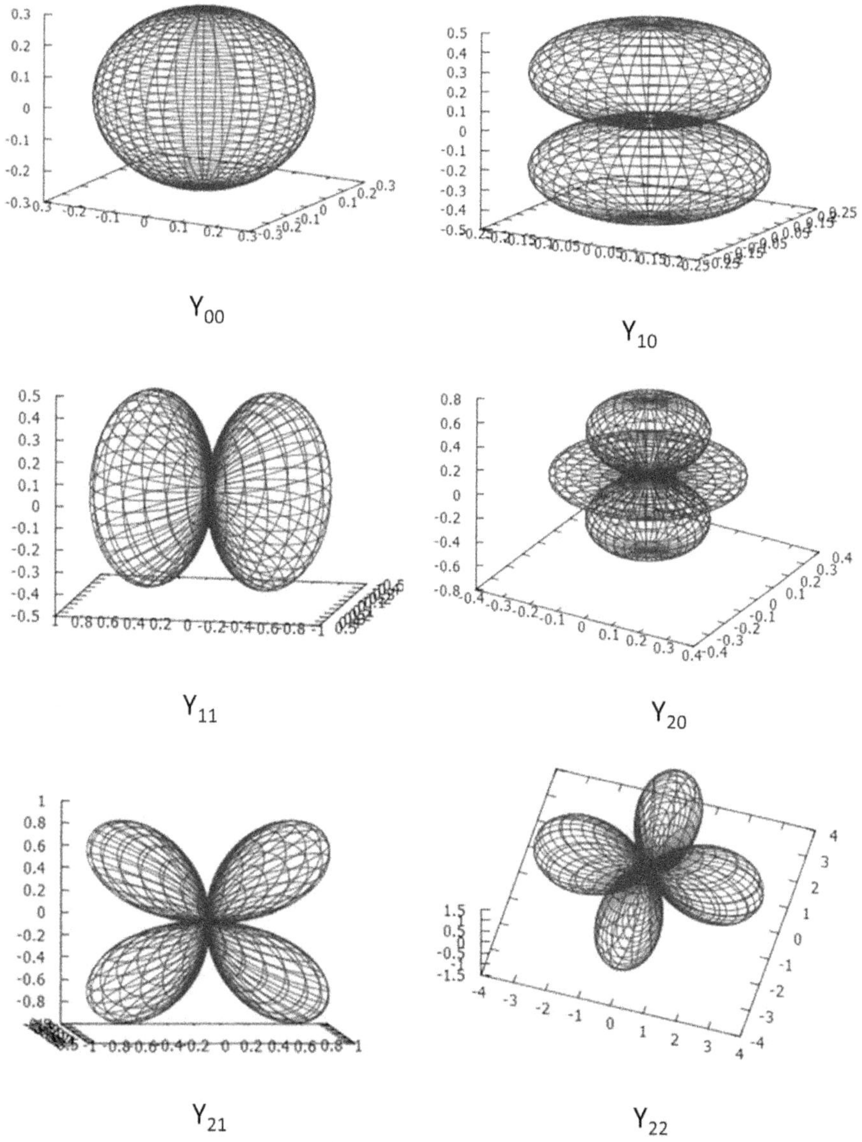

Fig. 17.1: Graphs of $Y_{0,0}$, $Y_{1,0}$, $Y_{0,1}$, $Y_{2,0}$, $Y_{2,1}$, and $Y_{2,2}$.

Let us multiply it by R(r)

$$\frac{d}{dr}\left(r^2\frac{d}{dr}\right)R(r)-\frac{2\mu r^2}{\hbar^2}\left[-\frac{KZe^2}{r}-E\right]R(r)-l(l+1)R(r)=0$$

By rearranging the second and third terms of R(r), we have:

$$\frac{d}{dr}\left(r^2\frac{d}{dr}\right)R(r)+\left[\frac{2\mu r^2}{\hbar^2}\frac{KZe^2}{r}+\frac{2\mu r^2}{\hbar^2}E-l(l+1)\right]R(r)=0$$

Let us undimensionalize partially the radial equation. It has two dimensional physical quantities: r and E. We have to change r into x and E into ε (in π m^{-1}).

Let us undimensionalize the length r. Table 17.1 shows the physical quantities and corresponding units (in SI) related to the radial equation for the hydrogen-like atom.

Let us replace R(r) with y(x) and r with x. Then, we intend to use y and x variables instead of R(r) and r variables in the radial equation.

By knowing that the volume element (dV) in spherical coordinates is given by three edges of an infinitesimal cube (see below) according to the Fig. 17.2.

$$edge1 : dr$$
$$edge2 : rd\theta$$
$$edge3 : r\sin\theta d\varphi$$

Where the edges 2 and 3 are related to the lengths of arc circumferences, l, from the relation below:

Length (m)	Angle (degrees)
2πr	360
l	α

$$l = \frac{\alpha \cdot 2\pi r}{360} = \frac{\alpha \cdot \pi r}{180} \quad or \quad l = \alpha_{rad} \cdot r$$

The probability to find the particle between r and r + dr from the center is given by:

$$\psi(r,\theta,\varphi) = R(r)Y(\theta,\varphi)$$
$$dV = rd\theta \cdot r\sin\theta d\varphi \cdot dr$$
$$then : \int \left|\psi(r,\theta,\varphi)\right|^2 rd\theta \cdot r\sin\theta d\varphi \cdot dr =$$
$$= \int \left|R(r)\right|^2 r^2 dr = \int \left|y(r)\right|^2 dr$$

Table 17.1: Physical quantities and corresponding units in SI.

Physical quantity	Unit (in SI)
Energy (E)	$kg.m^2.s^{-2}$
Length (r)	m
Mass (m)	kg
Planck's constant (h)	$kg.m^2.s^{-1}$

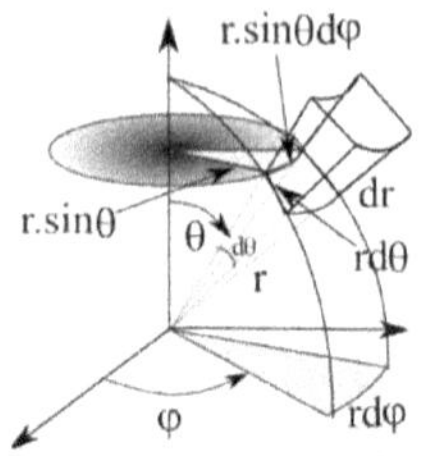

Fig. 17.2: Volume element in spherical coordinates.

Where we eliminated the angular variables from the integration above since the variation occurs only in the radial function.
Then, we have:

$$y(r) = rR(r) \quad \Rightarrow \quad R(r) = \frac{y(r)}{r}$$

Let us now make the proper substitution of R(r) in the first term of the radial equation:

$$\frac{d}{dr}\left(r^2 \frac{d}{dr}\right)R(r) = \frac{d}{dr}\left(r^2 \frac{d}{dr}\right)\left(r^{-1}\right)y(r) =$$

$$= \frac{d}{dr} r^2 \left[\left(-r^{-2}\right)y(r) + \left(r^{-1}\right)\frac{dy(r)}{dr}\right] =$$

$$= \frac{d}{dr}\left[-y(r) + r\frac{dy(r)}{dr}\right] = -\frac{dy(r)}{dr} + \frac{dy(r)}{dr} + r\frac{d^2 y(r)}{dr^2}$$

$$then: \frac{d}{dr}\left(r^2 \frac{d}{dr}\right)R(r) = r\frac{d^2 y(r)}{dr^2}$$

The radial equation becomes:

$$r\frac{d^2 y(r)}{dr^2} + \left[\frac{2\mu r^2 KZe^2}{r\hbar^2} + \frac{2\mu r^2 E}{\hbar^2} - l(l+1)\right]\frac{y(r)}{r} = 0$$

$$\Rightarrow r\frac{d^2 y(r)}{dr^2} + \left[\frac{2\mu r^2 KZe^2}{r^2\hbar^2} + \frac{2\mu r^2 E}{r\hbar^2} - \frac{l(l+1)}{r}\right]y(r) = 0$$

$\times 1/r$:

$$\Rightarrow \frac{d^2 y(r)}{dr^2} + \left[\frac{2\mu KZe^2}{r\hbar^2} + \frac{2\mu E}{\hbar^2} - \frac{l(l+1)}{r^2}\right]y(r) = 0$$

We have not yet changed r into x. We have only changed R(r) into y(r). Next, we will undimensionalize r. Let us make a second substitution:

$$-\left(\frac{\varepsilon}{2}\right)^2 = \frac{2\mu E}{\hbar^2}, \qquad \varepsilon = \frac{\sqrt{8\mu E}}{\hbar}$$

$$-[\varepsilon] = \frac{2\sqrt{2\left[kg \cdot m^2 \cdot s^{-2}\right][kg]}}{\left[kg \cdot m^2 \cdot s^{-1}(2\pi)^{-1}\right]} = \left[\frac{\pi}{m}\right]$$

The negative sign refers to bound states where $E < 0$.
Then, in order to undimensionalize the length r, we have:

$$x = r \cdot \varepsilon$$

$$[x] = [m] \cdot \frac{[\pi]}{[m]} = [\pi]$$

$$r = \frac{x}{\varepsilon}$$

As a consequence:

$$dr = \frac{dx}{\varepsilon}$$

$$\frac{d^2y(r)}{dr^2} = \frac{d}{dr}\frac{dy(r)}{dr}$$

$$1/dr = \varepsilon/dx, \quad y(r) \Rightarrow y(x)$$

$$then:$$

$$\frac{d^2y(r)}{dr^2} = \varepsilon\frac{d}{dx}\varepsilon\frac{dy(x)}{dx} = \varepsilon^2\frac{d^2y(x)}{dx^2}$$

The radial equation becomes:

$$\frac{d^2y(r)}{dr^2} + \left[\frac{2\mu KZe^2}{r\hbar^2} + \frac{2\mu E}{\hbar^2} - \frac{l(l+1)}{r^2}\right]y(r) = 0$$

$$\Rightarrow \varepsilon^2\frac{d^2y(x)}{dx^2} + \left[\frac{2\mu KZe^2\varepsilon}{x\hbar^2} - \frac{\varepsilon^2}{4} - \varepsilon^2\frac{l(l+1)}{x^2}\right]y(x) = 0$$

$$\times 1/\varepsilon^2$$

$$\frac{d^2y(x)}{dx^2} + \left[\frac{2\mu KZe^2}{x\varepsilon\hbar^2} - \frac{1}{4} - \frac{l(l+1)}{x^2}\right]y(x) = 0$$

The Bohr radius is given by the expressions:

$$a_0 = \frac{\hbar^2}{m_e e^2}[G.u.] = \frac{4\pi \in_0 \hbar^2}{m_e e^2}[SI] = 52.86\,pm \therefore$$

$$[a_0] = \frac{\left[C^2 \frac{s^2}{m\cdot kg} m^{-2}\right]\left[m^2 kg \cdot s^{-1}\right]^2}{[kg][C]^2} = [m]$$

$$as: \mu \approx m_e$$

$$then: a_0 = \frac{\hbar^2}{\mu e^2} = 52.9\,pm$$

Where pm means picometer (10^{-12} m), vacuum permitivity unit in SI unit might be $C^2 N^{-1} m^{-2}$ or $C V^{-1} m^{-1}$ and G.u. is Gaussian unit.
Then, let us use Bohr radius in the radial equation above:

$$\frac{d^2y(x)}{dx^2} + \left[\frac{2KZ}{x\varepsilon a_0} - \frac{1}{4} - \frac{l(l+1)}{x^2}\right]y(x) = 0$$

The equation above is equivalent to the differential equation below whose solution, $y_j^k(x)$, is also provided.

$$y_j^{k}{}''(x) + \left(-\frac{1}{4} + \frac{2j+k+1}{2x} - \frac{k^2-1}{4x^2}\right)y_j^k(x) = 0$$

$$y_j^k(x) = e^{-x/2}x^{(k+1)/2}L_j^k(x)$$

Where

$$l(l+1)=\frac{k^2-1}{4}$$

$$\frac{2KZ}{a_0\varepsilon}=\frac{2j+k+1}{2}$$

The above equation is a slightly different a type of the associated Laguerre equation (Chapter 4), See in the Appendix that the replacement of the solution $y_j^k(x)$ into the above equation along with some derivation yields to the associated Laguerre equation depicted below:

$$xL_j^k{}''(x)+(1-x+k)L_j^k{}'(x)+jL_j^k(x)=0$$

Then, by comparing:

$$\frac{d^2y(x)}{dx^2}+\left[\frac{2KZ}{x\varepsilon a_0}-\frac{1}{4}-\frac{l(l+1)}{x^2}\right]y(x)=0$$

And:

$$y_j^k{}''(x)+\left(-\frac{1}{4}+\frac{2j+k+1}{2x}-\frac{k^2-1}{4x^2}\right)y_j^k(x)=0$$

We see that:

$$k=2l+1$$

$$then:\frac{2j+k+1}{2}=\frac{2j+2l+1+1}{2}=j+l+1$$

$$and:n=j+l+1$$

Where n is the principal quantum number.
Then, we have the following equation:

$$\frac{2KZ}{a_0\varepsilon}=n$$

And we can obtain the equation for the total energy E with respect to the principal quantum number.

$$\varepsilon=\frac{2KZ}{na_0}\Rightarrow\varepsilon^2=\frac{4K^2Z^2}{n^2a_0^2}$$

As we have done to undimensionalize r, we obtained:

$$-\left(\frac{\varepsilon}{2}\right)^2=\frac{2\mu E}{\hbar^2}\Rightarrow\varepsilon^2=-\frac{8\mu E}{\hbar^2}$$

By comparing both equations:

$$-\frac{8\mu E}{\hbar^2}=\frac{4K^2Z^2}{n^2a_0^2}$$

$$E=-\frac{K^2Z^2\hbar^2}{2\mu n^2a_0^2}$$

The energy can be expressed in Rydberg (Ry):

$$a_0 = \frac{\hbar^2}{m_e e^2}[G.u.], \qquad a_0 = \frac{4\pi \in_0 \hbar^2}{m_e e^2}[SI]$$

$$\mu \cong m_e, \qquad K = \frac{1}{4\pi \in_0}$$

$$E = -\frac{K^2 Z^2 \hbar^2}{2m_e n^2 a_0^2} = -\frac{K^2 Z^2 \hbar^2 m_e^2 e^4}{2m_e n^2 \hbar^4}$$

$$E = -\frac{K^2 Z^2 m_e e^4}{2n^2 \hbar^2} = -\frac{Z^2}{n^2} Ry$$

$$Ry = \frac{K^2 m_e e^4}{2\hbar^2} = 13.6058 \text{ eV}$$

$$E_h = -\frac{Z^2}{2n^2}(a.u.)$$

$$2Ha = 1Ry$$

Where m_e is the mass of the electron ($m_e = 9.11 \times 10^{-31}$ kg), 1 electron-volt is equivalent to 1.602×10^{-19} J and E_h is energy in Hartree.

The atomic unit for energy is Hartree (Ha) instead of Rydberg. As one can see in the last equation above, two Hartrees is equivalent to one Rydberg. Then, the energy of the hydrogen atom in the ground state (n = 1, l = 0) is –0.5 Hartree. See Table 17.2. Let us check the relation between Ry and electron-volt.

$$Ry = \frac{K^2 m_e e^4}{2\hbar^2} = \frac{\left(9\times10^9\right)^2\left(9.109\times10^{-31}\right)\left(1.602\times10^{-19}\right)^4}{2\left(1.054\times10^{-34}\right)^2}$$

$$Ry = \frac{\left(81\times10^{18}\right)\left(9.109\times10^{-31}\right)\left(6.586\times10^{-76}\right)}{2.222\times10^{-68}}$$

$$Ry = \frac{4.859\times10^{-86}}{2.222\times10^{-68}} = 21.8676\times10^{-19} J$$

$$\frac{1eV}{1.602\times10^{-19} J}\times 21.8676\times10^{-19} J = 13.65eV$$

$$[Ry] = \frac{\left[Nm^2C^{-2}\right]^2 kg\left[C\right]^4}{\left[m^2 kg\cdot s^{-1}\right]^2} = \frac{N^2 m^4 C^{-4} kg C^4}{m^4 kg^2\cdot s^{-2}} = \frac{N^2}{kg\cdot s^{-2}}$$

$$[Ry] = \frac{\left[kg\cdot m\cdot s^{-2}\right]^2}{kg\cdot s^{-2}} = kg\cdot m^2\cdot s^{-2} = J$$

The difference between the calculated Rydberg value above and the correct one might be attributed to the rounding-off errors used for simplicity reasons in the calculations above.

Important to emphasize that Ry is the Rydberg unit of energy which corresponds to the ionization energy of the hydrogen atom and that R_∞ is the actual Rydberg

Table 17.2: Atomic energies of the hydrogen atom in Rydberg and Hartree according to the principal quantum number, n.

Atomic energies of the hydrogen atom		
level	Energy in Ry	Energy in Hartree
n = 1	–1	–0.5
n = 2	–0.25	–0.125
n = 3	–0.111	–0.0555
n = 4	–0.0625	–0.03125

constant related to the electromagnetic spectra of an atom whose value is $R_\infty = 1.0973 \times 10^7\ m^{-1}$.

$$R_\infty = \frac{m_e e^4}{\epsilon_0^2 (4\pi)^3 \hbar^3 c} = \frac{m_e e^4}{8 \epsilon_0^2 h^3 c}$$

$$Ry = hcR_\infty = hc\frac{m_e e^4}{8 \epsilon_0^2 h^3 c} = \frac{m_e e^4}{8 \epsilon_0^2 h^2}$$

$$Obs.: (4\pi)^2 2\hbar^2 = (4\pi)^2 2\left(\frac{h}{2\pi}\right)^2 = 8h^2$$

Let us come back to the equation of the energy of hydrogen-like atom:

$$E = -\frac{Z^2}{n^2} 13.6058 \text{ eV}$$

$$n = j + l + 1$$

From the last equation above, we see that:

$$n \geq l + 1$$

For a given l (where j = 0), n = l + 1, n = l + 2,..., so on. We also see that the energy depends only on the ***principal quantum number n***. However, the effective energy potential (V_{eff}) also depends on the ***secondary quantum number l*** (which divides the shells into smaller groups called subshells or types of atomic orbitals). **The secondary quantum number or the orbital quantum number is the quantum number of the total angular operator, L, and describes each type of subshell ranging from 0 to n–1. Each subshell (S, P, D, F) has a distinguished shape.**
Let us consider again the radial equation in the form:

$$\frac{d^2 y(r)}{dr^2} + \left[\frac{2\mu KZe^2}{r\hbar^2} + \frac{2\mu E}{\hbar^2} - \frac{l(l+1)}{r^2}\right] y(r) = 0$$

And multiply by $-\hbar^2/2\mu$:

$$-\frac{\hbar^2}{2\mu}\frac{d^2 y(r)}{dr^2} + \left[-\frac{KZe^2}{r} - E + \frac{\hbar^2 [l(l+1)]}{2\mu r^2}\right] y(r) = 0$$

$$\Rightarrow -\frac{\hbar^2}{2\mu}\frac{d^2 y(r)}{dr^2} + \left[-\frac{KZe^2}{r} + \frac{\hbar^2 [l(l+1)]}{2\mu r^2}\right] y(r) = Ey(r)$$

As a consequence, the **effective potential energy V_{eff}** is:

$$-\frac{\hbar^2}{2\mu}\frac{d^2y(r)}{dr^2}+V_{eff}\,y(r)=Ey(r)$$

$$V_C=-\frac{KZe^2}{r},V_L=\frac{\hbar^2\left[l(l+1)\right]}{2\mu r^2}$$

$$V_C(SI)=-\frac{2.3\times10^{-28}}{r},V_L(SI)=\frac{6.1\times10^{-39}\left[l(l+1)\right]}{r^2}$$

$$V_{eff}=V_C+V_L$$

Where ħ² is 1.11 × 10^{-68} ($m^2kg.s^{-1}$)²; the mass of the electron is 9.11 × 10^{-31} kg; V_C is the Coulomb potential and V_L is the potential dependent on the quantum number l (the centrifugal potential). See that the equation of classical V_L and the unit of V_L is also Joule (or $kg.m^2.s^{-2}$).

$$V_{L(classical)}=\frac{L^2}{mr^2},F_{cent}=\frac{L^2}{mr^3}$$

$$\left[V_L\right]=\frac{\left[m^2\cdot Kg\cdot s^{-1}\right]^2}{[kg][m]^2}=\left[m^2\cdot kg\cdot s^{-2}\right]or\left[J\right]$$

See the plots in Fig. 17.3 of the potential energy surface as a function of r position and the secondary quantum number, l.

Then, we have a degeneracy on the energy levels with the same n and different l values. By considering 2l + 1 (k = 2l + 1) possible values of the quantum number m, **the total number of states with same energy, or degeneracy (not considering relativistic effects and spin-orbit effects—see Section 7), g, is given below.** Note that all subshells in the same shell are degenerate in this scenario where other effects are not taken into account (see Table 17.3).

$$g=n^2=\sum_{l=0}^{n-1}(2l+1)$$

The degeneracy in each subshell (according to Madelung rule or diagonal rule) and the number of projections of angular momentum in each subshell are given by 2l + 1. The magnetic quantum number (or the quantum number of z-component of the angular momentum, L_z), m, specifies each projection in each subshell. It ranges from –l to +l.

The Fig. 17.3 indicates that there is no degeneracy for l = 0 because the effective potential energy depends solely on the Coulomb potential energy unlike l = 1 which have a stronger influence of the V_L.

As mentioned before, the solution for the radial equation is:

$$y_j^k(x)=e^{-x/2}x^{(k+1)/2}L_j^k(x)$$

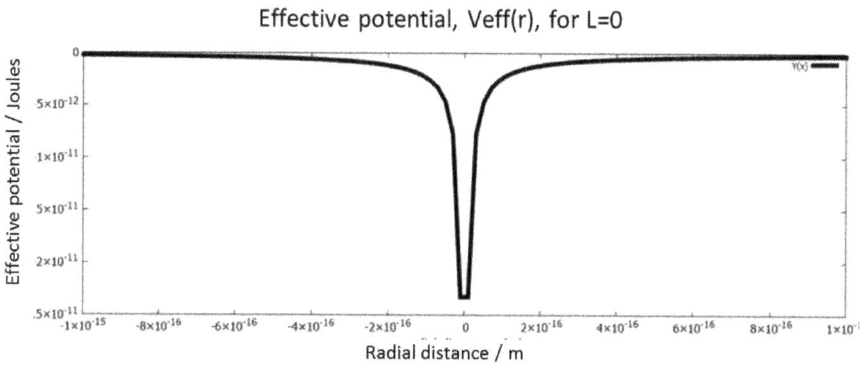

(A)

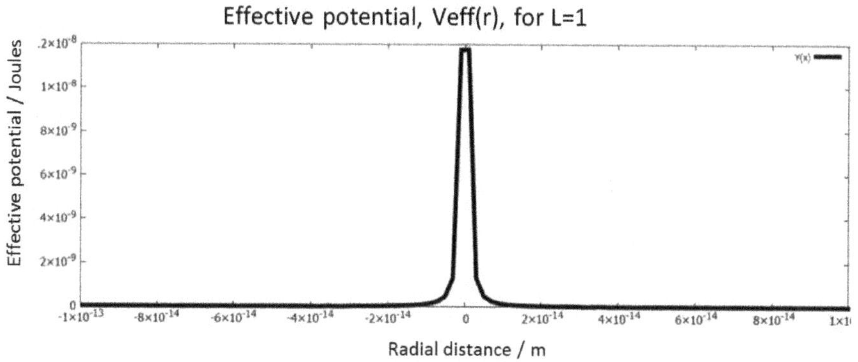

(B)

Fig. 17.3: Plots of the effective potential energy surface, V_{eff}, for l = 0,1,2,3.

Table 17.3: Number of subshells, projections of L, L_z, in each subshell and degeneracy of each shell.

Shell	Subshells/projections of L (m)									g
	S		P				D			
⋮	⋮	⋮	⋮	⋮	⋮	⋮	⋮	⋮	⋮	⋮
3	0	−1	0	1	−2	−1	0	1	2	9
2	0	−1	0	1						4
1	0									1

We can express the indices in terms of n and l, instead of k and j:

$$y_n^l(x) = e^{-x/2} x^{l+1} L_{n-l-1}^{2l+1}(x)$$

$$k = 2l+1, \qquad j = n-l-1$$

$$(k+1)/2 = (2l+1+1)/2 = l+1$$

And the independent variable x can be written in terms of the principal quantum number as well:

$$x = \varepsilon \cdot r$$

$$\varepsilon = \frac{2KZ}{a_0 n}$$

$$then:$$

$$x = \frac{2KZr}{a_0 n}$$

Where ***j* = *n*–*l*–*1* is the number of nodes of the radial wave function.**

$$j = nodes = n - l - 1$$

Which can be included in the solution of the radial equation so that we can change y(x) into y(r):

$$y_n^l(r) = e^{-KZr/na_0}\left(\frac{2KZr}{na_0}\right)^{l+1} L_{n-l-1}^{2l+1}\left(\frac{2KZr}{na_0}\right)$$

At last, let us replace y(r) with R(r):

$$rR_{n,l}(r) = e^{-KZr/na_0}\left(\frac{2KZr}{na_0}\right)^{l+1} L_{n-l-1}^{2l+1}\left(\frac{2KZr}{na_0}\right)$$

$$y(r) = rR(r)$$

Then, after division by r where the normalization term A incorporated the factor $2/na_0$ from the power term, we have:

$$rR_{n,l}(r) = e^{-KZr/na_0}\left(\frac{2KZr}{na_0}\right)\left(\frac{2KZr}{na_0}\right)^{l} L_{n-l-1}^{2l+1}\left(\frac{2KZr}{na_0}\right)$$

$$\times 1/r$$

$$R_{n,l}(r) = A\, e^{-KZr/na_0}\left(\frac{2KZ}{na_0}\right)\left(\frac{2KZr}{na_0}\right)^{l} L_{n-l-1}^{2l+1}\left(\frac{2KZr}{na_0}\right)$$

$$R_{n,l}(r) = A' e^{-KZr/na_0}\left(\frac{2KZr}{na_0}\right)^{l} L_{n-l-1}^{2l+1}(\rho)$$

$$where: \rho = \frac{2KZr}{na_0}, A' = A \cdot \left(\frac{2KZ}{na_0}\right)$$

For simplicity, A’ will be represented as A.
Where the formula for the associated Laguerre polynomial is:

$$L_{n-l-1}^{2l+1}(\rho) = \sum_{i=0}^{n-l-1} \frac{(-1)^i \left[(n+l)!\right]^2 \rho^i}{i!(n-l-1-i)!(2l+1+i)!}$$

$$where: \rho = \frac{2KZr}{na_0}$$

The normalization constant, A, is:

$$A=\sqrt{\left(\frac{2KZ}{na_0}\right)^3\frac{(n-l-1)!}{2n\left[(n+l)!\right]^3}}$$

Morse and Feshbach provided the following normalization condition (Morse and Feshbach 1953):

$$\int_0^\infty z^a e^{-z} L_b^a(z) L_b^a(z)dz=\delta_{b,c}\frac{\left[\Gamma(a+b+1)\right]^3}{\Gamma(a+b)}$$

$$\Gamma(j)=(j-1)!$$

The normalization condition for the radial wave function is:

$$\int_0^\infty \left(R_{n,l}(r)\right)^*\left(R_{n,l}(r)\right)r^2dr=1$$

$$\int_0^\infty A^2e^{-2KZr/na_0}\left(\left(\frac{2KZr}{na_0}\right)^l L_{n-l-1}^{2l+1}(\rho)\right)^*\left(\left(\frac{2KZr}{na_0}\right)^l L_{n-l-1}^{2l+1}(\rho)\right)r^2dr$$

$$\int_0^\infty A^2\left(\frac{2KZ}{na_0}\right)^2 e^{-2KZr/na_0}\left(\frac{2KZr}{na_0}\right)^{2l}\left(L_{n-l-1}^{2l+1}(\rho)\right)^2r^2dr=1$$

$$\int_0^\infty A^2\cdot e^{-2KZr/na_0}\left(\frac{2KZr}{na_0}\right)^{2l}\left(L_{n-l-1}^{2l+1}(\rho)\right)^2r^2dr=1$$

$$\rho=\frac{2KZr}{na_0},d\rho=\frac{2KZ}{na_0}dr,\qquad dr=\frac{na_0}{2KZ}d\rho$$

$$r=\frac{na_0}{2KZ}\rho,\qquad r^2=\left(\frac{na_0}{2KZ}\right)^2\rho^2$$

$$\int_0^\infty A^2\cdot e^{-2KZr/na_0}\left(\frac{2KZr}{na_0}\right)^{2l}\left(L_{n-l-1}^{2l+1}(\rho)\right)^2\left(\frac{na_0}{2KZ}\right)^3\left(\frac{2KZr}{na_0}\right)^2d\rho$$

$$A^2\left(\frac{na_0}{2KZ}\right)^3\int_0^\infty e^{-2KZr/na_0}\left(\frac{2KZr}{na_0}\right)^{2l+2}\left(L_{n-l-1}^{2l+1}(\rho)\right)^2d\frac{2KZr}{na_0}=1$$

For convenience, let us simply replace ρ to z and see the equation from Morse and Feshbach is nearly similar to the above equation.

$$1=|A|^2\left(\frac{na_0}{2KZ}\right)^3\int_0^\infty e^{-z}z^a\left[zL_b^az\right]L_b^azdz$$

$$z=\left(\frac{2KZr}{na_0}\right),\qquad a=2l+1\therefore b=n-l-1$$

After taking out one factor of the power of z (or ρ), we have now a similar expression to that from Morse and Feshbach's book.

$$(A')^2\left(\frac{na_0}{2KZ}\right)^3\int_0^\infty e^{-2KZr/na_0}\left(\frac{2KZr}{na_0}\right)^{2l+1}\left(\left(\frac{2KZr}{na_0}\right)L_{n-l-1}^{2l+1}(\rho)\right)L_{n-l-1}^{2l+1}(\rho)\,d\rho=1$$

Another relation which is important here from Morse and Feshbach's book (Morse and Feshbach 1953):

$$zL_b^a(z)=(a+2b+1)L_b^a(z)-\frac{b+1}{a+b+1}L_{b+1}^a(z)-(a+b)^2L_{b-1}^a(z)$$

Where taking the last term between brackets of the penultimate expression, it becomes:

$$\left(\frac{2KZr}{na_0}\right)L_{n-l-1}^{2l+1}(\rho)=2n\cdot L_{n-l-1}^{2l+1}(\rho)$$

After noticing that the Kronecker delta gives some integrals a zero value when the associated Laguerre indices are different. Finally, we have:

$$1=|A|^2\,2n\left(\frac{na_0}{2KZ}\right)^3\int_0^\infty e^{-z}z^{2l+1}L_{n-l-1}^{2l+1}zL_{n-l-1}^{2l+1}zdz$$

$$1=|A|^2\,2n\left(\frac{na_0}{2KZ}\right)^3\frac{\left[\Gamma(2l+1+n-l-1+1)\right]^3}{\Gamma(n-l-1+1)}$$

$$1=|A|^2\,2n\left(\frac{na_0}{2KZ}\right)^3\frac{\left[\Gamma(l+1+n)\right]^3}{\Gamma(n-l)}$$

$$1=|A|^2\,2n\left(\frac{na_0}{2KZ}\right)^3\frac{\left[(n+l)!\right]^3}{(n-l-1)!}$$

And the normalization constant becomes:

$$A=\sqrt{\left(\frac{2KZ}{na_0}\right)^3\frac{(n-l-1)!}{2n\left[(n+l)!\right]^3}}$$

Another solution for the normalization constant of the radial equation can be found in Pauling and Wilson's book (Pauling, L. and Wilson, E. B. 1935).
Then, the final solution to the radial part of the hydrogen-like atom is:

$$R_{n,l}(r)=\sqrt{\left(\frac{2KZ}{na_0}\right)^3\frac{(n-l-1)!}{2n\left[(n+l)!\right]^3}}\,e^{-KZr/na_0}\left(\frac{2KZr}{na_0}\right)^l L_{n-l-1}^{2l+1}(\rho)$$

$$\rho=\frac{2KZr}{na_0}$$

The formula of the radial distribution for the first states of the hydrogen-like atom are given below where K is removed from the equations for atomic units:

$$R_{10} = 2e^{-Zr/a_0}\left(\frac{Z}{a_0}\right)^{3/2}$$

$$R_{20} = \frac{1}{\sqrt{2}}e^{-Zr/2a_0}\left(\frac{Z}{a_0}\right)^{3/2}\left(1-\frac{Zr}{2a_0}\right)$$

$$R_{21} = \frac{1}{\sqrt{24}}e^{-Zr/2a_0}\left(\frac{Z}{a_0}\right)^{3/2}\left(\frac{Zr}{a_0}\right)$$

Let us derive the equation for hydrogen atom in the 2p state (n = 2, l = 1).

$$2l+1=2\cdot 1+1=3, \qquad n-l-1=2-1-1=0$$

$$L_{n-l-1}^{2l+1} = L_0^3(\rho) = \sum_0^0 \frac{(-1)^0\left[(2+1)!\right]^2\rho^0}{0!(2-1-1-0)!(2\cdot 1+1+0)!} = 6$$

$$\sqrt{\left(\frac{2}{na_0}\right)^3\frac{(n-l-1)!}{2n\left[(n+l)!\right]^3}} = \sqrt{\frac{8}{8a_0^3}\frac{(2-1-1)!}{2\cdot 2\left[(2+1)!\right]^3}} = \frac{a_0^{-3/2}}{12\sqrt{6}}$$

$$\left(\frac{2r}{na_0}\right)^l = \left(\frac{2r}{2a_0}\right)^1 = \frac{r}{a_0}$$

$$R_{21} = 6\cdot\frac{a_0^{-3/2}}{12\sqrt{6}}\cdot e^{-r/2a_0}\cdot\frac{r}{a_0} = \frac{a_0^{-3/2}}{\sqrt{24}}e^{-r/2a_0}\frac{r}{a_0}$$

Due to the importance of the atomic physics, there is a set of physical quantities which are used in atomic units (a.u.). They are shown in Table 17.4 along with the corresponding SI units conversion.

In atomic units, the radial equation for hydrogen-like atoms become:

$$\frac{d^2y(r)}{dr^2} + \left[\frac{2Z}{r} + 2E - \frac{l(l+1)}{r^2}\right]y(r) = 0$$

Table 17.4: Physical quantities in atomic units and SI units.

Physical quantity	Atomic units (expression in SI), [name]	SI units	Conversion
charge	e [elementary charge]	C	1.602×10^{-9}C
Mass	m_e	kg	9.109×10^{-31} kg
action	Ħ	$m^2 kg.s^{-1}$	1.054×10^{-34} $m^2 kg.s^{-1}$
permitivity	1/k ($1/4p\hat{I}_0$)	$C^2J^{-1}m^{-1}$	1.11×10^{-10} $C^2J^{-1}m^{-1}$
length	a_0 ($4pe_0\hbar^2/m_e a_0^2$) [Bohr]	m	0.5292 Å (52.92 pm)
energy	E_h ($\hbar^2/m_e a_0^2$) [Hartree]	J	4.36 x 10^{-18} J

5. Numerical analysis for radial equation of the hydrogen-like atom

5.1 Guess functions for Numerov integration

This subsection was partly based on Izaac and Wang's derivation of the Numerov method for radial wave function (Izaac and Wang 2018). Firstly, let us transform the radial equation below in order to determine the guess functions for the Numerov integration used to obtain the radial wave function of the hydrogen atom.

$$\frac{d^2y(r)}{dr^2}+\left[\frac{2\mu KZe^2}{r\hbar^2}+\frac{2\mu E}{\hbar^2}-\frac{l(l+1)}{r^2}\right]y(r)=0$$

$$\Rightarrow \frac{d^2y(r)}{dr^2}+\left\{\frac{2\mu}{\hbar^2}\left[E-V(r)\right]-\frac{l(l+1)}{r^2}\right\}y(r)=0$$

$$V(r)=-\frac{KZe^2}{r}$$

Let us first change the variables:

$$\rho=(1/\alpha)r,\qquad r=\alpha\rho$$

$$y(r)\to u(\rho)$$

Where α is a constant with length dimension. Then, we have:

$$\frac{1}{\alpha^2}\frac{d^2u(\rho)}{d\rho^2}+\left\{\frac{2\mu}{\hbar^2}\left[E-V(r)\right]-\frac{l(l+1)}{\alpha^2\rho^2}\right\}u(\rho)=0$$

$\times\alpha^2$:

$$\frac{d^2u(\rho)}{d\rho^2}+\left\{\frac{2\mu\alpha^2}{\hbar^2}\left[E-V(r)\right]-\frac{l(l+1)}{\rho^2}\right\}u(\rho)=0$$

Let us make the next substitutions in the equation above:

$$V(r)=-V_0f(r),\qquad V_0>0$$

$$\varepsilon=\frac{|E|}{V_0}\ \therefore E=-\varepsilon V_0,\qquad \varepsilon>0$$

Then, we have:

$$\frac{d^2u(\rho)}{d\rho^2}+\left\{\frac{2\mu\alpha^2}{\hbar^2}\left[-V_0\varepsilon+V_0f(r)\right]-\frac{l(l+1)}{\rho^2}\right\}u(\rho)=0$$

The next substitution is:

$$\gamma=\frac{2\mu\alpha^2V_0}{\hbar^2}$$

Then, we have:

$$\frac{d^2u(\rho)}{d\rho^2}+\gamma\left\{f(r)-\frac{1}{\gamma}\frac{l(l+1)}{\rho^2}-\varepsilon\right\}u(\rho)=0$$

$$\frac{d^2u(\rho)}{d\rho^2}+k(\rho)u(\rho)=0,\quad k(\rho)=\gamma\left\{f(r)-\frac{1}{\gamma}\frac{l(l+1)}{\rho^2}-\varepsilon\right\}$$

Let us now specify the boundary conditions of the function $f(r)$ in the spherical symmetrical potential and choose α as a (near the limit of the non-vanishing potential energy):

$$f(r)=\begin{cases}1 & for \quad r\leq a\\ -0 & for \quad r>a\end{cases}$$

$$\alpha=a$$

Then:

$$k(\rho)=\gamma\begin{cases}1-\frac{1}{\gamma}\frac{l(l+1)}{\rho^2}-\varepsilon, & r\leq a\\ -0-\frac{1}{\gamma}\frac{l(l+1)}{\rho^2}-\varepsilon, & r>a\end{cases}$$

The extremes of the wave function $u(\rho)$ will be used as guesses of the shooting method of the inward and outward integrations using Numerov method.
For small ρ, the equation:

$$\frac{d^2u(\rho)}{d\rho^2}+\gamma\left\{f(r)-\frac{1}{\gamma}\frac{l(l+1)}{\rho^2}-\varepsilon\right\}u(\rho)=0$$

reduces to:

$$\frac{d^2u(\rho)}{d\rho^2}-\frac{l(l+1)}{\rho^2}u(\rho)=0$$

since the term $l(l+1)/\rho^2$ will become very large as ρ^2 decreases and the solutions are:

$$u(\rho)=\rho^{l+1}, for:\rho\approx 0$$

Let us check this out:

$$\frac{d^2\rho^{l+1}}{d\rho^2}-\frac{l(l+1)}{\rho^2}\rho^{l+1}=0$$

$$l(l+1)\rho^{l-1}-l(l+1)\rho^{l-1}=0$$

For large ρ, the equation:

$$\frac{d^2u(\rho)}{d\rho^2}+\gamma\left\{f(r)-\frac{1}{\gamma}\frac{l(l+1)}{\rho^2}-\varepsilon\right\}u(\rho)=0$$

reduces to:

$$\frac{d^2u(\rho)}{d\rho^2} - \gamma\varepsilon u(\rho) = 0, \qquad \gamma > 0$$

since the term $l(l + 1)/\rho^2$ will become very small as ρ^2 increases and V(r) will be zero. The solutions are:

$$u(\rho) = \exp(-\gamma\varepsilon\rho), \qquad for : \rho >> 0$$

There are two very different solutions at the boundaries and can lead to totally wrong solutions if just a single integration is done throughout the r axis. Then, we need to divide the wave function into two parts which meet at the matching point (as it was done in Chapter fourteen).

As above mentioned, the algorithm generates two solutions: (1) one starting from small value of ρ and being integrated outwards and towards the matching point $\rho = \rho_{m,}$ whose function will be named $u < (\rho)$; (2) another one starting from very large value of ρ and being integrated inwards and towards the matching point whose function will be named $u > (\rho)$. Important to notice that the same reasoning is applied in the code of the Chapter fourteen.

Being the function u(ρ) analytic and finite everywhere, the logarithmic derivative at the matching point should be well defined.

$$\frac{\frac{du<(\rho)}{d\rho}}{u<(\rho)} = \frac{\frac{du>(\rho)}{d\rho}}{u>(\rho)}, at \qquad \rho = \rho_m$$

We can calculate the derivatives above by finite difference approximation (Chapter two):

$$forward : \frac{du>(\rho)}{d\rho} \approx \frac{u>(\rho_m + h) - u>(\rho_m)}{h}$$

$$backward : \frac{du<(\rho)}{d\rho} \approx \frac{u<(\rho_m) - u<(\rho_m - h)}{h}$$

Then, the criterion for a proper eigenfunction will be:

$$f = u>(\rho_m + h) - u<(\rho_m - h) = 0$$

The function f is obtained by the integration of the grid points of the pair rho-f for a specific energy E.

5.2 Code for potential energy versus radial distribution

Next, we provide the equations for the code of the all components of the potential energy in SI units.

This code is slightly inspired from Giannozzi, Ercolessi and Gironcoli (Giannozzi et al. 2013) for the radial grid.

Let us go back to the radial equation:

$$\frac{d^2y(r)}{dr^2}+\left\{\frac{2\mu}{\hbar^2}\left[E-V(r)\right]-\frac{l(l+1)}{r^2}\right\}y(r)=0$$

$$V(r)=-\frac{KZe^2}{r},\, y(r)=rR$$

In this code (HYDROGEN-POTENTIAL) and in the next one, it is used the exponential grid to deal with a variable-step grid since there is no symmetry in the corresponding wave function. Such grid becomes denser and denser as it approaches the origin. In the code of the Chapter fourteen, it was used constant-grid due to the symmetry in the wave function of the one-particle quantum harmonic oscillator.

```
Do i=0,mesh ! mesh=1000
  oldr = ximin + dr*i
  r(i)= exp(oldr)/zmesh
End do
```

Where zmesh = zeta. The zmesh variable is zeta variable as a real number while zeta is an integer.

In the Table 17.5, there are some selected values of i, r(i) and r^2(i). One can see that the equations above yield an exponential radial grid which is denser near the nucleus than away from it.

In the code HYDROGEN-POTENTIAL, we use the length unit in meters. Then, all the values of r(i) in Table 17.2 have to be multiplied by 5.29×10^{-11} since Bohr unit is $a_0 = 5.29 \times 10^{-11}$m.

The Coulomb potential energy (whose unit is Joule) is given by:

$$V(r)=-K\frac{Ze^2}{r} \therefore K=\frac{1}{4\pi\in_0}=9\times10^9\,Nm^2C^{-2}$$

$$V(r)=-2.3\times10^{-28}\,\frac{Z}{r}$$

Table 17.5: Selected values of i, r(i) and r^2(i) of the radial grid.

i	r(i)/Bohr	r^2(i)
0	3.4×10^{-4}	1.089×10^{-7}
100	9.1×10^{-4}	8.2×10^{-7}
500	0.0497	2.47×10^{-3}
750	0.606	0.367
1000	7.39	54.56

The Coulomb potential energy plays sole role only for secondary quantum magnetic l = 0. For l > 0, the potential energy dependent on l (the centrifugal potential) has to be included. As it was shown before, we have:

$$-\frac{\hbar^2}{2\mu}\frac{d^2y(r)}{dr^2}+V_{eff}y(r)=Ey(r)$$

$$V_C=-\frac{KZe^2}{r}, V_L=\frac{\hbar^2\left[l(l+1)\right]}{2\mu r^2}$$

$$V_C(SI)=-\frac{2.3\times10^{-28}}{r}, \quad Z=1$$

$$V_L(SI)=\frac{6.1\times10^{-39}\left[l(l+1)\right]}{r^2}$$

$$V_{eff}=V_C+V_L$$

```
!------------------------------------------------------------
!Program name: HYDROGEN-POTENTIAL
!Objective: plot the effective potential energy as a function of radial dis-
tance
!------------------------------------------------------------
!The Coulomb potential energy equation is:
! V(r) = -KZq²/r = -(Kq²)(Z/r)
! K(Coulomb constant)= 9*10⁹ Nm²C⁻²
! q = - 1.602 x 10⁻¹⁹C
! Kq²=2.3 x 10⁻²⁸
! The unit is Joule. We have to change into Rydberg.
!--------------------------------------------------
! The V(centrifugal) equation is:
! Vcent=6.1 x 10⁻³⁹ l(l+1)/r² (see text)
! The unit is Joule. We have to change into Rydberg
!------------------------------------------------------------
! The unit of the radial distance, r, is meter.
!------------------------------------------------------------
    Program hydrogen
    Implicit none
integer :: mesh, n, l, i, zeta
Real*8 rmin, dr, rmax
    double precision :: oldr, e, kq2, h2m
    double precision, allocatable :: r(:), r2(:), vpot(:), vc(:), vcent(:), veff(:)
!--------------------------------------------------
    zeta = 1
!---------------------------------------------------
! Initial data
    rmax= 100. ! Bohr
    rmin= 3.E-4 !Bohr
```

```
  dr= 0.005
    ! Number of points for the calculation (mesh)=1000 for Z=1
      mesh = 1000
      Allocate (r(mesh))
      Allocate (r2(mesh))
      Allocate (vc(mesh))
      Allocate (vcent(mesh))
      Allocate (vpot(mesh))
      Allocate (veff(mesh))
      !User entering data
      Write (*,*) 'Enter n and l. n > 0 and l=n-1'
      Read (*,*) n, l
      If (n<1) then
          Write (*,*) 'n < 1. Unphysical.'
          Stop
      Else if (n < l+1) then
          Write (*,*) 'n < l+1. It has to be n=l+1!'
          Stop
      Else if (l < 0) then
          Write (*,*) 'l < 0. Unphysical.'
          Stop
      End if
    ! Initialize exponential radial grid
      Do i=0,mesh
          oldr = rmin + dr*i !oldr is in Bohr unit
r(i)= (exp(oldr))*5.23E-11 !r(i) is changed to SI unit
          r2(i)= (r(i) * r(i))
      End do
!-----------------------------------------------------
! Initialize the potential energy grid
      kq2=2.3E-28
      h2m=6.1E-39
If (l==0) then
      Do i=0,mesh
      vc(i)= -(kq2*zeta)/r(i)
      veff(i)=vc(i) !veff is in Joule unit
!Let us change vpot unit to Rydberg
vpot(i)=veff(i)*0.458E+18 ! vpot is in Ry
       End do
Else if (l > 0) then
      Do i=0,mesh
      vc(i)= -(kq2*zeta)/r(i)
      vcent(i)=((h2m)*(l*(l+1))/r2(i))
      veff(i)= vc(i)+vl(i) !veff is in Joule unit
```

```
!Let us change vpot unit to Rydberg
vpot(i)=veff(i)*0.458E+18 ! vpot is in Ry
     End do
End if
!------------------------------------------------------------------
  !Printing results r(i) and vpot(i) in the 'fileout' (optional)
  open(7,file='potential.dat',status='replace')
  do i=0,mesh
          write (7,'(3e16.8,f12.6,3e16.8,f12.6)') r(i),vpot(i)
  End do
      write (7,'(/)')
      !Close(7)
  !----------------------------------------------------------------------
      Deallocate (r)
      Deallocate (r2)
      Deallocate (vc)
      Deallocate (vcent)
      deallocate (vpot)
      Deallocate (veff)
  End program
```

The results of the code above for hydrogen atom in the 1s, 3p and 3d states are shown in Fig. 17.4(A), 17.4(B) and 17.4(C), respectively.

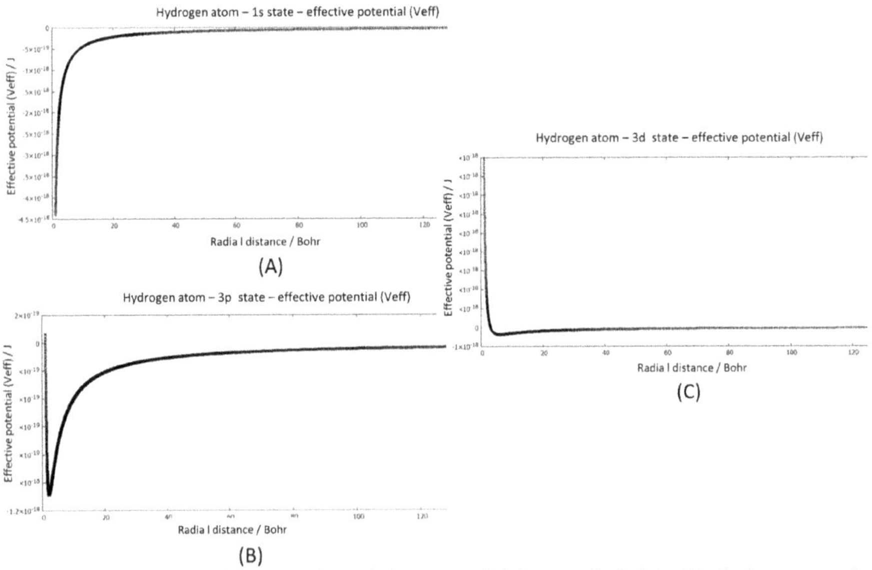

Fig. 17.4: Effective potential energy (in Joules) versus radial distance (in Bohr) of the hydrogen atom in (A) 1s state, (B) 3p state and (C) 3d state.

5.3 Code for the radial wave function versus radial distribution

The code named HYDROGEN-RADIAL is a very simple code that defines the radial grid and the potential grid in order to obtain the matching point and to calculate the integration of the Numerov wave function starting from the guess initial and final values of the function for all 1000 points of the grid. It does not use any resource of recursive iteration to improve the wave function.

Let us change the expression of radial equation of the hydrogen-like atoms below from SI units to atomic units:

$$-\frac{\hbar^2}{2\mu}\frac{d^2y(r)}{dr^2}+\left[-\frac{KZe^2}{r}+\frac{\hbar^2\left[l(l+1)\right]}{2\mu r^2}\right]y(r)=Ey(r)$$

Since e, m_e (where $\mu = m_e$), K, and ħ are one unit each in atomic units, then, we have:

$$-\frac{1}{2}\frac{d^2y(r)}{dr^2}+\left[V(r)-E+\frac{l(l+1)}{2r^2}\right]y(r)=0$$

$$V(r)=-\frac{Z}{r}, \qquad E=-\frac{Z}{2n^2}$$

The energy is given by:

$$E=-\frac{K^2Z^2m_ee^4}{2n^2\hbar^2}$$

$$E=-\frac{Z^2}{2n^2}(a.u.)$$

The effective potential used in the code is given by:

$$V_{eff}=V(r)+\frac{l(l+1)}{2r^2}$$

The second term is the centrifugal potential. Let us multiply the radial equation by –2:

$$\frac{d^2y(r)}{dr^2}+\left[-2V(r)+2E-\frac{l(l+1)}{r^2}\right]y(r)=0$$

Let us use the terminologies commonly applied to the Numerov method:

$$\frac{d^2y(r)}{dr^2}+K^2(r)y(r)=0$$

Where:

$$K^2(r)=-2V+2E-\frac{l(l+1)}{r^2}$$

However, the code below only works correctly with $(-1)K^2$, that is, $2V-2E+2V_{centr}$, where the implicit negative sign of V and E still exists. In the code we use V_{cent} instead of V_L.

From the Numerov method (see Chapter two), we have the following wave function:

$$\psi_{n+1} = \frac{(12-10f_n)\psi_n - f_{n-1}\psi_{n-1}}{f_{n+1}}$$

Where:

$$f_n = 1 + \frac{h^2}{12}K^2(r)$$

$$f_{n+1} = 1 + \frac{h^2}{12}K^2(r+h)$$

$$f_{n-1} = 1 + \frac{h^2}{12}K^2(r-h)$$

$$\psi_{n-1} = \psi(r-h), \quad \psi_{n+1} = \psi(r+h)$$

It is important to not confuse the h from Numerov method (which is an increment) with that from Planck constant. In our modified code, h is changed into dr.

The code below works nearly well for all states of the hydrogen atom. Other approaches can be found elsewhere such as Peng and Gong (Peng and Gong 2010) and Izaac and Wang (Izaac and Wang 2018) for instance.

```
!-------------------------------------------------------------
!Program name: HYDROGEN-RADIAL
!Radial equation of the hydrogen-like atom
!Objective: plot the radial wave function for each n,l pair
! The equation is atomic units
!-------------------------------------------------------------
!The Coulomb potential energy equation is:
! V(r) = -2Z/r
! The ceentrifugal potential equation is:
! VL=l(l+1)/r^2
! The energy is:
! E=-2*Z/n^2
!-------------------------------------------------------------
! Use of Numerov method to find the wave function
! Function fn from Numerov equation is
! Fn = 1 + K2*h^2/12 , where K2=-2(Veff - E)
! In the code, h is changed into dr
!-------------------------------------------------------------
  Program hydrogen
  Implicit none
integer :: mesh, n, l, i, j, zeta, m, maxiter=100
Real*8 rmin, dr, zmesh, y_out_m
  double precision :: e, fh12, norm, eps=1.0E-6
  double precision, allocatable :: r(:), r2(:), y(:), vc(:), vcent(:), &
     veff(:), k2(:), fn(:), f(:)
!--------------------------------------------------
```

```
  zeta = 1
!-----------------------------------------------------
! Initial data
  zmesh= zeta !zmesh is zeta as a real number
  rmin= 1.E-5
  dr= 0.005d0
!--------------------------------------------------------------
! Number of points for the calculation (mesh)=1000 for Z=1
mesh = 1000
  Allocate (r(mesh))
  Allocate (r2(mesh))
  Allocate (y(mesh))
  Allocate (vc(mesh))
  Allocate (f(mesh))
  Allocate (fn(mesh))
  Allocate (vcent(mesh))
  Allocate (veff(mesh))
  Allocate (k2(mesh))
!--------------------------------------------------------------
!User entering data
Write (*,*) 'Enter n and l. n > 0 and l=n-1'
Read (*,*) n, l
If (n<1) then
  Write (*,*) 'n < 1. Unphysical.'
  Stop
Else if (n < l+1) then
  Write (*,*) 'n < l+1. It has to be n=l+1!'
  Stop
Else if (l < 0) then
  Write (*,*) 'l < 0. Unphysical.'
  Stop
End if
!--------------------------------------------------------------
! Initialize exponential radial grid only for the potential
! This grid does not work for 1s and 2p states
Do i=0,mesh
     oldr = rmin + dr*i !oldr is in Bohr unit
r(i)= (exp(oldr))
     r2(i)= (r(i) * r(i))
End do
!---------------------------------------------------
! Initialize the potential energy grid for the y(i)
Do i=0,mesh
  vc(i)= -2*zmesh/r(i)
```

```
  vcent(i)=(l*(l+1.d0))/(r2(i))
veff(i)= vc(i)+vcent(i)
End do
!-----------------------------------------------------------------
! Initialize constant radial grid
dr= 0.02d0
  Do i=0,mesh
     r(i) = rmin + dr*i !oldr is in Bohr unit
r2(i)= (r(i) * r(i))
End do
!-----------------------------------------------------------------
!Set the eigenvalue
e= -zmesh/n**2
!-----------------------------------------------------------------
  !calculate part of Numerov function f, fh12 and K2
  fh12=dr*dr/12.0d0
  k2= vc-e+vcent
  f(0)= fh12*k2(0)
!-------------------------------------------------------------
!Start Numerov integration
!------------------------------------------------------------
!set boundary conditions
y(0)=r(0)**(l+1)
y(1)=r(1)**(l+1)
     y(mesh)=dr
m=0
!-----------------------------------------------------------
 !find the matching point m checking the change of sign of f
do i=1,mesh
        f(i)= fh12*k2(i)
if (f(i) == 0.d0) f(i)=1.E-20
        if (f(i) /= sign(f(i),f(i-1))) m=i
     end do
!----------------------------------------------------------
 ! With the values of f, obtain fn
 !obtain the y(mesh-1) for inward integration
        fn=1.d0-f
        y(mesh-1)=y(mesh)*(12.d0-10.d0*fn(mesh))/fn(mesh-1)
!----------------------------------------------------------
 !Start Numerov outward integration
     do i=1,m-1
        y(i+1)=y(i)*(12.d0-10.d0*fn(i))-(fn(i-1)*y(i-1))/fn(i+1)
        if (y(i) == 0.0d0) y(i)=1.d-20
        end do
     y_out_m = y(m)
     !print *, y(m)
 !------------------------------------------------------------
```

```
! Start Numerov inward integration
  do i = mesh-1,m+1,-1
     y(i-1)=y(i)*(12.d0-10.d0*fn(i))-(fn(i+1)*y(i+1))/fn(i-1)
     if (y(i-1) > 1.E10) then
        do j=mesh,i-1,-1
           y(j)=y(j)/y(i-1)
        end do
     end if
  end do
!------------------------------------------------------------
 ! rescale function to match at the turning point
  y(:m-1) = y(:m-1)/y_out_m
     y(m:) = y(m:)/y(m)
!------------------------------------------------------------
!Normalization process
     norm = sqrt(dot_product(y,y))
     y=y/norm
!------------------------------------------------------------
!Print the result of the wave function
do i=0,mesh
     print *, r(i),y(i)
End do
!------------------------------------------------------------
 !Printing results r(i) and y(i) in the 'radial'
open(7,file='radial.dat',status='replace')
do i=0,mesh
     write (7,'(3e16.8,f12.6,3e16.8,f12.6)') r(i),y(i)
End do
  write (7,'(/)')
  Close(7)
!---------------------------------------------------------------------
  Deallocate (r)
  Deallocate (r2)
  Deallocate (vc)
  Deallocate (vcent)
  Deallocate (veff)
  Deallocate (k2)
  Deallocate (y)
  Deallocate (f)
  Deallocate (fn)
 End program
```

The results of the code above for hydrogen atom in the 1s, 2s, 2p, 3s, 3p and 3d states are shown in Fig. 17.5(A), 17.5(B), 17.5(C), 17.5(D), 17.5(E) and 17.5(F), respectively. See that the radial wave function in 1s, 2p and 3d do not have node since $n–l–1 = 0$ for these states. As to 2s and 3p there is one node whereas 3s has two nodes.

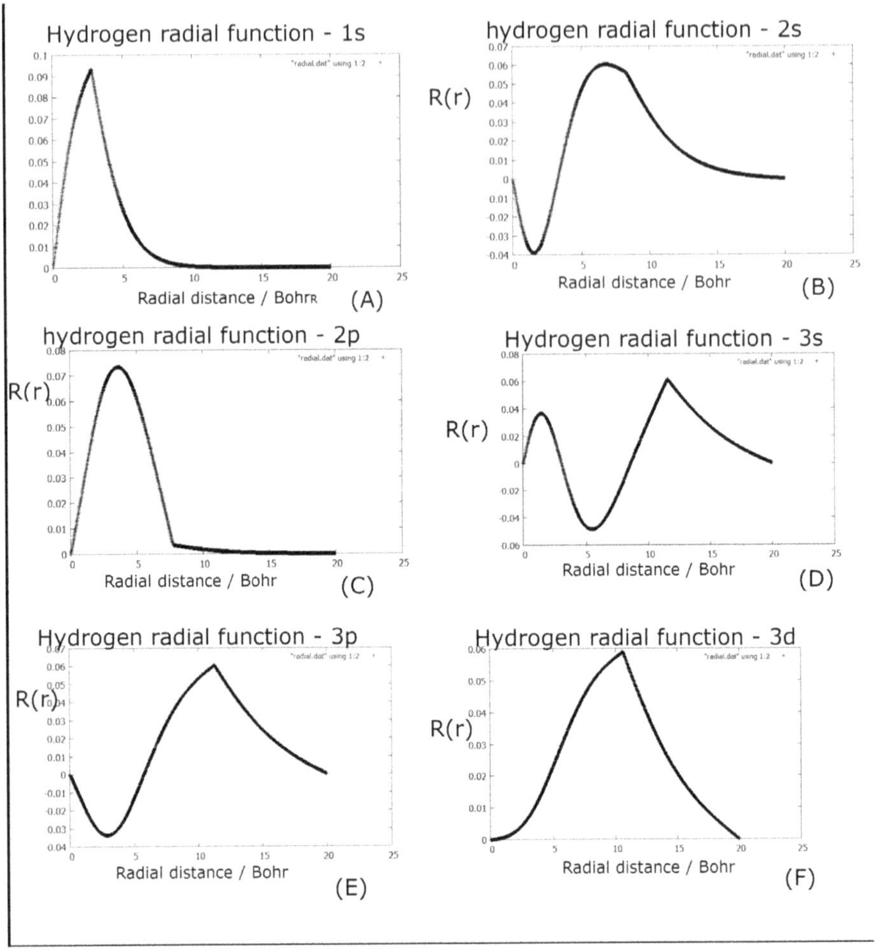

Fig. 17.5: Radial wave function versus radial distance (in Bohr) of the hydrogen atom in (A) 1s state, (B) 2s state, (C) 2p state and (D) 3s state.

6. Hydrogen atomic orbitals

The product of normalized the spherical harmonics, $Y_{l,m}(\theta,\varphi)$:

$$Y(\theta,\varphi) = (-1)^{m+|m|}\left[\frac{2l+1}{4\pi}\frac{(l-m)!}{(l+m)!}\right]^{1/2} P_l^m(\cos\theta)\,e^{im\varphi}$$

and the radial distribution, $R_{n,l}(r)$:

$$R_{n,l}(r) = \sqrt{\left(\frac{2KZ}{na_0}\right)^3 \frac{(n-l-1)!}{2n\left[(n+l)!\right]^3}}\,e^{-KZr/na_0}\left(\frac{2KZr}{na_0}\right)^l L_{n-l-1}^{2l+1}(\rho)$$

$$\rho = \frac{2KZr}{na_0}$$

gives the hydrogen atomic orbitals or hydrogen-like atomic orbitals:

$$\Psi_{n,l,m}(r,\theta,\varphi)=R_{n,l}(r)\cdot Y_{l,m}(\theta,\varphi)$$

Let us obtain, for example, the atomic orbital $3p_y$:

$$R_{3,1}(r)=\frac{8}{27\sqrt{6}}a_0^{-3/2}\left(1-\frac{r}{6a_0}\right)\frac{r}{a_0}e^{-r/3a_0}$$

$$Y_{1,-1}(\theta,\phi)=\sqrt{\frac{3}{8\pi}}\sin\theta\cdot e^{-i\phi}$$

$$\Psi_{3,1,-1}(r,\theta,\phi)=\left[\frac{8}{27\sqrt{6}}a_0^{-3/2}\left(1-\frac{r}{6a_0}\right)\frac{r}{a_0}e^{-r/3a_0}\right]\cdot\left[\sqrt{\frac{3}{8\pi}}\sin\theta\cdot e^{-i\phi}\right]$$

$$\Psi_{3,1,-1}(r,\theta,\phi)=\frac{8\sqrt{3}}{27\sqrt{6\cdot 8\pi}}\left[a_0^{-3/2}\left(1-\frac{r}{6a_0}\right)\frac{r}{a_0}e^{-r/3a_0}\right]\cdot\left[\sin\theta\cdot e^{-i\phi}\right]$$

$$\Psi_{3,1,-1}(r,\theta,\phi)=\frac{8\sqrt{3}a_0^{-3/2}}{27\cdot 4\sqrt{3}\sqrt{\pi}}\left[\left(1-\frac{r}{6a_0}\right)\frac{r}{a_0}e^{-r/3a_0}\right]\cdot\left[\sin\theta\cdot e^{-i\phi}\right]$$

$$Euler: e^{-i\phi}=\cos\phi-i\sin\phi \therefore i\sin\phi=0^{(*)}\Rightarrow e^{-i\phi}=\cos\phi$$

$$\Psi_{3,1,-1}(r,\theta,\phi)=\frac{2a_0^{-3/2}}{27\sqrt{\pi}}\left[\left(1-\frac{r}{6a_0}\right)\frac{r}{a_0}e^{-r/3a_0}\right]\cdot[\sin\theta\cdot\cos\phi]$$

$$a_0=1(SI),\Psi_{3,1,-1}(r,\theta,\phi)=\frac{1}{27\sqrt{\pi}}\left(2r-\frac{r^2}{3}\right)e^{-r/3}\sin\theta\cdot\cos\phi$$

(*) For the purpose of drawing the atomic orbitals (which have only real coordinates), we neglect the imaginary part of the Euler's equation, i.e., $i.\sin\varphi=0$.

The atomic orbitals of the hydrogen atom in atomic units (where $a_0 = 1$, $Z = 1$ and $K= 1$) for its first three shells in s and p sublevels are:

$$\psi_{1,0,0}=\frac{1}{\sqrt{\pi}}e^{-r}$$

$$\psi_{2,0,0}=\frac{1}{2\sqrt{2\pi}}\left(1-\frac{r}{2}\right)e^{-r/2}$$

$$\psi_{2,1,1}=\frac{1}{4\sqrt{2\pi}}r\sin\theta\cos\varphi\cdot e^{-r/2}$$

$$\psi_{2,1,-1}=\frac{1}{4\sqrt{2\pi}}r\sin\theta\sin\varphi\cdot e^{-r/2}$$

$$\psi_{2,1,0}=\frac{1}{4\sqrt{2\pi}}r\cos\theta\cdot e^{-r/2}$$

$$\psi_{3,0,0}=\frac{1}{81\sqrt{3\pi}}\left(27-18r+2r^2\right)\cdot e^{-r/3}$$

$$\psi_{3,1,1} = \frac{1}{27\sqrt{\pi}}\left(r - \frac{r^2}{3}\right)\sin\theta\cos\varphi \cdot e^{-r/3}$$

$$\psi_{3,1,-1} = \frac{1}{27\sqrt{\pi}}\left(r - \frac{r^2}{3}\right)\sin\theta\sin\varphi \cdot e^{-r/3}$$

$$\psi_{3,1,0} = \frac{1}{27\sqrt{\pi}}\left(r - \frac{r^2}{3}\right)\cos\theta \cdot e^{-r/3}$$

Notice that the atomic orbitals have four variables (r, θ, φ and Ψ). Then, a four-dimension plot is impracticable and it is necessary to parametrize, e.g., the θ as π/2 in order to obtain 3-D plots.

The plot of the probability density in the plane x, y of the 2s, 2p, 3p and 3d atomic orbitals is shown in Fig. 17.6. Check the do-it-yourself section to learn how to generate these figures.

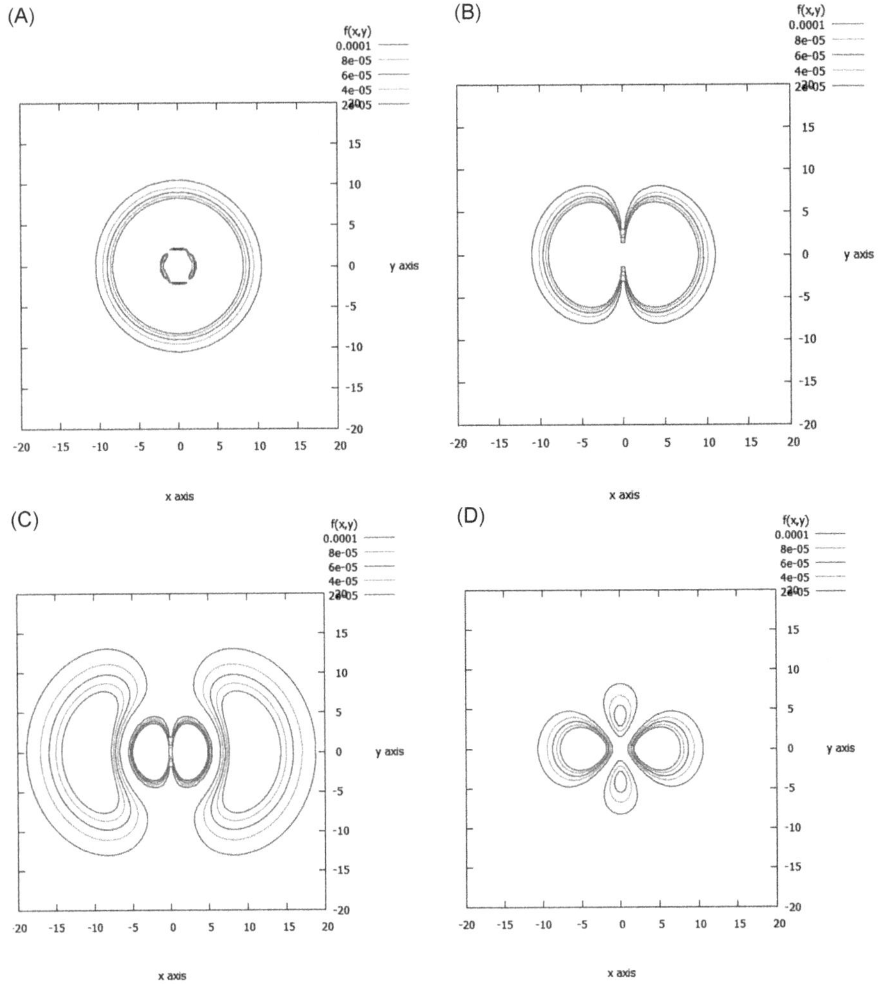

Fig. 17.6: Probability density in x,y plane of the hydrogen atomic orbitals (A) 2s, (B) 2p, (C) 3p and (D) 3d.

Important to add that any pair of s and p orbitals are orthogonal, that is, the integration of their product over the whole 3D space is zero. For example, let us use the same center for the 1s and $2p_x$ orbitals and apply the product of them. The result is a new p_x function bigger than $2p_x$. Its integration gives zero value due to the positive and negative lobes of p wave function.

$$\int 1s\cdot 2p_x d\tau = 0$$

Even two different pairs of s orbitals (for example, 2s and 3s) are orthogonal due to their nodes that change the sign of each s wave function.

7. Hydrogen fine structure-basics

The Halmitonian used in this chapter for the hydrogen-like atoms is known as $H^{(0)}$ because it does not include corrections such as relativistic effect, Darwin effect and spin-orbit effect for an hydrogen-like atom free of external potential (such as magnetic field). For a Hamiltonian that is non-relativistic, the speed of light is not included in the kinetic energy term.

Dirac found the Halmitonian that included these natural corrections (excluding external fields): relativistic effect (δH_{rel}), the spin-orbit effect ($\delta H_{spin\text{-}orbit}$) and Darwin effect ($\delta H_{Darwin}$).

$$H = \underbrace{\frac{p^2}{2m} + \mathbf{V}}_{H^{(0)}} - \underbrace{\frac{\mathbf{p}^4}{8m^3c^2}}_{\delta H_{rel}} + \underbrace{\frac{1}{2m^2c^2}\frac{1}{r}\frac{dV}{dr}S\cdot L}_{\delta H_{spin-orbit}} + \underbrace{\frac{\hbar^2}{8m^2c^2}\nabla^2 V}_{\delta H_{Darwin}}$$

Taking into account the relativistic effect, the degeneracy of l multiplets for a given n, according to g equation, is broken as one can see the equation of the energy of hydrogen atom with relativistic correction:

$$E^{(1)}_{n,l,m,s:rel} = -\frac{\left(E_n^{(0)}\right)^2}{2mc^2}\left[\frac{4n}{l+\frac{1}{2}} - 3\right]$$

$$where: E_n^{(0)} = -\frac{1}{n^2}Ry$$

When considering the relativistic correction (but not including the spin-orbit effect), the energy is dependent on n and l. Then, the degeneracy changes from the sum of 2l + 1 ranging from 0 to n – 1 into 2l + 1. The latter is similar to the diagram of energy level of the Madelung rule.

$$g^{(0)} = \sum_{l=0}^{n-1}(2l+1), \qquad g_{rel} = 2l+1$$

The energy of the hydrogen atom, taking into account the spin-orbit effect (which is zero for l = 0) is:

$$E^{(1)}_{nljm_j:spin-orbit} = \frac{\left(E_n^{(0)}\right)^2}{mc^2}\frac{n\left[j(j+1) - l(l+1) - \frac{3}{4}\right]}{l\left(l+\frac{1}{2}\right)(l+1)}, \qquad (l \neq 0)$$

$$j = l + s, \qquad s = 1/2$$

Where j is the quantum number for the operator J (sum of the angular momentum operator, L, and the spin angular momentum operator, S).
The combination of both energy corrections, after some derivation, give:

$$E^{(1)}_{nljm_j} = -\frac{\left(E^{(0)}_n\right)^2}{2mc^2}\left[\frac{4n}{j+\frac{1}{2}} - 3\right], \quad (l \neq 0)$$

As a consequence, the total energy of the hydrogen atom including relativistic and spin-orbit effects is:

$$E_{nljm_j} = -\frac{e^2}{2a_0}\frac{1}{n^2}\left[1+\frac{\alpha^2}{n^2}\left(\frac{n}{j+\frac{1}{2}} - \frac{3}{4}\right)\right]$$

Then, when taking into account both relativistic and spin-orbit effects, the degeneracy $g_{rel} = 2l + 1$ is broken. Now, the states $P_{1/2}$ and $P_{3/2}$ have different energies and the states $D_{3/2}$ and $D_{5/2}$ are not degenerate any more. The combination of both effects leads to the energy diagram depicted in Table 17.6 for the first three states of the hydrogen atom.

Table 17.6: Energy levels according to shell and subshell after the inclusion of the relativistic and spin-orbit effects.

Shell (n)	Subshell (l)/spectroscopic term		
	S	P	D
⋮	⋮	⋮	⋮
3			___ $3D_{5/2}$
		___ $3P_{3/2}$	___ $3D_{3/2}$
	___ $3S_{1/2}$	___ $3P_{1/2}$	
2		___ $2P_{3/2}$	
	___ $2S_{1/2}$	___ $2P_{1/2}$	
1	___ $1S_{1/2}$		

Do-it-yourself activity

(1) Plot the spherical harmonics in GNUPLOT.

Be aware that: (I) GNUPLOT does not recognize complex number i and the equation exp(i*φ) has to be replaced by sin(φ) or cos(φ) according to the Euler function and that we neglect the imaginary part; (II) variable u replaces θ and variable v replaces φ; (III) for the spherical harmonic functions, use the implicit function sqrt in GNUPLOT to represent square root function and pi to represent π.

Tip: type the following statements just once:

```
Set parametric
Set urange [0:pi]
Set vrange [0:2*pi]
Set isosample 36,36
Set ticslevel 0
Set size 0.65,1.0
Y(u,v)={spherical harmonics function}
Fx(u,v)=sin(u)*cos(v)*abs(Y(u,v))
Fy(u,v)=sin(u)*sin(v)*abs(Y(u,v))
Fz(u,v)=cos(u)*abs(Y(u,v))
Splot Fx(u,v),Fy(u,v),Fz(u,v)
```

Henceforth, to plot other spherical harmonics one needs only to retype the following statements:

```
Y(u,v)={spherical harmonics function}
Splot Fx(u,v),Fy(u,v),Fz(u,v)
```

(2) Draw the contour plot of the probability density of the orbitals 2s, 2p, 3p and 3d using GNUPLOT.

Tips:

(I) We have to change from polar coordinates to Cartesian coordinates.

(II) The probability density contour plot will be plotted in the xy plane (where $z = 0$). Then, we have to set $\theta = \pi/2$ and the probability density will be function of r and φ, i.e., $P(r,\varphi)$.

(III) you have to use the following general equations:

$$r = \sqrt{x^2 + y^2}$$

$$rho = (2./n) \cdot r$$

$$\varphi = \arctan\left(\frac{y}{x}\right)$$

$$\theta = \frac{\pi}{2}$$

(IV) There are specific functions for θ and φ. Each of them have part of the normalization constant of the spherical harmonics: N' and N'' derived from their own isolated functions.

(V) There is a specific function for radial distance ρ (named rad) containing the normalization constant of the radial function.

(VI) The final function, the probability function, is the square of the product of the functions for r, θ and φ.

(VII) There is a common part of the script for all cases:

```
Set contour
Unset surface
Set cntrparam levels 5
Set cntrparam levels incremental 2.e-5,2.e-5
Set isosamples 50
Set xrange [–20:20]
Set yrange [–20:20]
Set view 0,0,1.15,
Set size 0.62,1
```

(VIII) When it is possible, avoid functions containing θ. For example, choose $\Psi_{2,1,0}$ instead of $\Psi_{2,1,1}$ or $\Psi_{2,1,-1}$. Then, the function for θ will be only part of the normalization constant.

(IX) Let us see one example:

```
set contour
unset surface
set cntrparam levels 5
set cntrparam levels incremental 2.e-5,2.e-5
set isosamples 50
set xrange [-20:20]
set yrange [-20:20]
set view 0,0,1.15,
set size 0.62,1
rfun(x,y)=sqrt(x**2+y**2)
phi(x,y)=atan(y/x)
thetafun=1./sqrt(32.)
phifun(x,y)=(1./sqrt(pi))*cos(phi(x,y))
rad(x,y)=rfun(x,y)*exp(-rfun(x,y)/2.)
f(x,y)=(rad(x,y)*thetafun*phifun(x,y))**2
splot f(x,y)
```

Since n = 2 and ρ = r, we omitted the function rho in the script above.

$$\rho_{n=2} = (2/2)\cdot r = r$$

Exercises

(1) Plot the probability density functions for the radial wave functions shown in the Fig. 17.5 by using the code HYDROGEN-RADIAL incorporating the function P(r) for the probability density.

(2) Obtain the radial distribution of the hydrogen atom for n = 3 (l = 0,1,2) using the general formula:

$$R_{n,l}(r) = \sqrt{\left(\frac{2}{na_0}\right)^3 \frac{(n-l-1)!}{2n\left[(n+l)!\right]^3}} e^{-r/na_0} \left(\frac{2r}{na_0}\right)^l L_{n-l-1}^{2l+1}(\rho)$$

$$\rho = \frac{2r}{na_0}$$

$$L_{n-l-1}^{2l+1}(\rho) = \sum_{i=0}^{n-l-1} \frac{(-1)^i \left[(n+l)!\right]^2 \rho^i}{i!(n-l-1-i)!(2l+1+i)!}$$

Later, plot these radial distributions in the GNUPLOT.

Answer:

$$R_{3,0}(r) = \frac{2}{\sqrt{27}} a_0^{-3/2} \left(1 - \frac{2r}{3a_0} + \frac{2r^2}{27a_0^2}\right) e^{-r/3a_0}$$

$$R_{3,1}(r) = \frac{8}{27\sqrt{6}} a_0^{-3/2} \left(1 - \frac{r}{6a_0}\right) \frac{r}{a_0} e^{-r/3a_0}$$

$$R_{3,2}(r) = \frac{4}{81\sqrt{30}} a_0^{-3/2} \frac{r^2}{a_0^2} e^{-r/3a_0}$$

(3) Obtain the atomic wave function of the hydrogen atom (Z = 1) for 1s, 2s, 2p, 3s, 3p and 3d states.

References cited

Giannozzi, P., Ercolessi, F. and Gironcoli, S. 2013. Lecture notes: Numerical methods in quantum mechanics, Udine and Trieste, Italy (http://www.fisica.uniud.it/~giannozz/Didattica/MQ/mq.html).

Izaac, J. and Wang, J. 2018. Computational Quantum Mechanics. Springer Nature, Cham, Switzerland.

Morse, P.M. and Feshbach, H. 1953. Methods of Theoretical Physics, McGraw-Hill, New York.

Peng, L.-Y. and Gong, Q. 2010. An accurate Fortran code for computing hydrogenic continuum wave functions at a wide range of parameters. Comp. Phys. Comm. 181: 2098–2101.

Pauling, L. and Wilson, E.B. 1935. Introduction quantum mechanics with applications to chemistry. McGraw-Hill Inc. Tokyo, Japan.

Helium Atom, Variational Method and Perturbation Theory

18

1. Helium atom

After the hydrogen atom, helium is the simplest atom whose nucleus has two protons and two neutrons and the electrosphere has two electrons.
The operator of the kinetic energy in tridimensional cartesian coordinates is:

$$T = -\frac{\hbar^2}{2m}\left(\frac{\partial^2}{\partial x^2}+\frac{\partial^2}{\partial y^2}+\frac{\partial^2}{\partial z^2}\right) = -\frac{\hbar^2}{2m}\nabla^2$$

Then, the Hamiltonian of the hydrogen-like atom is given by:

$$H_H = -\frac{\hbar^2\nabla^2}{2m} - \frac{Ze^2}{r}$$

Where Z|e| is the charge of the hydrogen-like nucleus and r is the position vector of the electron.
Then, the Hamiltonian of the helium atom is:

$$H_{He} = -\frac{\hbar^2\nabla_1^2}{2m} - \frac{\hbar^2\nabla_2^2}{2m} - \frac{Ze^2}{r_1} - \frac{Ze^2}{r_2} + \frac{e^2}{\left|\vec{r_1} - \vec{r_2}\right|}$$

The last term in the above equation refers to the repulsion between the two electrons of the helium atom.

Then, we can state that the Hamiltonian of the helium atom can be expressed as the Hamiltonian of two hydrogen atoms, H_1 and H_2, plus the electronic repulsion term, V_{ee}.

$$H_{He} = H_{H1} + H_{H2} + V_{ee}$$

Since there is no exact solution for the above Hamiltonian, three approximate methods can be considered to solve the Schrödinger equation for the helium atom. They are: the model of distinguishable non-interacting particles, the variational method and the perturbation theory.

2. Model of distinguishable non-interacting particles

The model of distinguishable non-interacting particles is an approximation method to obtain the approximated wave function and energy of a multi-electron system.

The electrons (as well as protons and neutrons) are indistinguishable because their intrinsic physical properties (e.g., mass, electric charge, spin) are similar and the impossibility to track each individual trajectory due to their wave function nature.

In the model of distinguishable non-interacting particles, there are N electrons which do not interact among them and experience different potential, $V(\xi_i)$, and different mass, m_i.

The Hamiltonian of the single atom with N distinguishable non-interacting electrons is the sum of the N one-electron Hamiltonian.

$$H_N = \sum_{i=1}^{N} H_{Hi} = \sum_{i=1}^{N} \left[-\frac{\hbar^2 \nabla_i^2}{2m_i} + V(\xi_1) \right]$$

The wave function of the single atom with N distinguishable non-interacting electrons is the product of N one-electron hydrogen wave functions.

$$\Psi_{n_1,n_2,\ldots n_N}(\xi_1, \xi_2, \ldots \xi_N) = \Psi_{n_1}(\xi_1)\Psi_{n_2}(\xi_2)\ldots\Psi_{n_N}(\xi_N) = \prod_{i=1}^{N} \Psi_{n_i}(\xi_i)$$

Where n_i is the ith electron and ξ_i contain all the information of the ith electron.

The total energy of the single atom with N distinguishable non-interacting electrons is the sum of the N one-electron energy.

$$E_{n_1,n_2,\ldots n_N} = \varepsilon_{n_1} + \varepsilon_{n_2} + \ldots + \varepsilon_{n_N} = \sum_{i=1}^{N} \varepsilon_{n_i}$$

The wave function of the single atom with N distinguishable non-interacting electrons is the guess function to be used in the variational method.

The Schrödinger equation for the N distinguishable non-interacting electrons is:

$$H\Psi_{n_1,n_2,\ldots,n_N}(\xi_1, \xi_2, \ldots, \xi_N) = E_{n_1,n_2,\ldots,n_N} \Psi_{n_1,n_2,\ldots,n_N}(\xi_1, \xi_2, \ldots, \xi_N)$$

Which is separated into N equations:

$$\left[-\frac{\hbar^2 \nabla_i^2}{2m_i} + V_i(\xi_i) \right] \Psi_{n_i}(\xi_i) = \varepsilon_{n_i} \Psi_{n_i}(\xi_i)$$

Example: Find the ground state eigenvalue and wave function of a system containing four distinguishable non-interacting spinless electrons within an infinite, unidimensional potential well with length a.

Solution: Each electron moves in a potential whose boundary conditions are:

$V_i(x_i) = 0$ for $0 \leq x_i \leq a$

$V_i(x_i) = \infty$ for $x_i \geq 0$ and $x_i \geq a$

The Schrödinger equation for this system is:

$$\sum_{i=1}^{4}\left[-\frac{\hbar^2}{2m_i}\frac{d^2}{dx_i^2}\right]\Psi_{n_1,n_2,n_3,n_4}(x_1,x_2,x_3,x_4)=E_{n_1,n_2,n_3,n_4}\Psi_{n_1,n_2,n_3,n_4}(x_1,x_2,x_3,x_4)$$

Let us separate the above equation into:

$$-\frac{\hbar^2}{2m_i}\frac{d^2\Psi_{n_i}(x_i)}{dx_i^2}=\varepsilon_{n_i}\Psi_{n_i}(x_i), i=1,2,3,4$$

The energy and wave function of the ith electron is:

$$\varepsilon_{n_i}=\frac{\hbar^2\pi^2 n_i^2}{2m_i a^2},\quad \Psi_{n_i}(x_i)=\sqrt{\frac{2}{a}}\sin\left(\frac{n_i\pi}{a}x_i\right)$$

The total energy and total wave function are:

$$E_{n_1,n_2,n_3,n_4}=\frac{\hbar^2\pi^2}{2a^2}\left(\frac{n_1^2}{m_1}+\frac{n_2^2}{m_2}+\frac{n_3^2}{m_3}+\frac{n_4^2}{m_4}\right)$$

$$\Psi_{n_1,n_2,n_3,n_4}(x_1,x_2,x_3,x_4)=\frac{4}{a^2}\sin\left(\frac{n_1\pi}{a}x_1\right)\sin\left(\frac{n_2\pi}{a}x_2\right)\sin\left(\frac{n_3\pi}{a}x_3\right)\sin\left(\frac{n_4\pi}{a}x_4\right)$$

In the ground state, $n_1 = n_2 = n_3 = n_4 = 1$, then we have:

$$E_{1,1,1,1}=\frac{\hbar^2\pi^2}{2a^2}\left(\frac{1}{m_1}+\frac{1}{m_2}+\frac{1}{m_3}+\frac{1}{m_4}\right)$$

$$\Psi_{1,1,1,1}(x_1,x_2,x_3,x_4)=\frac{4}{a^2}\sin\left(\frac{\pi}{a}x_1\right)\sin\left(\frac{\pi}{a}x_2\right)\sin\left(\frac{\pi}{a}x_3\right)\sin\left(\frac{\pi}{a}x_4\right)$$

3. Variational method

The variational method is another approximation method to obtain the approximated wave function and energy of a multi-electron system.

The variational method starts with a guess function, named Φ, whose energy, E_i, is higher than the energy of the actual, exact wave function, Ψ, so called E_0 (the energy of the ground state), or both energies are equivalent (if the guess function is indeed the actual, exact wave function). Since the last possibility is nearly impossible for the guess function, one needs to change the starting guess function in order to obtain smaller and smaller energies until no change is found.

$$E_i=\langle\Phi|H|\Phi\rangle=\int\Phi^* H\Phi d\tau\geq E_0$$

With the aim to change the wave function, we have to use the expansion of a guess function using a discrete, complete basis set. The basis set is based on the hydrogen atomic orbitals, φ_k.

$$\Phi=\sum_k a_k\varphi_k$$

Where a_k is the expansion coefficient of kth hydrogen atomic orbital. Note that Schrödinger equation applied to: (1) Ψ gives the ground state energy, E_0; (2) Φ gives the energy E_i; and (3) φ_k gives the energy of each atomic orbital, ε_k.

$$H\Psi = E_0\Psi$$

$$H\Phi = E_i\Phi$$

$$H\varphi_k = \varepsilon_k\varphi_k$$

Let us use the expansion of the guess function in the integration to obtain the expectation value.

$$\int \Phi^* H\Phi d\tau = \int \sum_k a_k^*\varphi_k^* H \sum_j a_j\varphi_j d\tau =$$

$$= \int \sum_k a_k^*\varphi_k^* \sum_j a_j H\varphi_j d\tau = \int \sum_k a_k^*\varphi_k^* \sum_j a_j\varepsilon_j\varphi_j d\tau =$$

$$= \sum_k \sum_j a_k^* a_j \varepsilon_j \int \varphi_k^*\varphi_j d\tau$$

Since all hydrogen atomic orbitals are normalized:

$$\int_{-\infty}^{\infty} \varphi_i\varphi_i d\tau = 1$$

And two different hydrogen atomic orbitals are orthogonal (i.e., integrating the product of both functions yields a zero value), the Kronecker delta, δ_{kj}, can be used in the last expression as a orthonormality condition.

$$\int \Phi^* H\Phi d\tau = \sum_k \sum_j a_k^* a_j \varepsilon_j \int \varphi_k^*\varphi_j d\tau = \sum_k \sum_j a_k^* a_j \varepsilon_j \delta_{kj}$$

The Kronecker delta gives all the terms zero value except when k = j. In this case, we have:

$$\int \Phi^* H\Phi d\tau = \sum_k a_k^* a_k E_k = \sum_k |a_k|^2 E_k$$

When k = 1, the E_1 corresponds to the most low-lying eigenvalue of H or the ground state energy. Then, we have:

$$E_k \geq E_1, \quad k = 1, 2, 3, \ldots$$

$$then: |a_k|^2 E_k \geq |a_k|^2 E_1$$

$$hence: \sum_k |a_k|^2 E_k \geq \sum_k |a_k|^2 E_1$$

As a consequence, we have:

$$\int \Phi^* H\Phi d\tau = \sum_k |a_k|^2 E_k \geq \sum_k |a_k|^2 E_1 = E_1 \sum_k |a_k|^2$$

By using the normalization condition:

$$\int \Phi^*\Phi d\tau = 1, \quad \int \varphi^*\varphi d\tau = 1$$

And the expansion of the guess function:

$$\Phi = \sum_k a_k \varphi_k$$

We have:

$$1 = \int \left(\sum_k a_k^* \varphi_k^* \sum_j a_j \varphi_j \right) d\tau = \sum_k \sum_j a_k^* a_j \int \varphi_k^* \varphi_j d\tau =$$

$$1 = \sum_k \sum_j a_k^* a_j \delta_{kj} = \sum_k |a_k|^2$$

Then, let us use the above relation in the equation:

$$\int \Phi^* H \Phi d\tau = \sum_k |a_k|^2 E_k \geq E_1 \sum_k |a_k|^2$$

$$\text{since}: \sum_k |a_k|^2 = 1$$

$$then: \int \Phi^* H \Phi d\tau \geq E_1$$

Let us now suppose that the guess function Φ is not normalized. Then, it is needed to add the normalization constant in the above equation to give:

$$|N|^2 \int \Phi^* H \Phi d\tau \geq E_1$$

The equation for the normalization constant is given below:

$$\int (N\Phi)^* (N\Phi) d\tau = |N|^2 \int \Phi^* \Phi d\tau = 1$$

$$|N|^2 = \frac{1}{\int \Phi^* \Phi d\tau}$$

Then, the equation of the variational method for non-normalized guess function (the variational integral) is:

$$\frac{\int \Phi^* H \Phi d\tau}{\int \Phi^* \Phi d\tau} \geq E_1$$

With the aim to approach the ground state energy E_1, it is needed to test different guess functions in order to obtain lower and lower eigenvalues of E_k. The best guess function gives the value closest to E_1.

Example: Find the guess function for the electron in an unidimensional box with length l.

Solution: The boundary conditions are:

$\Psi = 0$ for $x = 0$ and $x = l$

And $\Phi = 0$ for $x = 0$ and $x = l$

In Chapter fifteen, we found that the wave function Ψ does not have nodes within the box in the ground state, then the guess function Φ should follow the same condition. As a consequence, the simplest function that satisfies these conditions is the parabolic function:

$$\Phi = x(l-x) = xl - x^2 \therefore 0 \leq x \leq l$$

The Hamiltonian within the box is:

$$H = -\frac{\hbar^2}{2m_i}\frac{d^2}{dx_i^2}$$

Since the guess function is non-normalized we have to use the variational equation for non-normalized guess functions. Let us find each term of the non-normalized variational integral. The numerator is:

$$\int \Phi^* H \Phi d\tau = -\frac{\hbar^2}{2m}\int_0^l \left(lx - x^2\right)\frac{d^2}{dx^2}\left(lx - x^2\right)dx$$

$$\frac{d}{dx}\left(lx - x^2\right) = l - 2x, \qquad \frac{d}{dx}(l-2x) = -2$$

$$-\frac{\hbar^2}{2m}\int_0^l \left(lx - x^2\right)(-2)\,dx = -\frac{\hbar^2}{m}\int_0^l \left(x^2 - lx\right)dx =$$

$$-\frac{\hbar^2}{m}\left[\frac{1}{3}x^3 \Big|_0^l - \frac{1}{2}lx^2 \Big|_0^l\right] = \frac{\hbar^2 l^3}{6m}$$

The denominator of the non-normalized variational integral is:

$$\int \Phi^* \Phi d\tau = \int_0^l x^2 (l-x)^2 dx =$$

$$\int_0^l \left(x^2 l^2 - 2x^3 l + x^4\right)dx =$$

$$\int_0^l \left(x^2 l^2\right)dx - \int_0^l \left(2x^3 l\right)dx + \int_0^l \left(x^4\right)dx$$

$$\frac{l^2 x^3}{3}\Big|_0^l - \frac{x^4 l}{2}\Big|_0^l + \frac{x^5}{5}\Big|_0^l =$$

$$\frac{10l^5 - 15l^5 + 6l^5}{30} = \frac{l^5}{30}$$

By replacing both terms in the non-normalized variational integral, we have:

$$\frac{\hbar^2 l^3/6m}{l^5/30} = \frac{5\hbar^2}{ml^2}, \qquad \hbar = h/2\pi, \qquad \frac{5h^2}{4\pi^2 l^2 m} \geq E_1$$

The error of this result is:

$$\frac{\left(\frac{5h^2}{4\pi^2 l^2 m}\right)-\left(\frac{h^2}{8l^2 m}\right)}{\left(\frac{h^2}{8l^2 m}\right)}\times 100\% = 1.3\%$$

The normalized guess wave function is:

$$\Phi = \left(30/l^5\right)^{1/2} x(l-x)$$

4. Perturbation theory (case: non-degenerate systems)

The perturbation theory is a mathematical method to find approximate solution for a problem without exact solution which has as a starting point the exact solution of the related problem. The perturbation theory is only applicable when it is possible to add a small term to the formulated problem.

The perturbation theory is based on the hypothesis that the formulated problem with non-exact solution is only slightly different (it has a small deviation) from that one having an exact solution. The perturbation theory can be used when the difference in solution between both problems is small.

As to the Hamiltonian of the formulated problem with non-exact solution, the contribution (the small difference) to the Hamiltonian, H_p, is added to the Hamiltonian which has an exact solution (H_0). Then, the perturbation theory builds approximate solutions (H) from exact solutions.

$$H = H_0 + H_p$$

From the perturbation theory, the H_p is very small compared to H_0 (Hamiltonian of the non-perturbated system).

Let us consider the system with exact H_0 non-degenerate eigenvalues, that is, each energy, $E_n^{(0)}$, has a distinct value which corresponds to only one eigenstate ϕ_n (wave function of the system with exact solution). The Schrödinger equation (using Dirac notation) of the system with exact solution is:

$$H_0\left|\phi_n\right\rangle = E_n^{(0)}\left|\phi_n\right\rangle$$

Where n is the principal quantum number. The Schrödinger equation of the formulated problem with non-exact solution is:

$$H\left|\psi_n\right\rangle = E_n\left|\psi_n\right\rangle$$

The main idea of the perturbation theory is to assume that the perturbated eigenvalues and eigenstates can be expanded in a power series (Taylor series—see Chapter two) over the parameter λ (real dimensionless parameter which delimits a very small contribution to the perturbation).

$$H_p = \lambda H', \quad \lambda \ll 1$$

$$H = H_0 + \lambda H'$$

Let us assume the wave function of the exact solution, ϕ_n, as the zeroth order wave function of the perturbated series (ψ^0_n):

$$|\phi_n\rangle = |\psi_n^{(0)}\rangle$$

Then, we have:

$$(H_0 + \lambda H')|\psi_n\rangle = E_n|\psi_n\rangle$$
$$E_n = E_n^{(0)} + \lambda E_n^{(1)} + \lambda^2 E_n^{(2)} + ... + \lambda^k E_n^{(k)}$$
$$|\psi_n\rangle = |\psi_n^{(0)}\rangle + \lambda|\psi_n^{(1)}\rangle + \lambda^2|\psi_n^{(2)}\rangle + ... + \lambda^k|\psi_n^{(k)}\rangle$$
$$|\psi_n^{(k)}\rangle = \frac{1}{k!}\frac{\partial^k \psi_n}{\partial \lambda^k}\bigg|_{\lambda\to 0}, E_n^{(k)} = \frac{1}{k!}\frac{\partial^k E_n}{\partial \lambda^k}\bigg|_{\lambda\to 0}, k = 1,2,3...$$

By using the expansion of E_n and ψ_n in the above Schrödinger equation, we have the Taylor series (Chapter two) of the perturbated system:

$$H_0|\psi_n^{(0)}\rangle + \lambda\left(H'|\psi_n^{(0)}\rangle + H_0|\psi_n^{(1)}\rangle\right) + \lambda^2\left(H'|\psi_n^{(1)}\rangle + H_0|\psi_n^{(2)}\rangle\right) + ... =$$
$$E_n^{(0)}|\psi_n^{(0)}\rangle + \lambda\left(E_n^{(1)}|\psi_n^{(0)}\rangle + E_n^{(0)}|\psi_n^{(1)}\rangle\right) +$$
$$+\lambda^2\left(E_n^{(2)}|\psi_n^{(0)}\rangle + E_n^{(1)}|\psi_n^{(1)}\rangle + E_n^{(0)}|\psi_n^{(2)}\rangle\right) + ...$$

It is important to add that there are cases where the perturbation is small but E_n and ψ_n are not expandable in the power of the parameter λ. Besides, there are cases where these series are not convergent, but the first terms provide a reasonable description of the system. Then, if we truncate the above expansion to one term (first order correction) or two terms (second order correction), the convergence problem is avoided.

The above series shows the zeroth, first and second order corrections. These corrections are respectively explicit as zeroth, first and second order corrections below.

$$H_0|\psi_n^{(0)}\rangle = E_n^{(0)}|\psi_n^{(0)}\rangle$$
$$H_0|\psi_n^{(1)}\rangle + H'|\psi_n^{(0)}\rangle = E_n^{(0)}|\psi_n^{(1)}\rangle + E_n^{(1)}|\psi_n^{(0)}\rangle$$
$$H'|\psi_n^{(1)}\rangle + H_0|\psi_n^{(2)}\rangle = E_n^{(2)}|\psi_n^{(0)}\rangle + E_n^{(1)}|\psi_n^{(1)}\rangle + E_n^{(0)}|\psi_n^{(2)}\rangle$$

4.1 First order perturbation theory

Let us take the explicit first order correction below (in terms of λ):

$$H_0|\psi_n^{(1)}\rangle + H'|\psi_n^{(0)}\rangle = E_n^{(0)}|\psi_n^{(1)}\rangle + E_n^{(1)}|\psi_n^{(0)}\rangle$$

And rearrange this equation as:

$$H_0|\psi_n^{(1)}\rangle - E_n^{(0)}|\psi_n^{(1)}\rangle = E_n^{(1)}|\psi_n^{(0)}\rangle - H'|\psi_n^{(0)}\rangle$$

Now, we multiply both terms of the above equation by the complex conjugate of ψ^0_n, which is $\langle\psi_m^{(0)}|$ in Dirac notation. Then, we have:

$$\left\langle\psi_m^{(0)}\middle|H_0\middle|\psi_n^{(1)}\right\rangle - E_n^{(0)}\left\langle\psi_m^{(0)}\middle|\psi_n^{(1)}\right\rangle =$$
$$= E_n^{(1)}\left\langle\psi_m^{(0)}\middle|\psi_n^{(0)}\right\rangle - \left\langle\psi_m^{(0)}\middle|H'\middle|\psi_n^{(0)}\right\rangle$$

By using the Hermetian property for the first term of the left side of the above equation, we have:

$$\left\langle\psi_m^{(0)}\middle|H_0\middle|\psi_n^{(1)}\right\rangle = \left\langle\psi_n^{(1)}\middle|H_0\middle|\psi_m^{(0)}\right\rangle^* = \left\langle\psi_n^{(1)}\middle|H_0\psi_m^{(0)}\right\rangle^* =$$
$$= \left\langle\psi_n^{(1)}\middle|E_m^{(0)}\psi_m^{(0)}\right\rangle^* = E_m^{(0)}\left\langle\psi_n^{(1)}\middle|\psi_m^{(0)}\right\rangle^* = E_m^{(0)}\left\langle\psi_m^{(0)}\middle|\psi_n^{(1)}\right\rangle$$
$$Where: H_0\left|\phi_n\right\rangle = E_n^{(0)}\left|\phi_n\right\rangle$$

Let us suppose that n corresponds to the ground state and m corresponds to the first excited state.
By replacing the resulting first term in the previous equation, we have:

$$E_m^{(0)}\left\langle\psi_m^{(0)}\middle|\psi_n^{(1)}\right\rangle - E_n^{(0)}\left\langle\psi_m^{(0)}\middle|\psi_n^{(1)}\right\rangle =$$
$$= E_n^{(1)}\left\langle\psi_m^{(0)}\middle|\psi_n^{(0)}\right\rangle - \left\langle\psi_m^{(0)}\middle|H'\middle|\psi_n^{(0)}\right\rangle$$

Which gives:

$$\left(E_m^{(0)} - E_n^{(0)}\right)\left\langle\psi_m^{(0)}\middle|\psi_n^{(1)}\right\rangle = E_n^{(1)}\left\langle\psi_m^{(0)}\middle|\psi_n^{(0)}\right\rangle - \left\langle\psi_m^{(0)}\middle|H'\middle|\psi_n^{(0)}\right\rangle$$

By knowing the Kronecker delta as:

$$\delta_{mn} = \left\langle\psi_m^{(0)}\middle|\psi_n^{(0)}\right\rangle$$

We have:

$$\left(E_m^{(0)} - E_n^{(0)}\right)\left\langle\psi_m^{(0)}\middle|\psi_n^{(1)}\right\rangle = E_n^{(1)}\delta_{mn} - \left\langle\psi_m^{(0)}\middle|H'\middle|\psi_n^{(0)}\right\rangle$$

4.1.1 m = n

If m = n, then left side of the above equation is zero and Kronecker delta is a unit. Then, we have the first order correction to the eigenvalue of the formulated problem.

$$E_n^{(1)} = \left\langle\psi_n^{(0)}\middle|H'\middle|\psi_n^{(0)}\right\rangle$$

If we truncate the series to the first order, then the total energy of the system with non-exact solution is:

$$E_n = E_n^{(0)} + \left\langle\psi_n^{(0)}\middle|H'\middle|\psi_n^{(0)}\right\rangle$$

Example (1): Find the first energy correction for the unidimensional non-harmonic oscillator with the following Hamiltonian:

$$H = -\frac{\hbar^2}{2m}\frac{d^2}{dx^2} + \frac{1}{2}kx^2 + cx^3 + dx^4$$

Where k, c, and d are constants.

Solution: The system with exact solution is the Hamiltonian of the harmonic oscillator:

$$H^{(0)} = -\frac{\hbar^2}{2m}\frac{d^2}{dx^2} + \frac{1}{2}kx^2$$

Remember that the ground state wave function (n = 0) for the unidimensional harmonic oscillator is:

$$\psi(x) = N_n H(y) e^{-\frac{\alpha x^2}{2}}$$

$$N_n = \frac{1}{2^n n!}(\alpha/\pi)^{1/4}$$

$$n = 0, \quad H(x) = 1$$

$$\psi(x) = (\alpha/\pi)^{1/4} e^{-\frac{\alpha x^2}{2}}$$

The Hamiltonian of the first order correction is the difference between the Hamiltonian of the formulated problem and the Hamiltonian of the zeroth order correction:

$$H' = H - H^{(0)}$$

$$H' = \left(-\frac{\hbar^2}{2m}\frac{d^2}{dx^2} + \frac{1}{2}kx^2 + cx^3 + dx^4\right) - \left(-\frac{\hbar^2}{2m}\frac{d^2}{dx^2} + \frac{1}{2}kx^2\right)$$

$$H' = cx^3 + dx^4$$

The first order correction is:

$$E_n^{(1)} = \langle \psi_n^{(0)} | H' | \psi_n^{(0)} \rangle = \int \psi_n^{(0)*} H' \psi_n^{(0)} d\tau$$

In the case of the harmonic oscillator, the wave function does not have imaginary component, then the equation for the first order energy correction is:

$$E_n^{(1)} = \int H' \left(\psi_n^{(0)}\right)^2 d\tau, \quad \psi_n^{(0)} = real$$

Then, we have the following unidimensional integral:

$$E^{(1)} = \frac{\alpha}{\pi} \int_{-\infty}^{\infty} e^{-x^2} \left(cx^3 + dx^4\right) dx$$

As one can see in Fig. 18.1, the integral of the function $cx^3\exp(-x^2)$ is zero in the region $-\infty$ to $+\infty$. Then, we use only the integral of the function $dx^4\exp(-x^2)$ which is non-zero in the region $-\infty$ to $+\infty$. (See Fig. 18.1). In Fig. 18.1, we have used the function $\exp(-x^2/2)$ but similar results exist for the function $\exp(-x^2)$.

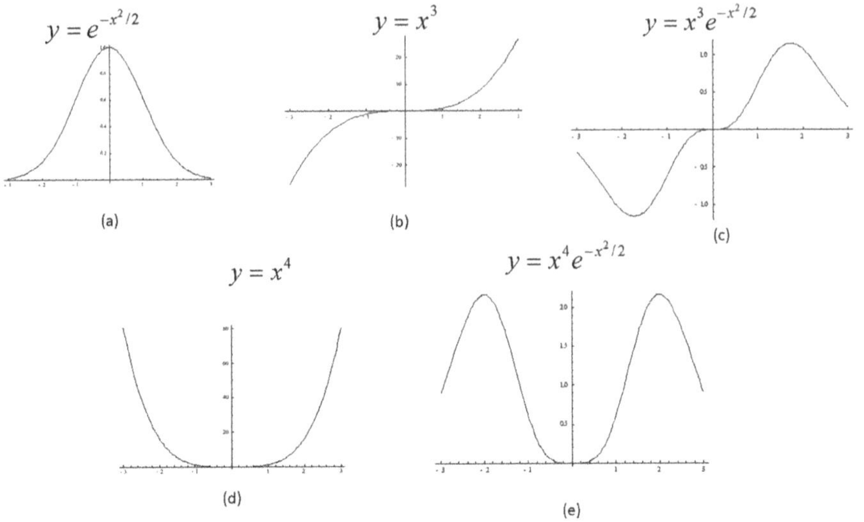

Fig. 18.1: Plots of the functions: (a) y = x^3; (b) y = exp($-x^2/2$); (c) y = x^3exp($-x^2/2$); (d) y = x^4; (e) y = x^4exp($-x^2/2$).

Then, the first order correction for the energy becomes:

$$E^{(1)} = 2d\frac{\alpha}{\pi}\int_0^{\infty} x^4 e^{-x^2}\,dx = \frac{3dh^2}{64\pi^2 v^2 m^2}$$

$$\alpha = \frac{2\pi m v}{\hbar}$$

Example (2): The Hamiltonian of the helium atom is:

$$H_{He} = -\frac{\hbar^2\nabla_1^2}{2m} - \frac{\hbar^2\nabla_2^2}{2m} - \frac{Ze^2}{r_1} - \frac{Ze^2}{r_2} + \frac{e^2}{\left|\vec{r_1} - \vec{r_2}\right|}$$

The Hamiltonian of the non-perturbated system is the sum of the Hamiltonian of two isolated hydrogen atoms H_1^0 and H_2^0 and the Hamiltonian of the perturbated system that is the potential energy operator between both electrons.

$$H_1^{(0)} = -\frac{\hbar^2\nabla_1^2}{2m} - \frac{Ze^2}{r_1}$$

$$H_2^{(0)} = -\frac{\hbar^2\nabla_2^2}{2m} - \frac{Ze^2}{r_2}$$

$$H^{(0)} = -\frac{\hbar^2\nabla_1^2}{2m} - \frac{Ze^2}{r_1} - \frac{\hbar^2\nabla_2^2}{2m} - \frac{Ze^2}{r_2}$$

$$H_p = \frac{e^2}{\left|\vec{r_1} - \vec{r_2}\right|}$$

$$H_{He} = H^{(0)} + H_p$$

The wave function of the non-perturbated system is the product of the wave function of two isolated hydrogen atoms.

$$\psi^{(0)}(r_1,\theta_1,\varphi_1,r_2,\theta_2,\varphi_2)=\chi_1(r_1,\theta_1,\varphi_1)\chi_2(r_2,\theta_2,\varphi_2)$$

The energies and Schrödinger equations of the non-perturbated system are:

$$E^{(0)}=E_1+E_2$$

$$H_1^{(0)}\chi_1=E_1\chi_1$$

$$H_2^{(0)}\chi_2=E_2\chi_2$$

The solutions of the above equations of the non-perturbated system is:

$$E_1=-\frac{Z^2}{n_1^2}\frac{e^2}{2a_0}$$

$$E_2=-\frac{Z^2}{n_2^2}\frac{e^2}{2a_0}$$

$$E^{(0)}=-Z^2\left(\frac{1}{n_1^2}+\frac{1}{n_2^2}\right)\frac{e^2}{2a_0}=-108.8eV$$

The experimental energy of the helium atom is –78,95 eV. Then, the error is 37,8%. When we take into account the first order correction of the Hamiltonian of the perturbated system, the corresponding first order corrected energy is:

$$E^{(1)}=\langle\psi^{(0)}|H'|\psi^{(0)}\rangle$$

$$E^{(1)}=\frac{Z^6e^2}{\pi^2a_0^6}\int_0^{2\pi}\int_0^{2\pi}\int_0^{\pi}\int_0^{\pi}\int_0^{\infty}\int_0^{\infty}e^{-2Zr_1/a_0}e^{-2Zr_2/a_0}\frac{1}{r_{12}}r_1^2\sin\theta_1 r_2^2\sin\theta_2 dr_1dr_2d\theta_1d\theta_2d\varphi_1d\varphi_2$$

Important to observe that the differential $d\tau$ in spherical coordinates is:

$$d\tau=(r\cdot d\theta)\cdot(dr)\cdot(d\phi\cdot r\cdot\sin\theta)$$

$$d\tau=r^2\cdot\sin\theta\cdot dr\cdot d\theta\cdot d\phi$$

According to Fig. 18.2

$$E^{(1)}=\frac{5Z}{8}\left(\frac{e^2}{a_0}\right)=34eV$$

The total energy of the helium is:

$$E_{helium}=E^{(0)}+E^{(1)}=-108.8+34=\ -74{,}8\ eV$$

The error in comparison with the experimental value (–78,95 EV) is 5.25%.

Example (3): Consider one electron in a unidimensional box having length a, an applied electric field, E, force, F, acting over the electron F = –eE, and the potential energy V(x) = eEx. Considering that eEa << E_0, where E_0 is the energy of the ground state of the electron in the absence of the perturbation, give the first order corrected energy of the electron.

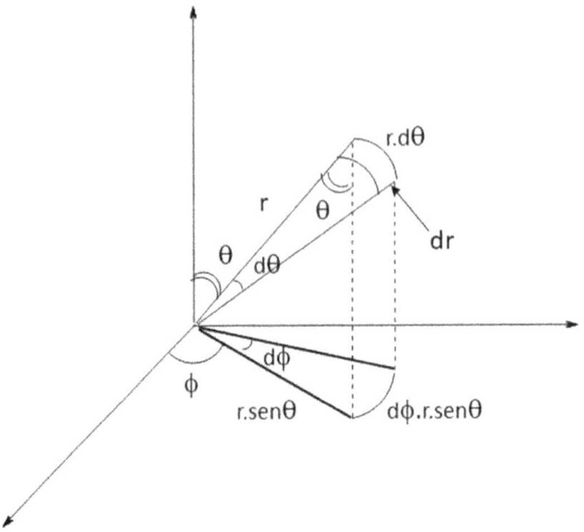

Fig. 18.2: The differential dτ in spherical coordinates

Solution: For an electron in a non-perturbated unidimensional box, with length a, the energy of the ground state and the wave function are:

$$E_n^{(0)} = \frac{n^2h^2}{8ma^2}, \qquad n = 1,2,3,...$$

$$\psi_n^{(0)}(x) = \left(\frac{2}{a}\right)^{1/2} \sin\left(\frac{n\pi x}{a}\right), 0 \le x \le a$$

The Hamiltonian of the perturbated system is:

$$H = H^{(0)} + H'$$

$$H' = eEx$$

The first order corrected energy:

$$E^{(1)} = \langle \psi^{(0)} | H' | \psi^{(0)} \rangle$$

$$E^{(1)} = \int_0^a \sqrt{\frac{2}{a}} \sin\frac{\pi x}{a} (eEx) \int_0^a \sqrt{\frac{2}{a}} \sin\frac{\pi x}{a} dx$$

$$E^{(1)} = \frac{2}{a}\int_0^a eEx \sin^2\frac{\pi x}{a} dx = \frac{2eE}{a}\int_0^a x\sin^2\frac{\pi x}{a} dx$$

Let us use the following integral property:

$$\int udv = uv - \int vdu$$

$$where: u = x, \qquad du = dx$$

$$dv = \sin^2\frac{\pi x}{a} dx, \qquad v = \frac{x}{2} - \frac{a}{4\pi}\sin\frac{2\pi x}{a}$$

$$E^{(1)} = \frac{2eE}{a}\left[x\left(\frac{x}{2}-\frac{a}{4\pi}\sin\frac{2\pi x}{a}\right)\right]_0^a - \int_0^a\left(\frac{x}{2}-\frac{a}{4\pi}\sin\frac{2\pi x}{a}\right)dx$$

$$E^{(1)} = \frac{2eE}{a}\left[\frac{x^2}{2}-\frac{ax}{4\pi}\sin\frac{2\pi x}{a}\right]_0^a - \left[\frac{x^2}{4}+\frac{a^2}{8\pi}\cos\frac{2\pi x}{a}\right]_0^a$$

$$E^{(1)} = \frac{2eE}{a}\left[\frac{x^2}{2}-\frac{x^2}{4}\right]_0^a - \frac{a^2}{4\pi}\sin\frac{2\pi a}{a} - \left[\frac{a^2}{8\pi}\cos\frac{2\pi x}{a}\right]_0^a$$

$$\sin 2\pi = 0$$

$$E^{(1)} = \frac{2eE}{a}\left[\frac{a^2}{2}-\frac{a^2}{4}-\frac{a^2}{8\pi}\cos\frac{2\pi a}{a}+\frac{a^2}{8\pi}\cos\frac{2\pi 0}{a}\right]$$

$$\cos 2\pi = \cos 0 = 1$$

$$E^{(1)} = \frac{2eE}{a}\left[\frac{a^2}{4}-\frac{a^2}{8\pi}+\frac{a^2}{8\pi}\right] = \frac{eEa}{2}$$

4.1.2 $m \neq n$

Now, let us come back to the equation:

$$\left(E_m^{(0)}-E_n^{(0)}\right)\left\langle\psi_m^{(0)}\middle|\psi_n^{(1)}\right\rangle = E_n^{(1)}\delta_{mn} - \left\langle\psi_m^{(0)}\middle|H'\middle|\psi_n^{(0)}\right\rangle$$

$$for: m \neq n, \quad \delta_{mn} = 0$$

$$then: \left(E_m^{(0)}-E_n^{(0)}\right)\left\langle\psi_m^{(0)}\middle|\psi_n^{(1)}\right\rangle = -\left\langle\psi_m^{(0)}\middle|H'\middle|\psi_n^{(0)}\right\rangle$$

If m≠n, then expand the wave function $\psi_n^{(1)}$ from a complete, orthonormal non-perturbated wave function $\psi_m^{(0)}$ of the H^0.

$$\psi_n^{(1)} = \sum_m a_m\psi_m^{(0)}$$

$$a_m = \left\langle\psi_m^{(0)}\middle|\psi_n^{(1)}\right\rangle$$

By replacing the expansion coefficient a_m in the previous equation, we have:

$$\left(E_m^{(0)}-E_n^{(0)}\right)a_m = -\left\langle\psi_m^{(0)}\middle|H'\middle|\psi_n^{(0)}\right\rangle$$

$$then: a_m = -\frac{\left\langle\psi_m^{(0)}\middle|H'\middle|\psi_n^{(0)}\right\rangle}{\left(E_m^{(0)}-E_n^{(0)}\right)}$$

Hence, we have the first order corrected wave function:

$$\psi_n^{(1)} = \sum_m a_m\psi_m^{(0)}$$

$$a_m = -\frac{\left\langle\psi_m^{(0)}\middle|H'\middle|\psi_n^{(0)}\right\rangle}{\left(E_m^{(0)}-E_n^{(0)}\right)} = \frac{\left\langle\psi_m^{(0)}\middle|H'\middle|\psi_n^{(0)}\right\rangle}{\left(E_n^{(0)}-E_m^{(0)}\right)}$$

$$\psi_n^{(1)} = \sum_{m\neq n}\frac{\left\langle\psi_m^{(0)}\middle|H'\middle|\psi_n^{(0)}\right\rangle}{\left(E_n^{(0)}-E_m^{(0)}\right)}\psi_m^{(0)}$$

4.2 Second order perturbation theory

Let us come back to the Taylor series of the perturbated system:

$$H_0\left|\psi_n^{(0)}\right\rangle+\lambda\left(H'\left|\psi_n^{(0)}\right\rangle+H_0\left|\psi_n^{(1)}\right\rangle\right)+\lambda^2\left(H'\left|\psi_n^{(1)}\right\rangle+H_0\left|\psi_n^{(2)}\right\rangle\right)+\ldots=$$

$$E_n^{(0)}\left|\psi_n^{(0)}\right\rangle+\lambda\left(E_n^{(1)}\left|\psi_n^{(0)}\right\rangle+E_n^{(0)}\left|\psi_n^{(1)}\right\rangle\right)+$$

$$+\lambda^2\left(E_n^{(2)}\left|\psi_n^{(0)}\right\rangle+E_n^{(1)}\left|\psi_n^{(1)}\right\rangle+E_n^{(0)}\left|\psi_n^{(2)}\right\rangle\right)+\ldots$$

Now, let us take the explicit second order correction below (in terms of λ^2):

$$H_0\left|\psi_n^{(2)}\right\rangle+H'\left|\psi_n^{(1)}\right\rangle=E_n^{(0)}\left|\psi_n^{(2)}\right\rangle+E_n^{(1)}\left|\psi_n^{(1)}\right\rangle+E_n^{(2)}\left|\psi_n^{(0)}\right\rangle$$

Let us now multiply both sides of the equation above by $\langle\psi_n^{(0)}|$ and integrate over the whole space:

$$\left\langle\psi_n^{(0)}\right|H_0\left|\psi_n^{(2)}\right\rangle+\left\langle\psi_n^{(0)}\right|H'\left|\psi_n^{(1)}\right\rangle=\left\langle\psi_n^{(0)}\right|E_n^{(0)}\left|\psi_n^{(2)}\right\rangle+\left\langle\psi_n^{(0)}\right|E_n^{(1)}\left|\psi_n^{(1)}\right\rangle+$$

$$+\left\langle\psi_n^{(0)}\right|E_n^{(2)}\left|\psi_n^{(0)}\right\rangle$$

$$\left\langle\psi_n^{(0)}\right|H_0\left|\psi_n^{(2)}\right\rangle+\left\langle\psi_n^{(0)}\right|H'\left|\psi_n^{(1)}\right\rangle=E_n^{(0)}\left\langle\psi_n^{(0)}\middle|\psi_n^{(2)}\right\rangle+E_n^{(1)}\left\langle\psi_n^{(0)}\middle|\psi_n^{(1)}\right\rangle+$$

$$+E_n^{(2)}\left\langle\psi_n^{(0)}\middle|\psi_n^{(0)}\right\rangle$$

According to the Schrödinger equation,

$$H^0\Psi_n^{(0)}=E_n^{(0)}\Psi_n^{(0)}$$

We have:

$$E_n^{(0)}\left\langle\psi_n^{(0)}\middle|\psi_n^{(2)}\right\rangle+\left\langle\psi_n^{(0)}\right|H'\left|\psi_n^{(1)}\right\rangle=E_n^{(0)}\left\langle\psi_n^{(0)}\middle|\psi_n^{(2)}\right\rangle+$$

$$+E_n^{(1)}\left\langle\psi_n^{(0)}\middle|\psi_n^{(1)}\right\rangle+E_n^{(2)}\left\langle\psi_n^{(0)}\middle|\psi_n^{(0)}\right\rangle$$

By considering the non-perturbated and the corrected functions as orthonormal, we have:

$$\left\langle\psi_n^{(0)}\middle|\psi_n^{(1)}\right\rangle=\left\langle\psi_n^{(0)}\middle|\psi_n^{(2)}\right\rangle=0$$

$$\left\langle\psi_n^{(0)}\middle|\psi_n^{(0)}\right\rangle=1$$

Then, the second order corrected energy is:

$$E_n^{(2)}=\left\langle\psi_n^{(0)}\right|H'\left|\psi_n^{(1)}\right\rangle$$

Let us use the expression for that we have obtained in the last section

$$\psi_n^{(1)}=\sum_m a_m\psi_m^{(0)}$$

$$a_m=-\frac{\left\langle\psi_m^{(0)}\right|H'\left|\psi_n^{(0)}\right\rangle}{\left(E_m^{(0)}-E_n^{(0)}\right)}=\frac{\left\langle\psi_m^{(0)}\right|H'\left|\psi_n^{(0)}\right\rangle}{\left(E_n^{(0)}-E_m^{(0)}\right)}$$

$$\Psi_n^{(1)}=\sum_{m\neq n}\frac{\left\langle\psi_m^{(0)}\right|H'\left|\psi_n^{(0)}\right\rangle}{E_n^{(0)}-E_m^{(0)}}\psi_m^{(0)}$$

In the second order corrected energy expression:

$$E_n^{(2)} = \left\langle \psi_n^{(0)} \middle| H' \middle| \sum_{m \neq n} \frac{\left\langle \psi_m^{(0)} \middle| H' \middle| \psi_n^{(0)} \right\rangle}{E_n^{(0)} - E_m^{(0)}} \psi_m^{(0)} \right\rangle$$

By removing the summation off the integral, we have:

$$E_n^{(2)} = \sum_{m \neq n} \frac{\left\langle \psi_m^{(0)} \middle| H' \middle| \psi_n^{(0)} \right\rangle}{E_n^{(0)} - E_m^{(0)}} \left\langle \psi_n^{(0)} \middle| H' \middle| \psi_m^{(0)} \right\rangle$$

From the Hermetian property:

$$\left\langle \psi_n^{(0)} \middle| H' \middle| \psi_m^{(0)} \right\rangle = \left\langle \psi_m^{(0)} \middle| H' \middle| \psi_n^{(0)} \right\rangle^*$$

By considering no imaginary terms in the wave function, we have:

$$E_n^{(2)} = \sum_{m \neq n} \frac{\left| \left\langle \psi_m^{(0)} \middle| H' \middle| \psi_n^{(0)} \right\rangle \right|^2}{E_n^{(0)} - E_m^{(0)}}$$

5. Helium spin function

Since fermions follow the anti-symmetry requirement of the Pauli exclusion principle and the spatial wave function of the helium atom is symmetrical, its spin function must be anti-symmetrical so that the whole wave function (product of spatial and spin functions) is anti-symmetrical. Let us use the permutation operator, P_{12}, in order to find a proper spin function for the helium atom. Firstly, remember the properties of the permutation operator (see Chapter three) and symmetry below:

$$\Theta(s_1, s_2) \xrightarrow{P_{12}} \Theta(s_2, s_1) = \Theta(s_1, s_2), \textit{symmetrical}$$

$$\Theta(s_1, s_2) \xrightarrow{P_{12}} \Theta(s_2, s_1) = -\Theta(s_1, s_2), \textit{anti-symmetrical}$$

Then, we have the possible spin functions of the helium, the resulting function after permutation operator and the indication of being symmetrical, anti-symmetrical or neither of them.

$\alpha(1)\alpha(2) \xrightarrow{P_{12}} \alpha(2)\alpha(1),$ symmetrical

$\beta(1)\beta(2) \xrightarrow{P_{12}} \beta(2)\beta(1),$ symmetrical

$\alpha(1)\beta(2) \xrightarrow{P_{12}} \alpha(2)\beta(1),$ not symmetrical and not anti-symmetrical

$\beta(1)\alpha(2) \xrightarrow{P_{12}} \beta(2)\alpha(1),$ not symmetrical and not anti-symmetrical

$$\frac{1}{\sqrt{2}}\{\alpha(1)\beta(2) + \beta(1)\alpha(2)\} \xrightarrow{P_{12}} \frac{1}{\sqrt{2}}\{\alpha(2)\beta(1) + \beta(2)\alpha(1)\} =$$

$$= \frac{1}{\sqrt{2}}\{\alpha(1)\beta(2) + \beta(1)\alpha(2)\}, \quad \text{symmetrical}$$

$$\frac{1}{\sqrt{2}}\{\alpha(1)\beta(2)-\beta(1)\alpha(2)\}\xrightarrow{P_{12}}\frac{1}{\sqrt{2}}\{\alpha(2)\beta(1)-\beta(2)\alpha(1)\}=$$

$$=-\frac{1}{\sqrt{2}}\{\alpha(1)\beta(2)-\beta(1)\alpha(2)\},\quad \text{anti-symmetrical}$$

$$\alpha = spin = +1/2, \qquad \beta = spin = -1/2$$

Then, since the spatial wave function of helium is symmetrical, we must use the following anti-symmetrical spin function:

$$\Theta(s_1,s_2)=\frac{1}{\sqrt{2}}\{\alpha(1)\beta(2)-\beta(1)\alpha(2)\}$$

6. Variational method for helium atom with effective nuclear charge

In the helium problem using first order perturbation theory, the helium nuclear charge was Z = 2, where no shielding effect was taken into account since one electron can shield the other electron leading to a decrease of the nuclear charge that the second electron is influenced to. In the next example, we will consider the effective nuclear charge, Z_{eff}, as a result of the shielding effect. Let us suppose that Z_{eff} is not known. We will use the variational parameter of the Z_{eff} called ξ to obtain Z_{eff} from the equation:

$$Z_{eff} = Z - \xi$$

We will use a mathematical artifact to find minimum of the function ξ and then obtain Z_{eff} of helium.

Solution: Based on the Hamiltonian of the helium:

$$H=-\frac{\hbar^2}{2m_e}\nabla_1^2-\frac{\hbar^2}{2m_e}\nabla_2^2-Z\frac{e^2}{r_1}-Z\frac{e^2}{r_2}+\frac{e^2}{r_{12}}$$

Let us add and subtract the variational parameter ξ in the Hamiltonian:

$$H=-\frac{\hbar^2}{2m_e}\nabla_1^2-\frac{\hbar^2}{2m_e}\nabla_2^2-Z\frac{e^2}{r_1}-Z\frac{e^2}{r_2}+\frac{e^2}{r_{12}}+\left(\xi\frac{e^2}{r_1}-\xi\frac{e^2}{r_1}+\xi\frac{e^2}{r_2}-\xi\frac{e^2}{r_2}\right)$$

Rearranging the above equation, we have:

$$H=\left[-\frac{\hbar^2}{2m_e}\nabla_1^2-\xi\frac{e^2}{r_1}-\frac{\hbar^2}{2m_e}\nabla_2^2-\xi\frac{e^2}{r_2}\right]+(\xi-Z)\frac{e^2}{r_1}+(\xi-Z)\frac{e^2}{r_2}+\frac{e^2}{r_{12}}$$

The term between the square brackets is the sum of Hamiltonians of two hydrogen atoms with nuclear charge ξ.

Now, let us use the Hamiltonian of two hydrogen atoms with nuclear charge Z in the Schrödinger equation with a guess function φ.

$$\left[-\frac{\hbar^2}{2m_e}\nabla_1^2-\frac{\hbar^2}{2m_e}\nabla_2^2-Z\frac{e^2}{r_1}-Z\frac{e^2}{r_2}\right]\varphi=-Z^2\left(\frac{1}{n_1^2}+\frac{1}{n_2^2}\right)\frac{e^2}{2a_0}\varphi$$

In the ground state (n = 1), we have

$$\left[-\frac{\hbar^2}{2m_e}\nabla_1^2-\frac{\hbar^2}{2m_e}\nabla_2^2-Z\frac{e^2}{r_1}-Z\frac{e^2}{r_2}\right]\varphi=-2Z^2\frac{e^2}{2a_0}\varphi$$

Let us replace Z with ξ.

$$\left[-\frac{\hbar^2}{2m_e}\nabla_1^2-\frac{\hbar^2}{2m_e}\nabla_2^2-\xi\frac{e^2}{r_1}-\xi\frac{e^2}{r_2}\right]\varphi=-2\xi^2\frac{e^2}{2a_0}\varphi=-\xi^2\frac{e^2}{a_0}\varphi$$

The variational integral of the normalized guess function is:

$$\int\varphi^*H\varphi.d\tau=\int\varphi^*\left\{\begin{matrix}\left[-\frac{\hbar^2}{2m_e}\nabla_1^2-\xi\frac{e^2}{r_1}-\frac{\hbar^2}{2m_e}\nabla_2^2-\xi\frac{e^2}{r_2}\right]\\+(\xi-Z)\frac{e^2}{r_1}+(\xi-Z)\frac{e^2}{r_2}+\frac{e^2}{r_{12}}\end{matrix}\right\}\varphi.d\tau$$

By replacing the term in the square brackets with the eigenvalue of the corresponding Schrödinger equation, we have:

$$\int\varphi^*H\varphi.d\tau=\int\varphi^*\left\{-\xi^2\frac{e^2}{a_0}\varphi+(\xi-Z)\frac{e^2}{r_1}+(\xi-Z)\frac{e^2}{r_2}+\frac{e^2}{r_{12}}\right\}\varphi.d\tau$$

Rearranging the above equation, we have the variational integral:

$$\int\varphi^*H\varphi.d\tau=-\xi^2\frac{e^2}{a_0}\int\varphi^*\varphi.d\tau+(\xi-Z).e^2\int\frac{\varphi^*\varphi}{r_1}d\tau+$$

$$+(\xi-Z).e^2\int\frac{\varphi^*\varphi}{r_2}d\tau+e^2\int\frac{\varphi^*\varphi}{r_{12}}d\tau$$

Let us assume that the guess function is a product of two 1s orbital, represented by f_1 and f_2. Next, let us evaluate each term of the above variational integral.

$$\int\varphi^*\varphi d\tau=\iint f_1^*f_2^*f_1f_2d\tau_1d\tau_2=\int f_1^*f_1d\tau_1\int f_2^*f_2d\tau_2=1$$

$$\int\frac{\varphi^*\varphi}{r_1}d\tau=\int\frac{f_1^*f_1}{r_1}d\tau_1\int f_2^*f_2d\tau_2=\int\frac{f_1^*f_1}{r_1}d\tau_1$$

$$d\tau=(r\cdot d\theta)\cdot(dr)\cdot(d\phi\cdot r\cdot\sin\theta)=r^2\cdot\sin\theta\cdot dr\cdot d\theta\cdot d\phi$$

$$\int\frac{f_1^*f_1}{r_1}d\tau_1=\frac{1}{\pi}\frac{\xi^3}{a_0^3}\int_0^\infty e^{-2\xi r_1/a_0}\frac{r_1^2}{r_1}dr_1\int_0^\pi\sin\theta_1d\theta_1\int_0^{2\pi}d\phi_1=\frac{\xi}{a_0}$$

$$\int\frac{\varphi^*\varphi}{r_1}d\tau=\int\frac{f_2^*f_2}{r_2}d\tau_1=\int\frac{f_1^*f_1}{r_1}d\tau_1=\frac{\xi}{a_0}$$

Then, the variational integral becomes:

$$\int \varphi^* H\varphi.d\tau = \left(\xi^2 - 2Z\xi + \frac{5}{8}\xi\right)\frac{e^2}{a_0}$$

Now, we will obtain the minimum of the variational integral as a function of ξ.

$$\frac{\partial}{\partial \xi}\int \varphi^* H\varphi.d\tau = \frac{\partial}{\partial \xi}\left\{\left(\xi^2 - 2Z\xi + \frac{5}{8}\xi\right)\frac{e^2}{a_0}\right\} = 0$$

$$\frac{\partial}{\partial \xi}\int \varphi^* H\varphi.d\tau = \left(2\xi - 2Z + \frac{5}{8}\right)\frac{e^2}{a_0} = 0$$

$$\xi = Z - \frac{5}{16}$$

By replacing ξ = Z – 5/16 in the variational integral, we have:

$$\int \varphi^* H\varphi.d\tau = \left(\xi^2 - 2Z\xi + \frac{5}{8}\xi\right)\frac{e^2}{a_0}$$

$$\xi^2 = \left(Z - \frac{5}{16}\right)^2 = Z^2 - \frac{10}{16}Z + \frac{25}{256}$$

$$-2Z\xi = -2Z\left(Z - \frac{5}{16}\right) = -2Z^2 + \frac{10}{16}Z$$

$$\frac{5}{8}\xi = \frac{5}{8}\left(Z - \frac{5}{16}\right) = \frac{5Z}{8} - \frac{25}{128}$$

$$\int \varphi^* H\varphi.d\tau = \left(Z^2 - \frac{10}{16}Z + \frac{25}{256} - 2Z^2 + \frac{10}{16}Z + \frac{5Z}{8} - \frac{25}{128}\right)\frac{e^2}{a_0}$$

$$\int \varphi^* H\varphi.d\tau = \left(-Z^2 + \frac{5}{8}Z - \frac{25}{256}\right)\frac{e^2}{a_0} = \left(Z - \frac{5}{16}\right)^2\frac{e^2}{a_0}$$

$$Helium: Z = 2$$

$$\int \varphi^* H\varphi.d\tau = \left(2 - \frac{5}{16}\right)^2\frac{e^2}{a_0} = -77.5eV$$

The error with respect the experimental value (–79.0 eV) is 1.9% which is much smaller than that using nuclear charge as Z (5.3%).

7. Exercises

(1) A particle of mass m in a unidimensional box having length a is perturbated as with a Hamiltonian H' from a/2 to a, calculate the first order corrected energy.

Data:

$$E_n^0 = \frac{n^2h^2}{8ma^2}, \varphi_n^0 = \sqrt{\frac{2}{a}}\sin\left(\frac{n\pi x}{a}\right), H' = \frac{h^2}{8ma^2}, a/2 \le x \le a$$

(2) The function $u = Ax(a^2 - x^2)$ is used as guess function for a particle having mass m in a unidimensional box where V = 0 for $0 \le x \le a$ and $V = \infty$ for other regions. Calculate the approximate eigenvalue and the relative error.

Data:

$$H = -\frac{1}{2}\frac{\hbar^2}{m}\frac{d^2}{dx^2}, \qquad E_1 = \frac{\hbar^2}{8ma^2}$$

Appendix

Special Case of Associated Laguerre Equation

The differential equation below:

$$y_j^{k\,''}(x)+\left(-\frac{1}{4}+\frac{2j+k+1}{2x}-\frac{k^2-1}{4x^2}\right)y_j^k(x)=0$$

having as solution:

$$y_j^k(x)=e^{-x/2}x^{(k+1)/2}L_j^k(x)$$

is a slightly different type of the associated Laguerre equation depicted below:

$$xL_j^k\,''(x)+(1-x+k)L_j^k\,'(x)+jL_j^k(x)=0$$

whose solution is also slightly different from that of the former equation:

$$y_j^k(x)=e^{-x/2}x^{k/2}L_j^k(x)$$

We will show that by replacing:

$$y_j^k(x)=e^{-x/2}x^{(k+1)/2}L_j^k(x)$$

In the former differential equation will lead to the latter differential equation (the associated Laguerre equation).

Firstly, let us simplify the notations so far used:

$$y_j^k(x)=y,\quad L_j^k(x)=v$$

and do the first derivative on:

$$y=e^{-x/2}x^{(k+1)/2}v$$

giving:

$$y' = -\frac{1}{2}e^{-x/2}x^{(k+1)/2}v + e^{-x/2}\left(\frac{k+1}{2}\right)x^{(k-1)/2}v + e^{-x/2}x^{(k+1)/2}v'$$

$$y' = -\frac{1}{2}e^{-x/2}x^{(k+1)/2}v + e^{-x/2}\left(\frac{k+1}{2}\right)x^{-1}x^{(k+1)/2}v + e^{-x/2}x^{(k+1)/2}v'$$

$$y' = -\frac{1}{2}e^{-x/2}x^{(k+1)/2}v + e^{-x/2}\left(\frac{k+1}{2x}\right)x^{(k+1)/2}v + e^{-x/2}x^{(k+1)/2}v'$$

$$y' = \left[-\frac{1}{2}v + \left(\frac{k+1}{2x}\right)v + v'\right]e^{-x/2}x^{(k+1)/2}$$

$$\Rightarrow \left(e^{x/2}x^{-(k+1)/2}\right)y' = -\frac{1}{2}v + \left(\frac{k+1}{2x}\right)v + v'$$

Let us derivative the equation:

$$y' = -\frac{1}{2}e^{-x/2}x^{(k+1)/2}v + e^{-x/2}\left(\frac{k+1}{2x}\right)x^{(k+1)/2}v + e^{-x/2}x^{(k+1)/2}v'$$

By derivating the first term, we have:

$$\frac{1}{4}e^{-x/2}x^{(k+1)/2}v - \frac{1}{2}e^{-x/2}\left(\frac{k+1}{2}\right)x^{-1}x^{(k+1)/2}v - \frac{1}{2}e^{-x/2}x^{(k+1)/2}v'$$

$$\Rightarrow e^{-x/2}x^{(k+1)/2}\left[\frac{1}{4}v - \frac{1}{2}\left(\frac{k+1}{2x}\right)v - \frac{1}{2}v'\right]$$

Let us rearrange the second term:

$$e^{-x/2}\left(\frac{k+1}{2x}\right)x^{(k+1)/2}v = e^{-x/2}\left(\frac{k+1}{2}\right)x^{(k-1)/2}v$$

And now let us derivate the second term:

$$e^{-x/2}\left(\frac{k+1}{2x}\right)x^{(k+1)/2}v = e^{-x/2}\left(\frac{k+1}{2}\right)x^{(k-1)/2}v$$

$$-\frac{1}{2}e^{-x/2}\left(\frac{k+1}{2x}\right)x^{(k+1)/2}v + e^{-x/2}\left(\frac{k+1}{2}\right)\left(\frac{k-1}{2}\right)x^{-1}x^{(k-1)/2}v +$$

$$+e^{-x/2}\left(\frac{k+1}{2x}\right)x^{(k+1)/2}v' = -\frac{1}{2}e^{-x/2}\left(\frac{k+1}{2x}\right)x^{(k+1)/2}v + e^{-x/2}\left(\frac{k+1}{2x}\right)\left(\frac{k-1}{2x}\right)x^{(k+1)/2}v$$

$$+e^{-x/2}\left(\frac{k+1}{2x}\right)x^{(k+1)/2}v'$$

$$\Rightarrow e^{-x/2}x^{(k+1)/2}\left[-\frac{1}{2}\left(\frac{k+1}{2x}\right)v + \left(\frac{k+1}{2x}\right)\left(\frac{k-1}{2x}\right)v + \left(\frac{k+1}{2x}\right)v'\right]$$

By derivating the third term, we have:

$$-\frac{1}{2}e^{-x/2}x^{(k+1)/2}v'+e^{-x/2}\left(\frac{k+1}{2x}\right)x^{(k+1)/2}v'+e^{-x/2}x^{(k+1)/2}v''$$

$$\Rightarrow e^{-x/2}x^{(k+1)/2}\left[-\frac{1}{2}v'+\left(\frac{k+1}{2x}\right)v'+v''\right]$$

By summing the derivatives of three terms of the original equation (y'), we have:

$$y''=e^{-x/2}x^{(k+1)/2}\left\{\begin{array}{l}\left[\frac{1}{4}v-\frac{1}{2}\left(\frac{k+1}{2x}\right)v-\frac{1}{2}v'\right]+\\ \left[-\frac{1}{2}\left(\frac{k+1}{2x}\right)v+\left(\frac{k+1}{2x}\right)\left(\frac{k-1}{2x}\right)v+\left(\frac{k+1}{2x}\right)v'\right]+\\ +\left[-\frac{1}{2}v'+\left(\frac{k+1}{2x}\right)v'+v''\right]\end{array}\right\}$$

And let us rearrange it to become:

$$\left(e^{x/2}x^{-(k+1)/2}\right)y''=\frac{1}{4}v-\frac{1}{2}\left(\frac{k+1}{2x}\right)v-\frac{1}{2}v'-\frac{1}{2}\left(\frac{k+1}{2x}\right)v$$

$$+\left(\frac{k+1}{2x}\right)\left(\frac{k-1}{2x}\right)v+\left(\frac{k+1}{2x}\right)v'-\frac{1}{2}v'+\left(\frac{k+1}{2x}\right)v'+v''$$

Let us take this equation above and the solution:

$$y=e^{-x/2}x^{(k+1)/2}v$$

and replace them in the special type of the associated Laguerre equation:

$$y''+\left(-\frac{1}{4}+\frac{2j+k+1}{2x}-\frac{k^2-1}{4x^2}\right)y=0$$

to give:

$$\left\{\begin{array}{l}\frac{1}{4}v-\frac{1}{2}\left(\frac{k+1}{2x}\right)v-\frac{1}{2}v'-\frac{1}{2}\left(\frac{k+1}{2x}\right)v\\ +\left(\frac{k+1}{2x}\right)\left(\frac{k-1}{2x}\right)v+\left(\frac{k+1}{2x}\right)v'-\frac{1}{2}v'+\left(\frac{k+1}{2x}\right)v'+v''\end{array}\right\}\left(e^{-x/2}x^{(k+1)/2}\right)+$$

$$+\left(-\frac{1}{4}+\frac{2j+k+1}{2x}-\frac{k^2-1}{4x^2}\right)\cdot\left(e^{-x/2}x^{(k+1)/2}v\right)=0$$

After dividing the equation above by $e^{-x/2}x^{(k+1)/2}$, the equation becomes:

$$\left\{\begin{array}{l}\frac{1}{4}v-\frac{1}{2}\left(\frac{k+1}{2x}\right)v-\frac{1}{2}v'-\frac{1}{2}\left(\frac{k+1}{2x}\right)v\\ +\left(\frac{k+1}{2x}\right)\left(\frac{k-1}{2x}\right)v+\left(\frac{k+1}{2x}\right)v'-\frac{1}{2}v'+\left(\frac{k+1}{2x}\right)v'+v''\end{array}\right\}+$$

$$+\left(-\frac{1}{4}+\frac{2j+k+1}{2x}-\frac{k^2-1}{4x^2}\right)\cdot v=0$$

$$\Rightarrow\left\{\begin{array}{l}\frac{1}{4}v-\frac{1}{2}\left(\frac{k+1}{2x}\right)v-\frac{1}{2}v'-\frac{1}{2}\left(\frac{k+1}{2x}\right)v\\+\frac{k^2-1}{4x^2}v+\left(\frac{k+1}{2x}\right)v'-\frac{1}{2}v'+\left(\frac{k+1}{2x}\right)v'+v''\end{array}\right\}+$$

$$-\frac{1}{4}v+\frac{2j+k+1}{2x}v-\frac{k^2-1}{4x^2}v=0$$

Where some terms are canceled out to give:

$$\left\{\begin{array}{l}-\frac{1}{2}\left(\frac{k+1}{2x}\right)v-\frac{1}{2}v'-\frac{1}{2}\left(\frac{k+1}{2x}\right)v\\+\left(\frac{k+1}{2x}\right)v'-\frac{1}{2}v'+\left(\frac{k+1}{2x}\right)v'+v''\end{array}\right\}+\frac{2j+k+1}{2x}v=0$$

Let us rearrange the equation above:

$$\left\{\begin{array}{l}-\left(\frac{k+1}{4x}\right)v-\frac{1}{2}v'-\left(\frac{k+1}{4x}\right)v\\+\left(\frac{k+1}{2x}\right)v'-\frac{1}{2}v'+\left(\frac{k+1}{2x}\right)v'+v''\end{array}\right\}+\frac{j}{x}v+\frac{k+1}{2x}v=0$$

And see that more terms are canceled out:

$$\left\{-\frac{1}{2}v'+\left(\frac{k+1}{2x}\right)v'-\frac{1}{2}v'+\left(\frac{k+1}{2x}\right)v'+v''\right\}+\frac{j}{x}v=0$$

More rearrangement in the equation yields:

$$\left\{-v'+\left(\frac{k+1}{x}\right)v'+v''\right\}+\frac{j}{x}v=0$$

$$\Rightarrow v''+\left(\frac{k+1}{x}\right)v'-v'+\frac{j}{x}v=0$$

$$\Rightarrow xv''+(k+1)v'-xv'+jv=0$$

$$\Rightarrow xv''+(k+1-x)v'+jv=0$$

Which becomes the associated Laguerre equation:

$$xL_j^k{}''(x)+(1-x+k)L_j^k{}'(x)+jL_j^k(x)=0$$

Index

F

G

H

I

J

K

L

M

N

O

P

For Product Safety Concerns and Information please contact our EU representative GPSR@taylorandfrancis.com
Taylor & Francis Verlag GmbH, Kaufingerstraße 24, 80331 München, Germany

www.ingramcontent.com/pod-product-compliance
Lightning Source LLC
Chambersburg PA
CBHW060743220326
41598CB00022B/2313
* 9 7 8 0 3 6 7 5 0 6 3 4 6 *